MEASURE AND INTEGRATION THEORY ON INFINITE-DIMENSIONAL SPACES

This is Volume 48 in
PURE AND APPLIED MATHEMATICS
A Series of Monographs and Textbooks
Editors: PAUL A. SMITH AND SAMUEL EILENBERG
A complete list of titles in this series appears at the end of this volume

Measure and Integration Theory on Infinite-Dimensional Spaces

ABSTRACT HARMONIC ANALYSIS

XIA DAO-XING

Tao-hsing Hsia

FUDAN UNIVERSITY
SHANGHAI

Translated by Elmer J. Brody

DEPARTMENT OF MATHEMATICS
THE CHINESE UNIVERSITY OF HONG KONG

ACADEMIC PRESS New York and London 1972

ACADEMIC PRESS, INC.
111 Fifth Avenue, New York, New York 10003

United Kingdom Edition published by
ACADEMIC PRESS, INC. (LONDON) LTD.
24/28 Oval Road, London NW1

LIBRARY OF CONGRESS CATALOG CARD NUMBER: 73-182639

PRINTED IN THE UNITED STATES OF AMERICA

CONTENTS

FOREWORD

The present book is a compendium of results which are mostly of fairly recent vintage, and the theory discussed herein is very much in a state of flux. Moreover, the original book seems to have been compiled and published rather hurriedly. Thus, there were a great number of inaccuracies in the original, ranging from typographical errors to very substantial gaps in the mathematical reasoning. I have made some effort to correct these inaccuracies; in many cases, I have altered and expanded proofs without burdening the reader with a tedious explanation of how and where the revised version deviates from the original. In many cases where doubts or difficulties still remain, I have called attention to these by footnotes. However, I cannot claim either completeness or consistency in this editorial work. Especially as regards Chapters III and IV, I feel dubious as to how much expenditure of effort would be justified in revising or developing the theory, at least until such time as more applications may be demonstrated; in this connection, the appearance of a subsequent volume, as indicated in the author's preface, would be most enlightening.

ELMER J. BRODY

PREFACE

The study of measures and integrals on infinite-dimensional spaces arose from the theory of stochastic processes, particularly the theory of Wiener processes. In recent years, the subject has been intimately connected with research on characteristic functionals, limit theorems, sample spaces, and generalized stochastic processes. Even more noteworthy is the fact that questions of integration on infinite-dimensional spaces have, during the last ten-odd years, appeared in many scientific fields, such as quantum mechanics, quantum field theory, statistical physics, thermodynamics of irreversible processes, turbulence theory, atomic reactor computations, and coding problems. However, the application of integration on infinite-dimensional spaces to these fields has encountered many profound difficulties, and a lack of adequate techniques. Thus, it seems that further study of this new subject is amply justified.

Heretofore, there have appeared no introductory books on this topic, either in this country or abroad. As far as the author knows, there has been only a volume of lecture notes, "Integration of Functionals," written by K. O. Friedrichs, H. N. Shapiro, *et al.*, 1957, and still unpublished. Moreover, except for Wiener integrals, the mathematical theory of measures and integrals on infinite-dimensional spaces largely began to develop only after 1956. As the mathematical background involved in the literature of this theory is rather extensive, the novice is likely to find the going somewhat difficult. Therefore, the author has been so bold as to write the present book with the hope of smoothing the way for Chinese comrades undertaking research in this direction.

This volume is primarily devoted to introducing abstract harmonic analysis. It essentially consists of three parts. The first part is concerned with the representation of positive functionals and operator rings (Chapter II), which constitutes the basis of abstract harmonic analysis. Although this topic cannot be regarded as lying entirely within the domain of infinite-dimensional measure and integration theory, the two are intimately related. The second part deals with abstract harmonic analysis on pseudo-invariant measure spaces (Chapters III and IV); except for just a few theorems, the results given here were, for the most part, obtained in China. This kind of harmonic analysis may provide tools for the further investigation of measure and integration on infinite-dimensional spaces. In the third part, we discuss a mathematical problem arising in quantum field theory, i.e., the representation of commutation relations in Bose–Einstein fields (Chapter VI); here, applications of the theory developed in the first two parts are given. In addition, one chapter (Chapter V) is devoted to another important example of measure theory on infinite-dimensional spaces, i.e., Gaussian measures.

In a subsequent volume, we shall deal with the so-called continual integral problems which appear frequently in the applications of integration theory on infinite-dimensional spaces, as well as functional variational equations and various other applications.

We assume that the reader is familiar with the treatise of Halmos [1], or its equivalent, and has the basic knowledge of functional analysis which may be found in ordinary textbooks on that subject. It is also expected that the reader has some acquaintance with the basic notions of topological spaces, topological groups, and linear topological spaces; in this connection, he may consult, for example, Guan Zhao-zhi [1]. Chapter I and Appendices I, II of the present book also provide some supplementary background material.

Owing to the author's limitations, and the rather short time taken to write this book, its shortcomings are undoubtedly numerous, and errors inevitable. The reader's criticisms will be welcomed.

Part of the manuscript of this book was read by Professor Zheng Ceng-tong of Zhongshan University, who offered valuable comments. The teachers and research students of the Functional Analysis Group, Function Theory Teaching and Research Section, Fudan University Mathematics Department, also offered valuable opinions, especially Comrade Yan Shao-zong. For these contributions, I hereby express my thanks.

Xia Dao-xing

CHAPTER

I

SOME SUPPLEMENTARY BACKGROUND IN MEASURE THEORY

The measure-theoretic concepts and results used in this book may, for the most part, be found in Halmos' *Measure Theory*, and will be directly applied in the sequel without additional explanation. However, certain supplementary measure-theoretic results, not included in Halmos' book, will be introduced in the present chapter; these results will also be essential in the subsequent chapters.

At some points in this book, we shall require the discussion of measures which are not σ-finite.[1] However, non-σ-finite measures in general are not well behaved (e.g., the Radon–Nikodyn theorem is not generally valid for such measures). Therefore, we shall in §1.2 investigate *localizable* measures, which are not necessarily σ-finite, but which do retain certain desirable properties of σ-finite measures. The measures ordinar-

[1] *Translator's note:* The term σ-finite, as used by the author, means *totally* σ-finite in the sense of Halmos [1]. This distinction is an important one in certain parts of this book. For example, according to the author's terminology, a Haar measure is pseudo-σ-finite, but not necessarily σ-finite.

ily used on groups are localizable, so that localizable measures, in fact, constitute a fairly broad class. Some rather deeper properties of localizable measures will be introduced in §2.4.

In §1.3 we shall introduce the Kolmogorov theorem. This is a fundamental theorem concerning the construction of measures on infinite-dimensional spaces from given measures on finite-dimensional spaces. We shall present this theorem in a very general form, related to the notion of a projective limit of locally convex linear topological spaces; in this form, it can be used for the construction of measures on locally convex linear topological spaces, starting from given measures on Banach spaces.

In §1.4, we introduce Kakutani inner measure, which plays an important role in the study of equivalence of measures on product spaces, as well as in the study of quasi-invariant measures.

§1.1. Some Measure-Theoretic Concepts

1° Extension and Restriction of Measures

We shall introduce certain generalizations of the usual notion (see Halmos [1]) of "measurable set."

Definition 1.1.1.[2] Let $(G, \mathfrak{B})$ be a measurable space. Let $A \subset G$, and suppose that, for every $B \in \mathfrak{B}$, we have $A \cap B \in \mathfrak{B}$. We then say that A is *measurable* with respect to $(G, \mathfrak{B})$. We denote the totality of such measurable sets by $\tilde{\mathfrak{B}}$.

Clearly, $\mathfrak{B} \subset \tilde{\mathfrak{B}}$, and $\tilde{\mathfrak{B}}$ is a σ-algebra on G. If $\mathfrak{B}$ is an algebra, then $\mathfrak{B} = \tilde{\mathfrak{B}}$.

Let f be a real (complex) function on G. If, for every Borel set A of the real line (complex plane), we have $\{g \mid f(g) \in A\} \in \tilde{\mathfrak{B}}$, we say that f is a *measurable function* on $(G, \mathfrak{B})$.

Definition 1.1.2. Let $(G, \mathfrak{B}, \mu)$ be a measure space. Define a set function $\tilde{\mu}$ on $(G, \tilde{\mathfrak{B}})$, as follows. For $A \in \tilde{\mathfrak{B}}$,

$$\tilde{\mu}(A) = \sup_{B \in \mathfrak{B}} \mu(A \cap B).$$

We call $\tilde{\mu}$ the *extension* of μ.

It is easily seen that, if $A \in \mathfrak{B}$, then $\tilde{\mu}(A) = \mu(A)$. Consequently, we shall, in the sequel, denote $\tilde{\mu}$ simply by μ, without danger of confusion.

[2] *Translator's note:* It should be recalled that, in Halmos' terminology, a σ-ring $\mathfrak{B}$ need not be a σ-algebra, that is, $\mathfrak{B}$ need not contain G itself.

In what follows, any measure space $(G, \mathfrak{B}, \mu)$ will, whenever necessary, be extended to $(G, \tilde{\mathfrak{B}}, \mu)$.

Again, extend $(G, \tilde{\mathfrak{B}}, \mu)$ to a complete measure space $(G, \mathfrak{B}^*, \mu^*)$. If f is measurable with respect to $(G, \mathfrak{B}^*)$, we say that f is a *measurable function* on $(G, \mathfrak{B}, \mu)$.

If $B \in \tilde{\mathfrak{B}}$ and $\mu(B) = 0$, we call B a *μ-null set*, or simply a *null set.*

Definition 1.1.3. Let $(G, \mathfrak{B}, \mu)$ be a measure space, $A \subset G$, and let

$$\mathfrak{B}_A = \{E \cap A \mid E \in \mathfrak{B}\}.$$

We call $\mathfrak{B}_A$ the *restriction* of $\mathfrak{B}$ to A.

If there exists a $C \in \mathfrak{B}$ such that the inner measure

$$\mu_*(C - A) = 0, \tag{1.1.1}$$

we may define a set function μ_A on $\mathfrak{B}_A$ as follows. For $E \in \mathfrak{B}$,

$$\mu_A(A \cap E) = \mu(E \cap C). \tag{1.1.2}$$

We call μ_A the *restriction*[3] of μ to A.

Lemma 1.1.1.[4] Let $(G, \mathfrak{B}, \mu)$ be a measure space, and let A be a subset of G satisfying condition (1.1.1) for a given $C \in \mathfrak{B}$. Then the restriction μ_A of μ to A is well defined, and $(A, \mathfrak{B}_A, \mu_A)$ is a measure space.

Proof. We need only prove that μ_A is well defined; the rest is obvious.

Let $E, F \in \mathfrak{B}$, with $A \cap E = A \cap F$. To justify the definition of μ_A, we need only show that

$$\mu(E \cap C) = \mu(F \cap C). \tag{1.1.3}$$

We may assume that $E \subset F$, for otherwise, we could replace F by $E \cup F$. Then, from $A \cap E = A \cap F$ and $E \subset F$, it follows that

$$A \cap (F - E) = 0,$$

whence $C - A \supset (C \cap F) - (C \cap E)$. But $\mu_*(C - A) = 0$, therefore $\mu((F \cap C) - (E \cap C)) = 0$, so that (1.1.3) holds.]

[3] *Translator's note:* Notice that μ_A depends upon the choice of C.

[4] See Halmos [1].

2° The Function Space $\mathfrak{L}_k^2(\Omega)$

We shall have occasion to use certain abstract functions taking values in a Hilbert space. We first introduce the following notions.

Definition 1.1.4. Let H be a Hilbert space, $\Omega = (G, B, \mu)$ a measure space, and f an abstract function on Ω such that (i) for every $g \in G$, $f(g) \in H$, (ii) for every $u \in H$, the numerical valued function $(f(g), u)$, $g \in G$, is a measurable function on Ω, and (iii) the range of values $\{f(g) \mid g \in G\}$ is contained in a separable subspace of H. We say that such an f is *measurable*, and denote the totality of such functions by $M(H, \Omega)$.

It is easily seen that $M(H, \Omega)$ forms a linear space with respect to ordinary addition of functions and multiplication by constants.

Lemma 1.1.2. Let $\{e_\lambda, \lambda \in \Lambda\}$ be a complete orthonormal system in the Hilbert space H. Then, a necessary and sufficient condition for f to belong to $M(H, \Omega)$ is that there exist a sequence $\{\lambda_n\} \subset \Lambda$ and a sequence of measurable functions f_{λ_n} on Ω such that

$$f(g) = \sum_{n=1}^{\infty} f_{\lambda_n}(g) e_{\lambda_n} . \tag{1.1.4}$$

PROOF. Assume that f satisfies the above condition. Then, the values of f are contained in the separable subspace spanned by $\{e_{\lambda_n}, n = 1, 2, \ldots\}$, and $(f(g), u) = \sum f_{\lambda_n}(g)(e_k, u)$ is measurable. Conversely, suppose that f is measurable, let M be a separable closed linear subspace containing the range of f, and let $\{\varphi_k\}$ be a complete orthonormal system in M. For each k, there is a sequence $\{\lambda_n^{(k)}\} \subset \Lambda$, such that

$$\varphi_k = \sum (\varphi_k, e_{\lambda_n^{(k)}}) e_{\lambda_n^{(k)}} .$$

Therefore, the range of f is contained in the separable closed linear subspace spanned by $\{e_{\lambda_n^{(k)}}, k, n = 1, 2, \ldots\}$. Since $(f, e_{\lambda_n^{(k)}})$ is a measurable function on Ω, and since

$$f(g) = \sum_{n,k=1}^{\infty} (f, e_{\lambda_n^{(k)}}) e_{\lambda_n^{(k)}} ,$$

the condition of the lemma is satisfied.]

Corollary 1.1.3. If $\varphi, f \in M(H, \Omega)$, then $(f(g), \varphi(g))$ is a measurable function on Ω. In particular, $\|f(g)\|^2$ is a measurable function on Ω.

PROOF. By Lemma 1.1.2, there is a sequence $\{e_{\lambda_n}\}$ such that (1.1.4) holds, therefore,

$$(f(g), \varphi(g)) = \sum (f(g), e_{\lambda_n})\overline{(\varphi(g), e_{\lambda_n})},$$

whence it follows at once that $(f(g), \varphi(g))$ is a measurable function.]

Definition 1.1.5. Let H be a Hilbert space, and let $\Omega = (G, \mathfrak{B}, \mu)$ be a measure space. Let $\mathfrak{L}^2(H, \Omega)$ be the totality of functions in $M(H, \Omega)$ which satisfy the condition

$$\int_G \|f(g)\|^2 \, d\mu(g) < \infty, \tag{1.1.5}$$

and define an inner product on $\mathfrak{L}^2(H, \Omega)$ as follows[5]:

$$(f, \varphi) = \int_G (f(g), \varphi(g)) \, d\mu(g). \tag{1.1.6}$$

We let $L_2(\Omega)$ (or $L^2(\Omega)$) denote the usual space of measurable quadratically integrable functions on Ω.

Theorem 1.1.4. Let $\{e_\lambda, \lambda \in \Lambda\}$ be a complete orthonormal system in H, and let $H_\lambda = \{f(g)\, e_\lambda \mid f \in L_2(\Omega)\}$. Then

$$\mathfrak{L}^2(H, \Omega) = \sum_{\lambda \in \Lambda} \oplus H_\lambda . \tag{1.1.7}$$

PROOF. Let $f \in \mathfrak{L}^2(H, \Omega)$. By Lemma 1.1.2, there is a sequence $\{\lambda_n\} \subset \Lambda$ such that (1.1.4) holds. Since $|f_{\lambda_k}(g)| \leqslant \|f(g)\|$, it follows that $f_{\lambda_k} \in L_2(\Omega)$, that is, $f_{\lambda_k}(g)\, e_{\lambda_k} \in H_{\lambda_k}$. Therefore, $f \in \sum_{k=1}^{\infty} \oplus H_{\lambda_k}$. This shows that f belongs to the right-hand side of (1.1.7).]

Notice that, if $f_{\lambda_k}(\cdot)\, e_{\lambda_k} \in H_{\lambda_k}$, $k = 1, 2, \ldots$, and if

$$\sum \|f_{\lambda_k}(\cdot) e_{\lambda_k}\|^2 < \infty,$$

then, forming $f \in M(H, \Omega)$ in accordance with (1.1.4), we have

$$\int_\Omega \|f(g)\|^2 \, d\mu(g) = \sum \int_\Omega |f_{\lambda_k}(g)|^2 \, d\mu(g) = \sum \|f_{\lambda_k}(\cdot) e_{\lambda_k}\|^2 < \infty.$$

Hence, $f \in \mathfrak{L}^2(H, \Omega)$. From this, we easily deduce the following result.

[5] By Corollary 1.1.3, $(f(g), \varphi(g))$, $g \in G$, is a measurable function on Ω. Moreover, by condition (1.1.5), $\int_G \|f(g)\| \, \|\varphi(g)\| \, d\mu(g) < \infty$, whence

$$\int_G |(f(g), \varphi(g))| \, d\mu(g) \leqslant \int_G \|f(g)\| \, \|\varphi(g)\| \, d\mu(g) < \infty,$$

so that (f, φ) is well defined. It is then easily verified that (f, φ) is an inner product on $\mathfrak{L}^2(H, \Omega)$.

Corollary 1.1.5. $\mathfrak{L}^2(H, \Omega)$ is a Hilbert space.[6]

Clearly, the concrete form of H has little bearing upon the properties of $\mathfrak{L}^2(H, \Omega)$; what is important is the dimension of H, that is, the cardinality of a complete orthonormal basis for H. If H is k-dimensional, we shall write H_k for H and denote $\mathfrak{L}^2(H, \Omega)$ by $\mathfrak{L}_k{}^2(\Omega)$. In particular, when $k = 1$, H_1 may be identified with the real line (or the complex plane), and $\mathfrak{L}_1{}^2(\Omega)$ is simply $L^2(\Omega)$.

As usual, $L^p(\Omega)$ (or $L_p(\Omega)$), $p \geqslant 1$, will denote the Banach space of all pth power integrable measurable real-valued (or complex-valued) functions on Ω, with the usual linear operations and the norm

$$\|f\|_p = \left(\int_G |f(x)|^p \, d\mu(x)\right)^{1/p}.$$

Also, $L_\infty(\Omega)$ (or $L^\infty(\Omega)$) will denote the Banach space consisting of all the essentially bounded measurable functions on Ω, with the usual operations and the norm

$$\|f\|_\infty = \inf_{\mu(E)=0} \sup_{x \in G-E} |f(x)|.$$

3° Determining Sets

Definition 1.1.6. Let $(G, \mathfrak{B})$ be a measurable space, where $\mathfrak{B}$ is a σ-algebra. Let $\mathfrak{D}$ be a family of measurable functions on $(G, \mathfrak{B})$, and suppose that there exists no σ-algebra $\mathfrak{B}_1 \subset \mathfrak{B}$, $\mathfrak{B}_1 \neq \mathfrak{B}$, such that $\mathfrak{D}$ constitutes a family of measurable functions on $(G, \mathfrak{B}_1)$. We then say that $\mathfrak{D}$ is a *determining set* (of functions) on $(G, \mathfrak{B})$, and that $\mathfrak{B}$ is the σ-algebra determined by $\mathfrak{D}$ on G.

It is easily seen that, if G is a set and $\mathfrak{D}$ is any family of functions on G, then there exists a unique σ-algebra $\mathfrak{B}$ determined by $\mathfrak{D}$. In fact, we need only let $\mathfrak{B}$ be the smallest σ-algebra which contains all sets of the form $f^{-1}(C)$, where $f \in \mathfrak{D}$ and C is a Borel set in the complex plane.

Definition 1.1.7. Let $\Omega = (G, \mathfrak{B}, \mu)$ be a measure space and $\mathfrak{D}$ a family of measurable functions on Ω. For any σ-finite set A of Ω, let $\mathfrak{B}_A$ denote the σ-algebra determined by $\mathfrak{D}$ on A. Suppose that, for every such A, every measurable set of Ω which is contained in A differs from some set of $\mathfrak{B}_A$ by a μ-null set. We then say that $\mathfrak{D}$ is a *determining set* (of functions) on Ω.

[6] *Translator's note:* Of course, Theorem 1.1.4 is meaningful only in the context of this corollary.

Clearly, if $\mathfrak{D}$ is a determining set on $(G, \mathfrak{B})$, then $\mathfrak{D}$ is also a determining set on $(G, \mathfrak{B}, \mu)$.

Lemma 1.1.6. Let $\mathfrak{D}$ be a family of bounded measurable real-valued functions on the measure space Ω, such that $\mathfrak{D}$, with respect to the usual operations, forms an algebra containing the unit element 1. Suppose also that $\mathfrak{D}$ is a determining set on Ω.

Choose any $\rho \in L^1(\Omega)$, $\rho \geqslant 0$, and let $L^2(\Omega, \rho)$ be the space consisting of all measurable real-valued functions f on Ω which satisfy the condition

$$\|f\| = \left(\int_G |f(g)|^2 \rho(g)\, d\mu(g)\right)^{1/2} < \infty.$$

Then $\mathfrak{D}$ is dense in $L^2(\Omega, \rho)$ with respect to the norm $\|f\|$.

PROOF. Let $\mathfrak{S}$ be the totality of sets of the form[7]

$$\bigcap_{j=1}^{n} \{x \mid f_j(x) \in (a_j\,, b_j]\}, \qquad f_j \in \mathfrak{D}, \tag{1.1.8}$$

and let $\mathfrak{F}$ be the collection of all finite unions of sets in $\mathfrak{S}$. Then $\mathfrak{F}$ is an algebra. In fact, it is obvious that $G \in \mathfrak{F}$, and that the union and intersection of any finite number of sets of $\mathfrak{F}$ also belongs to $\mathfrak{F}$. To show that $\mathfrak{F}$ is an algebra, it only remains to prove that the complement of any set of $\mathfrak{F}$ also belongs to $\mathfrak{F}$. It obviously suffices to prove that the complement of any set of $\mathfrak{S}$ belongs to $\mathfrak{F}$, but this fact follows at once from the formula

$$G - \bigcap_{j=1}^{n} \{x \mid f_j(x) \in (a_j\,, b_j]\}$$

$$= \bigcup_{j=1}^{n} \{x \mid f_j(x) \in (-\infty, a_j]\} \bigcup_{j=1}^{n} \{x \mid f_j(x) \in (b_j\,, \infty)\}.$$

Hence, $\mathfrak{F}$ is an algebra.

Let $\mathfrak{D}^0$ be the closure of $\mathfrak{D}$ in $L^2(\Omega, \rho)$. Then, since $\mathfrak{D}$ is linear, $\mathfrak{D}^0$ is a closed linear subspace. We now proceed to show that, for every $E \in \mathfrak{F}$, the characteristic function C_E of E belongs to $\mathfrak{D}^0$. Let $f_1\,,..., f_n \in \mathfrak{D}$. Then, there is a positive number ξ such that, for all $g \in G$, $|f_j(g)| \leqslant \xi$, $j = 1, 2,..., n$. On the interval $[-\xi, \xi]$, define the functions

$$\psi_j(x) = \begin{cases} 1, & x \in (a_j\,, b_j] \cap [-\xi, \xi], \\ 0, & x \in [-\xi, \xi] - (a_j\,, b_j]. \end{cases}$$

[7] Here, $a_j < b_j$, and, when $b_j = \infty$, then $(a_j\,, b_j]$ means the totality of real numbers greater than a_j.

It is easy to show that there exists a sequence of polynomials $\{p_{mj}\,;\, j = 1,..., n;\, m = 1, 2,...\}$, such that

$$\max_{|x|\leqslant\xi} |\, p_{mj}(x)| \leqslant 2, \tag{1.1.9}$$

and, for every j, and every $x \in [-\xi, \xi]$,

$$\lim_{m\to\infty} p_{mj}(x) = \psi_j(x). \tag{1.1.10}$$

If E is the set defined by (1.1.8), it is easily seen from (1.1.10) that

$$\lim_{m\to\infty} \prod_{j=1}^{n} p_{mj}(f_j(g)) = \prod_{j=1}^{n} \psi_j(f_j(g)) = C_E(g). \tag{1.1.11}$$

Since $\mathfrak{D}$ is an algebra containing the unit element 1, we have $\varphi_m(g) = \prod_{j=1}^{n} p_{mj}(f_j(g)) \in \mathfrak{D}$. By (1.1.9), $|\, \varphi_m(g)| \leqslant 2^n$. Hence, by the Lebesgue dominated convergence theorem, we obtain

$$\lim_{m\to\infty} \|\, \varphi_m - C_E \,\| = 0.$$

Thus, if $E \in \mathfrak{S}$, then $C_E \in \mathfrak{D}^0$. If E_1, $E_2 \in \mathfrak{S}$, then $E_1 \cap E_2 \in \mathfrak{S}$. Hence, by the formula

$$C_{E_1\cup E_2} = C_{E_1} + C_{E_2} - C_{E_1\cap E_2}\,,$$

and the linearity of $\mathfrak{D}^0$, it follows that $C_{E_1\cup E_2} \in \mathfrak{D}_0$. Consequently, $C_E \in \mathfrak{D}^0$ for every $E \in \mathfrak{F}$.

Suppose that $\mathfrak{D}^0 \neq L^2(\Omega, \rho)$. Then there is a nonzero vector $\varphi \in L^2(\Omega, \rho)$ such that $\varphi \perp \mathfrak{D}^0$. Hence, for every $E \in \mathfrak{F}$,

$$\int_E \varphi\rho\, d\mu = \int C_E\varphi\rho\, d\mu = 0. \tag{1.1.12}$$

By virtue of the countable additivity of the integral, it follows that (1.1.12) holds for every set E in the smallest σ-algebra $\mathfrak{F}_1$ containing $\mathfrak{F}$. Let $A = \{g \mid \rho(g) > 0\}$; then A is a σ-finite set of Ω. Since $\mathfrak{D}$ is a determining set, there exists, for every measurable set $F \subset A$, a set $E \in \mathfrak{F}_1$ such that $E \cap A$ differs from F by a μ-null set. Hence, by (1.1.12), we obtain

$$\int_F \varphi\rho\, d\mu = \int_{E\cap A} \varphi\rho\, d\mu = \int_E \varphi\rho\, d\mu = 0.$$

Therefore, $\varphi(g) = 0$ for almost all $g \in A$, that is, φ is the zero vector of $L^2(\Omega, \rho)$. This contradiction proves that $\mathfrak{D}^0 = L^2(\Omega, \rho)$.]

4° Measures on Product Spaces

Notice that, although Halmos [1] considers only products of countably many measure spaces, his method of treatment can also be used to define the product of arbitrarily many measure[8] spaces. We shall not give a detailed account of this construction here. In what follows, the product of the family of measure spaces $\{\Omega_\alpha = (G_\alpha, \mathfrak{B}_\alpha, \mu_\alpha), \alpha \in \mathfrak{A}\}$ will be denoted by $\mathsf{X}_{\alpha\in\mathfrak{A}}\,\Omega_\alpha$, or by $(\mathsf{X}_{\alpha\in\mathfrak{A}}\,G_\alpha, \mathsf{X}_{\alpha\in\mathfrak{A}}\,\mathfrak{B}_\alpha, \mathsf{X}_{\alpha\in\mathfrak{A}}\,\mu_\alpha)$. We proceed to mention a few obvious facts concerning such products.

Let $(G, \mathfrak{B}, \mu_k)$, $k = 1, 2$, and $(H, \mathfrak{F}, \nu_k)$, $k = 1, 2$, be measure spaces. Then, for the measure $\mu_1 \times \nu_1$ on $(G \times H, \mathfrak{B} \times \mathfrak{F})$ to be absolutely continuous with respect to $\mu_2 \times \nu_2$, it is necessary[9] and sufficient that $\mu_1 \ll \mu_2$, $\nu_1 \ll \nu_2$. In this case, we have[10]

$$\frac{d\mu_1 \times \nu_1(g, h)}{d\mu_2 \times \nu_2(g, h)} = \frac{d\mu_1(g)}{d\mu_2(g)}\,\frac{d\nu_1(h)}{d\nu_2(h)}\,. \tag{1.1.13}$$

Let $\{\Omega_n = (G_n, \mathfrak{B}_n, \mu_n),\ n = 1, 2, \ldots\}$ be a sequence of probability measure spaces, and let $\Omega = (G, \mathfrak{B}, \mu) = \mathsf{X}_{n=1}^{\infty}\,\Omega_n$. Let $\Omega^{(n)} = \mathsf{X}_{\nu=1}^{n}\,\Omega_\nu = (G^{(n)}, \mathfrak{B}^{(n)}, \mu^{(n)})$. For every $f \in L^2(\Omega^n)$, define a function on Ω by means of the correspondence $g \to f(g^{(n)})$, where $g = \{g_1, g_2, \ldots, g_n, \ldots\} \in G$ and $g^{(n)} = \{g_1, \ldots, g_n\} \in G^{(n)}$. This function clearly belongs to $L^2(\Omega)$. In this way, $L^2(\Omega^{(n)})$ is imbedded as a closed linear subspace of $L^2(\Omega)$. Let P_n be the operator projecting $L^2(\Omega)$ onto $L^2(\Omega^{(n)})$. We then have the following lemma.

Lemma 1.1.7. $\{P_n\}$ converges strongly to the identity operator I.

PROOF. Since we obviously have $P_1 \leqslant P_2 \leqslant \cdots \leqslant P_n \leqslant \cdots$, we need only prove that $Q = \bigcup_{n=1}^{\infty} L^2(\Omega^{(n)})$ is dense in $L^2(\Omega)$. Let $\mathfrak{D}$ denote the totality of bounded measurable real-valued functions in Q; clearly $\mathfrak{D}$ is a real algebra containing the unit element 1. Now, for any n, and any n-dimensional Borel set E, let

$$\tilde{E} = \{g \mid g = \{g_1, g_2, \ldots\} \in G, \quad \{g_1, \ldots, g_n\} \in E\}.$$

Then, the characteristic function $C_{\tilde{E}} \in L^2(\Omega^{(n)})$, hence $C_{\tilde{E}} \in \mathfrak{D}$. Moreover, the totality of such sets $\tilde{E}$ generates $\mathfrak{B}$. Consequently, $\mathfrak{D}$ is a determining set on Ω. Therefore, by Lemma 1.1.6, any real-valued function in $L^2(\Omega)$ can be approximated by elements of $\mathfrak{D}$. Since $\mathfrak{D} \subset Q$, it follows easily that, in either the real or complex case, Q is dense in $L^2(\Omega)$.]

[8] *Translator's note:* If the number of measure spaces is infinite, then, clearly, one must require that all but a finite number of them be probability spaces, that is, $\mu_\alpha(\Omega_\alpha) = 1$.

[9] *Translator's note:* If, say, μ_1 is identically zero, we can draw no conclusion regarding ν_1.

[10] *Translator's note:* Presumably, all the measures concerned are assumed to be σ-finite.

5° Direct Sums of Measures

Definition 1.1.8. Let $\Omega_\alpha = (G_\alpha, \mathfrak{B}_\alpha, \mu_\alpha)$, $\alpha \in \mathfrak{A}$ be a family of measure spaces, where $\{G_\alpha, \alpha \in \mathfrak{A}\}$ is a family of pairwise disjoint sets. Let $G = \bigcup_{\alpha \in \mathfrak{A}} G_\alpha$. Let $\mathfrak{B}$ be the totality of sets of the form

$$A = \bigcup_{\nu=1}^{\infty} A_{\alpha_\nu}, \qquad A_{\alpha_\nu} \in \mathfrak{B}_{\alpha_\nu}, \quad \{\alpha_1, \alpha_2, ..., \alpha_n, ...\} \subset \mathfrak{A}. \tag{1.1.14}$$

Define the set function μ on $(G, \mathfrak{B})$ as follows: if A is of the form (1.1.14), then

$$\mu(A) = \sum_{\nu=1}^{\infty} \mu_{\alpha_\nu}(A_{\alpha_\nu}).$$

Then we say that $\Omega = (G, \mathfrak{B}, \mu)$ is the *direct sum* of the measure spaces $\{\Omega_\alpha, \alpha \in \mathfrak{A}\}$. If, moreover, $\mu_\alpha(G_\alpha) < \infty$ for all $\alpha \in \mathfrak{A}$, we say that $\{G_\alpha, \alpha \in \mathfrak{A}\}$ is a *partition* of Ω.

Clearly, the Ω described in Definition 1.1.8 is a measure space.

Lemma 1.1.8. Let $\Omega = (G, \mathfrak{B}, \mu)$ be the direct sum of the family of measure spaces $\{(G_\alpha, \mathfrak{B}_\alpha, \mu_\alpha), \alpha \in \mathfrak{A}\}$. Then $B \in \tilde{\mathfrak{B}}$ if and only if, for every $\alpha \in \mathfrak{A}$,

$$B \cap G_\alpha \in \tilde{\mathfrak{B}}_\alpha, \tag{1.1.15}$$

and, in that case,

$$\mu(B) = \sum_{\alpha \in \mathfrak{A}} \mu_\alpha(B \cap G_\alpha). \tag{1.1.16}$$

PROOF. If B satisfies condition (1.1.15), then, for any A of the form (1.1.14), we obviously have

$$A \cap B = \bigcup_{\nu=1}^{\infty} (A_{\alpha_\nu} \cap B),$$

whence, by (1.1.15), $A_{\alpha_\nu} \cap B \in \mathfrak{B}_{\alpha_\nu} \subset \mathfrak{B}$, and therefore $B \in \tilde{\mathfrak{B}}$. Conversely, if $B \in \tilde{\mathfrak{B}}$, $A_\alpha \in \mathfrak{B}_\alpha$, then

$$(B \cap G_\alpha) \cap A_\alpha = B \cap A_\alpha \in \mathfrak{B},$$

but $B \cap A_\alpha \subset G_\alpha$, hence $B \cap A_\alpha \in \mathfrak{B}_\alpha$, thus, (1.1.15) holds.

It remains to prove (1.1.16). Let $A \in \mathfrak{B}$ be of the form (1.1.14). Then

$$\mu(B \cap A) = \sum \mu_{\alpha_i}(B \cap A_{\alpha_i}) \leqslant \sum \mu_\alpha(B \cap G_\alpha),$$

whence we get

$$\mu(B) = \sup_{A \in \mathfrak{B}} \mu(B \cap A) \leqslant \sum \mu_\alpha(B \cap G_\alpha). \tag{1.1.17}$$

If the right-hand side of (1.1.17) is a finite number, then there are at most countably many indices α such that $\mu_\alpha(G_\alpha \cap B) > 0$; denoting these indices by $\alpha_1, \ldots, \alpha_n, \ldots$, we have

$$\sum_\alpha \mu_\alpha(G_\alpha \cap B) = \sum \mu_{\alpha_i}(G_{\alpha_i} \cap B). \tag{1.1.18}$$

If the right-hand side of (1.1.17) is ∞, one can also find indices $\alpha_1, \ldots, \alpha_n, \ldots \in \mathfrak{A}$ such that (1.1.18) holds. However, $\mu_{\alpha_i}(G_{\alpha_i} \cap B) = \mu(G_{\alpha_i} \cap B)$, hence

$$\mu(B) \geqslant \sum_{\alpha_i} \mu(G_{\alpha_i} \cap B) \geqslant \sum \mu_{\alpha_i}(G_{\alpha_i} \cap B). \tag{1.1.19}$$

Combining (1.1.17), (1.1.18), and (1.1.19), we obtain (1.1.16).]

Corollary 1.1.9. Under the conditions of Lemma 1.1.8, let $B \in \tilde{\mathfrak{B}}$; then B is a μ-null set if and only if, for all $\alpha \in \mathfrak{A}$, $\mu_\alpha(B \cap G_\alpha) = 0$.

Example 1.1.1. Let $\Omega = (G, \mathfrak{B}, \mu)$ be a measure space. If there is a sequence G_n, $n = 1, 2, \ldots$, of disjoint sets of $\mathfrak{B}$, such that $G = \bigcup_{n=1}^{\infty} G_n$, then $\{G_n, n = 1, 2, \ldots\}$ is a partition of Ω.

In general, let $\Omega = (G, \mathfrak{B}, \mu)$ be a measure space, and let $\{G_\alpha, \alpha \in \mathfrak{A}\} \subset \mathfrak{B}$ be a family of disjoint sets such that $G = \bigcup_{\alpha \in \mathfrak{A}} G_\alpha$. If $\mathfrak{A}$ is not countable, then $\{G_\alpha, \alpha \in \mathfrak{A}\}$ is not necessarily a partition of Ω. For example, if G is the interval $[0, 1]$, and μ is the ordinary Lebesgue measure on G, then $\{\{\alpha\}, \alpha \in [0, 1]\}$ is not a partition of $[0, 1]$.

Let $\Omega_\alpha = (G_\alpha, \mathfrak{B}_\alpha, \mu_\alpha)$, $\alpha \in \mathfrak{A}$ be a family of measure spaces, and let $\Omega = (G, \mathfrak{B}, \mu)$ be the direct sum of this family. For each $\alpha \in \mathfrak{A}$, let f_α be a given measurable function on $\Omega_\alpha = (G_\alpha, \mathfrak{B}_\alpha, \mu_\alpha)$. Define a function f on G as follows: if $g \in G_\alpha$, then

$$f(g) = f_\alpha(g).$$

For any Borel set A and any $\alpha \in \mathfrak{A}$, the set

$$\{g \mid f(g) \in A\} \cap G_\alpha = \{g \mid f_\alpha(g) \in A\}$$

is measurable, hence f is measurable. Moreover, it is easy to prove the following lemma.

Lemma 1.1.10. Let $\Omega = (G, \mathfrak{B}, \mu)$ be the direct sum of the family of measure spaces $\{\Omega_\alpha = (G_\alpha, \mathfrak{B}_\alpha, \mu_\alpha), \alpha \in \mathfrak{A}\}$. Extend each $f_\alpha \in L^2(\Omega_\alpha)$ to a function on G by defining its values on $G - G_\alpha$ to be zero. Then,

$$L^2(\Omega) = \sum_{\alpha \in \mathfrak{A}} \oplus L^2(\Omega_\alpha).$$

6⁰ Measures on Groups

In the sequel, we shall frequently make use of the following type of measure on a group.

Definition 1.1.9. Let G be a group, $\mathfrak{B}$ a σ-ring of subsets of G, $\Omega = (G, \mathfrak{B}, \mu)$ a measure space. Suppose there is a subgroup G_0 of G such that each of the left (right) cosets $\{G_\alpha, \alpha \in \mathfrak{A}\}$ of G_0 in G belongs to $\mathfrak{B}$ and is σ-finite. Let $\mathfrak{B}_\alpha, \mu_\alpha$ be the restriction of $\mathfrak{B}$, μ to G_α, and suppose that Ω is the direct sum of the $\{(G_\alpha, \mathfrak{B}_\alpha, \mu_\alpha), \alpha \in \mathfrak{A}\}$. We then say that Ω is *pseudo-σ-finite.*

Definition 1.1.10. Let G be a topological space, $(G, \mathfrak{B}, \mu)$ a measure space. If, for every $x_0 \in G$, there is a neighborhood V of x_0 such that $V \in \mathfrak{B}$ and $\mu(V) < \infty$, then $(G, \mathfrak{B}, \mu)$ is said to be *locally finite.*

For example, if G is a locally compact group, the Haar measure space on G is locally finite.

If $(G, \mathfrak{B}, \mu)$ is a locally finite measure space, then, for any compact subset C of G, there is an open set $O \in \mathfrak{B}$ such that $C \subset O$ and $\mu(O) < \infty$.

In fact, for each $x \in C$ there is a $V_x \in \mathfrak{B}$ such that $\mu(V_x) < \infty$. By the compactness of C, there exist $x_1, \ldots, x_n$ such that

$$O = V_{x_1} \cup \cdots \cup V_{x_n} \supset C.$$

Then $O \in \mathfrak{B}$ and $\mu(O) \leqslant \sum_{\nu=1}^{n} \mu(V_{x_\nu}) < \infty$.

Lemma 1.1.11. Let G be a locally compact group, $\mathfrak{B}$ the σ-ring generated by the totality of compact subsets of G. If $(G, \mathfrak{B}, \mu)$ is a locally finite measure space, then it is also pseudo-σ-finite.

PROOF. Choose any neighborhood V of the identity e of G, such that the closure C of V is compact, and $C = C^{-1}$. Let C^i denote the set

$$\overbrace{CC \cdots C}^{i},$$

and form the subgroup

$$G_0 = \bigcup_{i=1}^{\infty} C^i.$$

Since C is compact, and C^2 is the image, under the continuous mapping $(x, y) \to xy$, of the compact subset $C \times C$ of the product space $G \times G$, it follows that C^2 is compact. Similarly, every C^i is compact, so that the measure of C^i is finite, and hence G_0 is σ-finite.

Let $\{G_\alpha, \alpha \in \mathfrak{A}\}$ be the left coset system of G_0 in G. Clearly, each G_α is also the union of countably many compact sets, and hence is

σ-finite. We shall now prove that $(G, \mathfrak{B}, \mu)$ is the direct sum of the measure spaces $\{(G_\alpha, \mathfrak{B}_\alpha, \mu_\alpha), \alpha \in \mathfrak{A}\}$, where $\mathfrak{B}_\alpha, \mu_\alpha$ denotes the restriction of $\mathfrak{B}, \mu$ to G_α. Since $\mathfrak{B}$ is generated by the totality of compact subsets of G, we need only prove that, for any compact set $K \subset G$,

$$\mu(K) = \sum_\alpha \mu_\alpha(K \cap G_\alpha).$$

Since $\mu_\alpha(K \cap G_\alpha) = \mu(K \cap G_\alpha)$, it suffices to prove that there are only countably many α such that $\mu(K \cap G_\alpha) > 0$. In fact, since K is compact, there exists a finite set $x_1, \ldots, x_n$ such that

$$\bigcup_{\nu=1}^{n} (x_\nu V) \supset K.$$

Let $x_\nu \in G_{\alpha_\nu}$. Since $C = C^{-1}$, $G_{\alpha_\nu} C = G_{\alpha_\nu}$, therefore

$$K \subset \bigcup_{\nu=1}^{n} G_{\alpha_\nu}.$$

Hence, if $\alpha \neq \alpha_1, \ldots, \alpha_n$, then $K \cap G_\alpha = 0$, so that $\mu(K \cap G_\alpha) = 0$.]

Corollary 1.1.12. The Haar measure on a locally compact topological group is pseudo-σ-finite.

We shall next introduce some results concerning integrals with respect to Haar measures on locally compact groups. In the following three propositions, we assume that G is a locally compact group, and that $\Omega = (G, \mathfrak{B}, \mu)$ is the left invariant Haar measure space on G.

Lemma 1.1.13. Let $a \in L^1(\Omega)$, $\xi \in L^1(\Omega)$ (or $L^2(\Omega)$). Let

$$(a * \xi)(g) = \int_G a(g_1)\, \xi(g_1^{-1} g)\, d\mu(g_1). \tag{1.1.20}$$

Then $a * \xi \in L^1(\Omega)(L^2(\Omega))$. Moreover, if G is commutative, then $a * \xi = \xi * a$.

Proof. If $\xi \in L^1(\Omega)$, then, by the Fubini theorem,

$$\begin{aligned}
&\int_G \left(\int_G | a(g_1)\, \xi(g_1^{-1} g)|\, d\mu(g_1) \right) d\mu(g) \\
&= \int_G \left(\int_G | a(g_1)\, \xi(g_1^{-1} g)|\, d\mu(g) \right) d\mu(g_1) \\
&= \int_G | a(g_1)|\, d\mu(g_1) \int_G | \xi(g_1)|\, d\mu(g_1),
\end{aligned} \tag{1.1.21}$$

hence $a * \xi \in L^1(\Omega)$, and $\| a * \xi \|_1 \leqslant \| a \|_1 \| \xi \|_1$.

If $\xi \in L^2(\Omega)$, then, by the Cauchy inequality,

$$\left(\int_G | a(g_1)\, \xi(g_1^{-1}g)|\, d\mu(g_1)\right)^2$$
$$\leqslant \int_G | a(g_1)|\, d\mu(g_1) \int_G | a(g_1)|\, | \xi^2(g_1^{-1}g)|\, d\mu(g_1). \tag{1.1.22}$$

Also, $\xi^2 \in L^1(\Omega)$, hence, substituting ξ^2 for ξ in (1.1.21), we get

$$\int_G \left(\int_G | a(g_1)|\, | \xi^2(g_1^{-1}g)|\, d\mu(g_1)\right) d\mu(g) \leqslant \| a \|_1 \| \xi \|_2^2 . \tag{1.1.23}$$

Integrating both sides of (1.1.22) with respect to g, and using (1.1.23), we obtain

$$\int_G \left(\int_G | a(g_1)\, \xi(g_1^{-1}g)|\, d\mu(g_1)\right)^2 d\mu(g) \leqslant \| a \|_1^2 \| \xi \|_2^2 .$$

Consequently, $a * \xi \in L^2(\Omega)$, and $\| a * \xi \|_2 \leqslant \| a \|_1 \| \xi \|_2$.

If G is commutative, then μ is also right invariant, and $d\mu(g_1) = d\mu(g_1^{-1})$. Setting $g' = gg_1^{-1}$, (1.1.20) becomes

$$(a * \xi)(g) = \int a(g'^{-1}g)\, \xi(g')\, d\mu(g'),$$

whence $a * \xi = \xi * a$.]

Lemma 1.1.14. Let $\xi, \eta \in L^2(\Omega)$. Then there exist $\{a_n\} \subset L^1(\Omega) \cap L^2(\Omega)$ such that

$$\lim_{n\to\infty} \| a_n * \xi - \xi \|_2 = \lim_{n\to\infty} \| a_n * \eta - \eta \|_2 = 0. \tag{1.1.24}$$

PROOF. Let U be a neighborhood of the identity in G such that $\mu(U) < \infty$. Form the function

$$Z_U(g) = \begin{cases} \dfrac{1}{\mu(U)}, & g \in U, \\ 0, & g \bar{\in} U. \end{cases}$$

Then $Z_U \in L^1(\Omega) \cap L^2(\Omega)$, and $\int_G Z_U(g)\, d\mu(g) = 1$, hence

$$(Z_U * \xi)(g) - \xi(g) = \int Z_U(g_1)(\xi(g_1^{-1}g) - \xi(g))\, d\mu(g_1).$$

Write $\xi(g_1^{-1}g) = \xi_{g_1}(g)$. Substituting Z_U for a and replacing $\xi(g_1^{-1}g)$ by

$\xi(g_1^{-1}g) - \xi(g)$ in (1.1.20), then integrating with respect to g, we get

$$\| Z_U * \xi - \xi \|_2^2 \leqslant \int_G \left(\int_G Z_U(g_1) \mid \xi(g_1^{-1}g) - \xi(g) \mid d\mu(g_1)\right)^2 d\mu(g)$$

$$\leqslant \| Z_U \|_1 \int_G \mid Z_U(g_1) \mid \| \xi_{g_1} - \xi \|_2^2 \, d\mu(g)$$

$$\leqslant \sup_{g_1 \in U} \| \xi_{g_1} - \xi \|_2^2 .$$

Now, given any positive number ϵ, there exists[11] a continuous function ξ', with a compact support, such that $\| \xi' - \xi \|_2 < \epsilon/3$. Furthermore, there exists[12] a neighborhood V_ϵ of the identity such that, if $U \subset V_\epsilon$, $g^1 \in U$, then

$$\| \xi'_{g_1} - \xi' \|_2 < \epsilon/3.$$

Therefore, for such a U, using $\| Z_U \|_1 = 1$, we get

$$\| Z_U * \xi - \xi \|_2 \leqslant \| Z_U * (\xi - \xi') \|_2 + \| \xi - \xi' \|_2 + \| Z_U * \xi' - \xi' \|_2$$

$$\leqslant 2 \| \xi - \xi' \|_2 + \max_{g_1 \in U} \| \xi'_{g_1} - \xi' \|_2 < \epsilon.$$

Similarly, for $\eta \in L^2(\Omega)$ and $\epsilon > 0$, there exists a neighborhood W_ϵ of the identity such that, if $U \subset W_\epsilon$, then

$$\| Z_U * \eta - \eta \|_2 < \epsilon.$$

Taking $a_n = Z_{V_{1/n} \cap W_{1/n}}$, we have

$$\| a_n * \xi - \xi \|_2 < \frac{1}{n}, \qquad \| a_n * \eta - \eta \|_2 < \frac{1}{n}.$$

Thus, we obtain (1.1.24).]

Corollary 1.1.15. Let $\xi \in L^2(\Omega)$. If, for every $a \in L^1(\Omega) \cap L^2(\Omega)$, $a * \xi = 0$, then $\xi = 0$.

PROOF. Choose $\{a_n\} \subset L^1(\Omega) \cap L^2(\Omega)$ such that (1.1.24) holds. Then $a_n * \xi = 0$, whence, by (1.1.24), we obtain $\xi = 0$.]

7° Regular Measures on Topological Spaces

In the present book, we shall often restrict our considerations to regular measures on topological spaces (not necessarily locally compact).

11 *Translator's note:* See Naimark [1], p. 137.
12 *Translator's note:* See Naimark [1], p. 373.

Definition 1.1.11. Let G be a topological space, $(G, \mathfrak{B}, \mu)$ a measure space. If, for every $E \in \mathfrak{B}$,

$$\mu(E) = \inf\{\mu(U) \mid U \in \mathfrak{B}, \quad U \supset E, \quad U \text{ open}\},$$

then $(G, \mathfrak{B}, \mu)$ is said to be *outer regular*. If, for every $E \in \mathfrak{B}$,

$$\mu(E) = \sup\{\mu(C) \mid C \in \mathfrak{B}, \quad C \subset E, \quad C \text{ closed compact}\},$$

then $(G, \mathfrak{B}, \mu)$ is said to be *inner regular*. $(G, \mathfrak{B}, \mu)$ is regular if it is both inner regular and outer regular.

We proceed to establish some conditions for the regularity of measure spaces.

Lemma 1.1.16. Let G be a topological space, $(G, \mathfrak{B}, \mu)$ a finite measure space.[13] Then $(G, \mathfrak{B}, \mu)$ is outer regular if and only if, for every $E \in \mathfrak{B}$,

$$\mu(E) = \sup\{\mu(F) \mid F \in \mathfrak{B}, \quad F \subset E, \quad F \text{ closed}\}. \tag{1.1.25}$$

Proof. We know that $(G, \mathfrak{B}, \mu)$ is outer regular if and only if, for every $E \in \mathfrak{B}$, there is a sequence of open sets $\{O_n\} \subset \mathfrak{B}$ such that $O_n \supset E$, $\mu(O_n) \to \mu(E)$. Letting $G - E = E'$, this condition is equivalent to the existence of $F_n(= G - O_n) \subset E'$ such that $\mu(F_n) \to \mu(E')$. But this is just condition 1.1.25.]

Lemma 1.1.17. Let G be a separable metric space, $(G, \mathfrak{B}, \mu)$ a finite measure space, and suppose that every ball belongs to $\mathfrak{B}$. Then, for any positive number ϵ, there exists a totally bounded set $C \in \mathfrak{B}$, such that $\mu(G - C) < \epsilon$.

Proof. Let $\{x_n\}$ be a sequence of points dense in G. Let $S(x, a)$ denote the closed ball with center x and radius a. Then, for any natural number k,

$$\bigcup_{n=1}^{\infty} S\left(x_n, \frac{1}{k}\right) = G.$$

Consequently, there is a natural number $p(k)$ such that

$$\mu\left(\bigcup_{n=1}^{p(k)} S\left(x_n, \frac{1}{k}\right)\right) > \mu(G) - \frac{\epsilon}{2^k}. \tag{1.1.26}$$

Form the set

$$C = \bigcap_{k=1}^{\infty} \bigcup_{n=1}^{p(k)} S\left(x_n, \frac{1}{k}\right) \in \mathfrak{B}.$$

[13] Here, $\mathfrak{B}$ is assumed to be an algebra.

From (1.1.26), it is easily seen that

$$\mu(G - C) \leqslant \sum_{k=1}^{\infty} \mu\left(G - \bigcup_{n=1}^{p(k)} S\left(x_n, \frac{1}{k}\right)\right) < \epsilon.$$

But, by its very construction, C is totally bounded. In fact, given any $\delta > 0$, choose a natural number k such that $1/k < \delta$; then $x_1, \ldots, x_{p(k)}$ is a δ-net for C.]

Theorem 1.1.18. Let G be a complete separable metric space, $\mathfrak{B}$ the σ-algebra generated by the totality of closed sets of G, and μ a finite measure on $(G, \mathfrak{B})$. Then $(G, \mathfrak{B}, \mu)$ is a regular measure space.

PROOF. By virtue of Lemma 1.1.17, for every natural number n, there exists a compact set C_n such that

$$\mu(G - C_n) < \frac{1}{n}. \tag{1.1.27}$$

Since C_n is a compact Baire set, the restriction of μ to C_n is regular. Hence, for any $E \in \mathfrak{B}$, there is a compact set $F_n \in \mathfrak{B}$ such that $F_n \subset E \cap C_n$ and

$$\mu((E \cap C_n) - F_n) < \frac{1}{n}. \tag{1.1.28}$$

Since $E - F_n \subset G - C_n + [(E \cap C_n) - F_n]$, it follows from (1.1.27) and (1.1.28) that

$$\mu(E) < \mu(F_n) + \frac{2}{n}.$$

Thus, we obtain a sequence of closed compact sets $F_n \subset E$, such that $\mu(F_n) \to \mu(F)$. This proves that $(G, \mathfrak{B}, \mu)$ is inner regular. By Lemma 1.1.16, μ is also outer regular.]

8° Some Probabilistic Terminology

We now briefly present some terminology from probability theory which will be used in the sequel. Let $S = (\Omega, \mathfrak{B}, P)$ be a probability measure space, and let $X(\omega)$, $\omega \in \Omega$, be a measurable function on S. We call $X(\omega)$ a *random variable*[14] on S, and

$$E(e^{iX(\cdot)t}) = \int_{\Omega} e^{iX(\omega)t}\, dP(\omega), \qquad -\infty < t < \infty$$

[14] In the present book, we shall, in general, consider only the case of real-valued functions.

the *characteristic function* of X. If $X(\cdot) \in L^1(S)$, then

$$E(X) = \int_\Omega X(\omega)\, dP(\omega)$$

is called the *mean value* or *mathematical expectation* of the random variable X. If $X \in L^2(S)$, then

$$\sigma(X) = E((X - E(X))^2) = \int_\Omega (X(\omega) - E(X))^2\, dP(\omega)$$

is known as the *variance* of X. If X, Y are two random variables such that $X(\cdot)$, $Y(\cdot) \in L^2(S)$, we call

$$E(XY) = \int X(\omega)\, Y(\omega)\, dP(\omega)$$

the *correlation number* of X and Y. Let $X_1, \ldots, X_n$ be n random variables on S. Form a Borel measure μ on n-dimensional real space R_n, as follows: if E is a Borel set in R_n, define

$$\mu(E) = P(\{\omega \mid (X_1(\omega), \ldots, X_n(\omega)) \in E\}).$$

This μ is a probability measure, and is called the *joint (probability) distribution* of $X_1, \ldots, X_n$. Sometimes, the function

$$F(x_1, \ldots, x_n) = P(\{\omega \mid X_\nu(\omega) \leqslant x_\nu, \quad \nu = 1, 2, \ldots, n\})$$

is called the *joint distribution function* of $X_1, \ldots, X_n$. In particular, when $n = 1$, $X = X_1$, μ (or the corresponding F) is known as the *probability distribution* (or *distribution function*) of the random variable X. It is easily seen that, if μ is the joint distribution of $X_1, \ldots, X_n$, then, for any bounded Baire function f, we have

$$E(f(X_1, \ldots, X_n)) = \int f(X_1(\omega), \ldots, X_n(\omega))\, dP(\omega)$$

$$= \int_{-\infty}^{\infty} \cdots \int_{-\infty}^{\infty} f(x_1, \ldots, x_n)\, d\mu(x_1, \ldots, x_n).$$

9° Absolute Continuity and Singularity Relations among Measures

In this subsection, we shall prove a theorem concerning limits of Radon–Nikodym derivatives. For this purpose, we first introduce the following concept.

Definition 1.1.12. Let $S = (\Omega, \mathfrak{B}, P)$ be a probability measure space, $\mathfrak{B}_1, \ldots, \mathfrak{B}_n$ a finite sequence of σ-algebras such that $\mathfrak{B}_1 \subset \mathfrak{B}_2 \subset \cdots \subset \mathfrak{B}_n \subset \mathfrak{B}$. Let $x_1(\omega), \ldots, x_n(\omega)$ be (real) random variables on S such that x_i is measurable with respect to $(\Omega, \mathfrak{B}_i)$. If, for any i, $1 \leqslant i < n$, and any $\Lambda \in \mathfrak{B}_i$,

$$\int_\Lambda x_i(\omega)\, dP(\omega) \leqslant \int_\Lambda x_{i+1}(\omega)\, dP(\omega), \tag{1.1.29}$$

then $\{x_i, \mathfrak{B}_i, i = 1, \ldots, n\}$ is said to be a *semimartingale* on S.

Clearly, if $\{x_i, \mathfrak{B}_i, i = 1, \ldots, n\}$ is a semimartingale, then, for any real number c, and any $\Lambda \in \mathfrak{B}_i$, we have

$$\begin{aligned}\int_\Lambda \max(x_i(\omega) - c, 0)\, dP(\omega) &= \int_{\Lambda\{\omega \mid x_i(\omega) \geqslant c\}} (x_i(\omega) - c)\, dP(\omega)\\ &\leqslant \int_{\Lambda\{\omega \mid x_i(\omega) \geqslant c\}} (x_{i+1}(\omega) - c)\, dP(\omega)\\ &\leqslant \int_\Lambda \max(x_{i+1}(\omega) - c, 0)\, dP(\omega).\end{aligned}$$

Consequently, $\{\max(x_i(\omega) - c, 0), \mathfrak{B}_i, i = 1, \ldots, n\}$ is also a semimartingale.

We shall also make use of certain quantities defined as follows. Let $\xi_1, \ldots, \xi_n$ be a finite sequence of real numbers, and let r_1, r_2 be a pair of real numbers with $r_1 < r_2$. Let ν_1 be the smallest index j such that $\xi_j \leqslant r_1$, let ν_2 be the smallest index j such that $\xi_j \geqslant r_2$, $j > \nu_1$, let ν_3 be the smallest index j such that $\xi_j \leqslant r_1$, $j > \nu_2$, and so on. Continuing in this manner, we get

$$\xi_{\nu_1} \leqslant r_1, \ \xi_{\nu_2} \geqslant r_2, \ \xi_{\nu_3} \leqslant r_1, \ \xi_{\nu_4} \geqslant r_2, \ldots .$$

Let m be the greatest integer i such that ν_i can be defined in this manner, and let β denote the greatest integer not exceeding $m/2$. We call β the number of times the sequence $\xi_1, \ldots, \xi_n$ *crosses* the interval $[r_1, r_2]$.

Lemma 1.1.19. Let $\{x_i(\omega), B_i, i = 1, \ldots, n\}$ be a semimartingale on the probability measure space $S = (\Omega, \mathfrak{B}, P)$. Let r_1, r_2 be two real numbers, $r_1 < r_2$. For each $\omega \in \Omega$, let $\beta(\omega)$ denote the number of times that $x_1(\omega), \ldots, x_n(\omega)$ crosses $[r_1, r_2]$. Then

$$\begin{aligned}E(\beta) &\leqslant \frac{1}{r_2 - r_1} E(\max(x_n - r_1, 0))\\ &\leqslant \frac{1}{r_2 - r_1} (E(|x_n|) + |r_1|).\end{aligned} \tag{1.1.30}$$

PROOF. The second inequality in (1.1.30) is obvious; we proceed to prove the first. Firstly, set

$$r_1 = 0, \qquad x_i(\omega) \geqslant 0, \qquad i = 1,\ldots, n, \quad \omega \in \Omega. \tag{1.1.31}$$

For each $\omega \in \Omega$, we take $x_1(\omega),\ldots, x_n(\omega)$ as the $\xi_1 ,\ldots, \xi_n$ in the procedure described above, thereby obtaining numbers $\nu_1(\omega)(= 1),\ldots, \nu_{m(\omega)}(\omega)$ and $\beta(\omega)$. It is easily seen that $\beta(\omega)$ is a measurable function. For those indices j such that $m(\omega) < j \leqslant n$, we shall, for convenience, define $\nu_j(\omega) = n + 1$. The $\nu_1(\omega),\ldots, \nu_n(\omega)$ so obtained are also measurable functions on S. We further define measurable functions $u_2 ,\ldots, u_n$ and x as follows:

$$u_j(\omega) = \begin{cases} 1, & j \leqslant \nu_1(\omega), \\ 1, & \nu_i(\omega) < j \leqslant \nu_{i+1}(\omega), \quad i \text{ even}, \\ 0, & \nu_i(\omega) < j \leqslant \nu_{i+1}(\omega), \quad i \text{ odd}, \end{cases}$$

$$x(\omega) = x_1(\omega) + \sum_{j=2}^{n} u_j(\omega)(x_j(\omega) - x_{j-1}(\omega)).$$

Then,

$$\begin{aligned} \Lambda_{j-1} &= \{\omega \mid u_j(\omega) = 1\} \\ &= \bigcup_{\substack{i\text{even} \\ i \leqslant j-1}} (\{\omega \mid \nu_i(\omega) < j\} - \{\omega \mid \nu_{i+1}(\omega) < j\}). \end{aligned} \tag{1.1.32}$$

The set on the right-hand side of (1.1.32) is determined by $x_1(\omega),\ldots, x_{j-1}(\omega)$. Consequently, $\Lambda_{j-1} \in \mathfrak{B}_{j-1}$. Using (1.1.29), we get

$$\int_{\Lambda_{j-1}} (x_j(\omega) - x_{j-1}(\omega))\, dP(\omega) \geqslant 0,$$

and hence

$$\begin{aligned} &\int_\Omega x(\omega)\, dP(\omega) \\ &\quad = \int_\Omega x_1(\omega)\, dP(\omega) + \sum_{j=2}^{n} \int_{\Lambda_{j-1}} (x_j(\omega) - x_{j-1}(\omega))\, dP(\omega) \geqslant 0. \end{aligned}$$

On the other hand,

$$x(\omega) \leqslant x_n(\omega) - r_2\beta(\omega),$$

therefore,

$$\begin{aligned} 0 &\leqslant \int_\Omega x(\omega)\, dP(\omega) \\ &\leqslant \int_\Omega x_n(\omega)\, dP(\omega) - r_2 \int_\Omega \beta(\omega)\, dP(\omega). \end{aligned} \tag{1.1.33}$$

Thus, (1.1.30) holds under the assumptions (1.1.31). As for the general case, we need only consider

$$x_j'(\omega) = \max(x_j(\omega) - r_1, 0).$$

As we pointed out above, $\{x_j', \mathfrak{B}_j, j = 1,..., n\}$ is also a semimartingale. The number of times that the sequence $x_1',..., x_n'$ crosses $[0, r_2 - r_1]$ is equal to the number of times that $x_1,..., x_n$ crosses $[r_1, r_2]$. But $x_1',..., x_n'$ and $[0, r_2 - r_1]$ satisfy conditions (1.1.31), hence, by (1.1.33), we have

$$0 \leqslant \int_\Omega x_n'(\omega)\, dP(\omega) - (r_2 - r_1) \int_\Omega \beta(\omega)\, dP(\omega),$$

which is just (1.1.30).]

Theorem 1.1.20. Let Ω be a set, and let $\mathfrak{B}_1,..., \mathfrak{B}_n,...$ and $\mathfrak{B}$ be σ-algebras in Ω such that

$$\mathfrak{B}_1 \subset \mathfrak{B}_2 \subset \cdots \subset \mathfrak{B}_n \subset \cdots \subset \mathfrak{B},$$

and such that $\mathfrak{B}$ is generated by $\bigcup_{n=1}^\infty \mathfrak{B}_n$. Let P, Q be two probability measures on $(\Omega, \mathfrak{B})$, with $Q \ll P$.[15] Let P_n, Q_n be the probability measures on $(\Omega, \mathfrak{B}_n)$ defined by

$$P_n(E) = P(E), \qquad Q_n(E) = Q(E), \qquad E \in \mathfrak{B}_n.$$

Obviously, $Q_n \ll P_n$. Let $dQ_n(\omega)/dP_n(\omega)$ and $dQ(\omega)/dP(\omega)$ denote the appropriate Radon–Nikodym derivatives. Then, for almost all $\omega \in \Omega$ (with respect to P), we have

$$\lim_{n\to\infty} \frac{dQ_n(\omega)}{dP_n(\omega)} = \frac{dQ(\omega)}{dP(\omega)}.$$

PROOF. Set $x_n(\omega) = dQ_n(\omega)/dP_n(\omega)$, $x(\omega) = dQ(\omega)/dP(\omega)$. Then x_n is measurable with respect to $\mathfrak{B}_n$, moreover, for any $\Lambda \in \mathfrak{B}_n$, we have $Q_n(\Lambda) = Q_{n+1}(\Lambda) = Q(\Lambda)$ and

$$\int_\Lambda x_n(\omega)\, dP(\omega) = \int_\Lambda x_n(\omega)\, dP_n(\omega) = Q_n(\Lambda),$$

whence

$$\int_\Lambda x_n(\omega)\, dP(\omega) = \int_\Lambda x_{n+1}(\omega)\, dP(\omega)$$
$$= \int_\Lambda x(\omega)\, dP(\omega) = Q(\Lambda). \tag{1.1.34}$$

[15] That is, Q is absolutely continuous with respect to P.

It follows from (1.1.34) that, for any n, $\{x_i, \mathfrak{B}_i, i = 1,\ldots, n\}$ is a semimartingale. We shall now prove that $\lim_{n\to\infty} x_n(\omega)$ exists almost everywhere. Let

$$x^*(\omega) = \overline{\lim_{n\to\infty}}\, x_n(\omega), \qquad x_*(\omega) = \underline{\lim_{n\to\infty}}\, x_n(\omega).$$

Now, for any $c > 0$, we have

$$P(\{\omega \mid x_n(\omega) \geqslant c\}) \leqslant \frac{1}{c}\int_\Omega x_n(\omega)\, dP(\omega) = \frac{1}{c} Q_n(\Omega) = \frac{1}{c}. \tag{1.1.35}$$

Consider the set

$$\Lambda_\infty \equiv \{\omega \mid \lim_{n\to\infty} x_n(\omega) = \infty\} = \bigcap_{i=1}^{\infty} \bigcup_{m=1}^{\infty} \bigcap_{n=m}^{\infty} \{\omega \mid x_n(\omega) > i\}.$$

Since, by (1.1.35), $P(\{\omega \mid x_n(\omega) > i\}) \leqslant 1/i$, it follows that $P(\Lambda_\infty) = 0$. Consequently, if $\lim_{n\to\infty} x_n(\omega)$ did not exist almost everywhere, there would exist positive numbers $r_1 < r_2$ such that the set

$$A = \{\omega \mid x^*(\omega) > r_2 > r_1 > x_*(\omega)\}$$

has positive probability measure. Let $\beta_n(\omega)$ be the number of times that $x_1(\omega),\ldots, x_n(\omega)$ crosses $[r_1, r_2]$. If $\omega \in A$, then $\lim_{n\to\infty} \beta_n(\omega) = \infty$. But, by Lemma 1.1.19,

$$E(\beta_n) \leqslant \frac{1}{r_2 - r}(1 + r_1). \tag{1.1.36}$$

Moreover, since $\beta_1(\omega) \leqslant \beta_2(\omega) \leqslant \cdots \leqslant \beta_n(\omega) \leqslant \cdots$, it follows by the Levi lemma and (1.1.36) that $\lim_{n\to\infty} \beta_n(\omega)$ is finite almost everywhere. This is a contradiction. Therefore, there exists a finite-valued measurable function $x_\infty(\omega)$ on Ω such that

$$\lim_{n\to\infty} x_n(\omega) = x_\infty(\omega) \tag{1.1.37}$$

holds almost everywhere.

We shall now prove that the sequence of integrable functions $\{x_n(\omega)\}$ has uniformly absolutely continuous indefinite integrals on Ω. In fact, for any positive number ϵ, there is a positive number δ such that, when $\Lambda \in \mathfrak{B}$, $P(\Lambda) \leqslant \delta$,

$$Q(\Lambda) = \int_\Lambda x(\omega)\, dP(\omega) < \frac{\epsilon}{2}.$$

On the other hand, by (1.1.35), we have $P(\{\omega \mid x_n(\omega) \geqslant 1/\delta\}) \leqslant \delta$. Hence, if $P(\Lambda) \leqslant \epsilon\delta/2$, then

$$\int_\Lambda x_n(\omega)\, dP(\omega)$$

$$\leqslant \int_{\Lambda \cap \{\omega \mid x_n(\omega) < 1/\delta\}} x_n(\omega)\, dP(\omega) + \int_{\{\omega \mid x_n(\omega) \geqslant 1/\delta\}} x_n(\omega)\, dP(\omega)$$

$$< \frac{1}{\delta} P(\Lambda) + Q\left(\left\{\omega \,\middle|\, x_n(\omega) \geqslant \frac{1}{\delta}\right\}\right) \leqslant \epsilon.$$

This proves that $\{x_n(\omega)\}$ has uniformly absolutely continuous indefinite integrals. Therefore, by (1.1.37) and the well-known Vitali theorem, it follows that $x_\infty(\omega) \in L^1(\Omega)$, and, for every $\Lambda \in \mathfrak{B}$,

$$\lim_{n\to\infty} \int_\Lambda x_n(\omega)\, dP(\omega) = \int_\Lambda x_\infty(\omega)\, dP(\omega).$$

However, by (1.1.34), we have, for $\Lambda \in \mathfrak{B}_n$, $m \geqslant n$,

$$\int_\Lambda x_m(\omega)\, dP(\omega) = \int_\Lambda x(\omega)\, dP(\omega).$$

Consequently, for any $\Lambda \in \bigcup_{n=1}^\infty \mathfrak{B}_n$,

$$\int_\Lambda x_\infty(\omega)\, dP(\omega) = \int_\Lambda x(\omega)\, dP(\omega). \tag{1.1.38}$$

But $\mathfrak{B}$ is the σ-algebra generated by $\bigcup_{n=1}^\infty \mathfrak{B}_n$, hence (1.1.38) also holds for any $\Lambda \in \mathfrak{B}$. This proves that $x_\infty(\omega)$ is equal to $x(\omega)$ almost everywhere.]

We next establish a condition for the mutual singularity of two measures.

Lemma 1.1.21. Let P_1, P_2 be two measures on the measurable space $(G, \mathfrak{B})$. If, for any positive number ϵ, there exist disjoint sets $A_\epsilon \in \mathfrak{B}$, $B_\epsilon \in \mathfrak{B}$, such that $G = A_\epsilon \cup B_\epsilon$ and

$$P_1(A_\epsilon) < \epsilon, \qquad P_2(B_\epsilon) < \epsilon,$$

then P_1 and P_2 are mutually singular.

PROOF. Form the sets

$$A = \bigcap_{m=1}^\infty \bigcup_{n=m}^\infty A_{1/2^n}, \qquad B = \bigcup_{m=1}^\infty \bigcap_{n=m}^\infty B_{1/2^n}.$$

Then,

$$A \cup B = G, \qquad A \cap B = \text{empty set},$$

moreover,

$$P_1(A) = \lim_{m\to\infty} P_1\left(\bigcup_{n=m}^{\infty} A_{1/2^n}\right) \leqslant \lim_{m\to\infty} \sum_{n=m}^{\infty} \frac{1}{2^n} = 0,$$

$$P_2(B) = \lim_{m\to\infty} P_2\left(\bigcap_{n=m}^{\infty} B_{1/2^n}\right) \leqslant \lim_{m\to\infty} P_2(B_{1/2^m}) = 0.$$

Thus, P_1 and P_2 are mutually singular.]

§1.2. Localizable Measure Spaces

1° A Partial Ordering in the Collection of Measurable Sets

Definition 1.2.1. Let $\Omega = (G, \mathfrak{B}, \mu)$ be a measure space, $\tilde{\mathfrak{B}}$ the totality of measurable sets of Ω, $\mathfrak{B}_0$ the totality of null sets. If $E, F \in \tilde{\mathfrak{B}}$ and $(E - F) \cup (F - E) \in \mathfrak{B}_0$, we say that E and F are *equivalent*, and write $E \sim F$. If $E, F \in \tilde{\mathfrak{B}}$ and $E - F \in \mathfrak{B}_0$, we write $E \prec F$.

Clearly, the relations $\sim$ and $\prec$ have the following properties (for $E, F, H \in \tilde{\mathfrak{B}}$):

(i) $E \sim E$; (ii) if $E \sim F$, then $F \sim E$; (iii) if $E \sim F$, $F \sim H$, then $E \sim H$; (iv) if $E \sim F$, then $E \prec F$; (v) if $E \prec F$, $F \prec E$, then $E \sim F$; (vi) if $E \prec F$, $F \prec H$, then $E \prec H$.

Consequently, if equivalent elements of $\tilde{\mathfrak{B}}$ are identified, then $\tilde{\mathfrak{B}}$ becomes a partially ordered set with respect to the relation $\prec$. We now introduce the usual notions of upper and lower bounds, least upper bound and greatest lower bound in this partially ordered set.

Definition 1.2.2. Let $\mathfrak{F} \subset \tilde{\mathfrak{B}}$, $K \in \tilde{\mathfrak{B}}$. If, for every $E \in \mathfrak{F}$, we have $E \prec K$, we call K an *upper bound* for $\mathfrak{F}$. If K_0 is an upper bound for $\mathfrak{F}$ and, for any upper bound K for $\mathfrak{F}$, we have $K_0 \prec K$, then we call K_0 a *least upper bound* for $\mathfrak{F}$, and write $K_0 = \bigvee_{E\in\mathfrak{F}} E$. In a similar fashion, we define *lower bound* and *greatest lower bound*; we denote any greatest lower bound for $\mathfrak{F}$ by $\bigwedge_{E\in\mathfrak{F}} E$.

It is easily seen that, if $\mathfrak{F}$ is a finite or countable collection

$$\bigvee_{E\in\mathfrak{F}} E \sim \bigcup_{E\in\mathfrak{F}} E, \qquad \bigwedge_{E\in\mathfrak{F}} E \sim \bigcap_{E\in\mathfrak{F}} E.$$

Let $\mathfrak{F}$ be a subcollection of $\mathfrak{B}$; if $K \in \mathfrak{B}$, then $\Omega - K \in \mathfrak{B}$, hence, it is easily seen that

$$\Omega - \bigvee_{E \in \mathfrak{F}} E \sim \bigwedge_{E \in \mathfrak{F}} (\Omega - E), \tag{1.2.1}$$

$$\Omega - \bigwedge_{E \in \mathfrak{F}} E \sim \bigvee_{E \in \mathfrak{F}} (\Omega - E). \tag{1.2.2}$$

Lemma 1.2.1. Let E be a set of finite measure, $\mathfrak{F}$ a family of measurable subsets of E. Then, $\mathfrak{F}$ has a least upper bound, and there exists a sequence of sets $E_n \in \mathfrak{F}$, $n = 1, 2, \ldots$, such that $\bigcup_{n=1}^{\infty} E_n$ is a least upper bound for $\mathfrak{F}$.

PROOF. Let

$$\mathfrak{F}_1 = \left(\bigcup_{n=1}^{\infty} F_n \mid F_n \in \mathfrak{F}, \quad n = 1, 2, \ldots \right).$$

Then, any countable union of sets of $\mathfrak{F}_1$ also belongs to $\mathfrak{F}_1$. Let

$$\alpha = \sup_{F \in \mathfrak{F}_1} \mu(F). \tag{1.2.3}$$

Then $\alpha \leqslant \mu(E) < \infty$. Hence, there is a sequence $\lambda_n \in \mathfrak{F}_1$, $n = 1, 2, \ldots$, such that

$$\alpha = \lim_{n \to \infty} \mu(\lambda_n). \tag{1.2.4}$$

Let $E_0 = \bigcup_{n=1}^{\infty} \lambda_n \in \mathfrak{F}_1$; then, $\alpha \geqslant \mu(E_0) \geqslant \mu(\lambda_n)$, and by (1.2.4), we have

$$\mu(E_0) = \alpha.$$

We shall now prove that $E_0 = \bigvee_{F \in \mathfrak{F}} F$. Obviously, if $F \in \mathfrak{F}$, then $F \in \mathfrak{F}_1$, hence $E_0 \cup F \in \mathfrak{F}_1$. But, by (1.2.3), we have $\mu(E_0 \cup F) \leqslant \alpha = \mu(E_0)$, whence $E_0 \cup F - E_0$ is a null set. Thus, E_0 is an upper bound for $\mathfrak{F}$. On the other hand, if K is any upper bound for $\mathfrak{F}$, then, since E_0 is the union of countably many sets $E_n \in \mathfrak{F}$, $n = 1, 2, \ldots$,

$$E_0 - K \subset \bigcup_{n=1}^{\infty} (E_n - K)$$

is a μ-null set. Thus, $E_0 \prec K$, so that E_0 is a least upper bound for $\mathfrak{F}$.]

Corollary 1.2.2. Let E be a σ-finite set, $\mathfrak{F}$ a collection of measurable subsets of E. Then $\mathfrak{F}$ has a least upper bound, and there exists a sequence $E_n \in \mathfrak{F}$, $n = 1, 2, \ldots$, such that $\bigcup_{n=1}^{\infty} E_n = \bigvee_{F \in \mathfrak{F}} F$.

PROOF. Decompose E into the union of a sequence of mutually disjoint sets $E^{(n)}$ having finite measure, $n = 1, 2, \ldots$. Let

$$\mathfrak{F}^{(n)} = \{A \cap E^{(n)} \mid A \in \mathfrak{F}\}.$$

By Lemma 1.2.1, there exists, for every n, a sequence $F_{kn} \in \mathfrak{F}$, $k = 1, 2, \ldots$, such that

$$\bigcup_{k=1}^{\infty} (F_{kn} \cap E^{(n)}) = \bigvee_{A \in \mathfrak{F}^{(n)}} A.$$

It is easy to show that

$$\bigcup_{k,n=1}^{\infty} F_{kn} = \bigvee_{A \in \mathfrak{F}} A. \quad]$$

Lemma 1.2.3. Let $\Omega = (G, \mathfrak{B}, \mu)$ be a measure space, and let $\mathfrak{F}$ be a directed collection of sets of $\tilde{\mathfrak{B}}$, that is, for any $A, B \in \mathfrak{F}$, there exists $C \in \mathfrak{F}$ such that $A \prec C$, $B \prec C$. If E is any least upper bound for $\mathfrak{F}$, then

$$\mu(E) = \sup_{A \in \mathfrak{F}} \mu(A).$$

PROOF. If $A \in \mathfrak{F}$, then, since $A \prec E$, we have $\mu(A) \leqslant \mu(E)$. Hence, it suffices to prove that $\mu(E) \leqslant \sup_{A \in \mathfrak{F}} \mu(A)$ whenever $\sup_{A \in \mathfrak{F}} \mu(A) = \alpha < \infty$. Now, in this case, there are sets $A_n \in \mathfrak{F}$, $n = 1, 2, \ldots$, such that $\mu(A_n) \to \alpha$. Since $\mathfrak{F}$ is directed, we may assume that

$$A_1 \prec A_2 \prec \cdots \prec A_n \prec \cdots.$$

Let $A = \bigcup_{n=1}^{\infty} A_n$; then, $\mu(A) = \alpha$. We shall now prove that A is an upper bound for $\mathfrak{F}$. If it were not, then there would exist $K \in \mathfrak{F}$ such that $\mu(K - A) > 0$. Choosing $B_n \in \mathfrak{F}$ such that

$$K \prec B_n, \qquad A_n \prec B_n,$$

and letting $n \to \infty$, we get

$$\begin{aligned} \alpha \geqslant \mu(B_n) \geqslant \mu(K + A_n) &= \mu(K - A_n) + \mu(A_n) \\ &\to \mu(K - A) + \alpha > \alpha, \end{aligned}$$

a contradiction.]

Lemma 1.2.4. Let $\Omega = (G, \mathfrak{B}, \mu)$ be a measure space, and let

$\mathfrak{F} \subset \mathfrak{B}$ be such that, for any two distinct sets F, F' of $\mathfrak{F}$, we have $\mu(F \cap F') = 0$. If E is a least upper bound for $\mathfrak{F}$, then

$$\mu(E) = \sum_{F \in \mathfrak{F}} \mu(F). \tag{1.2.5}$$

PROOF. If the right-hand side of (1.2.5) is ∞, then one can find $\{F_n \mid n = 1, 2, \ldots\} \subset \mathfrak{F}$ such that $\sum_{n=1}^{\infty} \mu(F_n) = \infty$. But since E is a least upper bound for $\mathfrak{F}$, it is an upper bound for $\{F_n\}$, hence

$$\mu(E) \geqslant \mu\left(\bigcup_{n=1}^{\infty} F_n\right) = \sum_{n=1}^{\infty} \mu(F_n) = \infty.$$

If $\sum_{F \in \mathfrak{F}} \mu(F) < \infty$, then there are at most countably many sets of positive measure in $\mathfrak{F}$; denote these by $F_1, F_2, \ldots$. It is easily seen that $\bigcup_n F_n$ is a least upper bound for $\mathfrak{F}$, therefore, E and $\bigcup_n F_n$ are equivalent, whence we obtain (1.2.5).]

2° Measurable Sets and Projection Operators

Let $L^2(\Omega)$ be the Hilbert space formed by the totality of quadratically integrable (real or complex-valued) functions on Ω (see §1.1). For each $E \in \mathfrak{B}$, let C_E denote the characteristic function of E:

$$C_E(g) = \begin{cases} 1 & \text{if } g \in E, \\ 0 & \text{if } g \in G - E. \end{cases}$$

Let H_E be the closed linear subspace formed by those functions in $L^2(\Omega)$ which vanish almost everywhere (a.e.) outside of E, and let P_E denote the projection from $L^2(\Omega)$ onto H_E . Then

$$(P_E f)(g) = C_E(g) f(g), \qquad f \in L^2(\Omega).$$

Note that the closed linear subspaces of any Hilbert space H form a partially ordered set with respect to the inclusion relation $\subset$. Any family $\mathfrak{F}$ of closed linear subspaces has a greatest lower bound (g.l.b.) with respect to this ordering, namely, the intersection of all the subspaces in $\mathfrak{F}$; we denote this by $\bigwedge_{M \in \mathfrak{F}} M$. Also, $\mathfrak{F}$ has a least upper bound (l.u.b.), that is, the intersection of all those closed linear subspaces which contain all the subspaces in $\mathfrak{F}$; alternatively, it may be described as the closure of the linear subspace consisting of all linear combinations of vectors in $\bigcup_{L \in \mathfrak{F}} L$. We denote the l.u.b. of $\mathfrak{F}$ by $\bigvee_{M \in \mathfrak{F}} M$.

Correspondingly, the (orthogonal) projection operators in H are partially ordered by the relation $\leqslant$. There is a natural one-to-one correspondence between projection operators and closed linear subspaces, and this correspondence is order-preserving, that is, if P, Q are projection operators in H, then $PH \subset QH$ is equivalent to $P \leqslant Q$. Consequently, any family $\mathfrak{F}_1$ of projections in H has a l.u.b. P, defined by the relation

$$PH = \bigvee_{Q \in \mathfrak{F}_1} QH.$$

Denote this l.u.b. by $P = \bigvee_{Q \in \mathfrak{F}_1} Q$. Similarly, $\mathfrak{F}_1$ has a g.l.b., denoted by $\bigwedge_{M \in \mathfrak{F}_1} M$.

Note that, if $E, F \in \tilde{\mathfrak{B}}$, then $E \sim F$ if and only if $H_E = H_F$ (or $P_E = P_F$), and $E \prec F$ if and only if $H_E \subset H_F$ (or $P_E \leqslant P_F$).

In the following lemma, we assume that, for any $E \in \mathfrak{B}$ such that $\mu(E) > 0$, there exists $F \in \mathfrak{B}$, $F \subset E$, such that $0 < \mu(F) < \infty$. This is one of the conditions for localizability (see Definition 1.2.3).

Lemma 1.2.5. Let $\mathfrak{F} \subset \tilde{\mathfrak{B}}$, $E \in \tilde{\mathfrak{B}}$. Then $E = \bigvee_{F \in \mathfrak{F}} F$ if and only if

$$P_E = \bigvee_{F \in \mathfrak{F}} P_F. \tag{1.2.6}$$

PROOF. Suppose $E = \bigvee_{F \in \mathfrak{F}} F$. Then, if $F \in \mathfrak{F}$, we have $F \prec E$, hence $P_F \leqslant P_E$. Therefore,

$$P_E \geqslant \bigvee_{F \in \mathfrak{F}} P_F. \tag{1.2.7}$$

Let $f \in H_E$, and $\mathfrak{M}_f = \{g \mid f(g) \neq 0\}$. Then $\mathfrak{M}_f$ is σ-finite and $\mathfrak{M}_f - E$ is a null set. Let $\mathfrak{F}_f = \{A \cap \mathfrak{M}_f \mid A \in \mathfrak{F}\}$. By Corollary 1.2.2, there is a sequence $A_n \in \mathfrak{F}$, $n = 1, 2, \ldots$, such that

$$\bigcup_{n=1}^{\infty} (A_n \cap \mathfrak{M}_f) \sim \bigvee_{A \in \mathfrak{F}} (A \cap \mathfrak{M}_f).$$

But it is easy to verify that

$$\bigvee_{A \in \mathfrak{F}} (A \cap \mathfrak{M}_f) \sim \Big(\bigvee_{A \in \mathfrak{F}} A\Big) \cap \mathfrak{M}_f = E \cap \mathfrak{M}_f \sim \mathfrak{M}_f.$$

Consequently, as $N \to \infty$,

$$\|f - C_{\cup_{n=1}^{N} A_n} f\|^2 = \int_{\mathfrak{M}_f - \cup_{n=1}^{N} A_n} |f(g)|^2 \, d\mu(g) \to 0. \tag{1.2.8}$$

Moreover, it is easily seen that

$$C_{\cup_{n=1}^{N} A_n} f \in H_{\cup_{n=1}^{N} A_n} \subset \bigvee_{F \in \mathfrak{F}} H_F.$$

Since $\bigvee_{F\in\mathfrak{F}} H_F$ is a closed set, it follows from (1.2.8) that $f \in \bigvee_{F\in\mathfrak{F}} H_F$. Therefore, we have

$$H_E \subset \bigvee_{F\in\mathfrak{F}} H_F,$$

whence $P_E \leqslant \bigvee_{F\in\mathfrak{F}} P_F$. Combining this inequality with (1.2.7), we obtain (1.2.6). The proof of sufficiency is left to the reader.]

Lemma 1.2.5 shows the relationship between the l.u.b. of a family of sets and the l.u.b. of the corresponding family of projection operators.

3° Definition of Localizable Measure Spaces, and Some of Their Properties

Definition 1.2.3. Let $\Omega = (G, \mathfrak{B}, \mu)$ be a measure space satisfying the following two conditions: (i) if $B \in \mathfrak{B}$, $\mu(B) > 0$, then there exists $E \in \mathfrak{B}$, $E \subset B$, such that $0 < \mu(E) < \infty$; (ii) if $\tilde{\mathfrak{B}}$ denotes the totality of measurable sets of Ω, then, for any $\mathfrak{F} \subset \tilde{\mathfrak{B}}$, there exists a l.u.b. $\bigvee_{A\in\mathfrak{F}} A \in \tilde{\mathfrak{B}}$. Then, the measure space Ω is said to be *localizable.*

By Corollary 1.2.2, we have the following example.

Example 1.2.1. Any σ-finite (or finite) measure space is localizable.

The following theorem indicates an important method of constructing localizable measure spaces.

Theorem 1.2.6. The direct sum of a family of localizable measure spaces is localizable.

PROOF. Let $\Omega = (G, \mathfrak{B}, \mu)$ be the direct sum of the family of localizable measure spaces $\Omega_\xi = (G_\xi, \mathfrak{B}_\xi, \mu_\xi)$, $\xi \in \Xi$. Let $\mathfrak{F} \subset \tilde{\mathfrak{B}}$. We proceed to prove that $\mathfrak{F}$ has a l.u.b. in $\tilde{\mathfrak{B}}$.

Let $\mathfrak{F}_\xi = \{F \cap G_\xi \mid F \in \mathfrak{F}\}$. By hypothesis, each $\mathfrak{F}_\xi$ has a l.u.b. $E_\xi \in \tilde{\mathfrak{B}}_\xi$. Form the set $E = \bigcup_{\xi\in\Xi} E_\xi$; this is an element of $\tilde{\mathfrak{B}}$, and is a l.u.b. for the family of sets $\{E_\xi, \xi \in \Xi\}$. We shall prove that E is a l.u.b. for $\mathfrak{F}$.

If $F \in \mathfrak{F}$, then, for each $\xi \in \Xi$, $(F - E) \cap G_\xi = F \cap G_\xi - E_\xi$ is a null set, hence $F - E \in \mathfrak{B}_0$. Thus, E is an upper bound for $\mathfrak{F}$. On the other hand, if K is any upper bound for $\mathfrak{F}$, then, for each $F \in \mathfrak{F}$ and each $\xi \in \Xi$, $(F - K) \cap G_\xi = (F \cap G_\xi) - (K \cap G_\xi)$ is a null set, that is, $K \cap G_\xi$ is an upper bound for $\mathfrak{F}_\xi$, hence

$$E_\xi - (K \cap G_\xi) = (E - K) \cap G_\xi$$

is a null set. Therefore, $E - K \in \mathfrak{B}_0$. Moreover, it is obvious that any set of infinite measure must contain a set of finite positive measure. Thus, Ω is localizable.]

Example 1.2.2. Any quasi-σ-finite measure on a group, in particular, any Haar measure on a locally compact group, is localizable.

The following concept will also be used in the sequel.

Definition 1.2.4. Let $(G, \mathfrak{B}, \mu)$ be a measure space, let $E \in \tilde{\mathfrak{B}}$, and let $[E]$ denote $\{F \mid F \sim E, F \in \tilde{\mathfrak{B}}\}$, that is, the equivalence class of E with respect to the relation $\sim$ in $\tilde{\mathfrak{B}}$. Let $\mathfrak{U}$ be the totality of such equivalence classes. If $[E], [F] \in \mathfrak{U}$, let $[E] \vee [F] = [E \vee F]$, $[E] \wedge [F] = [E \wedge F]$. We call $\mathfrak{U}$ the *measurable ring* associated with the measure space $(G, \mathfrak{B}, \mu)$. Define a function μ on $\mathfrak{U}$ by

$$\mu([E]) = \mu(E).$$

μ is called a *measure* on the ring $\mathfrak{U}$, and the pair $(\mathfrak{U}, \mu)$ is called the *measure ring* associated with the measure space $(G, \mathfrak{B}, \mu)$.

Let $\Omega = (G, \mathfrak{B}, \mu)$ and $\Omega' = (G', \mathfrak{B}', \mu')$ be two measure spaces, $(\mathfrak{U}, \mu)$ and $(\mathfrak{U}', \mu')$ the associated measure rings. Suppose there exists a one-to-one mapping φ from $\mathfrak{U}$ onto $\mathfrak{U}'$, satisfying the following conditions:

$$\begin{gathered} \varphi([\text{empty set}]) = [\text{empty set}], \qquad \varphi([G]) = [G], \\ \varphi(\xi \wedge \eta) = \varphi(\xi) \wedge \varphi(\eta), \qquad \varphi(\xi \vee \eta) = \varphi(\xi) \vee \varphi(\eta), \end{gathered} \tag{1.2.9}$$

moreover, $\mu(\xi) = 0$ if and only if $\mu'(\varphi(\xi)) = 0$. Then, we say that the measure spaces Ω and Ω' are *equivalent*.

In particular, if $G = G'$, $\mathfrak{B} = \mathfrak{B}'$, and $\varphi([E]) = [E]$, then this notion of equivalence reduces to the ordinary definition of equivalence between two measures μ, μ' on a single measurable space.

§1.3. The Kolmogorov Theorem

Products of measure spaces are introduced in Chapter 7 of Halmos [1]. It is clear that the treatment presented there can easily be extended to products of families of probability measure spaces indexed by sets of arbitrary cardinality. Accordingly, in what follows, all our discussion of product spaces will refer to products of an arbitrary number of spaces.

1° Projective Limits of Families of Measure Spaces

We shall now effect some generalizations of the concept of a product measure space.

Definition 1.3.1. Let Λ be a directed set, and let $(\Gamma_\lambda, \mathfrak{B}_\lambda)$, $\lambda \in \Lambda$, be a family of measurable spaces. Let $P_\lambda^{\lambda'}: \Gamma_{\lambda'} \to \Gamma_\lambda$, $\lambda, \lambda' \in \Lambda$, $\lambda \prec \lambda'$, be a family of mappings such that, if $A \in \mathfrak{B}_\lambda$, then $(P_\lambda^{\lambda'})^{-1}(A) \in \mathfrak{B}_{\lambda'}$ [i.e., $P_\lambda^{\lambda'}$ is a measurable mapping from $(\Gamma_{\lambda'}, \mathfrak{B}_{\lambda'})$ to $(\Gamma_\lambda, \mathfrak{B}_\lambda)$], and such that the following consistency condition is satisfied: if $\lambda \prec \lambda'$, $\lambda' \prec \lambda''$, then

$$P_\lambda^{\lambda'} P_{\lambda'}^{\lambda''} = P_\lambda^{\lambda''}. \tag{1.3.1}$$

Then, we say that $\{P_\lambda^{\lambda'}, \lambda \prec \lambda', \lambda, \lambda' \in \Lambda\}$ is a *consistent family of projections* in the family of measurable spaces $\{(\Gamma_\lambda, \mathfrak{B}_\lambda), \lambda \in \Lambda\}$.

Let Γ be a set and let $P_\lambda: \Gamma \to \Gamma_\lambda$, $\lambda \in \Lambda$, be a family of mappings such that $P_\lambda(\Gamma) = \Gamma_\lambda$, and, when $\lambda \prec \lambda'$,

$$P_\lambda^{\lambda'} P_{\lambda'} = P_\lambda. \tag{1.3.2}$$

Let $\mathfrak{B}_\lambda{}^* = \{P_\lambda^{-1}(A) \mid A \in \mathfrak{B}_\lambda\}$. The set $P_\lambda^{-1}(A)$, $A \in \mathfrak{B}_\lambda$, is called a *Borel cylinder* in Γ, with *base* A, belonging to the index λ. Let $\mathfrak{B}_0$ denote the totality of Borel cylinders in Γ, that is, $\mathfrak{B}_0 = \bigcup_{\lambda \in \Lambda} \mathfrak{B}_\lambda{}^*$, and $\mathfrak{B}$ the smallest σ-ring in Γ containing $\mathfrak{B}_0$. The pair $(\Gamma, \mathfrak{B}_0)$ is called a *projective limit* (with respect to the family of projections $\{P_\lambda^{\lambda'}\}$) of the family of measurable spaces $\{(\Gamma_\lambda, \mathfrak{B}_\lambda), \lambda \in \Lambda\}$.

Notice that, if $\lambda \prec \lambda'$, then $\mathfrak{B}_\lambda{}^* \subset \mathfrak{B}_{\lambda'}^*$, moreover, for each $\lambda \in \Lambda$, $\mathfrak{B}_\lambda{}^*$ is a σ-ring. It follows easily that $\mathfrak{B}_0$ is a ring; however, $\mathfrak{B}_0$ need not be a σ-ring.

Definition 1.3.2. Let μ_0 be a nonnegative set function on $\mathfrak{B}_0$, such that, for each $\lambda \in \Lambda$, the restriction of μ_0 to $\mathfrak{B}_\lambda{}^*$ is a measure. Then μ_0 is called a *cylinder measure* on $(\Gamma, \mathfrak{B}_0)$.

Clearly, μ_0 is finitely additive on $\mathfrak{B}_0$, but need not be countably additive on $\mathfrak{B}_0$.

Definition 1.3.3. Let $(\Gamma_\lambda, \mathfrak{B}_\lambda, \mu_\lambda)$, $\lambda \in \Lambda$ be a family of measure spaces, $\{P_\lambda^{\lambda'}, \lambda \prec \lambda', \lambda, \lambda' \in \Lambda\}$ a consistent family of projections in the family of measurable spaces $\{(\Gamma_\lambda, \mathfrak{B}_\lambda), \lambda \in \Lambda\}$. If, whenever $\lambda \prec \lambda'$ and $A \in \mathfrak{B}_\lambda$, we have

$$\mu_\lambda(A) = \mu_{\lambda'}((P_\lambda^{\lambda'})^{-1}A), \tag{1.3.3}$$

then the family of measures μ_λ, $\lambda \in \Lambda$ [or the family of measure spaces $(\Gamma_\lambda, \mathfrak{B}_\lambda, \mu_\lambda)$, $\lambda \in \Lambda$] is said to be *consistent* with the family of projections $\{P_\lambda^{\lambda'}\}$.

Definition 1.3.4. Let $(\Gamma, \mathfrak{B}_0)$ be a projective limit of the family of measurable spaces $\{(\Gamma_\lambda, \mathfrak{B}_\lambda), \lambda \in \Lambda\}$, $P_\lambda: \Gamma \to \Gamma_\lambda$, $\lambda \in \Lambda$ the associated mappings, and μ_0 a cylinder measure on $(\Gamma, \mathfrak{B}_0)$. Let μ_λ be the measure on $(\Gamma_\lambda, \mathfrak{B}_\lambda)$ defined by

$$\mu_\lambda(A) = \mu_0(P_\lambda^{-1}(A)) \tag{1.3.4}$$

for all $A \in \mathfrak{B}_\lambda$. We then say that $(\Gamma, \mathfrak{B}_0, \mu_0)$ is a *projective limit* of the family of measure spaces $(\Gamma_\lambda, \mathfrak{B}_\lambda, \mu_\lambda)$, $\lambda \in \Lambda$.

Lemma 1.3.1. Let $(\Gamma, \mathfrak{B}_0)$ be a projective limit of the family of measurable spaces $(\Gamma_\lambda, \mathfrak{B}_\lambda)$, $\lambda \in \Lambda$, and, for each $\lambda \in \Lambda$, let μ_λ be a measure on $(\Gamma_\lambda, \mathfrak{B}_\lambda)$. Then, the consistency of the family of measure spaces $\{(\Gamma_\lambda, \mathfrak{B}_\lambda, \mu_\lambda), \lambda \in \Lambda\}$ with the projections $\{P_\lambda^{\lambda'}\}$ is a necessary and sufficient condition for the existence of a cylinder measure μ_0 on $(\Gamma, \mathfrak{B}_0)$ such that $(\Gamma, \mathfrak{B}_0, \mu_0)$ is a projective limit of the family $\{(\Gamma_\lambda, \mathfrak{B}_\lambda, \mu_\lambda), \lambda \in \Lambda\}$.

PROOF. Assume that $\{(\Gamma_\lambda, \mathfrak{B}_\lambda, \mu_\lambda), \lambda \in \Lambda\}$ is consistent. Define a set function μ_0 on $\mathfrak{B}_0$ as follows: for $A \in \mathfrak{B}_\lambda$, $P_\lambda^{-1}(A) \in \mathfrak{B}_\lambda{}^*$, let

$$\mu_0(P_\lambda^{-1}(A)) = \mu_\lambda(A). \tag{1.3.5}$$

First, we show that μ_0 is well-defined, that is, if $A \in \mathfrak{B}_\lambda$, $A' \in \mathfrak{B}_{\lambda'}$, and $P_\lambda^{-1}(A) = P_{\lambda'}^{-1}(A')$, then

$$\mu_{\lambda'}(A') = \mu_\lambda(A). \tag{1.3.6}$$

In fact, there exists $\lambda'' \in \Lambda$ such that $\lambda < \lambda''$ and $\lambda' < \lambda''$. By (1.3.2), we have $(P_{\lambda'}^{\lambda''})^{-1}A' = P_{\lambda''}(P_{\lambda'}^{-1}A') = P_{\lambda''}(P_\lambda^{-1}A) = (P_\lambda^{\lambda''})^{-1}A$, whence by (1.3.3) we get

$$\mu_{\lambda'}(A') = \mu_{\lambda''}((P_{\lambda'}^{\lambda''})^{-1}A') = \mu_{\lambda''}((P_\lambda^{\lambda''})^{-1}A) = \mu_\lambda(A),$$

that is, (1.3.6) holds. The countable additivity of μ_0 on $(\Gamma, \mathfrak{B}_\lambda{}^*)$ follows easily from the countable additivity of μ_λ. Thus, μ_0 is a cylinder measure on $(\Gamma, \mathfrak{B}_0)$.

Conversely, let $(\Gamma, \mathfrak{B}_0, \mu_0)$ be a projective limit of the family $\{(\Gamma_\lambda, \mathfrak{B}_\lambda, \mu_\lambda), \lambda \in \Lambda\}$. Take any $\lambda, \lambda' \in \Lambda$, $\lambda < \lambda'$, and any $A \in \mathfrak{B}_\lambda$. By (1.3.2), we have $P_\lambda^{-1}(A) = P_{\lambda'}^{-1}((P_\lambda^{\lambda'})^{-1}A)$. Hence, by (1.3.4), we get

$$\mu_\lambda(A) = \mu_0(P_\lambda^{-1}(A)) = \mu_0(P_{\lambda'}^{-1}(P_\lambda^{\lambda'})^{-1}A) = \mu_{\lambda'}((P_\lambda^{\lambda'})^{-1}A).$$

Thus, the family of measures $\{\mu_\lambda, \lambda \in \Lambda\}$ is consistent with the projections.]

We now give an important example.

Example 1.3.1. Let $(\Gamma_\alpha, \mathfrak{B}_\alpha)$, $\alpha \in \mathfrak{A}$ be a family of measurable spaces, and let Λ be the totality of nonempty finite subsets of $\mathfrak{A}$. We define the following partial ordering (called the *natural ordering*) in

Λ: if $\lambda, \lambda' \in \Lambda$, and $\lambda \subset \lambda'$, then $\lambda \prec \lambda'$. Obviously, Λ is directed by this ordering. For each $\lambda \in \Lambda$, form the product measurable space $(\Gamma_\lambda, \mathfrak{B}_\lambda) = (\mathsf{X}_{\alpha\in\lambda}\, \Gamma_\alpha, \mathsf{X}_{\alpha\in\lambda}\, \mathfrak{B}_\alpha)$. In particular, when λ contains but the single element α, we identify Γ_λ with Γ_α. If $\lambda = \{\alpha_1, ..., \alpha_n\}$, we denote the points of Γ_λ by $\omega_\lambda = (\omega_{\alpha_1}, ..., \omega_{\alpha_n})$, where $\omega_{\alpha_\nu} \in \Gamma_{\alpha_\nu}$. If $\lambda \prec \lambda'$, we may suppose that $\lambda' = \{\alpha_1, ..., \alpha_m\}$, $m \geqslant n$, and define the mapping $P_\lambda^{\lambda'}: \Gamma_{\lambda'} \to \Gamma_\lambda$ as follows:

$$P_\lambda^{\lambda'}(\omega_{\alpha_1}, ..., \omega_{\alpha_m}) = (\omega_{\alpha_1}, ..., \omega_{\alpha_n}).$$

It is easily seen that this family of mappings $\{P_\lambda^{\lambda'}\}$ satisfies the consistency condition (1.3.1). We call $\{P_\lambda^{\lambda'}\}$ the family of *natural projections*.

Let Γ be the product space $\mathsf{X}_{\alpha\in\mathfrak{A}}\, \Gamma_\alpha$. For each $\omega \in \Gamma$, let ω_α denote the αth coordinate of ω, and for each $\lambda \in \Lambda$, $\lambda = \{\alpha_1, ..., \alpha_n\}$, let P_λ be the mapping defined by

$$\omega \to \omega_\lambda = (\omega_{\alpha_1}, ..., \omega_{\alpha_n}).$$

P_λ is called the *natural projection* of Γ onto Γ_λ; in particular, when λ is a singleton $\{\alpha\}$, we denote P_λ by P_α. Clearly, $\{P_\lambda\}$ and $\{P_\lambda^{\lambda'}\}$ satisfy the relation (1.3.2). Let $\mathfrak{B}_0$ be the totality of Borel cylinders in Γ. We call $(\Gamma, \mathfrak{B}_0)$ the *canonical projective limit* of the family of measurable spaces $(\Gamma_\lambda, \mathfrak{B}_\lambda)$, $\lambda \in \Lambda$. In this case, $(\Gamma, \mathfrak{B})$ is merely the product of the family of measurable spaces $(\Gamma_\alpha, \mathfrak{B}_\alpha)$, $\alpha \in \mathfrak{A}$.

For each $\alpha \in \mathfrak{A}$, let μ_α be a probability measure on $(\Gamma_\alpha, \mathfrak{B}_\alpha)$, and for each $\lambda \in \Lambda$, $\lambda = \{\alpha_1, ..., \alpha_n\}$, let μ_λ be the product measure $\mathsf{X}_{k=1}^{n}\, \mu_{\alpha_k}$ on $(\Gamma_\lambda, \mathfrak{B}_\lambda)$. Then the family of measure spaces $(\Gamma_\lambda, \mathfrak{B}_\lambda, \mu_\lambda)$ is consistent. Now there exists (see Halmos [1]) a product measure μ on $(\Gamma, \mathfrak{B})$, that is, $(\Gamma, \mathfrak{B}, \mu) = (\mathsf{X}_{\alpha\in\mathfrak{A}}\, \Gamma_\alpha, \mathsf{X}_{\alpha\in\mathfrak{A}}\, \mathfrak{B}_\alpha, \mathsf{X}_{\alpha\in\mathfrak{A}}\, \mu_\alpha)$. Restricting μ to $\mathfrak{B}_0$, we obtain a cylinder measure μ_0, and $(\Gamma, \mathfrak{B}_0, \mu_0)$ is the projective limit of the family of measure spaces $\{(\Gamma_\lambda, \mathfrak{B}_\lambda, \mu_\lambda), \lambda \in \Lambda\}$. In this case, μ_0 is countably additive on $\mathfrak{B}_0$.

Regarding projective limits of consistent families of measure spaces, we shall, in general, be primarily interested in the question of whether the cylinder measure μ_0 on $(\Gamma, \mathfrak{B}_0)$ is countably additive. If μ_0 is countably additive, then it can be extended to a measure μ on the σ-ring $\mathfrak{B}$. For the simple case described in the above example, μ_0 is always countably additive.

In the following, we shall investigate the countable additivity of μ_0 when the Γ_λ are topological spaces.

Definition 1.3.5. Let Λ be a directed set, and $(\Gamma_\lambda, \mathfrak{T}_\lambda)$, $\lambda \in \Lambda$ a family of topological spaces. Let $P_\lambda^{\lambda'}: \Gamma_{\lambda'} \to \Gamma_\lambda$, $\lambda, \lambda' \in \Lambda$, $\lambda \prec \lambda'$ be a family of continuous mappings satisfying the consistency condition

(1.3.1). Then, $\{P_\lambda^{\lambda'}\}$ is said to be a *consistent family of projections* in the family of topological spaces $\{(\Gamma_\lambda, \mathfrak{T}_\lambda), \lambda \in \Lambda\}$.

Let Γ be a set and let $P_\lambda: \Gamma \to \Gamma_\lambda$, $\lambda \in \Lambda$, be a family of mappings such that $P_\lambda(\Gamma) = \Gamma_\lambda$ and, whenever $\lambda, \lambda' \in \Lambda$, $\lambda \prec \lambda'$, relation (1.3.2) holds.

Define a topology in Γ as follows. For each $\lambda \in \Lambda$, let

$$\mathfrak{T}_\lambda{}^* = \{P_\lambda^{-1}(A) \mid A \in \mathfrak{T}_\lambda\}.$$

Form $\mathfrak{T}_0 = \bigcup_{\lambda \in \Lambda} \mathfrak{T}_\lambda{}^*$.[16] Let $\mathfrak{T}$ be the weakest topology containing $\mathfrak{T}_0$. Then $(\Gamma, \mathfrak{T})$ is said to be a[17] *projective limit* (with respect to the family of projections $\{P_\lambda^{\lambda'}\}$) of the family of topological spaces $\{(\Gamma_\lambda, \mathfrak{T}_\lambda), \lambda \in \Lambda\}$.

The projective limit $(\Gamma, \mathfrak{T})$ is said to be *projectively complete* if the following condition is satisfied: given any sequence $\{\lambda_n\} \subset \Lambda$,

$$\lambda_1 \prec \lambda_2 \prec \cdots \prec \lambda_n \prec \cdots, \tag{1.3.7}$$

and any sequence of nonempty sets $\{E_n\}$ such that E_n is closed and compact with respect to the topology $\mathfrak{T}_{\lambda_n}^*$,

$$E_1 \supset E_2 \supset \cdots \supset E_n \supset \cdots, \tag{1.3.8}$$

then $\bigcap_{n=1}^{\infty} E_n$ is nonempty.

Lemma 1.3.2. Let $(\Gamma, \mathfrak{T})$ be a projective limit of the family of Hausdorff topological spaces $\{(\Gamma_\lambda, \mathfrak{T}_\lambda), \lambda \in \Lambda\}$. Suppose that, for each sequence $\{\lambda_n\} \subset \Lambda$ satisfying condition (1.3.7), and each sequence of elements $\{\xi_n, n = 1, 2,...\}$ satisfying the condition

$$\xi_n \in \Gamma_{\lambda_n}, \qquad P_{\lambda_n}^{\lambda_m} \xi_m = \xi_n \qquad \text{for} \quad m \geqslant n, \tag{1.3.9}$$

there exists $\xi \in \Gamma$ such that, for every n,

$$P_{\lambda_n} \xi = \xi_n .$$

Then $(\Gamma, \mathfrak{T})$ is projectively complete.

Proof. Take any sequence $\{\lambda_n\} \subset \Lambda$ satisfying condition (1.3.7), and any descending sequence of nonempty sets $\{E_n\}$ such that each E_n is

[16] The elements of $\mathfrak{T}_0$ are called *open cylinders* in Γ.

[17] *Translator's note:* It should be noted that the limit of a projective system, as defined here, is not unique: given any set Γ' and mapping P from Γ' onto Γ, one can obtain another projective limit $(\Gamma', \mathfrak{T}')$ in the obvious way, that is, $\mathfrak{T}' = P^{-1}(\mathfrak{T})$. Moreover, the condition that the projections P_λ (and hence also $P_\lambda^{\lambda'}$) be *onto* means that such a limit may fail to exist, even if all the given projections $P_\lambda^{\lambda'}$ are onto. Thus, this definition is far too restrictive for topological work, but it seems to suffice for the purposes of this book. Similar remarks apply to Definition 1.3.1.

closed and compact in the topology $\mathfrak{T}^*_{\lambda_n}$. We shall prove that $\bigcap_{n=1}^{\infty} E_n$ is nonempty.

Since $P_{\lambda_m} E_m$ is compact in $(\Gamma_{\lambda_m}, \mathfrak{T}_{\lambda_m})$, and, for any $n \leqslant m$, $P^{\lambda_n}_{\lambda_m}$ is continuous, it follows that $P_{\lambda_n} E_m = P^{\lambda_m}_{\lambda_n} P_{\lambda_m} E_m$ is compact (and hence closed) in $(\Gamma_{\lambda_n}, \mathfrak{T}_{\lambda_n})$. Thus, the intersection of the descending sequence of nonempty closed compact sets $\{P_{\lambda_1} E_m, m = 1, 2, \ldots\}$ in $(\Gamma_{\lambda_1}, \mathfrak{T}_{\lambda_1})$ contains a point ξ_1.

Since $\xi_1 \in P_{\lambda_1} E_m$, $m = 1, 2, \ldots$, the set $P^{-1}_{\lambda_1} \xi_1 \cap E_m$ is nonempty. Moreover, since each $(\Gamma_\lambda, \mathfrak{T}_\lambda)$ is Hausdorff, and $P^{-1}_{\lambda_1} \xi_1 = P^{-1}_{\lambda_m} (P^{\lambda_m}_{\lambda_1})^{-1} \xi_1$, it follows that $P^{-1}_{\lambda_1} \xi_1$ is closed with respect to $\mathfrak{T}^*_{\lambda_m}$. Hence, $P^{-1}_{\lambda_1} \xi_1 \cap E_m$ is a nonempty closed compact set of $(\Gamma, \mathfrak{T}^*_{\lambda_m})$. If we apply the foregoing argument to the sequence of sets

$$P^{-1}_{\lambda_1} \xi_1 \cap E_2 \supset \cdots \supset P^{-1}_{\lambda_1} \xi_1 \cap E_m \supset \cdots,$$

we conclude that the sets $\{P_{\lambda_2}(P^{-1}_{\lambda_1} \xi_1 \cap E_m), m = 2, 3, \ldots\}$ contain a common point ξ_2. We then have

$$P^{\lambda_2}_{\lambda_1} \xi_2 = \xi_1, \qquad \xi_2 \in P_{\lambda_2} E_m, \qquad m = 2, \ldots .$$

Repeating this argument, we obtain a sequence of elements $\{\xi_n, n = 1, 2, \ldots\}$ having the property

$$P^{\lambda_m}_{\lambda_n} \xi_m = \xi_n \in P_{\lambda_n} E_m \qquad \text{for} \quad m \geqslant n, \tag{1.3.10}$$

that is, $\{\xi_n\}$ satisfies condition (1.3.9). By hypothesis, there exists $\xi \in \Gamma$ such that, for each n, $P_{\lambda_n} \xi = \xi_n$, whence, by (1.3.10), we have

$$P_{\lambda_n} \xi \in P_{\lambda_n} E_n, \qquad n = 1, 2, \ldots .$$

But E_n is closed with respect to $\mathfrak{T}^*_{\lambda_n}$, hence $\xi \in E_n$, $n = 1, 2, \ldots$, that is, $\bigcap_{n=1}^{\infty} E_n$ is nonempty.]

Example 1.3.2. Let $(\Gamma_\alpha, \mathfrak{T}_\alpha)$, $\alpha \in \mathfrak{A}$, be a family of nonempty topological spaces, and let Λ be the totality of nonempty finite subsets of $\mathfrak{A}$. As in Example 1.3.1, we order Λ by inclusion, and, for each $\lambda \in \Lambda$, form the topological product $(\Gamma_\lambda, \mathfrak{T}_\lambda) = (\times_{\alpha \in \lambda} \Gamma_\alpha, \times_{\alpha \in \lambda} \mathfrak{T}_\alpha)$. In particular, when λ is a singleton $\{\alpha\}$, we identify Γ_λ with Γ_α. Let $(\Gamma, \mathfrak{T})$ be the topological product of the family of spaces $\{(\Gamma_\alpha, \mathfrak{T}_\alpha), \alpha \in \mathfrak{A}\}$. Define the projections $\{P^{\lambda'}_\lambda\}$ and $\{P_\lambda\}$ as in Example 1.3.1. It is easily seen that $(\Gamma, \mathfrak{T})$ is a projective limit of the family of topological spaces $\{(\Gamma_\lambda, \mathfrak{T}_\lambda), \lambda \in \Lambda\}$ with respect to the projections $\{P^{\lambda'}_\lambda\}$. We call it the *canonical projective limit.*

Lemma 1.3.3. Any canonical projective limit is projectively complete.

PROOF. We need only verify that the condition of Lemma 1.3.2 is satisfied. Take any sequence $\{\lambda_n\} \subset \Lambda$ satisfying (1.3.7). We may assume that

$$\lambda_n = \{\alpha_1, \alpha_2, \ldots, \alpha_{k_n}\} \subset \mathfrak{A}.$$

If $\{\xi_n\}$ is a sequence satisfying condition (1.3.9), then there are elements $\eta_{\alpha_m} \in \Gamma_{\alpha_m}$ such that

$$\xi_n = \{\eta_{\alpha_1}, \ldots, \eta_{\alpha_{k_n}}\}.$$

Define an element $\xi \in \times_{\alpha \in \mathfrak{A}} \Gamma_\alpha$ as follows: the α_mth coordinate of ξ is η_{α_m}, $m = 1, 2, \ldots$, and the other coordinates of ξ are chosen arbitrarily (invoking the axiom of choice). Obviously, $P_{\lambda_n}\xi = \xi_n$ for all n. Therefore, by Lemma 1.3.2, the canonical projective limit $(\Gamma, \mathfrak{T})$ is projectively complete.]

In §4.3 we shall encounter another example in which the condition of Lemma 1.3.2 is satisfied.

2° Countable Additivity of Cylinder Measures

We now present the basic theorem concerning the construction of measures on infinite-dimensional spaces, stated in its most general form.

Theorem 1.3.4. Let $(\Gamma_\lambda, \mathfrak{T}_\lambda)$, $\lambda \in \Lambda$, be a family of topological spaces, and, for each $\lambda \in \Lambda$, let $(\Gamma_\lambda, \mathfrak{B}_\lambda, \mu_\lambda)$ be a regular probability measure space. Let $P_\lambda^{\lambda'}: \Gamma_{\lambda'} \to \Gamma_\lambda$ be a family of continuous projections, consistent with respect to the family of measures $\{\mu_\lambda\}$. Let $(\Gamma, \mathfrak{T})$ be a projectively complete projective limit of the family $\{(\Gamma, \mathfrak{T}_\lambda)\}$, with associated projections $P_\lambda: \Gamma \to \Gamma_\lambda$. Suppose also that $(\Gamma, \mathfrak{B}_0, \mu_0)$ is a projective limit of the family of measure spaces $\{(\Gamma_\lambda, \mathfrak{B}_\lambda, \mu_\lambda)\}$, with the same associated projections P_λ. Then the cylinder measure μ_0 is countably additive.

PROOF. By Theorem F, §9 of Halmos [1], it suffices to prove that, for any sequence of cylinders

$$B_1 \supset B_2 \supset \cdots \supset B_n \supset \cdots$$

in $\mathfrak{B}_0$ such that

$$\lim_{n\to\infty} \mu_0(B_n) = L > 0, \tag{1.3.11}$$

the intersection $\bigcap_{n=1}^{\infty} B_n$ is nonempty.

Now, for each n, there is a $\lambda_n \in \Lambda$ such that $B_n \in \mathfrak{B}^*_{\lambda_n}$. We may assume that

$$\lambda_1 \prec \lambda_2 \prec \lambda_3 \prec \cdots \prec \lambda_n \prec \cdots. \tag{1.3.12}$$

In fact, since Λ is directed, there is an index λ_2' such that $\lambda_1 \prec \lambda_2'$ and $\lambda_2 \prec \lambda_2'$. Then $B_2 \in \mathfrak{B}^*_{\lambda_2} \subset \mathfrak{B}^*_{\lambda'_2}$, and we may replace λ_2 by λ_2'. Proceeding in the same manner for $n = 3, 4, \ldots$, we obtain a sequence of indices satisfying (1.3.12). For each n, let $B_n = P^{-1}_{\lambda_n}(A_n)$, $A_n \in \mathfrak{B}_{\lambda_n}$. By the regularity of μ_{λ_n}, there exists a closed compact subset C_n of Γ_{λ_n}, with $C_n \in \mathfrak{B}_{\lambda_n}$, such that $C_n \subset A_n$ and

$$\mu_{\lambda_n}(A_n - C_n) < \frac{L}{2^{n+1}}.$$

Hence, by (1.3.4), we have

$$\mu_0(B_n - (P^{-1}_{\lambda_n}C_n)) < \frac{L}{2^{n+1}}. \tag{1.3.13}$$

Let $E_n = \bigcap_{k=1}^n (P^{-1}_{\lambda_k}C_k)$. Since $P^{-1}_{\lambda_k}C_k \subset B_k$, and

$$B_n - E_n = \bigcup_{k=1}^n (B_n - P^{-1}_{\lambda_k}C_k) \subset \bigcup_{k=1}^n (B_k - P^{-1}_{\lambda_k}C_k),$$

it follows from (1.3.13) that $\mu_0(B_n - E_n) < L/2$. Hence, by (1.3.11), we get

$$\mu_0(E_n) = \mu_0(B_n) - \mu_0(B_n - E_n) > \frac{L}{2} > 0.$$

Therefore, E_n is nonempty. Moreover, E_n is closed and compact with respect to $\mathfrak{T}^*_{\lambda_n}$, and

$$E_1 \supset E_2 \supset \cdots \supset E_n \supset \cdots. \tag{1.3.14}$$

By virtue of the projective completeness of $(\Gamma, \mathfrak{T})$, it follows from (1.3.12) and (1.3.14) that $\bigcap_{n=1}^\infty E_n$ is nonempty. But $\bigcap_{n=1}^\infty B_n \supset \bigcap_{n=1}^\infty E_n$, hence $\bigcap_{n=1}^\infty E_n$ is nonempty.]

The following theorem (due to Kolmogorov) deals with the case of greatest interest.

Corollary 1.3.5.[18] Let $\mathfrak{A}$ be an arbitrary indexing set, and let Λ be the totality of nonempty finite subsets of $\mathfrak{A}$, directed by the natural ordering. For each $\lambda = \{\alpha_1, \ldots, \alpha_n\} \in \Lambda$, let R_λ be an n-dimensional euclidean space, $\mathfrak{B}_\lambda$ the σ-algebra of all Borel sets in R_λ. Denote the points of R_λ by $x_\lambda = \{x_{\alpha_1}, \ldots, x_{\alpha_n}\}$. For each $\lambda \in \Lambda$, let μ_λ be a probability measure on $(R_\lambda, \mathfrak{B}_\lambda)$, and suppose that the measures $\{\mu_\lambda, \lambda \in \Lambda\}$ are consistent in the following sense: given any pair $\lambda, \lambda' \in \Lambda$, with

[18] Regarding the terminology used here, see Example 1.3.1.

$\lambda = \{\alpha_1, \alpha_2, ..., \alpha_n\} \subset \{\beta_1, \beta_2, ..., \beta_m\} = \lambda'$, and any $E \in \mathfrak{B}_\lambda$, we have

$$\mu_{\lambda'}(\{x_{\lambda'} \mid \{x_{\alpha_1}, ..., x_{\alpha_n}\} \in E\}) = \mu_\lambda(E). \tag{1.3.15}$$

Let $\Gamma = \mathsf{X}_{\alpha\in\mathfrak{A}} R_\alpha$, $\mathfrak{B} = \mathsf{X}_{\alpha\in\mathfrak{A}} \mathfrak{B}_\alpha$, and denote the elements of Γ by $x = \{x_\alpha, \alpha \in \mathfrak{A}\}$; then, there is a unique probability measure μ on $(\Gamma, \mathfrak{B})$, satisfying the following condition: for any $\lambda = \{\alpha_1, ..., \alpha_n\} \in \Lambda$, and any $E \in \mathfrak{B}_\lambda$,

$$\mu(\{x \mid \{x_{\alpha_1}, ..., x_{\alpha_n}\} \in E\}) = \mu_\lambda(E). \tag{1.3.16}$$

(The measures μ_λ, or the distribution functions on the respective spaces R_λ determined by the μ_λ, are usually known as the finite-dimensional probability distributions of μ.)

PROOF. Each μ_λ, being a Borel measure, is regular. We form the families of projections $\{P_\lambda^{\lambda'}\}$, $\{P_\lambda\}$ as in Example 1.3.1. Then (1.3.15) shows that the family of measures $\{\mu_\lambda, \lambda \in \Lambda\}$ is consistent in the sense of Definition 1.3.3. By Lemma 1.3.1, there exists a cylinder measure μ_0 on $(\Gamma, \mathfrak{B}_0)$, satisfying condition (1.3.16), and, clearly, μ_0 is uniquely determined by (1.3.16). Let $\mathfrak{T}_\lambda$ be the Euclidean topology on Γ_λ, and let $\mathfrak{T}$ be the product topology $\mathsf{X}_{\alpha\in\mathfrak{A}} \Gamma_\alpha$ on Γ. Then $(\Gamma, \mathfrak{T})$ is the canonical projective limit of $\{(\Gamma_\lambda, \mathfrak{T}_\lambda), \lambda \in \Lambda\}$ (see Example 1.3.2). By Lemma 1.3.3, $(\Gamma, \mathfrak{T})$ is projectively complete. Therefore, it follows by Theorem 1.3.4 that μ_0 is countably additive. It is easily seen that $\mathfrak{B}$ is the smallest σ-algebra containing $\mathfrak{B}_0$, hence μ_0 can be extended to a probability measure μ on $(\Gamma, \mathfrak{B})$. The uniqueness of μ follows immediately from the uniqueness of μ_0.]

Corollary 1.3.5 can be stated in another useful form. (In the following, the characteristic function[19] of a probability distribution on a finite dimensional Euclidean space, will be referred to simply by the term "characteristic function.")

Corollary 1.3.5′. Let $\mathfrak{A}$ be an arbitrary indexing set, and let Ξ denote the totality of finite ordered[20] sequences $\xi = (\alpha_1, ..., \alpha_n)$

[19] According to the Bochner–Khinchin theorem, a function $\phi(x)$ of n variables $x = (x_1, ..., x_n)$ is the characteristic function of some probability distribution if and only if the following three conditions are satisfied: (i) $\phi(x)$ is continuous; (ii) $\phi(0) = 1$; (iii) ϕ is positive definite, that is, for any sequence of points $x^{(1)}, ..., x^{(m)}$ and arbitrary complex numbers $\zeta_1, ..., \zeta_m$,

$$\sum_{k,l=1}^{m} \varphi(x^{(k)} - x^{(l)})\, \zeta_k \bar{\zeta}_l \geqslant 0.$$

[20] This means that, for any nontrivial permutation $\alpha_1', \alpha_2', ..., \alpha_n'$ of the elements $\alpha_1, \alpha_2, ..., \alpha_n$, the sequences $(\alpha_1, \alpha_2, ..., \alpha_n)$ and $(\alpha_1', \alpha_2', ..., \alpha_n')$ are regarded as distinct.

($\alpha_i \in \mathfrak{A}$, $i = 1,\ldots, n$; $n = 1, 2,\ldots$). Let $\{F_\xi , \xi \in \Xi\}$ be a family of characteristic functions, having the following properties: (i) if $\xi' = (\alpha_1',\ldots, \alpha_n')$ is a permutation of $\xi = (\alpha_1 ,\ldots, \alpha_n)$, then

$$F_{\xi'}(t_{\alpha_1'} ,\ldots, t_{\alpha_n'}) = F_\xi(t_{\alpha_1} ,\ldots, t_{\alpha_n}); \tag{1.3.17}$$

(ii) for $m \geqslant n$,

$$F_{(\alpha_1,\ldots,\alpha_m)}(t_{\alpha_1} ,\ldots, t_{\alpha_n} , 0, 0,\ldots, 0) = F_{(\alpha_1,\ldots,\alpha_n)}(t_{\alpha_1} ,\ldots, t_{\alpha_n}). \tag{1.3.18}$$

For each $\alpha \in \mathfrak{A}$, let R_α be a 1-dimensional Euclidean space, $\mathfrak{B}_\alpha$ the totality of Borel sets in R_α , $\Gamma = \mathsf{X}_{\alpha\in\mathfrak{A}} R_\alpha$, $\mathfrak{B} = \mathsf{X}_{\alpha\in\mathfrak{A}} \mathfrak{B}_\alpha$, and denote the points of Γ by $x = \{x_\alpha , \alpha \in \mathfrak{A}\}$. Then, there is a unique probability measure μ on $(\Gamma, \mathfrak{B})$ such that, for any $\xi = (\alpha_1 ,\ldots, \alpha_n) \in \Xi$,

$$F_\xi(t_{\alpha_1} ,\ldots, t_{\alpha_n}) = \int_\Gamma \exp i(t_{\alpha_1}x_{\alpha_1} + \cdots + t_{\alpha_n}x_{\alpha_n})\, d\mu(x). \tag{1.3.19}$$

PROOF. Let λ be any nonempty finite subset of $\mathfrak{A}$, and let the elements of λ be ordered in an arbitrary sequence $\xi = (\alpha_1 , \alpha_2 ,\ldots, \alpha_n)$. Since F_ξ is a characteristic function, there exists a Borel measure μ_ξ on $R_\xi = R_{\alpha_1} \times R_{\alpha_2} \times \cdots \times R_{\alpha_n}$ such that

$$F_\xi(t_{\alpha_1} ,\ldots, t_{\alpha_n}) = \int_{R_\xi} \exp i(t_{\alpha_1}x_{\alpha_1} + \cdots + t_{\alpha_n}x_{\alpha_n})\, d\mu_\xi(x_{\alpha_1} ,\ldots, x_{\alpha_n}). \tag{1.3.20}$$

If the indices comprising ξ are permuted, it follows from condition (1.3.17) that, after the corresponding permutation of the coordinate axes, we obtain the same measure. Moreover, (1.3.15) can be deduced from (1.3.18). In fact, if $\xi' = \{\alpha_1 ,\ldots, \alpha_m\} \supset \{\alpha_1 ,\ldots, \alpha_n\}$, form the measure

$$\mu_\xi'(E) = \mu_{\xi'}\{(x_{\alpha_1} ,\ldots, x_{\alpha_m}) \mid (x_{\alpha_1} ,\ldots, x_{\alpha_n}) \in E\}.$$

Then

$$\begin{aligned} F_{\xi'}(t_{\alpha_1} ,\ldots, t_{\alpha_n} , 0,\ldots, 0) &= \int \cdots \int \exp i(t_{\alpha_1}x_{\alpha_1} + \cdots + t_{\alpha_n}x_{\alpha_n})\, d\mu_{\xi'}(x_{\alpha_1} ,\ldots, x_{\alpha_m}) \\ &= \int \cdots \int \exp i(t_{\alpha_1}x_{\alpha_1} + \cdots + t_{\alpha_n}x_{\alpha_n})\, d\mu_\xi'(x_{\alpha_1} ,\ldots, x_{\alpha_n}). \end{aligned} \tag{1.3.21}$$

By (1.3.18), (1.3.20), (1.3.21), and the uniqueness of the measure determined by a characteristic function, it follows that

$$\mu_{\xi'} = \mu_\xi .$$

Thus, (1.3.15) holds. Using Corollary 1.3.5, we obtain a unique measure μ on $(\Gamma, \mathfrak{B})$, satisfying condition (1.3.16).

Since the P_λ, $\lambda = \{\alpha_1, \ldots, \alpha_n\}$, are measurable mappings, (1.3.19) follows immediately from (1.3.16) and (1.3.20).]

3° Sample Measure Spaces

Definition 1.3.6. Let $\mathfrak{S} = (\Omega, \mathfrak{F}, P)$ be a probability measure space, $\{x_\alpha(\ ), \alpha \in \mathfrak{A}\}$ a family of (real) random variables on $\mathfrak{S}$ (one sometimes calls $\{x_\alpha(\ ), \alpha \in \mathfrak{A}\}$ a *stochastic process* on $\mathfrak{S}$). Also, suppose that $\{x_\alpha(\ ), \alpha \in \mathfrak{A}\}$ constitutes a determining set of functions on $(\Omega, \mathfrak{F}, P)$. For each $\alpha \in \mathfrak{A}$, let R_α be a copy of the real line, and $\mathfrak{B}_\alpha$ the totality of Borel sets in R_α. Let $\Gamma \subset \times_{\alpha \in \mathfrak{A}} R_\alpha$, and let $\mathfrak{B}$ be the restriction of $\times_{\alpha \in \mathfrak{A}} \mathfrak{B}_\alpha$ to Γ. Denote the αth coordinate of $x \in \Gamma$ by $x_\alpha = x_\alpha(x)$. Suppose that μ is a probability measure on $(\Gamma, \mathfrak{B})$ such that, for any finite set of indices $\alpha_1, \ldots, \alpha_n$, and any Borel set E in n-dimensional Euclidean space R^n, we have

$$\mu(\{x \mid (x_{\alpha_1}, \ldots, x_{\alpha_n}) \in E\}) = P(\{\omega \mid (x_{\alpha_1}(\omega), \ldots, x_{\alpha_n}(\omega)) \in E\}). \tag{1.3.22}$$

Then, $(\Gamma, \mathfrak{B}, \mu)$ is said to be a *sample probability measure space* associated with the stochastic process $\{x_\alpha(\omega), \alpha \in \mathfrak{A}\}(\omega \in \Omega)$ on $\mathfrak{S}$, and $\{x_\alpha(x), \alpha \in \mathfrak{A}\}$ the corresponding *sample process*.

Lemma 1.3.6. For any stochastic process, there exists an associated sample probability measure space.

PROOF. We use the notation of Definition 1.3.6. For each nonempty finite set of indices $\{\alpha_1, \ldots, \alpha_n\} \subset \mathfrak{A}$, define a measure on R^n, as follows:

$$\mu_{\{\alpha_1, \ldots, \alpha_n\}}(E) = P(\{\omega \mid (x_{\alpha_1}(\omega), \ldots, x_{\alpha_n}(\omega)) \in E\}). \tag{1.3.23}$$

It is easily verified that the family of measures $\{\mu_{\{\alpha_1, \ldots, \alpha_n\}}, \{\alpha_1, \ldots, \alpha_n\} \subset \mathfrak{A}\}$ is consistent. Let $\Gamma = \times_{\alpha \in \mathfrak{A}} R_\alpha$. By Corollary 1.3.5, there exists a probability measure μ on $(\Gamma, \mathfrak{B})$ such that (1.3.16) holds. Combining (1.3.16) and (1.3.23), we get (1.3.22).]

We conclude this section by proving a lemma, required later, which provides a criterion for the equivalence or singularity of probability measures in terms of the equivalence or singularity of their associated sample probability measures. First, we introduce a convenient modification of Definition 1.1.7.[21]

[21] *Translator's note:* Definition 1.3.7 was introduced by the translator in order to prove Lemma 1.3.7 (which was unprovable in its original form) and to fill a gap in the author's proof of Theorem 5.2.5.

Definition 1.3.7. Let $\Omega_\lambda = (G, \mathfrak{B}, \mu_\lambda)$, $\lambda \in \Lambda$, be a family of measure spaces, and $\mathfrak{D}$ a family of measurable functions on $(G, \mathfrak{B})$. Let $\mathfrak{B}_{\mathfrak{D}}$ denote the σ-algebra determined by $\mathfrak{D}$ on G. Suppose that, for every set $A \in \mathfrak{B}$, there exists a set $B_A \in \mathfrak{B}_{\mathfrak{D}}$ such that

$$\mu_\lambda(A \triangle B_A) = 0$$

for all $\lambda \in \Lambda$. We then say that $\mathfrak{D}$ is a *joint determining set* (of functions) for the family of measure spaces Ω_λ.

In particular, if $\mathfrak{D}$ is a determining set of functions on $(G, \mathfrak{B})$, then $\mathfrak{D}$ is a joint determining set for any family of measure spaces $(G, \mathfrak{B}, \mu_\lambda)$.

Lemma 1.3.7. Let $(\Gamma, \mathfrak{B}, \mu_k)$, $k = 1, 2$, be sample probability measure spaces (associated with the same stochastic process $\{x_\alpha(\cdot), \alpha \in \mathfrak{A}\}$) belonging to the respective probability measure spaces $\mathfrak{S}_k = (\Omega, \mathfrak{F}, P_k)$, $k = 1, 2$. Suppose also that $\{x_\alpha(\cdot), \alpha \in \mathfrak{A}\}$ constitutes a joint determining set for the measure spaces $\mathfrak{S}_k$, $k = 1, 2$. Then $\mu_1 \ll \mu_2$ if and only if $P_1 \ll P_2$, while $\mu_1 \perp \mu_2$ if and only if $P_1 \perp P_2$.

Proof. Consider the mapping φ from Ω to Γ defined by $\varphi\colon \omega \to x = \{x_\alpha(\omega), \alpha \in \mathfrak{A}\} \in \Gamma$. Clearly, φ is a measurable mapping from $(\Omega, \mathfrak{F})$ to $(\Gamma, \mathfrak{B})$. From (1.3.22), it is easily proved that, for any $B \in \mathfrak{B}$,

$$\mu_k(B) = P_k(\varphi^{-1}(B)), \qquad k = 1, 2. \tag{1.3.24}$$

Since $\{x_\alpha(\cdot), \alpha \in \mathfrak{A}\}$ is a joint determining set for $\mathfrak{S}_k$, $k = 1, 2$, and $\{\varphi^{-1}(B) \mid B \in \mathfrak{B}\}$ is the smallest σ-algebra in Ω with respect to which all the functions $\{x_\alpha(\cdot), \alpha \in \mathfrak{A}\}$ are measurable, it follows that, for any $E \in \mathfrak{F}$, there is a $B \in \mathfrak{B}$ such that

$$P_k((E - \varphi^{-1}(B)) \cup (\varphi^{-1}(B) - E)) = 0, \qquad k = 1, 2. \tag{1.3.25}$$

Assume that $\mu_1 \ll \mu_2$, $P_2(E) = 0$; we must prove that $P_1(E) = 0$. Let $B \in \mathfrak{B}$ be such that (1.3.25) holds. Then, by (1.3.24) and (1.3.25),

$$\mu_2(B) = P_2(E) = 0.$$

Hence, $\mu_1(B) = 0$. Again, by (1.3.24) and (1.3.25), we get

$$P_1(E) = \mu_1(B) = 0.$$

Therefore, $P_1 \ll P_2$. The remainder of the proof is left to the reader.]

§1.4. Kakutani Distance

1° Elementary Properties of Kakutani Distance

We now introduce a distance function which is useful in dealing with finite measure spaces.

Definition 1.4.1. Let Γ be a set, $\mathfrak{B}$ a σ-algebra consisting of certain subsets of Γ, and $\mathfrak{M}(\Gamma, \mathfrak{B})$ the totality of finite measures on $(\Gamma, \mathfrak{B})$. For any two measures $m, n \in \mathfrak{M}(\Gamma, \mathfrak{B})$, choose some $r \in \mathfrak{M}(\Gamma, \mathfrak{B})$ such that both m and n are absolutely continuous[22] with respect to r. Define

$$\rho(m, n) = \int_\Gamma \left(\frac{dm(\omega)}{dr(\omega)} \frac{dn(\omega)}{dr(\omega)}\right)^{1/2} dr(\omega), \tag{1.4.1}$$

where dm/dr, dn/dr denote the respective Radon–Nikodym derivatives of m and n with respect to r. The number $\rho(m, n)$ is called the *Kakutani inner product* of m and n. Also, define

$$d(m, n) = \left(\int \left[\left(\frac{dm(\omega)}{dr(\omega)}\right)^{1/2} - \left(\frac{dn(\omega)}{dr(\omega)}\right)^{1/2}\right]^2 dr(\omega)\right)^{1/2}.$$

$d(m, n)$ is called the *Kakutani distance* of m and n.

First, we note that $\rho(m, n)$ and $d(m, n)$ are, in fact, independent of the choice of r. Indeed, for any measurable function $f(\omega)$ on $(\Gamma, \mathfrak{B})$, if the integral

$$\int f(\omega) \left(\frac{dm(\omega)}{dr(\omega)}\right)^{1/2} \left(\frac{dn(\omega)}{dr(\omega)}\right)^{1/2} dr(\omega) \tag{1.4.2}$$

exists, then its value is independent of r. For, if there is another r' such that

$$\int f(\omega) \left(\frac{dm(\omega)}{dr'(\omega)}\right)^{1/2} \left(\frac{dn(\omega)}{dr'(\omega)}\right)^{1/2} dr'(\omega) \tag{1.4.3}$$

exists, let $r'' = r + r'$. Then it is easy to calculate that both (1.4.2) and (1.4.3) are equal to

$$\int f(\omega) \left(\frac{dm(\omega)}{dr''(\omega)}\right)^{1/2} \left(\frac{dn(\omega)}{dr''(\omega)}\right)^{1/2} dr''(\omega).$$

Therefore, (1.4.2) is independent of r, and will, accordingly, be denoted by

$$\int f(\omega)(dm(\omega)\, dn(\omega))^{1/2}.$$

Furthermore, it is easily seen that d and ρ are related by the formula

$$d(m, n)^2 = m(\Gamma) + n(\Gamma) - 2\rho(m, n). \tag{1.4.4}$$

[22] Such an r always exists, e.g., $r = m + n$.

In particular, when m and n are probability measures, (1.4.4) becomes

$$d(m, n)^2 = 2(1 - \rho(m, n)).$$

Moreover, $d(m, n)$ is, in fact, a metric on the set $\mathfrak{M}(\Gamma, \mathfrak{B})$.

Lemma 1.4.1. Two finite measures m, n on $(\Gamma, \mathfrak{B})$ are mutually singular if and only if their Kakutani inner product vanishes.

PROOF. If $\rho(m, n) = 0$, then

$$\frac{dm(\omega)}{dr(\omega)} \cdot \frac{dn(\omega)}{dr(\omega)} = 0 \tag{1.4.5}$$

for almost all ω (with respect to the measure r). Let

$$A = \left\{ \omega \,\middle|\, \frac{dm(\omega)}{dr(\omega)} = 0 \right\}.$$

Then

$$m(A) = \int_A \frac{dm(\omega)}{dr(\omega)}\, dr(\omega) = 0.$$

Since $dm(\omega)/dr(\omega) \neq 0$ on $\Gamma - A$, it follows from (1.4.5) that $dn(\omega)/dr(\omega)$ vanishes almost everywhere on $\Gamma - A$. This implies that $n(\Gamma - A) = 0$. Therefore, m and n are mutually singular. The proof of necessity is similar.]

2° Another Expression for the Kakutani Inner Product

We shall express the Kakutani distance in a form which does not involve Radon–Nikodym derivatives. This expression will prove useful in Chapters III and IV.

Lemma 1.4.2. Let μ, ν be two probability measures on the measurable space $(G, \mathfrak{B})$, and let $\mathfrak{F}$ denote the totality of countable partitions[23] of G. Then the Kakutani inner product of μ and ν is given by

$$\rho(\mu, \nu) = \inf_{\{E_k\} \in \mathfrak{F}} \sum_k (\mu(E_k)\, \nu(E_k))^{1/2}. \tag{1.4.6}$$

If G is a topological space and μ, ν are both regular measures, then $\mathfrak{F}$

[23] That is, the elements of $\mathfrak{F}$ are sequences $\{E_k\}$ of mutually disjoint sets in $\mathfrak{B}$, such that $\mu(G - \sum_k E_k) = \nu(G - \sum_k E_k) = 0$.

in (1.4.6) can be replaced by $\mathfrak{F}_0$, where $\mathfrak{F}_0$ denotes the totality of countable partitions consisting of closed compact sets in $\mathfrak{B}$.

PROOF. Take any partition $\{E_k\} \in \mathfrak{F}$. By the Schwarz inequality,

$$\int_{E_k} (d\mu(g)\, d\nu(g))^{1/2} \leqslant (\mu(E_k)\, \nu(E_k))^{1/2},$$

whence it follows at once that

$$\rho(\mu, \nu) \leqslant \sum_k (\mu(E_k)\, \nu(E_k))^{1/2}. \tag{1.4.7}$$

Consider the probability measure $r = \frac{1}{2}(\mu + \nu)$ on $(G, \mathfrak{B})$. The derivatives $d\mu(g)/dr(g)$, $d\nu(g)/dr(g)$ are nonnegative measurable functions, hence, for any number $a > 1$, and any integers k, l, the set

$$E_{k,l} = \left\{g \,\middle|\, a^{k-1} \leqslant \left(\frac{d\mu(g)}{dr(g)}\right)^{1/2} < a^k, \qquad a^{l-1} \leqslant \left(\frac{d\nu(g)}{dr(g)}\right)^{1/2} < a^l\right\}$$

is measurable[24]. We then have

$$\int_{E_{k,l}} \left(\frac{d\mu(g)}{dr(g)}\right)^{1/2} \left(\frac{d\nu(g)}{dr(g)}\right)^{1/2} dr(g) \geqslant a^{k+l-2} r(E_{k,l}), \tag{1.4.8}$$

$$\mu(E_{k,l}) = \int_{E_{k,l}} \frac{d\mu(g)}{dr(g)}\, dr(g) \leqslant a^{2k} r(E_{k,l}), \tag{1.4.9}$$

$$\nu(E_{k,l}) \leqslant a^{2l} r(E_{k,l}). \tag{1.4.10}$$

From (1.4.8)–(1.4.10), we get

$$\int_{E_{k,l}} (d\mu(g)\, d\nu(g))^{1/2} \geqslant \frac{1}{a^2} (\mu(E_{k,l})\, \nu(E_{k,l}))^{1/2}. \tag{1.4.11}$$

But $\{E_{k,l}\} \in \mathfrak{F}$. Hence, it follows from (1.4.11) that

$$\rho(\mu, \nu) \geqslant \frac{1}{a^2} \sum_{k,l} (\mu(E_{k,l})\, \nu(E_{k,l}))^{1/2} \geqslant \frac{1}{a^2} \inf_{\{E_k\} \in \mathfrak{F}} \sum_k (\mu(E_k)\, \nu(E_k))^{1/2}.$$

Letting $a \to 1$ in the above relation, and using (1.4.7), we obtain (1.4.6).

Now, suppose that G is a topological space, and μ, ν are regular measures. Then, we notice that, for any $E \in \mathfrak{B}$, there is a sequence of mutually disjoint closed compact sets $\{F_l,\ l = 1, 2, \ldots\}$, such that $F_l \subset E$ and

[24] *Translator's note:* To form a partition, one must also include the two sets $\{g \mid (d\mu/dr) = 0, (d\nu/dr) > 0\}$ and $\{g \mid (d\mu/dr) > 0, (d\nu/dr) = 0\}$. However, the proof still remains valid.

$F_0 = E - \sum_l F_l$ is a null set with respect to both μ and ν. By the Schwarz inequality,

$$\sum_{l \geqslant 0} (\mu(F_l))^{1/2} (\nu(F_l))^{1/2} \leqslant \Big(\sum_{l \geqslant 0} \mu(F_l)\Big)^{1/2} \Big(\sum_{l \geqslant 0} \nu(F_l)\Big)^{1/2}$$

$$= (\mu(E)\, \nu(E))^{1/2}. \tag{1.4.12}$$

Hence, if $\{E_k\} \in \mathfrak{F}$, there is an $\{F_{kl}\} \in \mathfrak{F}_0$ such that $F_{kl} \subset E_k$, and, by (1.4.12),

$$\sum_k (\mu(E_k)\, \nu(E_k))^{1/2} \geqslant \sum_{kl} (\mu(F_{kl})\, \nu(F_{kl}))^{1/2}.$$

Consequently,

$$\inf_{\{E_k\} \in \mathfrak{F}} \sum_k (\mu(E_k)\, \nu(E_k))^{1/2} \geqslant \inf_{\{F_k\} \in \mathfrak{F}_0} \sum_k (\mu(F_k)\, \nu(F_k))^{1/2}. \tag{1.4.13}$$

But $\mathfrak{F}_0 \subset \mathfrak{F}$, therefore, the inequality in (1.4.13) can be replaced by an equality.]

3° Kakutani Inner Products of Product Measures

We first consider products of finitely many finite measures.

Lemma 1.4.3. Let Γ_k, $k = 1, 2, \ldots, l$ be sets, and, for each k, let $\mathfrak{B}_k$ be a σ-algebra of sets in Γ_k, and let μ_k, ν_k be finite measures on $(\Gamma_k, \mathfrak{B}_k)$. Let $(\Gamma_l^*, \mathfrak{B}_l^*)$ denote the product measurable space $(\times_{k=1}^{l} \Gamma_k, \times_{k=1}^{l} \mathfrak{B}_k)$, and let μ_l^*, ν_l^* denote the product measures $\times_{k=1}^{l} \mu_k$, $\times_{k=1}^{l} \nu_k$, respectively. Then

$$\rho(\mu_l^*, \nu_l^*) = \prod_{k=1}^{l} \rho(\mu_k, \nu_k). \tag{1.4.14}$$

PROOF. Let r_k be a finite measure on $(\Gamma_k, \mathfrak{B}_k)$, such that $\mu_k \ll r_k$, $\nu_k \ll r_k$. Form the product measure $r_l^* = \times_{k=1}^{l} r_k$ on $(\Gamma_l^*, \mathfrak{B}_l^*)$. Since $\mu_k \ll r_k$ for each k, it follows that $\mu_l^* \ll r_l^*$. Similarly, $\nu_l^* \ll r_l^*$. Then, writing $\omega_l^* = (\omega_1, \ldots, \omega_l)$, $\omega_k \in \Gamma_k$, $\omega \in \Gamma$, we have

$$\frac{d\mu_l^*(\omega_l^*)}{dr_l^*(\omega_l^*)} = \frac{d\mu_1(\omega_1)}{dr_1(\omega_1)} \cdots \frac{d\mu_l(\omega_l)}{dr_l(\omega_l)}, \tag{1.4.15}$$

and a similar formula for ν_l^*. Using these two formulas, we easily calculate (1.4.14).]

Theorem 1.4.4. Let $(\Gamma_k, \mathfrak{B}_k)$, $k = 1, 2, \ldots$, be a sequence of measurable spaces, and, for each k, let μ_k, ν_k be probability measures on

$(\Gamma_k, \mathfrak{B}_k)$. Let $(\Gamma, \mathfrak{B}) = (\mathsf{X}_{k=1}^{\infty} \Gamma_k, \mathsf{X}_{k=1}^{\infty} \mathfrak{B}_k)$, $\mu = \mathsf{X}_{k=1}^{\infty} \mu_k$, $\nu = \mathsf{X}_{k=1}^{\infty} \nu_k$. Suppose that, for each k, μ_k and ν_k are equivalent. Then μ and ν are either equivalent or mutually singular. A necessary and sufficient condition for the equivalence (mutual singularity) of μ and ν is $\prod_{k=1}^{\infty} \rho(\mu_k, \nu_k) > 0$ $(= 0)$. Moreover, in either case,

$$\rho(\mu, \nu) = \prod_{k=1}^{\infty} \rho(\mu_k, \nu_k). \tag{1.4.16}$$

PROOF. Consider the measurable spaces $(\Gamma_l^*, \mathfrak{B}_l^*)$ and measures μ_l^*, ν_l^*, r_l^*, as defined in Lemma 1.4.3. By virtue of (1.4.14), and the fact that $\rho(\mu_k, \nu_k) \leqslant 1$, we know that $\rho(\mu_l^*, \nu_l^*)$, $l = 1, 2, \ldots$, is a monotonic decreasing sequence. Hence, the infinite product

$$\prod_{k=1}^{\infty} \rho(\mu_k, \nu_k) = \lim_{l\to\infty} \rho(\mu_l^*, \nu_l^*) \tag{1.4.17}$$

either converges to a positive number or diverges to zero. We consider these two cases separately.

(i) $\lim_{l\to\infty} \rho(\mu_l^*, \nu_l^*) = 0$.

In this case, for any positive number ϵ, there exists an integer l such that

$$\rho(\mu_l^*, \nu_l^*) < \epsilon.$$

Consider the set

$$A_l^* = \left\{ \omega_l^* \;\middle|\; \frac{d\mu_l^*(\omega_l^*)}{dr_l^*(\omega_l^*)} < \frac{d\nu_l^*(\omega_l^*)}{dr_l^*(\omega_l^*)} \right\}$$

in $\mathfrak{B}_l^*$. We have

$$\begin{aligned}
\mu_l^*(A_l^*) &= \int_{A_l^*} \frac{d\mu_l^*(\omega_l^*)}{dr_l^*(\omega_l^*)}\, dr_l^*(\omega_l^*) \\
&\leqslant \int_{A_l^*} \left(\frac{d\mu_l^*(\omega_l^*)}{dr_l^*(\omega_l^*)} \frac{d\nu_l^*(\omega_l^*)}{dr_l^*(\omega_l^*)} \right)^{1/2} dr_l^*(\omega_l^*) \\
&\leqslant \rho(\mu_l^*, \nu_l^*) < \epsilon.
\end{aligned} \tag{1.4.18}$$

Similarly, since

$$\Gamma_l^* - A_l^* = \left\{ \omega_l^* \;\middle|\; \frac{d\mu_l^*(\omega_l^*)}{dr_l^*(\omega_l^*)} \geqslant \frac{d\nu_l^*(\omega_l^*)}{dr_l^*(\omega_l^*)} \right\},$$

we obtain

$$\nu_l^*(\Gamma_l^* - A_l^*) \leqslant \int_{\Gamma_l^* - A_l^*} \left(\frac{d\mu_l^*(\omega_l^*)}{dr_l^*(\omega_l^*)} \frac{d\nu_l^*(\omega_l^*)}{dr_l^*(\omega_l^*)} \right)^{1/2} dr_l^*(\omega_l^*) < \epsilon. \tag{1.4.19}$$

Now, in Γ, form the cylinder A_l with base A_l^*, that is,

$$A_l = \{\omega \mid (\omega_1, \ldots, \omega_l) \in A_l^*\} \in \mathfrak{B},$$

where ω_m denotes the mth coordinate of ω. By the definition of product measure, we have $\mu(A_l) = \mu_l^*(A_l^*)$, $\nu(A_l) = \nu_l^*(A_l^*)$. By (1.4.18) and (1.4.19), we get

$$\mu(A_l) < \epsilon, \qquad \nu(\Gamma - A_l) < \epsilon.$$

By Lemma 1.1.21, we conclude that μ and ν are mutually singular, moreover, by Lemma 1.4.1, $\rho(\mu, \nu) = 0$, so that (1.4.16) holds.

(ii) $\lim_{l\to\infty} \rho(\mu_l^*, \nu_l^*) > 0$.

Since μ_k and ν_k are equivalent, we may choose $r_k = \nu_k$. We note that, for every l, $d\mu_l^*(\omega_l^*)/d\nu_l^*(\omega_l^*)$ may be regarded as a function of ω. Moreover, if we define

$$\psi_l(\omega) = \left(\frac{d\mu_l^*(\omega_l^*)}{d\nu_l^*(\omega_l^*)}\right)^{1/2},$$

then $\psi_l \in L^2(\Gamma, \mathfrak{B}, \nu)$. For $k \leqslant l$, the functions ψ_k, ψ_l are measurable with respect to $\mathfrak{B}_l^*$, and it is easy to calculate that

$$\int_\Gamma |\psi_l(\omega) - \psi_k(\omega)|^2 \, d\nu(\omega)$$
$$= 2\left(1 - \int_{\Gamma_l^*} \left(\frac{d\mu_l^*(\omega_l^*)}{d\nu_l^*(\omega_l^*)} \frac{d\mu_k^*(\omega_k^*)}{d\nu_k^*(\omega_k^*)}\right)^{1/2} d\nu_l^*(\omega_l^*)\right). \tag{1.4.20}$$

Furthermore, by (1.4.15), we have

$$\int_{\Gamma_l^*} \left(\frac{d\mu_l^*(\omega_l^*)}{d\nu_l^*(\omega_l^*)} \frac{d\mu_k^*(\omega_k^*)}{d\nu_k^*(\omega_k^*)}\right)^{1/2} d\nu_l^*(\omega_l^*)$$
$$= \int_{\Gamma_l^*} \prod_{p=1}^{k} \frac{d\mu_p(\omega_p)}{d\nu_p(\omega_p)} \prod_{p=k+1}^{l} \left(\frac{d\mu_p(\omega_p)}{d\nu_p(\omega_p)}\right)^{1/2} d\nu_l^*(\omega_l^*)$$
$$= \prod_{p=k+1}^{l} \rho(\mu_p, \nu_p). \tag{1.4.21}$$

Since we have assumed that (1.4.17) is positive,

$$\lim_{l \geqslant k\to\infty} \prod_{p=k+1}^{l} \rho(\mu_p, \nu_p) = 1.$$

Therefore, it follows from (1.4.20) and (1.4.21) that $\{\psi_l\}$, $l = 1, 2, \ldots$, is a fundamental sequence in $L^2(\Gamma, \mathfrak{B}, \nu)$, and hence converges in $L^2(\Gamma, \mathfrak{B}, \nu)$ to a limit function ψ. Let $A_m^* \in \mathfrak{B}_m^*$, let

$$A = \{\omega \mid (\omega_1, ..., \omega_m) \in A_m^*\},$$

and, for $l \geqslant m$, let $A_l^* = \{(\omega_1, ..., \omega_l) \mid (\omega_1, ..., \omega_m) \in A_m^*\}$. Then

$$\mu(A) = \lim_{\substack{l \geqslant m \\ l \to \infty}} \mu_l^*(A_l^*) = \lim_{\substack{l \geqslant m \\ l \to \infty}} \int_{A_l^*} \frac{d\mu_l^*(\omega_l^*)}{d\nu_l^*(\omega_l^*)} \, d\nu_l^*(\omega_l^*)$$

$$= \lim_{\substack{l \geqslant m \\ l \to \infty}} \int_A \frac{d\mu_l^*(\omega_l^*)}{d\nu_l^*(\omega_l^*)} \, d\nu(\omega) = \int_A \psi^2(\omega) \, d\nu(\omega).$$

From this, it is easily verified that, for any $A \in \mathfrak{B}$, $\mu(A) = \int_A \psi^2(\omega) \, d\nu(\omega)$. Hence, μ is absolutely continuous with respect to ν, and

$$\frac{d\mu(\omega)}{d\nu(\omega)} = \psi^2(\omega).$$

Similarly, ν is absolutely continuous with respect to μ. Thus, μ and ν are equivalent, moreover,

$$\rho(\mu, \nu) = \int_\Gamma \psi(\omega) \, d\nu(\omega)$$

$$= \lim_{l \to \infty} \int_\Gamma \left(\frac{d\mu_l^*(\omega_l^*)}{d\nu_l^*(\omega_l^*)} \right)^{1/2} d\nu(\omega) = \lim_{l \to \infty} \int_{\Gamma_l^*} \left(\frac{d\mu_l^*(\omega_l^*)}{d\nu_l^*(\omega_l^*)} \right)^{1/2} d\nu_l^*(\omega_l^*)$$

$$= \lim_{l \to \infty} \rho(\mu_l^*, \nu_l^*),$$

hence, (1.4.16) holds.]

CHAPTER

II

REPRESENTATION OF POSITIVE FUNCTIONALS AND OPERATOR RINGS

As is well known, the theory of integral representations of positive functionals on normed algebras is a basic tool of harmonic analysis on locally compact groups, and in fact, the integral representation of positive functionals is in itself an important problem of abstract harmonic analysis. In this chapter, we shall use measures on infinite-dimensional products to prove representation theorems for positive functionals on some rather broad classes of linear topological algebras.

In what follows, we shall discuss groups which are not locally compact. Such a group lacks the group algebra which frequently serves as a convenient tool of harmonic analysis, hence, the usual normed algebra techniques used in the study of locally compact groups are ineffective. Consequently, one must resort to the use of commutative weakly closed operator algebras on Hilbert spaces. We shall, in §2.4, show the connection between such algebras and multiplication algebras on localizable measure spaces, thus throwing further light on the properties of localizable measure spaces. The results obtained will find application in subsequent chapters.

§2.1. Topological Algebras with Involution: Fundamental Concepts

1° Seminormed Algebras

First, we recall the meaning of the term "algebra."

Definition 2.1.1. Let R be a set, with elements denoted by $x, y, z, \ldots$. Suppose that linear operations and pairwise multiplication have been defined in R, in such a way that the following conditions are satisfied.

(i) R forms a linear space (over the real or complex numbers).

(ii) For any number λ and elements $x, y, z \in R$, we have $\lambda(xy) = (\lambda x)y = x(\lambda y)$, $(xy)z = x(yz)$, $(x + y)z = xz + yz$, $z(x + y) = zx + zy$.

Then R is called an *algebra.* In deference to established usage, we shall sometimes use the term "ring" in referring to certain types of algebras. If, for every $x, y \in R$, we have $xy = yx$, then R is called a *commutative algebra.* If there is an $e \in R$ such that, for every $x \in R$, $xe = x$, then e is called a *unit* of R.

Let R be an algebra, and let $|x|$, $x \in R$, be a real-valued function on R, not identically zero, such that, for any number λ and elements x, $y \in R$, we have

$$\text{(i)}\quad |x| \geqslant 0, \qquad \text{(ii)}\quad |x + y| \leqslant |x| + |y|,$$

$$\text{(iii)}\quad |\lambda x| = |\lambda|\,|x|, \qquad \text{(iv)}\quad |xy| \leqslant |x|\,|y|,$$

then $|x|$ is called a *seminorm*[1] on R.

Let R be an algebra, and $\{|x|_\alpha, \alpha \in \mathfrak{A}\}$ (here, $\mathfrak{A}$ is an arbitrary indexing set) a family of seminorms on R. If, for each nonzero element $x \in R$, there is an $\alpha \in \mathfrak{A}$ such that $|x|_\alpha > 0$, then R is said to be a *seminormed algebra* with respect to $\{|x|_\alpha, \alpha \in \mathfrak{A}\}$ (or, in some of the literature, a locally multiplicatively convex algebra).

In particular, if $\mathfrak{A}$ contains but a single index, then R is called a *normed algebra.* If $\mathfrak{A}$ is denumerable, for instance, if $\mathfrak{A}$ is the totality of natural numbers, we call R a *countably seminormed algebra* with respect to $\{|x|_n, n = 1, 2, \ldots\}$. If R has a unit e, we shall always assume that $|e|_\alpha = 1$ for all $\alpha \in \mathfrak{A}$.

[1] *Translator's note:* In the terminology of Naimark [1], a "symmetric pseudonorm."

We note that a seminorm is, in particular, a pseudonorm (see §I.1), hence a seminormed algebra R is a linear topological space with respect to the family of pseudonorms $\{|x|_\alpha, \alpha \in \mathfrak{A}\}$, moreover, multiplication of elements is continuous in this topology. We shall not discuss this topology in detail; for our present purposes, we need only consider the convergence of sequences in R.

Definition 2.1.2. Let R be a seminormed algebra with respect to the family of seminorms $\{|x|_\alpha, \alpha \in \mathfrak{A}\}$. Let $\{x_m\}$ be a sequence of elements of R. If there is an $x \in R$ such that

$$\lim_{m\to\infty} |x_m - x|_\alpha = 0,$$

for every $\alpha \in \mathfrak{A}$, we say that $\{x_m\}$ *converges* to x, and write $x_m \to x$. If $\{x_n\}$ is a sequence in R such that

$$\lim_{n,m\to\infty} |x_n - x_m|_\alpha = 0$$

for every $\alpha \in \mathfrak{A}$, we say that $\{x_n\}$ is a *fundamental sequence* in R. If each fundamental sequence in R converges to some element in R, we say that R is complete.[2] A complete normed algebra is called a *Banach algebra*.

Let f be a linear functional on R. If there is a nonempty finite set of indices $\alpha_1, \ldots, \alpha_n \in \mathfrak{A}$, and a positive number M, such that, for every $x \in R$,

$$|f(x)| \leqslant M \max(|x|_{\alpha_1}, \ldots, |x|_{\alpha_n}),$$

then f is said to be a *continuous linear functional* on R.

If R is countably seminormed with respect to $\{|x|_n, n = 1, 2, \ldots\}$, we form the sequence of seminorms $\{|x|_n'\}$ as follows:

$$|x|_n' = \max(|x|_1, \ldots, |x|_n).$$

Then $|x|_1' \leqslant |x|_2' \leqslant \cdots$, moreover, this sequence of seminorms induces the same topology as $\{|x|_n\}$. Hence, we may always assume that the family of seminorms $\{|x|_n\}$ satisfies the condition

$$|x|_1 \leqslant |x|_2 \leqslant \cdots \leqslant |x|_n \leqslant \cdots.$$

Furthermore, it is easy to prove that a linear functional f on a countably seminormed algebra R is continuous if and only if, for every convergent sequence $\{x_n\}$ in R, $x_n \to x$ implies $f(x_n) \to f(x)$.

[2] In the theory of linear topological spaces, this property is known as *sequential completeness*.

2° Involutions and Positive Functionals

Definition 2.1.3. Let R be an algebra, and let $x \to x^*$ be a one-to-one mapping of R onto R, such that, for any numbers λ, μ, and elements $x, y \in R$,

$$\text{(i)}\quad (\lambda x + \mu y)^* = \bar{\lambda} x^* + \bar{\mu} y^*, \qquad \text{(ii)}\quad (xy)^* = y^* x^*, \qquad \text{(iii)}\quad (x^*)^* = x.$$

Then the mapping $x \to x^*$ is said to be an *involution* on R, and x^* is called the *conjugate* of x. If $x = x^*$, we say that x is *self-conjugate*.

For any $x \in R$, the element x^*x is self-conjugate. If R has a unit element e, then e must be self-conjugate. In fact, since $e^* = e^*e$ is self-conjugate, we have $e = (e^*)^* = e^*$. Furthermore, $(ex)^* = x^*e^* = x^*e = x^*$, hence $(ex)^{**} = x^{**}$, that is, $ex = x$, so that e also acts as a right-sided unit. For any $x \in R$, the elements

$$x_1 = \frac{x + x^*}{2}, \qquad x_2 = \frac{x - x^*}{2i}$$

are self-conjugate. Therefore, any element of R can be expressed as a linear combination of self-conjugate elements:

$$x = x_1 + ix_2 .$$

Definition 2.1.4. Let R be an algebra with an involution. If f is a linear functional on R such that

$$f(x^*) = \overline{f(x)} \tag{2.1.1}$$

for every $x \in R$, then f is said to be *symmetric*. If f is a linear functional on R satisfying the condition

$$f(x^*x) \geqslant 0, \qquad x \in R, \tag{2.1.2}$$

then f is called a *positive functional*.

Let R be an algebra with an involution, f a positive functional on R. For any $x, y \in R$, and any complex number λ, we have, by (2.1.2),

$$\begin{aligned} &f(x^*x) + \lambda f(x^*y) + \bar{\lambda} f(y^*x) + \lambda\bar{\lambda} f(y^*y) \\ &\quad = f((x + \lambda y)^*(x + \lambda y)) \geqslant 0. \end{aligned} \tag{2.1.3}$$

Since $f(x^*x) \geqslant 0, f(y^*y) \geqslant 0$, it follows that $\lambda f(x^*y) + \bar{\lambda} f(y^*x)$ is real. Hence, by suitably choosing the value of λ, we get

$$f(x^*y) = \overline{f(y^*x)}. \tag{2.1.4}$$

Also, since the form (2.1.3) is positive semi-definite, it follows that, for any $x, y \in R$,

$$|f(x^*y)| \leqslant (f(x^*x))^{1/2} (f(y^*y))^{1/2}. \tag{2.1.5}$$

This is known as the *Schwarz inequality* for positive functionals.

If R has a unit e, then by setting $y = e$ in (2.1.4), we get (2.1.1). Thus, on an algebra with an involution and a unit, every positive functional is symmetric.

Definition 2.1.5. Let R be a seminormed algebra with respect to $\{| x |_\alpha, \alpha \in \mathfrak{A}\}$. Suppose there is given an involution $x \to x^*$ on R such that, for every $\alpha \in \mathfrak{A}$, $x \in R$, we have

$$| x^* |_\alpha = | x |_\alpha . \tag{2.1.6}$$

Then, we call R a *symmetric seminormed algebra*.

Notice that (2.1.6) is, essentially, the condition that the involution $x \to x^*$ be continuous. For, if (2.1.6) is satisfied, then obviously $x \to x^*$ is continuous; conversely, if $x \to x^*$ is continuous, we may replace each $| x |_\alpha$ by the seminorm

$$| x |_\alpha' = \max(| x |_\alpha , | x^* |_\alpha),$$

without altering the topology of R, and obviously $| x^* |_\alpha' = | x |_\alpha'$.

3° Multiplicative Linear Functionals

Definition 2.1.6. Let R be an algebra with unit e, and let f be a linear functional on R such that

(i) for any $x, y \in R, f(xy) = f(x)f(y)$;

(ii) $f(e) = 1$.

Then f is said to be a *multiplicative linear functional* on R.

Thus, a multiplicative linear functional is just a homomorphism from an algebra R into the field of complex numbers.

In this book, we shall be interested in symmetric multiplicative linear functionals on algebras with involutions. If R is an algebra with an involution, and f is a symmetric multiplicative linear functional on R, then, for any $x \in R$,

$$f(x^*x) = f(x^*)f(x) = \overline{f(x)}f(x) \geqslant 0.$$

Therefore, any symmetric multiplicative linear functional is positive. Later on, we shall represent positive functionals, defined on certain types of countably seminormed algebras, in terms of symmetric multiplicative linear functionals.

In the ensuing discussion, the letter $\mathfrak{M}$ will denote the totality of symmetric multiplicative linear functionals on a given algebra with an involution.

We shall represent symmetric multiplicative linear functionals by points of an infinite dimensional product space. First, we begin by using such points to represent real linear functionals.

Let Φ be a linear space, and let $\{x_\beta, \beta \in B\}$ be a system of linearly independent vectors in Φ such that every vector in Φ is a finite linear combination of vectors in $\{x_\beta, \beta \in B\}$. Such a system does exist (although it is not unique). In fact, let $\mathfrak{F}$ denote the family of all linearly independent sets ξ of vectors in Φ. Define a partial ordering in $\mathfrak{F}$ as follows: if $\xi, \eta \in \mathfrak{F}$ and $\xi \subset \eta$, then $\xi \prec \eta$. Using Zorn's lemma, one can prove that $\mathfrak{F}$ contains a maximal element, and this maximal element satisfies our requirements. Such a system of vectors $\{x_\beta, \beta \in B\}$ is called a *linear basis* for Φ.

Having selected a linear basis $\{x_\beta, \beta \in B\}$ for a linear space Φ, we obtain, for each linear functional f on Φ, a system of numbers $\{f(x_\beta), \beta \in B\}$. Conversely, given any system of numbers $\{u_\beta, \beta \in B\}$, there is a unique linear functional f on Φ, such that $f(x_\beta) = u_\beta, \beta \in B$; this functional is given by the formula

$$f(c_{\beta_1}x_{\beta_1} + \cdots + c_{\beta_n}x_{\beta_n}) = c_{\beta_1}u_{\beta_1} + \cdots + c_{\beta_n}u_{\beta_n} . \tag{2.1.7}$$

Thus, we get a one-to-one correspondence between linear functionals and systems of numbers $\{u_\beta, \beta \in B\}$.

In particular, let R be an algebra with an involution and a unit e, and let Φ be the totality of self-conjugate elements of R. Then Φ is a linear space over the real numbers, and $e \in \Phi$. For each symmetric linear functional f on R, the restriction $f \mid \Phi$ is a real linear functional on Φ. Conversely, any real linear functional f on Φ can be extended to a symmetric linear functional on R by defining

$$f(x) = f\left(\frac{x + x^*}{2}\right) + if\left(\frac{x - x^*}{2i}\right), \qquad x \in R.$$

Choose a linear basis $\{x_\beta, \beta \in B\}$ for Φ; we may assume that $0 \in B$ and $x_0 = e$ (the unit element). Let $A = B - \{0\}$. For each index $\beta \in A$, let E_β be a copy of the space of real numbers, and let E_0 be the singleton containing only the real number 1. Form the Cartesian product

$$E = \mathop{\times}_{\beta \in B} E_\beta , \tag{2.1.8}$$

and denote the points of E by $u = \{u_\beta , \beta \in B\}$, where $u_\beta \in E_\beta$ is the βth coordinate of u. In accordance with (2.1.7), we have a one-to-one correspondence between points $u \in E$ and symmetric linear functionals f on R which satisfy the condition $f(e) = 1$. By identifying every f with the corresponding u, we may regard the totality $\mathfrak{M}$ of symmetric multiplicative linear functionals as a subset of E.

Consider any two basis elements x_β , $x_{\beta'}$, $\beta, \beta' \in B$; since x_β , $x_{\beta'} \in R$, the product $x_\beta x_{\beta'}$ is of the form $y + iz$, where $y, z \in \Phi$. Hence, we have a system of complex numbers $\{c^k_{\beta,\beta'} , k \in B\}$, of which only finitely many are nonzero, such that

$$x_\beta x_{\beta'} = \sum_k c^k_{\beta\beta'} x_k . \tag{2.1.9}$$

Accordingly, for each pair of indices $\beta, \beta' \in B$, we form the subset $\mathfrak{M}_{\beta,\beta'}$ of E, consisting of all points $\{u_\beta , \beta \in B\}$ which satisfy the condition

$$u_\beta u_{\beta'} = \sum_k c^k_{\beta\beta'} u_k . \tag{2.1.10}$$

In particular, $\mathfrak{M}_{0,0} = E$.

Lemma 2.1.1. $\mathfrak{M} = \bigcap_{\beta,\beta' \in B} \mathfrak{M}_{\beta,\beta'}$.

PROOF. Let $u \in \bigcap_{\beta,\beta' \in B} \mathfrak{M}_{\beta,\beta'}$. Let f be the linear functional corresponding to u [see (2.1.7)]. For any two elements $x, y \in R$, we have two systems of numbers, $\{\lambda_\beta , \beta \in B\}$, $\{\xi_\beta , \beta \in B\}$ (only finitely many of them being nonzero) such that

$$x = \sum_\beta \lambda_\beta x_\beta , \qquad y = \sum_{\beta'} \xi_{\beta'} x_{\beta'} .$$

Using (2.1.9), we get

$$xy = \sum_{\beta,\beta',k} c^k_{\beta\beta'} \lambda_\beta \xi_{\beta'} x_k .$$

Since we have assumed that u satisfies (2.1.10) for all β, β', it follows by (2.1.7) that

$$f(xy) = \sum_{\beta,\beta',k} c^k_{\beta\beta'} u_k \lambda_\beta \xi_{\beta'} = \sum_{\beta,\beta'} u_\beta \lambda_\beta u_{\beta'} \xi_{\beta'} = f(x)f(y).$$

Thus, f is a symmetric multiplicative linear functional such that $f(e) = 1$, that is, $u \in \mathfrak{M}$.

Conversely, if $u \in \mathfrak{M}$, then the corresponding functional f is multiplicative. Applying f to both sides of (2.1.9), we get (2.1.10). Thus, $u \in \mathfrak{M}_{\beta,\beta'}$ for all β, $\beta' \in B$, hence $\mathfrak{M} \subset \bigcap_{\beta,\beta'} \mathfrak{M}_{\beta,\beta'}$.]

Now, let each E_β have the Euclidean topology (i.e., the topology induced by the Euclidean metric) and let E have the product topology.

Then,

$$\mathfrak{M} \text{ is a closed subspace of } E. \tag{2.1.11}$$

In fact, for each $\beta \in B$, the projection $u \rightarrow u_\beta$ is a continuous function on E, hence the correspondence $u \rightarrow u_\beta u_{\beta'} - \sum_{k \in B} c^k_{\beta\beta'} u_k$ is also a continuous function, and so its null space $\mathfrak{M}_{\beta,\beta'}$ is closed. Hence, by virtue of Lemma 2.1.1, we get (2.1.11).

The relative topology in $\mathfrak{M}$, as a subset of E, is known as the *weak topology*. Clearly, this topology is Hausdorff. Henceforth, unless otherwise stated, we shall always suppose that $\mathfrak{M}$ is given the weak topology.

The idea of imbedding a linear space in its second conjugate space is well known. In a similar fashion, we shall regard the elements of R as functions on $\mathfrak{M}$. More precisely, for each $x \in R$, we form the function $\hat{x}(f)$, $f \in \mathfrak{M}$, defined by

$$\hat{x}(f) = f(x).$$

Let $\hat{R}$ be the totality of such functions. $\hat{R}$ is an algebra with respect to the ordinary linear operations and multiplication of functions. We also introduce an involution $\hat{x} \rightarrow \hat{x}^*$ on $\hat{R}$, as follows:

$$(\hat{x})^*(f) = \overline{\hat{x}(f)}, \qquad f \in \mathfrak{M}.$$

The correspondence $x \rightarrow \hat{x}$ is then an isomorphism of R onto $\hat{R}$, moreover, it is consistent with the involutions in R and $\hat{R}$:

$$(\hat{x})^* = \hat{x}^*.$$

For convenience, we shall henceforth write $x(f)$ in place of $\hat{x}(f)$.

When $\mathfrak{M}$ is given the weak topology, every element $\hat{x}$ of $\hat{R}$ is a continuous function on $\mathfrak{M}$. In fact, given any $x \in R$, there is a finite set of indices $\beta_1, \ldots, \beta_n \in B$, and a set of complex numbers $c_{\beta_1}, \ldots, c_{\beta_n}$, such that $x = \sum_{\nu=1}^n c_{\beta_\nu} x_{\beta_\nu}$, hence

$$x(f) = \sum_{\nu=1}^{n} c_{\beta_\nu} u_{\beta_\nu}, \qquad u_{\beta_\nu} = f(x_{\beta_\nu}).$$

But this is a continuous function of $u_{\beta_1}, u_{\beta_2}, \ldots, u_{\beta_n}$. Therefore, the function $x(f)$, $f \in \mathfrak{M}$, is continuous in the weak topology.

4° Some Examples

We next give some examples which will be used in the sequel.

Example 2.1.1. Let A_n be the totality of entire[3] functions $x(z)$ of n complex variables $z = (z_1, \dots, z_n)$. With respect to the ordinary linear operations and multiplication of functions, A_n is a commutative algebra with the unit element 1. For each $x \in A_n$, we define the function x^* as follows:

$$x^*(z_1, \dots, z_n) = \overline{x(\bar{z}_1, \dots, \bar{z}_n)}.$$

It is easily verified that the correspondence $x \to x^*$ is an involution in A_n. We now proceed to find the totality of symmetric multiplicative linear functionals on A_n.

For an arbitrary but fixed n-tuple of real numbers $z^0 = (z_1{}^0, \dots, z_n{}^0)$, consider the functional on A_n defined by

$$f_{z^0}: \quad x \to x(z^0).$$

It is easily seen that f_{z^0} is a symmetric multiplicative linear functional on A_n. Conversely, let f be any symmetric multiplicative linear functional on A_n. For the functions $z_\nu \in A_n$, $\nu = 1, 2, \dots, n$, we have $z_\nu{}^* = z_\nu$, hence, if

$$z_\nu{}^0 = f(z_\nu),$$

then the $z_\nu{}^0$, $\nu = 1, 2, \dots, n$, are real numbers. Now, for each function $x(z)$ in A_n, form the power series expansion

$$x(z) = \sum_{\nu_1, \dots, \nu_n} a_{\nu_1 \cdots \nu_n} (z_1 - z_1{}^0)^{\nu_1} \cdots (z_n - z_n{}^0)^{\nu_n},$$

which may also be written in the form

$$x(z) = x(z_1{}^0, \dots, z_n{}^0) + \sum_{k=1}^{n} g_k(z)(z_k - z_k{}^0),$$

where

$$g_k(z) = \sum_{\nu_k \geqslant 1} a_{0 \cdots 0 \nu_k \cdots \nu_n} (z_k - z_k{}^0)^{\nu_k - 1} \cdots (z_n - z_n{}^0)^{\nu_n}, \qquad k = 1, 2, \dots$$

are entire functions, i.e., $g_k \in A_n$. Hence, since $f \in \mathfrak{M}$ and $f(z_k) = z_k{}^0$, we get

$$f(x) = x(z_1{}^0, \dots, z_n{}^0) + \sum_{k=1}^{n} f(g_k) f(z_k - z_k{}^0) = f_{z^0}(x),$$

[3] That is to say, analytic on the entire n-dimensional complex number space.

where $z^0 = (z_1{}^0, \ldots, z_n{}^0)$. Therefore, the correspondence

$$z^0 \to f_{z^0}$$

defines a one-to-one mapping from real n-dimensional space R_n onto $\mathfrak{M}$. If we identify R_n and $\mathfrak{M}$ by identifying z^0 and f_{z^0}, then, when $f = f_{z^0}$, we have

$$x(f) = x(z^0).$$

We define a sequence of seminorms[4] $\{| x |_\alpha, \alpha = 1, 2, \ldots\}$ on A_n, as follows

$$| x |_\alpha = \max_{|z_\nu| \leqslant \alpha, \nu=1,2,\ldots,n} | x(z) |.$$

Then, A_n is a countably seminormed algebra with respect to $\{| x |_\alpha, \alpha = 1, 2, \ldots\}$. Moreover, a sequence $\{x_k\}$ in A_n converges to x if and only if the functions $\{x_k(z)\}$ converge uniformly to $x(z)$ on every compact subset of n-dimensional complex number space. It is easily proved that A_n is symmetric and complete.

We note that A_n is a universal commutative symmetric complete seminormed algebra, in the following sense. Let R be an arbitrary commutative symmetric complete seminormed algebra with a unit. Choose any $y_1, \ldots, y_n \in R$. Let $\varphi \in A_n$, and expand φ in a power series

$$\varphi(z) = \sum_{\nu_1, \ldots, \nu_n} a_{\nu_1 \cdots \nu_n} z_1^{\nu_1} \cdots z_n^{\nu_n}.$$

Since φ is an entire function, it is easily shown that, for any set of n positive numbers $c_1, \ldots, c_n$, we have

$$\sum_{\nu_1, \ldots, \nu_n} | a_{\nu_1 \cdots \nu_n} | c_1^{\nu_1} \cdots c_n^{\nu_n} < \infty.$$

Hence, for any $\alpha \in \mathfrak{A}$,

$$\sum_{\nu_1, \ldots, \nu_n} | a_{\nu_1 \cdots \nu_n} | \, | y_1 |_\alpha^{\nu_1} \cdots | y_n |_\alpha^{\nu_n} < \infty.$$

By virtue of the completeness of R, it follows that the series

$$\sum_{\nu_1, \ldots, \nu_n} a_{\nu_1 \cdots \nu_n} y_1^{\nu_1} \cdots y_n^{\nu_n}$$

[4] In this case, the seminorms $| x |_\alpha$ are, in fact, norms.

converges to some element of R, which we denote by

$$\varphi(y_1, \ldots, y_n).$$

We then have

$$\varphi(y_1, \ldots, y_n)^* = \varphi^*(y_1^*, \ldots, y_n^*).$$

In particular, when $y_1, \ldots, y_n$ are self-conjugate elements of R, the mapping $\varphi \to \varphi(y_1, \ldots, y_n)$ is an involution-preserving homomorphism from A_n into R. We shall denote the image of this homomorphism by $R_{y_1, \ldots, y_n}$.

Example 2.1.2. Let H be an inner product space, and $\mathscr{B}(H)$ the totality of bounded linear operators in H. Then $\mathscr{B}(H)$ is an algebra with respect to the ordinary addition, scalar multiplication and composition of operators, and the identity operator I is the unit of $\mathscr{B}(H)$. For any $A \in \mathscr{B}(H)$, let A^* denote the operator conjugate to A. Then $A \to A^*$ is an involution in $\mathscr{B}(H)$. Given this involution and the ordinary operator norm, $\mathscr{B}(H)$ becomes a symmetric normed algebra. For any $\xi \in H$,

$$F_\xi(A) = (A\xi, \xi), \qquad A \in \mathscr{B}(H)$$

is a continuous positive functional on $\mathscr{B}(H)$.

We shall refer to $\mathscr{B}(H)$ and its subalgebras as *operator algebras* (or, sometimes, *operator rings*) over the space H. The norm and involution in such algebras will always be as stated in Example 2.1.2.

The most important type of positive functional to be considered in this book is that described in the next example.

Example 2.1.3. Let S be a compact topological space, $C(S)$ the totality of continuous complex-valued functions on S. With respect to the ordinary linear operations, the norm

$$|x| = \max_{s \in S} |x(s)|, \qquad x \in C(S),$$

and the involution $x \to x^*$ defined by $x^*(s) = \overline{x(s)}$, $C(S)$ is a commutative symmetric Banach algebra with the unit element 1. Let $\mathfrak{B}$ be the σ-algebra generated by the totality of closed subsets of S, and let μ be a (nonnegative) finite measure on $(S, \mathfrak{B})$. Then

$$f(x) = \int_S x(s)\, d\mu(s), \qquad x \in C(S)$$

is a continuous positive functional on $C(S)$. In actual fact, all positive functionals on $C(S)$ are of this form.

Example 2.1.4. Let R be an algebra with an involution and a unit. Let $\mathfrak{M}$ be the topological space formed by the totality of symmetric multiplicative linear functionals on R, with the weak topology. Let $\mathfrak{D}$ be a compact subset of $\mathfrak{M}$, $\mathfrak{B}$ the σ-algebra in $\mathfrak{D}$ generated by the closed subsets of $\mathfrak{D}$, and μ a (nonnegative) finite measure on $(\mathfrak{D}, \mathfrak{B})$. Then the functional

$$f(x) = \int_{\mathfrak{D}} x(M)\, d\mu(M), \qquad x \in R \tag{2.1.12}$$

is well defined. In fact, for any $x \in R$, the function $x(M)$ is continuous on $\mathfrak{M}$, hence continuous on $\mathfrak{D}$, and therefore bounded on $\mathfrak{D}$. Consequently, $x(M)$, restricted to $\mathfrak{D}$, is a bounded measurable function on $(\mathfrak{D}, \mathfrak{B})$, so that the integral on the left of (2.1.12) exists. Obviously,

$$f(x^*x) = \int_{\mathfrak{D}} |\, x(M)|^2\, d\mu(M) \geqslant 0, \qquad x \in R.$$

Therefore, f is a positive functional on R. In the following section, we shall prove that, under certain conditions, (2.1.12) is the general form for positive functionals on R.

Example 2.1.5. Let $\Omega = (G, \mathfrak{B}, \mu)$ be a measure space, $L^\infty(\Omega)$ the totality of essentially bounded measurable functions on Ω. Then $L^\infty(\Omega)$ is an algebra with respect to the ordinary linear operations and multiplication of functions. For any $x \in L^\infty(\Omega)$, let

$$\| x \| = \operatorname*{essential\ supremum}_{g \in G} |\, x(g)|; \qquad x^*(g) = \overline{x(g)}, \quad g \in G.$$

Given this norm and involution, $L^\infty(\Omega)$ becomes a symmetric Banach algebra with a unit.

5° Algebras of Continuous Functions

First, we introduce the following notation. If f and g are two real functions, we write

$$f \vee g = \max(f, g), \qquad f \wedge g = \min(f, g).$$

Next, we state and prove a general approximation theorem for spaces of continuous functions.

Theorem 2.1.2. (Stone–Weierstrass). Let $\mathfrak{M}$ be a compact topological space, and let $R(\mathfrak{M})$ be the Banach algebra formed by the totality of real continuous functions on $\mathfrak{M}$ with the norm $\| \varphi \| = \max_{M \in \mathfrak{M}} |\, \varphi(M)|$.

Suppose that Q is a closed subalgebra of $R(\mathfrak{M})$, satisfying the following condition[5]: for any two distinct points M, N of $\mathfrak{M}$, there is a $\varphi \in Q$ such that $\varphi(M) = 0$ and $\varphi(N) = 1$. Then, $Q = R(\mathfrak{M})$.

PROOF. (1) First, we prove that if $\varphi \in Q$, then $|\varphi| \in Q$. Let ϵ be an arbitrary positive number. There exists a polynomial P such that

$$||t| - P(t)| < \frac{\epsilon}{2} \tag{2.1.13}$$

whenever $|t| \leqslant \|\varphi\|$. Let $P_0(t) = P(t) - P(0)$. Then P_0 is a polynomial without a constant term. Since Q is an algebra, $P_0(\varphi) \in Q$. By (2.1.13), we have

$$\max_{|t| \leqslant \|\varphi\|} ||t| - P_0(t)| < \frac{\epsilon}{2} + |P(0)| < \epsilon.$$

Consequently, since $|\varphi(M)| \leqslant \|\varphi\|$, we get

$$||\varphi|(M) - P_0(\varphi)(M)| < \epsilon$$

for all $M \in \mathfrak{M}$, whence $\||\varphi| - P_0(\varphi)\| < \epsilon$. But Q is closed, hence $|\varphi| \in Q$.

It now follows that, if $\varphi, \psi \in Q$, then $\varphi \vee \psi = \frac{1}{2}(\varphi + \psi + |\varphi - \psi|) \in Q$, and similarly $\varphi \wedge \psi \in Q$.

(2) For any two distinct points M, N of $\mathfrak{M}$, there exists, by hypothesis, an element $\varphi \in Q$ such that $\varphi(M) = 0$, $\varphi(N) = 1$. Reversing the roles of M and N, we can also choose $\psi \in Q$ so that $\psi(N) = 0$, $\psi(M) = 1$. Thus, for any two real numbers a, b, we have $a\varphi + b\psi \in Q$, and $(a\varphi + b\psi)(M) = b$, $(a\varphi + b\psi)(N) = a$.

Now, take any $x(M) \in R(\mathfrak{M})$. Then, for any $M, N \in \mathfrak{M}$, there is an $x_{M,N} \in Q$ such that $x_{M,N}(M) = x(M)$, $x_{M,N}(N) = x(N)$. Hence, each of the open sets

$$U_{M,N} = \{u \mid x_{M,N}(u) < x(u) + \epsilon\},$$

$$V_{M,N} = \{u \mid x_{M,N}(u) > x(u) - \epsilon\}$$

contains both M and N. Thus, $\{U_{M,N} \mid M \in \mathfrak{M}\}$ is an open covering of $\mathfrak{M}$. By the compactness of $\mathfrak{M}$, there exist $M_1, \ldots, M_n \in \mathfrak{M}$ such that $U_{M_1,N} + \cdots + U_{M_n,N} \supset \mathfrak{M}$. By the result proved in part (1) above, we have

$$x_N = x_{M_1,N} \wedge x_{M_2,N} \wedge \cdots \wedge x_{M_n,N} \in Q,$$

[5] It can be proved that this condition is equivalent to the following. For any two distinct points M, N of $\mathfrak{M}$, there is a $\varphi \in Q$ such that $\varphi(M) \neq \varphi(N)$, and, for any $M \in \mathfrak{M}$, there is a $\psi \in Q$ such that $\psi(M) \neq 0$.

and, for every $u \in \mathfrak{M}$, $x_N(u) < x(u) + \epsilon$. Let $V_N = \bigcap_{\nu=1}^{n} V_{M_\nu, N}$. Then V_N is a neighborhood of N, and $x_N(u) > x(u) - \epsilon$ for every $u \in V_N$. Again, by compactness, there exist $N_1, \ldots, N_m \in \mathfrak{M}$ such that $V_{N_1} \cup \cdots \cup V_{N_m} \supset \mathfrak{M}$, and, by part (1)

$$\tilde{x} = x_{N_1} \vee x_{N_2} \vee \cdots \vee x_{N_m} \in Q.$$

Moreover,

$$x(u) - \epsilon < \tilde{x}(u) < x(u) + \epsilon, \qquad u \in \mathfrak{M},$$

whence $\| \tilde{x} - x \| < \epsilon$. Since Q is closed, and ϵ is arbitrary, it follows that $x \in Q$.]

Corollary 2.1.3. Let $\mathfrak{M}$ be a compact topological space, and let $C(\mathfrak{M})$ be the symmetric normed algebra formed by the totality of continuous complex-valued functions on $\mathfrak{M}$ (see Example 2.1.3). Suppose that S is a closed symmetric[6] subalgebra of $C(\mathfrak{M})$, with the property that, for any two distinct points M, N of $\mathfrak{M}$, there is a $\varphi \in S$ such that $\varphi(M) = 0$, $\varphi(N) = 1$. Then $S = C(\mathfrak{M})$.

PROOF. Let Q be the totality of real functions in S. Clearly, Q is a closed subalgebra of $R(\mathfrak{M})$. Given any two distinct points $M, N \in \mathfrak{M}$, there is, by hypothesis, a function $\varphi \in S$ such that $\varphi(M) = 0$, $\varphi(N) = 1$. Since S is symmetric, $\bar{\varphi} \in S$, hence $\psi = \frac{1}{2}(\varphi + \bar{\varphi}) \in Q$. Also, $\psi(M) = 0$, $\psi(N) = 1$. Thus, Q satisfies the conditions of Theorem 2.1.2, therefore, $Q = R(\mathfrak{M})$. For any $x \in C(\mathfrak{M})$, the elements $x_1 = (x + \bar{x})/2$, $x_2 = (x - \bar{x})/2i$ both belong to $R(\mathfrak{M})$ (i.e., belong to Q), hence $x = x_1 + ix_2 \in S$. Thus, $S = C(\mathfrak{M})$.]

§2.2. Representation of Positive Functionals on Seminormed Algebras

We shall first prove that any positive functional, defined on a complete symmetric countably seminormed algebra with unit, is continuous. Here, continuity of a linear functional f means, as usual, that, for any sequence $\{x_n\}$ converging to the zero element, the values $f(x_n)$ converge to zero.

Lemma 2.2.1. Let R be a complete symmetric countably seminormed algebra with unit e. Let f be a positive functional on R. Then f is continuous.

PROOF. If $f(e) = 0$, it follows easily from (2.1.5) that f is identically zero. Consequently, we may assume that $f(e) = 1$. Suppose that f is not

[6] That is, if $x \in S$, then $x^* \in S$.

continuous. Then, there is a sequence $\{x_n\} \subset R$ such that $x_n \to 0$ and $f(x_n) \not\to 0$. Taking a subsequence of $\{x_n\}$, if necessary, we may suppose that

$$\epsilon = \inf_n |f(x_n)| > 0. \qquad (2.2.1)$$

Setting $x = x_n$, $y = e$ in the Schwarz inequality (2.1.5), we get

$$|f(x_n)| \leqslant (f(x_n{}^*x_n))^{1/2},$$

whence it follows by (2.2.1) that $f(x_n{}^*x_n) \geqslant \epsilon^2$. Also, since $x_n \to 0$, we have $x_n{}^* \to 0$. Let $y_n = x_n{}^*x_n/f(x_n{}^*x_n)$. Then

$$y_n{}^* = y_n, \qquad y_n \to 0, \qquad f(y_n) = 1.$$

For any $x \in R$, let $N_n(x) = \max(|x|_n, 1)$.[7] Since $y_n \to 0$, we can pick a subsequence $\{z_n\}$ of $\{y_n\}$ such that

$$\begin{aligned} Q_n = {} & N_n(z_1) + (N_n(z_2) + (N_n(z_3) + (\cdots + (N_n(z_{n-1}) + |z_n|_n^2)^2 \cdots)^2 \\ & - \{N_n(z_1) + (N_n(z_2) + (N_n(z_3) + (\cdots + (N_n(z_{n-2}) \\ & + N_n(z_{n-1})^2)^2 \cdots)^2\} < 1/2^n. \end{aligned} \qquad (2.2.2)$$

In fact, suppose that $z_1, \ldots, z_{l-1}$ have already been chosen, $z_\nu = y_{k_\nu}$, and that (2.2.2) holds for $n = 1, 2, \ldots, l-1$. Then, we need only choose $k_l > k_{l-1}$ so that $|y_{k_l}|_l$ is sufficiently small, take $z_l = y_{k_l}$, and (2.2.2) will also hold for $n = l$ (since $\lim_{n\to\infty} |y_n|_l = 0$, such an index k_l can certainly be found). Thus, we obtain the required subsequence $\{z_n\}$.

Consider the element

$$\lambda_n = z_1 + (z_2 + (z_3 + (\cdots + z_n{}^2)^2 \cdots)^2 \qquad (2.2.3)$$

of R. Observe that λ_n is a polynomial in $z_1, \ldots, z_n$, with positive coefficients. Let $P_n(t_1, \ldots, t_n)$ denote the polynomial obtained by replacing every z_ν in (2.2.3) by the complex variable t_ν. It is easily seen that $\lambda_n - \lambda_{n-1}$ is also a polynomial in $z_1, \ldots, z_n$, with positive coefficients; replacing each z_ν by t_ν, we denote this polynomial by $S_n(t_1, \ldots, t_n)$. Then we have

$$\begin{aligned} |\lambda_n - \lambda_{n-1}|_n & \leqslant S_n(|z_1|_n, \ldots, |z_n|_n) \\ & \leqslant S_n(N_n(z_1), \ldots, N_n(z_{n-1}), |z_n|_n) \\ & = P_n(N_n(z_1), \ldots, N_n(z_{n-1}), |z_n|_n) \\ & \quad - P_{n-1}(N_n(z_1), \ldots, N_n(z_{n-1})) \\ & = Q_n < 1/2^n. \end{aligned} \qquad (2.2.4)$$

[7] Here, $\{|x|_n, n = 1, 2, \ldots\}$ denotes the given family of seminorms on R.

It follows that $\{\lambda_n\}$ is a fundamental sequence in R, hence, it converges to an element χ_1 of R. Since $z_\nu = z_\nu^*$, $\lambda_n = \lambda_n^*$, it follows that $\chi_1 = \chi_1^*$.

Let $\lambda_n' = z_2 + (z_3 + (z_4 + (\cdots + (z_{n-1} + z_n^2)^2 \cdots)^2$. Proceeding as we did in the derivation of (2.2.4), we can likewise show that

$$\begin{aligned} | \lambda_n' - \lambda_{n-1}' |_n &\leqslant N_n(z_2) + (N_n(z_3) + (\cdots + (N_n(z_{n-1}) + | z_n |_n^2)^2 \cdots)^2 \\ &\quad - \{N_n(z_2) + (N_n(z_3) + (\cdots + (N_n(z_{n-2}) \\ &\quad + N_n(z_{n-1})^2)^2 \cdots)^2\} \\ &= Q_n\{N_n(z_2) + (N_n(z_3) + (\cdots \\ &\quad + (N_n(z_{n-1}) + | z_n |_n^2)^2 \cdots)^2 \\ &\quad + N_n(z_2) + (N_n(z_3) + (\cdots \\ &\quad + (N_n(z_{n-2}) + N_n(z_{n-1})^2)^2 \cdots)^2\}^{-1} \\ &< \frac{1}{2^n} . \end{aligned}$$

The final step above follows from the fact that $N_n(x) \geqslant 1$. Again, by completeness, there is a $\chi_2 \in R$ such that $\lambda_n' \to \chi_2$, and $\chi_2 = \chi_2^*$. Moreover, since $\lambda_n = z_1 + \lambda_n'^2$, we have

$$\chi_1 = z_1 + \chi_2^2.$$

Continuing in this manner, we obtain a sequence of elements $\{\chi_n\}$ in R, such that

$$\chi_n^* = \chi_n , \qquad \chi_n = z_n + \chi_{n+1}^2 , \quad n = 1, 2, \ldots . \tag{2.2.5}$$

Since $\chi_n^* = \chi_n$, $f(e) = 1$, the Schwarz inequality (2.1.5) yields

$$f(\chi_n)^2 \leqslant f(\chi_n^2). \tag{2.2.6}$$

From (2.2.5) and (2.2.6), since $f(z_n) = 1$, we conclude that

$$\begin{aligned} f(\chi_1) &\geqslant 1 + f(\chi_2)^2 \geqslant 1 + (1 + f(\chi_3)^2)^2 \geqslant \cdots \\ &\geqslant 1 + (1 + (1 + \cdots + (1 + f(\chi_n)^2)^2 \cdots)^2 \geqslant n \end{aligned}$$

for all n. But this is impossible. Hence, f is continuous.]

Corollary 2.2.2. Under the hypotheses of Lemma 2.2.1, if the seminorms $\{| x |_\alpha , \alpha = 1, 2, \ldots\}$ of R satisfy the condition $| x |_1 \leqslant | x |_2 \leqslant \cdots$, then there is a natural number α and a positive number c such that

$$|f(x)| \leqslant c \, | x |_\alpha \tag{2.2.7}$$

for all $x \in R$.

PROOF. If this were not so, then $\sup_{x,\alpha} |f(x)|/|x|_\alpha = \infty$. Consequently, for every n, there would be some $x_n \in R$ such that $f(x_n) = 1$, $|x_n|_n < 1/n$. But obviously, such a sequence $\{x_n\}$ converges to zero. This contradicts the continuity of f.]

There are examples which show that this lemma cannot be extended to algebras with uncountable families of seminorms.

Inequality (2.2.7) can be made more precise, that is,

$$|f(x)| \leqslant f(e)\, |x|_\alpha . \tag{2.2.8}$$

In fact, by repeated application of the Schwarz inequality (2.1.5), and using (2.2.7), we get

$$|f(x)| \leqslant f(e)^{1/2} f(x^*x)^{1/2} \leqslant f(e)^{\frac{1}{2}+\frac{1}{4}} f((x^*x)^2)^{1/4}$$
$$\leqslant f(e)^{\frac{1}{2}+\frac{1}{4}+\cdots+1/2^m} f((x^*x)^{2^{m-1}})^{1/2^m} \leqslant f(e)^{1-(1/2^m)} c^{1/2^m} (|(x^*x)^{2^{m-1}}|_\alpha)^{1/2^m}.$$

Letting $m \to \infty$, we get

$$|f(x)| \leqslant f(e) \lim_{m\to\infty} (|(x^*x)^{2^{m-1}}|_\alpha)^{1/2^m}. \tag{2.2.9}$$

However, $|(x^*x)^{2^{m-1}}|_\alpha \leqslant |x^*|_\alpha^{2^{m-1}} |x|_\alpha^{2^{m-1}} = |x|_\alpha^{2^m}$, whence (2.2.8) follows at once from (2.2.9).

In particular, when R is a normed algebra, (2.2.8) reduces to

$$|f(x)| \leqslant f(e)\, |x|, \qquad x \in R, \tag{2.2.10}$$

where $|x|$ denotes the norm in R. In this case, a simple proof of (2.2.10) can be given, as follows (see Naimark [1]).

Let x be self-conjugate and $|x| \leqslant 1$. Consider the binomial expansion

$$(1 - \lambda)^{1/2} = 1 - \frac{\lambda}{2} - \frac{1}{2!}\frac{\lambda^2}{2^2} - \cdots. \tag{2.2.11}$$

The right-hand side of (2.2.11) converges for $\lambda = 1$. It follows easily that the series

$$y = e - \frac{x}{2} - \frac{x^2}{2!\,2^2} - \cdots - \frac{1 \cdot 3 \cdots (2n-3)}{n!\,2^n} x^n - \cdots \in R \tag{2.2.12}$$

converges in the norm to an element y of R. Since x is self-conjugate, so are the partial sums in (2.2.12), hence, since $|z| = |z^*|$ for all $z \in R$, it follows that y is self-conjugate. Moreover, it is easily verified that $y^2 = e - x$. By the positiveness of f, we have $f(y^2) = f(y^*y) \geqslant 0$, hence

$$f(x) \leqslant f(e).$$

It follows from this that (2.2.10) holds for all self-conjugate elements of R. Then, by virtue of the Schwarz inequality and the self-conjugacy of x^*x, we have, for any $x \in R$,

$$|f(x)|^2 \leqslant f(e) f(x^*x) \leqslant f(e)^2 \mid x^*x \mid \leqslant f(e)^2 \mid x \mid^2,$$

that is, (2.2.10) holds.

However, this method of proof does not generalize to countably seminormed algebras.

Lemma 2.2.3. Let A_n be the algebra of entire functions of n variables (see Example 2.1.1), and let f be a positive function on A_n, with $f(1) = 1$. Then, there is a bounded closed subset D of n-dimensional real space R_n, and a probability measure μ on D, such that, for any $\varphi \in A_n$,

$$f(\varphi) = \int_D \varphi(z)\, d\mu(z). \tag{2.2.13}$$

Moreover, the μ satisfying (2.2.13) is unique.

PROOF. Let $\{| \varphi |_\alpha, \alpha = 1, 2, \ldots\}$ be the sequence of seminorms specified in Example 2.1.1. Since A_n satisfies the conditions of Corollary 2.2.2, there is a positive number c and an integer α such that inequality (2.2.7) holds for all $x \in A_n$.

For any real n-tuple $t = (t_1, \ldots, t_n)$, and any complex n-tuple $z = (z_1, \ldots, z_n)$, we write

$$t \cdot z = t_1 z_1 + \cdots + t_n z_n .$$

Consider the function $\psi(t) = f(\exp(it \cdot z))$. The continuity of f implies that $\psi(t)$ is continuous with respect to t. Also, $\psi(0) = 1$. Let $t^{(l)}$, $l = 1, 2, \ldots, k$, be any k points of R_n, let $\xi_1, \ldots, \xi_k$ be any k complex numbers, and let

$$x(z) = \sum_{l=1}^{k} \xi_l \exp(it^{(l)} \cdot z).$$

Then

$$\sum_{l,m} \psi(t^{(l)} - t^{(m)})\, \xi_l \bar{\xi}_m = f(x^*x) \geqslant 0,$$

where $t^{(l)} - t^{(m)} = (t_1^{(l)} - t_1^{(m)}, \ldots, t_n^{(l)} - t_n^{(m)})$. Thus, ψ is a positive definite function. By the Bochner–Khinchin theorem, there exists a unique probability measure μ on R_n, such that

$$f(e^{it \cdot z}) = \int_{R_n} e^{it \cdot x}\, d\mu(x). \tag{2.2.14}$$

We shall now prove that μ is concentrated on the subset

$$D = \{z \mid |z_\nu| \leqslant \alpha + 1, \nu = 1, 2, \ldots, n\}$$

of R_n. Form the sequence of functions

$$g_m(t_1) = \frac{1}{(2\pi)^{1/2}} \exp\left\{-\frac{(t_1 + im)^2}{2} - \alpha m\right\}, \qquad m = 1, 2, \ldots .$$

One easily calculates that

$$h_m(z) = \int_{-\infty}^{\infty} \exp(it_1 z_1)\, g_m(t_1)\, dt_1 = \exp\left\{-\frac{z_1^2}{2} + m(z_1 - \alpha)\right\}, \quad (2.2.15)$$

which belongs to A_n. Notice that the integral in (2.2.15) converges in the topology of A_n. Thus, by taking $t = (t_1, 0, \ldots, 0)$ in (2.2.14), then multiplying both sides of (2.2.14) by $g_m(t_1)$, integrating with respect to t_1, using the continuity of f on the left-hand side, and interchanging the order of integration on the right-hand side, we get

$$f(h_m) = \int_{R_n} h_m(x)\, d\mu(x) \geqslant \int_{x_1 \geqslant \alpha+1} \exp\left(-\frac{x_1^2}{2} + m\right) d\mu(x_1). \quad (2.2.16)$$

Since $| h_m |_\alpha \leqslant \exp \alpha^2/2$, it follows from (2.2.7) that the left-hand side of (2.2.16) remains bounded as $m \to \infty$. But the right-hand side of (2.2.16) cannot be bounded as $m \to \infty$ unless

$$\int_{z_1 \geqslant \alpha+1} \exp(-z_1^2/2)\, d\mu(z_1) = 0.$$

Consequently, the μ-measure of the halfspace $z_1 \geqslant \alpha + 1$ is zero. Applying the same reasoning to the other coordinates, we conclude that the measure μ is concentrated on D.

Therefore, we may replace R_n by D in the integral (2.2.14). We now apply the operator

$$\frac{\partial^{k_1 + \cdots + k_n}}{\partial t_1^{k_1} \cdots \partial t_n^{k_n}}$$

to both sides of (2.2.14). It is easily shown that this operator may be taken within the parentheses on the left of (2.2.14), and under the integral sign on the right of (2.2.14). Then, setting $t = (0, \ldots, 0)$, we see that (2.2.13) holds for any function of the form $\varphi(z) = \prod_{l=1}^{n} z_l^{k_l}$, and therefore holds for any polynomial. Since the totality of polynomials is dense in A_n, it follows by the continuity of f that (2.2.13) holds for all

$\varphi \in A_n$. The uniqueness of the measure μ can be deduced from the uniqueness of the measure in the Bochner–Khinchin theorem.]

Using the description of symmetric multiplicative linear functionals given in §2.1, and the Kolmogorov theorem, we can extend Lemma 2.2.3 to more general algebras. This is done in the following theorem, which is, at present, the most general abstractly formulated result of this kind.

Theorem 2.2.4. Let R be a symmetric commutative complete seminormed algebra with a unit e, and let $\mathfrak{M}$ be the topological space formed by the totality of symmetric multiplicative linear functionals on R. Let f be a positive functional on R, with $f(e) = 1$. Then, there exists a compact subset $\mathfrak{M}_f$ of $\mathfrak{M}$, and a unique measure μ on $\mathfrak{M}_f$, such that

$$f(x) = \int_{\mathfrak{M}_f} x(M)\, d\mu(M) \tag{2.2.17}$$

for all $x \in R$.

PROOF. As in §2.1, choose a linear basis $\{x_\beta, \beta \in B\}$ for the self-conjugate elements of R. For any fixed set of indices $\beta_1, \ldots, \beta_n$ in B, and any function $\varphi(z_1, \ldots, z_n)$ in A_n, form the element $\varphi(x_{\beta_1}, \ldots, x_{\beta_n})$ in R. Consider the functional on A_n defined by

$$\varphi \to f(\varphi(x_{\beta_1}, \ldots, x_{\beta_n})). \tag{2.2.18}$$

Since the mapping $\varphi \to \varphi(x_{\beta_1}, \ldots, x_{\beta_n})$ is an involution-preserving homomorphism from A_n into the ring $R_{x_{\beta_1}, \ldots, x_{\beta_n}}$ (see §2.1 for the definition of the latter), it follows that (2.2.18) is a positive functional on A_n. By Lemma 2.2.3, there exists a closed bounded subset $D_{\beta_1, \ldots, \beta_n}$ of the n-dimensional real space $E_{\beta_1} \times \cdots \times E_{\beta_n}$, and a unique probability measure $\mu_{\beta_1, \ldots, \beta_n}$, concentrated on $D_{\beta_1, \ldots, \beta_n}$, such that, for any $\varphi \in A_n$,

$$f(\varphi(x_{\beta_1}, \ldots, x_{\beta_n})) = \int_{D_{\beta_1, \ldots, \beta_n}} \varphi(u_{\beta_1}, \ldots, u_{\beta_n})\, d\mu_{\beta_1, \ldots, \beta_n}(u_{\beta_1}, \ldots, u_{\beta_n}). \tag{2.2.19}$$

When $m \geqslant n$, we have

$$\begin{aligned}\int \exp i(t_1 u_{\beta_1} + \cdots + t_n u_{\beta_n})\, d\mu_{\beta_1, \ldots, \beta_n}(u_{\beta_1}, \ldots, u_{\beta_n}) &= f(\exp i(t_1 x_{\beta_1} + \cdots + t_n x_{\beta_n})) \\ &= f(\exp i(t_1 x_{\beta_1} + \cdots + t_n x_{\beta_n} + 0 \cdot x_{\beta_{n+1}} + \cdots + 0 \cdot x_{\beta_m})) \\ &= \int \exp i(t_1 u_{\beta_1} + \cdots + t_n u_{\beta_n})\, d\mu_{\beta_1, \ldots, \beta_m}(u_{\beta_1}, \ldots, u_{\beta_m}),\end{aligned}$$

hence $\{\mu_{\beta_1,\ldots,\beta_n}, \beta_1,\ldots, \beta_n \in B\}$ forms a consistent family of finite-dimensional probability measures. We may assume that the set $\mathfrak{D}_{\beta_1}$ in (2.2.19) (taking the case $n = 1$) is a closed interval of the form $|u_{\beta_1}| \leqslant q_{\beta_1}$, and similarly for the remaining indices. By consistency, it follows easily that the measure $u_{\beta_1,\ldots,\beta_n}$ is concentrated on the product set $|u_{\beta_\nu}| \leqslant q_{\beta_\nu}$, $\nu = 1,\ldots, n$. Now, by the Kolmogorov theorem (Corollary 1.3.5′), there exists a unique probability measure μ on $E = \mathsf{X}_{\beta\in B} E_\beta$, such that, for any Borel cylinder $\tilde{Q}$ with base Q, corresponding to the indices $\beta_1,\ldots, \beta_n$, we have $\mu(\tilde{Q}) = \mu_{\beta_1,\ldots,\beta_n}(Q)$. Furthermore, if we let $\mathfrak{D} = \mathsf{X}_{\beta\in B} \mathfrak{D}_\beta$, then it is easily seen that the outer measure $\mu^*(D) = 1$. Also, from (2.2.19), we get

$$f(\varphi(x_{\beta_1},\ldots, x_{\beta_n})) = \int_E \varphi(u_{\beta_1},\ldots, u_{\beta_n})\, d\mu(u). \tag{2.2.20}$$

For any $\beta, \beta' \in B$, let $\{c^k_{\beta\beta'}\}$ be the system of numbers appearing in (2.1.9). By (2.1.9) and (2.2.20), we have

$$\int_E \left(u_\beta u_{\beta'} - \sum c^k_{\beta\beta'} u_k\right)^2 d\mu(u) = f\left(\left(x_\beta x_{\beta'} - \sum c^k_{\beta\beta'} x_k\right)^2\right) = 0. \tag{2.2.21}$$

But the measurable set $E - \mathfrak{M}_{\beta,\beta'}$ (see §2.1) is just the totality of points u such that the integrand in (2.2.21) is positive. Hence,

$$\mu(E - \mathfrak{M}_{\beta,\beta'}) = 0. \tag{2.2.22}$$

Regarding $\mathfrak{M}$ as a subset of E, we proceed to prove that the outer measure $\mu^*(\mathfrak{D} \cap \mathfrak{M}) = 1$. It suffices (see Halmos [1]) to prove that, if $\Gamma = \bigcup_{n=1}^\infty \Gamma_n$ is any countable union of open cylinders such that

$$\Gamma \supset \mathfrak{D} \cap \mathfrak{M}, \tag{2.2.23}$$

then $\mu(\Gamma) = 1$. Now, by the Tychonoff theorem (see, e.g., Guan Zhaozhi [1]), $\mathfrak{D}$, being a product of compact intervals, is compact, and by (2.1.11), $\mathfrak{M}$ is closed in E. Hence $\mathfrak{D} \cap \mathfrak{M}$ is compact, and we may assume that Γ itself is an open cylinder. Then, $\Lambda = E - \Gamma$ is a closed cylinder, and, by (2.2.23), we have

$$\mathfrak{D} \cap \mathfrak{M} \cap \Lambda = 0. \tag{2.2.24}$$

We assert that there is a finite sequence of sets $\mathfrak{M}_{\beta_1,\beta_1'},\ldots, \mathfrak{M}_{\beta_l,\beta_l'}$ such that

$$\mathfrak{D} \cap \Lambda \cap \bigcap_{\nu=1}^{l} \mathfrak{M}_{\beta_\nu,\beta_\nu'} = 0. \tag{2.2.25}$$

For otherwise, any finite number of the sets $\{\mathfrak{D} \cap \Lambda \cap \mathfrak{M}_{\beta,\beta'}, \beta, \beta' \in B\}$, which are closed in $\mathfrak{D}$, would have a nonempty intersection. Since $\mathfrak{D}$ is compact, this implies that $\mathfrak{D} \cap \Lambda \cap \bigcap_{\beta,\beta'} \mathfrak{M}_{\beta,\beta'}$ is nonempty, which, in view of Lemma 2.1.1, contradicts (2.2.24). Hence (2.2.25) holds for some finite subcollection of $\{\mathfrak{M}_{\beta,\beta'}\}$. But, by (2.2.22), we have $\mu(E - \bigcap_{\nu=1}^{l} \mathfrak{M}_{\beta_\nu,\beta_\nu'}) = 0$, and since $\Lambda \cap \mathfrak{D} \subset E - \bigcap_{\nu=1}^{l} \mathfrak{M}_{\beta_\nu,\beta_\nu'}$, it follows that

$$\mu^*(\Lambda \cap \mathfrak{D}) = 0.$$

But $\mu^*(\mathfrak{D}) = 1$, hence $\mu(\Lambda) = \mu^*(\Lambda \cap \mathfrak{D}) = 0$, and therefore $\mu(\Gamma) = 1$. This proves that $\mu^*(\mathfrak{D} \cap \mathfrak{M}) = 1$. Now, using the method described in §1.1, 1°, we concentrate μ on the compact set $\mathfrak{M}_f = \mathfrak{D} \cap \mathfrak{M}$. Thus, from (2.2.20), we obtain (2.2.17). The uniqueness of μ follows directly from the uniqueness of the measures $\mu_{\beta_1,\ldots,\beta_n}$.]

By formula (2.2.17), it follows that the positive functional f in Theorem 2.2.4 satisfies the inequality

$$|f(x)| \leqslant f(e) \sup_{M \in \mathfrak{M}_f} |x(M)|. \tag{2.2.26}$$

Next, we shall use inequality (2.2.10) to derive a result concerning the space of symmetric multiplicative linear functionals on a symmetric Banach algebra.

Theorem 2.2.5. Let R be a symmetric Banach algebra with a unit, and $\mathfrak{M}$ the totality of symmetric multiplicative linear functionals on R. Then $\mathfrak{M}$ is compact in the weak topology.

PROOF. Using the procedure of §2.1, imbed $\mathfrak{M}$ in the space E. As we remarked in §2.1, every element of $\mathfrak{M}$ is also a positive functional on R. Hence, it follows by (2.2.10) that, for every $\beta \in B$,

$$| u_\beta | \leqslant | x_\beta |. \tag{2.2.27}$$

Let C_β denote the compact subset $\{u_\beta \mid | u_\beta | \leqslant | x_\beta |\}$ of E_β. By the Tychonoff theorem, $C = \times_{\beta \in B} C_\beta$ is a compact subset of E, and, by (2.2.27), $\mathfrak{M} \subset C$. By (2.1.11), $\mathfrak{M}$ is closed in E, hence $\mathfrak{M}$, as a closed subset of a compact set, is itself compact.]

Next, we use Theorems 2.2.4 and 2.2.5 to characterize the structure of commutative symmetric operator rings on Hilbert spaces.

Theorem 2.2.6. Let H be a Hilbert space, and let R be a commutative symmetric complete algebra of bounded linear operators on H, such that the unit operator $I \in R$. Let $\mathfrak{M}$ denote the compact Hausdorff space formed by the totality of symmetric multiplicative linear functionals on R, and let $C(\mathfrak{M})$ be the ring of all continuous complex-valued functions on $\mathfrak{M}$. Then, the mapping

$$A \to A(M), \qquad M \in \mathfrak{M} \tag{2.2.28}$$

is a symmetric norm-preserving isomorphism from R onto $C(\mathfrak{M})$.

PROOF. If $A \in R$, then the function $A(M)$, $M \in \mathfrak{M}$, is an element of $C(\mathfrak{M})$; let $\hat{R}$ denote the totality of such elements. For any $x \in H$, define a positive functional on R, as follows:

$$f_x(A) = (Ax, x), \qquad A \in R.$$

By (2.2.26), we have

$$f_x(A^*A) \leqslant \max_{M \in \mathfrak{M}} | A(M)|^2 f_x(I),$$

hence

$$\| A \|^2 = \| A^*A \| = \sup_{x \neq 0} \frac{f_x(A^*A)}{f_x(I)} \leqslant \max_{M \in \mathfrak{M}} | A(M)|^2.$$

On the other hand, by (2.2.10), we have $| A(M)| \leqslant \| A \|$ for all $M \in \mathfrak{M}$. Hence,

$$\| A \| = \max_{M \in \mathfrak{M}} | A(M)|.$$

Therefore, (2.2.28) is a norm-preserving isomorphism from R onto $\hat{R}$. Since R is symmetric and complete, $\hat{R}$ is also symmetric and complete, and therefore constitutes a symmetric closed subalgebra of $C(\mathfrak{M})$. For any two distinct points M_0 and N_0 in $\mathfrak{M}$, there exists an $A \in R$ such that $A(M_0) \neq A(N_0)$. Let

$$\varphi = \frac{A(M_0)I - A}{A(M_0) - A(N_0)}.$$

Then $\varphi \in R$, the function $\varphi(M) \in \hat{R}$, and $\varphi(M_0) = 0$, $\varphi(N_0) = 1$. Hence, by Corollary 2.1.3, $\hat{R} = C(\mathfrak{M})$.]

§2.3. Weakly Closed Operator Algebras: Fundamental Concepts

1° Various Topologies on the Algebra $\mathscr{B}(H)$

Let H be a Hilbert space, and let $\mathscr{B}(H)$ be the algebra formed by the totality of bounded linear operators from H to H, with respect to the usual operations. The correspondence $A \to A^*$, where A^* denotes the adjoint of A, is an involution on $\mathscr{B}(H)$. We proceed to introduce various topologies on $\mathscr{B}(H)$, as follows:

I. Weak topology. For any $A_0 \in \mathscr{B}(H)$, we arbitrarily choose a positive number ϵ and elements $f_1, \ldots, f_n, \varphi_1, \ldots, \varphi_n$ of H. We call the set

$$U(A_0; f_1, \ldots, f_n, \varphi_1, \ldots, \varphi_n, \epsilon)$$
$$= \{A \mid |((A - A_0)f_k, \varphi_k)| < \epsilon, k = 1, \ldots, n\}$$

a *weak neighborhood* of A_0. Taking the totality of sets of this form as a neighborhood basis, we obtain a topology on $\mathscr{B}(H)$, which we call the *weak topology*. In general, this topology does not satisfy the first axiom of countability. For any $S \subset \mathscr{B}(H)$, we shall denote the closure of S, in the weak topology, by S^1. It is easily seen that, given the weak topology, $\mathscr{B}(H)$ becomes a linear topological space. For any fixed $B \in \mathscr{B}(H)$, the multiplication operation $A \to BA$, $A \in \mathscr{B}(H)$ (and $A \to AB$, $A \in \mathscr{B}(H)$) is continuous; however, the operation $(A, B) \to AB$, from $\mathscr{B}(H) \times \mathscr{B}(H)$ to $\mathscr{B}(H)$, need not be continuous with respect to the weak topology (see Example 2.3.1). On the other hand, since

$$|((A - A_0)f_k, \varphi_k)| = |(f_k, (A^* - A_0^*)\varphi_k)|,$$

the involution $A \to A^*$ is continuous in the weak topology.

II. Strong topology. For any $A_0 \in \mathscr{B}(H)$, we arbitrarily choose a positive number ϵ and finitely many vectors $f_1, \ldots, f_n$ in H. We call the set

$$V(A_0; f_1, \ldots, f_n; \epsilon) = \{A \mid \|(A - A_0)f_k\| < \epsilon, k = 1, 2, \ldots, n\}$$

a *strong neighborhood* of A_0. Taking the totality of sets of this form as a neighborhood basis, we obtain a topology on $\mathscr{B}(H)$, which we call the *strong topology*. In general, this topology does not satisfy the first axiom of countability. For any $S \subset \mathscr{B}(H)$, we shall denote the closure of S, in the strong topology, by S^2. $\mathscr{B}(H)$ is also a linear topological space with respect to the strong topology. Again, the multiplication operations $A \to AB$ and $A \to BA$ are continuous for any fixed B, but the operation $(A, B) \to AB$ from $\mathscr{B}(H) \times \mathscr{B}(H)$ to $\mathscr{B}(H)$ need not be continuous in the strong topology (see Example 2.3.1). Moreover, in the strong topology, the involution $A \to A^*$ need not be continuous (see Example 2.3.1).

III. Uniform topology. We have already mentioned (in §2.1) that $\mathscr{B}(H)$, given the ordinary operator norm, forms a symmetric Banach algebra. This norm determines a topology on $\mathscr{B}(H)$, such that, for any $A_0 \in \mathscr{B}(H)$, the sets

$$\{A \mid \| A - A_0 \| < \epsilon\}$$

form a neighborhood basis at A_0. This is called the *uniform topology*. The uniform topology satisfies the first axiom of countability, furthermore, in this topology, the addition, multiplication, and involution operations are all continuous. For any $S \subset \mathscr{B}(H)$, we shall denote the closure of S, in the uniform topology, by S^3.

Clearly, the weak topology on $\mathscr{B}(H)$ is weaker than the strong topology, and the strong topology is weaker than the uniform topology. In general, these three topologies are all distinct (see the following example). For any $S \subset \mathscr{B}(H)$, we obviously have

$$S^3 \subset S^2 \subset S^1.$$

Example 2.3.1. Let H be a separable Hilbert space, $\{e_\nu\}$, $\nu = 1, 2,\ldots$, any complete orthonormal system in H. Define an operator U as follows. For any $x \in H$, let

$$Ux = \sum_{\nu=1}^{\infty} (x, e_{\nu+1})e_\nu .$$

Obviously, $\| U \| \leqslant 1$. Let $A_n = U^n$. Then, for any $x \in H$,

$$A_n x = \sum_{\nu=1}^{\infty} (x, e_{\nu+n})e_\nu ,$$

hence $\| A_n x \|^2 = \sum_{k=n+1}^{\infty} |(x, e_k)|^2$. Since $\sum_{k=1}^{\infty} |(x, e_k)|^2 = \| x \|^2 < \infty$, it follows that, for every $x \in H$, we have

$$\lim_{n\to\infty} \| A_n x \| = 0,$$

that is, $\{A_n\}$ converges to zero in the strong topology. However, one easily calculates that the adjoint of A_n is

$$A_n{}^* x = \sum_{\nu=1}^{\infty} (x, e_\nu)e_{\nu+n} ,$$

hence $\| A_n{}^* x \| = \| x \|$, and so $\{A_n{}^*\}$ does not converge to zero in the strong topology. This shows that (i) the mapping $A \to A^*$ is not continuous in the strong topology, and (ii) the strong and weak topologies on $\mathscr{B}(H)$ are distinct.

Since $\{A_n\}$ also converges to zero in the weak topology, it follows that $\{A_n{}^*\}$ converges to zero in the weak topology (since the mapping $A \to A^*$ is always continuous in the weak topology). But $A_n A_n{}^* = I$, the identity operator, so that $\{A_n A_n{}^*\}$ does not converge to zero. This shows that the multiplication operation $(A, B) \to AB$ is not continuous in the weak topology.

Next, we proceed to show that the multiplication operation $(A, B) \to AB$ is not continuous in the strong topology. Choose any $f \in H$, $\|f\| = 1$, and any positive number $\epsilon < 1$. Consider the strong neighborhood $V(0; f, \epsilon)$ in $\mathscr{B}(H)$. It suffices to prove that, for any two neighborhoods of zero,

$$V(0; f_1, \ldots, f_p, \epsilon_1) \qquad \text{and} \qquad V(0; \varphi_1, \ldots, \varphi_q, \epsilon_2),$$

one can find operators $A \in V(0; f_1, \ldots, f_p, \epsilon_1)$ and $B \in V(0; \varphi_1, \ldots, \varphi_q, \epsilon_2)$, such that $AB \overline{\in} V(0; f, \epsilon)$. Choose a positive number $\delta < \epsilon_2/\max_k \| \varphi_k \|$. Then $\| \delta A_n{}^* \varphi_k \| = \delta \| \varphi_k \| < \epsilon_2$, $k = 1, \ldots, q$, hence,

$$\delta A_n{}^* \in V(0; \varphi_1, \ldots, \varphi_q, \epsilon_2), \qquad n = 1, 2, \ldots .$$

Choose n_0 sufficiently large, so that

$$\left\| \frac{1}{\delta} A_{n_0} f_k \right\| < \epsilon_1, \qquad k = 1, 2, \ldots, p.$$

Then,

$$\frac{1}{\delta} A_{n_0} \in V(0; f_1, \ldots, f_p, \epsilon_1).$$

But

$$I = \left(\frac{1}{\delta} A_{n_0}\right) \cdot (\delta A_{n_0}^*) \overline{\in} V(0; f, \epsilon).$$

This proves our assertion.

2° von Neumann Algebras

Definition 2.3.1. A subalgebra R of $\mathscr{B}(H)$ which is symmetric (i.e., $A \in R$ implies $A^* \in R$), weakly closed, and contains the identity operator I is called a *weakly closed operator ring* (or *weakly closed operator algebra*)[8] over H.

If $S \subset \mathscr{B}(H)$, the smallest weakly closed operator ring containing S is called the weakly closed operator ring *generated* by S, and will be denoted by $R(S)$.

Since the intersection of any collection of weakly closed operator rings is also a weakly closed operator ring, it is clear that $R(S)$ is just the intersection of all the weakly closed operator rings which contain S.

Let $S \subset \mathscr{B}(H)$, and let $S^* = \{A^* \mid A \in S\}$. Let S' denote the totality of bounded linear operators which commute with all the operators in

[8] In some of the literature, such an R is called a *von Neumann algebra*. Also, some authors use the term weakly closed operator ring (or *W^*-algebra*) in reference to a weakly closed symmetric operator algebra R (without requiring that $I \in R$).

$S \cup S^*$. It is easy to verify that S' is a weakly closed operator ring, and that

(i) if $S_1 \subset S_2$, then $S_2' \subset S_1'$;

(ii) $S \subset (S')'$.

Writing S'' for $(S')'$, S''' for $(S'')'$, and so on, we have

(iii) $S' = S''' = S^{\text{V}} = \cdots$, $S'' = S^{\text{IV}} = S^{\text{VI}} = \cdots$.

In fact, setting $S_1 = S$ and $S_2 = S''$ in (i), we get $S''' \subset S'$. Substituting S' for S in (ii), we get $S' \subset S'''$. Combining these results, we get $S' = S'''$. Continuing in this fashion, we obtain (iii).

We use $R_*(S)$ to denote the smallest symmetric, uniformly closed[9] operator ring which contains S and the identity operator. Starting from the elements of $S \cup S^*$ and the identity operator, and performing all possible finite sequences of algebraic operations, we obtain an operator algebra; the closure of this algebra, in the uniform topology, is $R_*(S)$. It is easily proved that

(iv) $S' = (R_*(S))' = (R(S))'$.

By virtue of (ii), S'' is a weakly closed operator ring containing S, hence

(v) $R(S) \subset S''$.

Theorem 2.3.1. Let $S \subset \mathscr{B}(H)$. Then, $R(S)$ is the closure of $R_*(S)$ in both the strong and weak topologies. Moreover,

$$R(S) = S''. \tag{2.3.1}$$

Proof. Choose any $A_0 \in S''$. We shall prove that $A_0 \in (R_*(S))^2$. That is, we shall prove that any strong neighborhood of A_0, say, $V(A_0; f_1^0, \ldots, f_n^0, \epsilon)$, contains an element of $R_*(S)$. Consider the direct sum[10]

[9] That is, closed in the uniform topology.

[10] In general, let H be a Hilbert space, and let n be an arbitrary cardinal number. Take any set Λ of cardinality n. Let $x = \{x_\lambda, \lambda \in \Lambda\}$ be any vector function on Λ with values in H (i.e., $x_\lambda \in H$) such that

$$\sum_{\lambda \in \Lambda} \| x_\lambda \|^2 < \infty.$$

Let H_n denote the totality of such functions x. Define linear operations in H_n as follows: if α, β are scalars, then $\alpha\{x_\lambda, \lambda \in \Lambda\} + \beta\{y_\lambda, \lambda \in \Lambda\} = \{\alpha x_\lambda + \beta y_\lambda, \lambda \in \Lambda\}$. Then H_n becomes a linear space. Define an inner product in H_n by the formula

$$(\{x_\lambda, \lambda \in \Lambda\}, \{y_\lambda, \lambda \in \Lambda\}) = \sum_\lambda (x_\lambda, y_\lambda).$$

It is easily verified that H_n forms a Hilbert space with respect to the linear operations and inner product defined above; it is called the *direct sum* of n copies of H. When n is a finite number, we write $\Lambda = (1, 2, \ldots, n)$ and $\{x_\lambda, \lambda \in \Lambda\} = \{x_1, \ldots, x_n\}$.

$H_n = H \oplus H \oplus \cdots \oplus H$. For each $B \in R_*(S)$, form the vector

$$f_B = \{Bf_1^0, Bf_2^0, \ldots, Bf_n^0\}$$

in H_n. The totality of such vectors f_B, $B \in R_*(S)$, forms a linear subspace E of H_n. Let E^0 be the closure of E (with respect to the norm of H_n); E^0 is a closed linear subspace of H_n. Let P be the projection of H_n onto E^0. Form an operator P_{ml} on H, as follows. For $f \in H$, let $P_{ml}f$ be the lth component of the vector

$$P\{\underbrace{0, 0, \ldots, 0}_{m-1}, f, 0, \ldots, 0\}.$$

It is easily seen that $P_{ml} \in \mathscr{B}(H)$. Next, we shall show that $P_{ml} \in S'$.
Choose any $C \in R_*(S)$, and define an operator $\tilde{C}$ on H_n, as follows:

$$\tilde{C}\{f_1, \ldots, f_n\} = \{Cf_1, \ldots, Cf_n\}.$$

Then, for any $C, B \in R_*(S)$, we have $\tilde{C}f_B = f_{CB} \in E$. Thus, the subspace E is invariant under $\tilde{C}$, hence, E^0 is also invariant under $\tilde{C}$, and therefore

$$P\tilde{C}P = \tilde{C}P.$$

On the other hand, C^* also belongs to $R_*(S)$, and $\tilde{C}^* = \widetilde{C^*}$, hence, if we substitute C^* for C in the above relation, and take the adjoint of both sides, we get $P\tilde{C}P = P\tilde{C}$. Consequently, whenever $C \in R_*(S)$, we have

$$\tilde{C}P = P\tilde{C},$$

and hence

$$\begin{aligned}&\{CP_{m1}f, CP_{m2}f, \ldots, CP_{mn}f\}\\ &\quad = \tilde{C}P\{\underbrace{0, \ldots, 0}_{m-1}, f, 0, \ldots, 0\} = P\tilde{C}\{0, \ldots, 0, f, 0, \ldots, 0\}\\ &\quad = P\{0, \ldots, 0, Cf, 0, \ldots, 0\} = \{P_{m1}Cf, \ldots, P_{mn}Cf\}.\end{aligned}$$

Thus, we conclude that $P_{ml}C = CP_{ml}$ for all $C \in R_*(S)$, that is, $P_{ml} \in (R_*(S))' = S'$. Since $A_0 \in S''$, it follows that $P_{ml}A_0 = A_0P_{ml}$.
Now, $I \in R_*(S)$, $f^0 = f_I \in E^0$, hence, $Pf^0 = f^0$, that is, for every l,

$$f_l^0 = \sum_{m=1}^{n} P_{ml}f_m^0.$$

Applying the operator A_0, we get

$$A_0 f_l^0 = \sum_{m=1}^{n} A_0 P_{ml} f_m^0 = \sum_{m=1}^{n} P_{ml} A_0 f_m^0$$

for every l, whence $Pf_{A_0} = f_{A_0}$, that is, $f_{A_0} \in E^0$. Consequently, there is a $B \in R_*(S)$, such that

$$\sum_{l=1}^{n} \|(A_0 - B) f_l^0\|^2 = \|f_B - f_{A_0}\|^2 < \epsilon^2.$$

This proves that $A_0 \in (R_*(S))^2 \subset (R_*(S))^1 \subset R(S)$. Therefore, we have

$$S'' \subset R(S),$$

whence, by virtue of property (v), we get $S'' = R(S)$. Since $S'' \subset (R_*(S))^2 \subset (R_*(S))^1 \subset R(S)$, it follows that $(R_*(S))^2 = (R_*(S))^1 = R(S)$.]

Corollary 2.3.2. Let R be a symmetric subring of $\mathscr{B}(H)$ such that $I \in R$. Then $R^1 = R^2$.

PROOF. By Theorem 2.3.1, $R^1 = (R_*(R))^2 = (R^3)^2 = R^2$.]

Corollary 2.3.3. If M is a weakly closed operator algebra, then $M = M''$.

PROOF. Since M is a weakly closed operator algebra, $R(M) = M$. The conclusion then follows from (2.3.1).]

Corollary 2.3.4. Let A be a bounded self-adjoint operator on a Hilbert space H and let $\{P_\lambda\}$ denote the resolution of the identity associated with A. If M is a weakly closed operator algebra over H, then $A \in M$ implies that $P_\lambda \in M$, $-\infty < \lambda < \infty$. If M is an operator algebra which is closed in the uniform topology, then $P_\lambda \in M$, $-\infty < \lambda < \infty$, implies that $A \in M$.

PROOF. If M is weakly closed, $A \in M$, $B \in M'$, then since B commutes with A, it follows that B commutes with every P_λ, hence, by Corollary 2.3.3, $P_\lambda \in M'' = M$. If M is closed in the uniform topology, and $P_\lambda \in M$ for all λ, then, since there is a sequence of partitions of the real line, $\{\lambda_0^{(n)}, \lambda_1^{(n)}, \ldots, \lambda_{m_n}^{(n)}\}$, such that the sequence of operators

$$A_n = \sum_{\nu=1}^{m_n} \lambda_\nu^{(n)} (P_{\lambda_\nu^{(n)}} - P_{\lambda_{\nu-1}^{(n)}}) \in M$$

converges uniformly to A, it follows that $A \in M$.]

For any weakly closed operator algebra M, we denote by M^P the totality of projection operators belonging to M. As an immediate consequence of Corollary 2.3.4, we have:

Corollary 2.3.5. If M and N are weakly closed operator algebras, then $M = N$ is equivalent to $M^P = N^P$, moreover, $M = R(M^P)$.

For any weakly closed operator algebra M, we denote by M^U the totality of unitary operators belonging to M.

Corollary 2.3.6. If M is a weakly closed operator algebra, then $R(M^U) = M$. If N is also a weakly closed operator algebra, then $M^U = N^U$ is equivalent to $M = N$.

PROOF. Obviously, $R(M^U) \subset M$. If $P \in M^P$, then

$$I - 2P = (I - P) + (-P) \in M^U.$$

But $I \in M^U$, hence $P \in R(M^U)$. Thus, $M^P \subset R(M^U)$. Consequently, $M = R(M^P) \subset R(M^U)$, whence we conclude that $R(M^U) = M$. The remaining assertion then follows trivially.]

Let M be a family of linear operators on H, and let G be a linear subspace of H. If, for every $\xi \in G$ and every $A \in M$, we have $A\xi \in G$, then the subspace G is said to be *invariant* under M, or simply an *M-invariant subspace*. If $M \subset \mathscr{B}(H)$, and G is an M-invariant subspace, then, clearly, the closure G^o of G is also an M-invariant subspace.

Lemma 2.3.7. Let G be a closed linear subspace of H, and let $P: H \to G$ be the corresponding projection operator. Then, G is M-invariant if and only if $P \in M'$.

PROOF. If $P \in M'$, then, for any $A \in M$, $AP = PA$, hence $AP = PAP$, which shows that G is invariant under A. Conversely, if G is M-invariant, then, for any $A \in M$, we have $AP = PAP$. But since $A^* \in M$, we also have $A^*P = PA^*P$, whence, by taking complex conjugates, we get $PA = PAP$. Therefore, $AP = PA$.]

3° Center, Commutative Operator Algebras, and Factors

Let M be a weakly closed operator algebra. Then, $\mathfrak{Z}(M) = M \cap M'$ is called the *center* of M; it consists of those operators in M which commute with every operator in M. The following two cases will be of special interest to us. (1) $\mathfrak{Z}(M)$ is maximal, that is, $\mathfrak{Z}(M) = M$. This means that $M \subset M'$, in other words, M is a commutative weakly closed operator algebra. (2) $\mathfrak{Z}(M)$ is minimal, that is, $\mathfrak{Z}(M)$ contains only the operators $\{\alpha I \mid \alpha \in F\}$, where F denotes the coefficient field of H; in this case, we say that M is a *factor*.

Notice that, if M is a commutative weakly closed operator algebra, then M' is not necessarily commutative. However, if M is a factor, then M' is also a factor, and conversely. This fact follows immediately from the relation

$$\mathfrak{Z}(M) = M \cap M' = M' \cap M'' = \mathfrak{Z}(M').$$

Example 2.3.2. $(\mathscr{B}(H))' = \{\alpha I \mid \alpha \in F\}$, $\{\alpha I \mid \alpha \in F\}' = \mathscr{B}(H)$. In fact, $\{\alpha I \mid \alpha \in F\}$ is clearly a weakly closed operator algebra, and, obviously, $\{\alpha I\}' = \mathscr{B}(H)$. Therefore,

$$(\mathscr{B}(H))' = \{\alpha I\}'' = \{\alpha I\}.$$

Consequently, the largest weakly closed operator ring $\mathscr{B}(H)$ and the smallest weakly closed operator ring $\{\alpha I\}$ are both factors.

Section §2.4 will be devoted exclusively to the discussion of commutative weakly closed operator rings.

4° Unitary Equivalence; Restriction

Definition 2.3.2. Let H and G be Hilbert spaces, U a unitary operator from H onto G. If M is a weakly closed operator algebra over H, then

$$N = UMU^{-1} = \{UAU^{-1} \mid A \in M\}$$

is a weakly closed operator algebra over G. We say that M and N are *unitarily equivalent.* In this case, we clearly have $N' = UM'U^{-1}$, $\mathfrak{Z}(N) = U\mathfrak{Z}(M)\,U^{-1}$. Consequently, if one of two unitarily equivalent operator algebras is commutative, then the other is also commutative; if one is a factor, then the other is also a factor.

Let M be a weakly closed operator ring over H, and let G be a closed linear subspace of H. Let M_G denote the totality of operators in M which are reduced by G. Then M_G may be regarded as a ring of operators on G, which we call the *contraction* of M to G. It can be proved that the contraction of M to G is a weakly closed operator ring over G.

§2.4. Representation of Commutative Weakly Closed Operator Rings

Throughout the present section, H will denote a Hilbert space and $\mathscr{B}(H)$ will denote the algebra of all bounded linear operators on H. First, we prove a lemma which will be frequently used in the sequel.

Lemma 2.4.1. Let $S \subset \mathscr{B}(H)$, and suppose that $S^* = S \subset S'$. Then $R(S)$ is commutative.

PROOF. The totality of elements obtainable from S and I by a finite sequence of algebraic operations constitutes a commutative symmetric operator ring E. Furthermore, $R(S) = R_*(S)^1 = (E^3)^1 = E^1$. However, it is easily proved that the weak closure of a commutative operator ring is also commutative, hence, $R(S)$ is commutative.]

1° Maximal Commutative Weakly Closed Operator Rings

Definition 2.4.1. Let M be a commutative weakly closed operator ring over H. If the only commutative weakly closed operator ring over H, which contains M, is M itself, then M is said to be *maximal commutative weakly closed.*

Lemma 2.4.2. An operator ring M is maximal commutative weakly closed if and only if

$$M = M'. \tag{2.4.1}$$

PROOF. Let M be a maximal commutative weakly closed operator ring. Since M is commutative, $M \subset M'$. We shall now prove that $M' \subset M$. Let A be any self-adjoint operator in M', and consider the set $S = M \cup \{A\}$. Since S is symmetric and commutative, it follows by Lemma 2.4.1 that $R(S)$ is commutative. But $R(S) \supset M$, hence, by the maximality of M, we must have $R(S) = M$, so that $A \in M$. Using Corollary 2.3.5, we deduce that $M' \subset M$, and therefore $M = M'$.

Conversely, assume that (2.4.1) holds, and let N be a commutative weakly closed operator ring containing M. Then, by property (i) of S' in §2.3, we have

$$N' \subset M'. \tag{2.4.2}$$

Since N is commutative, $N \subset N'$, hence, from (2.4.1) and (2.4.2), we obtain

$$N \subset M,$$

that is, $N = M$. Therefore, M is maximal.]

Definition 2.4.2. Let M be a commutative weakly closed operator ring. Suppose there exists a family $\{H_\lambda \mid \lambda \in \Lambda\}$ of M-invariant closed linear subspaces of H such that

$$H = \sum_{\lambda \in \Lambda} \oplus H_\lambda ,$$

the restriction M_λ of M to H_λ is maximal commutative weakly closed for every $\lambda \in \Lambda$, and all the M_λ are unitarily equivalent. We then say that M has *uniform multiplicity* k, where k is the cardinal number of Λ.

It can be proved that the cardinal number k is uniquely determined, that is, if M satisfies the above definition for both of the cardinal numbers k and k', then $k = k'$. In particular, a maximal commutative weakly closed operator ring is just a commutative weakly closed operator ring having uniform multiplicity 1.

I. E. Segal has proved the following decomposition theorem for commutative weakly closed operator rings. Since this theorem will be used only once in the present volume, the proof, which is rather lengthy, will not be given here; we refer the reader to Segal [2].

Theorem 2.4.3. Let M be a commutative weakly closed operator ring over H. Then, there is a correspondence which assigns an M-invariant closed linear subspace H_n of H to every cardinal number[11] n, such that

$$H = \sum_n \oplus H_n$$

and, when $H_n \neq \{0\}$, the restriction of M to H_n has uniform multiplicity n. Moreover, the family of subspaces $\{H_n\}$ is uniquely determined; more precisely, if $\{H_n'\}$ is another family of M-invariant closed linear subspaces satisfying the above conditions, then $H_n = H_n'$ for every n.

Suppose that M and N are two unitarily equivalent weakly closed operator rings. If M is maximal commutative weakly closed (i.e., $M = M'$), then we see from §2.3, 4°, that $N = N'$, that is, N is also maximal commutative weakly closed. Furthermore, if M is commutative and has uniform multiplicity n, then N clearly has the same property.

Example 2.4.1. An important type of commutative operator ring is the *multiplication algebra*, defined as follows.

Definition 2.4.3. Let $\Omega = (G, \mathfrak{B}, \mu)$ be a measure space, and let $\mathfrak{L}_k^2(\Omega)$ be the Hilbert space formed by the totality of quadratically integrable vector-valued functions on Ω, taking values in a k-dimensional Hilbert space H_k (see §1.1, 2°). If $f(g)$, $g \in G$, is any bounded measurable function on Ω, then the mapping

$$T_f: \quad \xi(g) \to f(g)\, \xi(g), \qquad g \in G, \quad \xi \in \mathfrak{L}_k^2(\Omega)$$

is a bounded linear operator in $\mathfrak{L}_k^2(\Omega)$; we call T_f the *multiplication operator* corresponding to the function f. The set of all such multiplica-

[11] *Translator's note:* Of course, $H_n = \{0\}$ for all sufficiently large cardinal numbers n. It is clear that logical difficulties can be avoided by a more meticulous wording of the theorem.

tion operators forms an operator ring $\mathfrak{M}_k(\Omega)$, which we call the *multiplication algebra* over $\mathfrak{L}_k^2(\Omega)$. It is easily seen that the correspondence $f \to T_f$ is a norm-preserving[12] isomorphism from $L^\infty(\Omega)$[13] onto $\mathfrak{M}_k(\Omega)$.

In particular, when $k = 1$, $\mathfrak{M}_1(\Omega)$ is just the multiplication algebra over $L^2(\Omega)$; we shall sometimes write simply $\mathfrak{M}(\Omega)$ in place of $\mathfrak{M}_1(\Omega)$.

In what follows, we shall prove that, if Ω is a localizable measure space, then $\mathfrak{M}_k(\Omega)$ has uniform multiplicity k, and, in fact, that all commutative weakly closed operator rings are essentially of this type.

Lemma 2.4.4. Let $\Omega = (G, \mathfrak{B}, \mu)$ be a measure space, $\mathfrak{E}$ a determining set of bounded measurable functions on Ω. Let $\tilde{\mathfrak{E}} = \{T_f \mid f \in \mathfrak{E}\}$. Then $R(\tilde{\mathfrak{E}}) \supset \mathfrak{M}_k(\Omega)$.

PROOF. Let $\mathfrak{D}$ be the smallest symmetric closed subalgebra of $L^\infty(\Omega)$ containing $\mathfrak{E}$ and 1; obviously, $\mathfrak{D}$ is also a determining set. It is easily seen that $R_*(\tilde{\mathfrak{E}}) = \{T_\varphi \mid \varphi \in \mathfrak{D}\}$. Choose any $\varphi \in L^\infty(\Omega)$, $f_1, \ldots, f_n \in \mathfrak{L}_k^2(\Omega)$, and form the function

$$\rho(g) = \sum_{l=1}^{n} \|f_l(g)\|^2$$

in $L^1(\Omega)$. Then $\varphi \in L^2(\Omega, \rho)$. By Lemma 1.1.6, for any positive number δ, there exists $\psi \in \mathfrak{D}$ such that

$$\int_G |\varphi(g) - \psi(g)|^2 \rho(g) \, d\mu(g) < \delta^2$$

Thus, $T_\psi \in V(T_\varphi; f_1, \ldots, f_n, \delta)$. This shows that $T_\varphi \in R_*(\tilde{\mathfrak{E}})^2$. Hence, by Theorem 2.3.1, we have $T_\varphi \in R_*(\tilde{\mathfrak{E}})^1 = R(\tilde{\mathfrak{E}})$.]

2° Representation of Maximal Commutative Weakly Closed Operator Rings

Let H be a Hilbert space, $\mathfrak{A}$ a maximal commutative weakly closed ring of bounded linear operators on H, and Γ the compact Hausdorff space consisting of all symmetric multiplicative linear functionals on $\mathfrak{A}$. Let $\mathfrak{B}$ be the σ-algebra generated by the totality of closed subsets of Γ. Given a measure μ on $(\Gamma, \mathfrak{B})$, we shall write $\Omega = (\Gamma, \mathfrak{B}, \mu)$, and $\mathfrak{M}(\Omega)$ will denote the multiplication algebra over $L^2(\Omega)$. With this notation, we have the following theorem.

[12] *Translator's note:* This correspondence is norm-preserving if and only if condition (i) of Definition 1.2.3 is satisfied.

[13] Regarding $L^\infty(\Omega)$ as an algebra (see Example 2.1.5).

Theorem 2.4.5. There exists a localizable[14] measure μ on $(\Gamma, \mathfrak{B})$ and a unitary operator φ from H onto $L^2(\Omega)$ such that

$$(\varphi A \varphi^{-1}) f(\gamma) = A(\gamma) f(\gamma), \qquad f \in L^2(\Omega)$$

for all $A \in \mathfrak{A}$, and the correspondence

$$A \to \varphi A \varphi^{-1}$$

is an isomorphism from $\mathfrak{A}$ onto $\mathfrak{M}(\Omega)$.

To prove this theorem, we require the following lemma.

Lemma 2.4.6. Let $\mathfrak{A}$ be a symmetric operator ring over a Hilbert space H, such that $I \in \mathfrak{A}$. Then H permits an orthogonal sum decomposition

$$H = \sum_{\xi \in \Xi} \oplus H_\xi \,, \tag{2.4.3}$$

where H_ξ is the smallest $\mathfrak{A}$-invariant closed linear subspace of H containing ξ. (H_ξ is called a *cyclic subspace* of H relative to $\mathfrak{A}$, and ξ is said to be a *cyclic element* of H_ξ relative to $\mathfrak{A}$.)

Proof. (1) Assuming that $H \neq \{0\}$, choose any element $\xi_1 \in H$, $\xi_1 \neq 0$. Let H_{ξ_1} denote the closure of the set

$$\{A\xi_1 \mid A \in \mathfrak{A}\}. \tag{2.4.4}$$

Then H_{ξ_1} is a closed linear subspace of H, and since (2.4.4) is $\mathfrak{A}$-invariant, so is H_{ξ_1}. Thus, ξ_1 is a cyclic element of H_{ξ_1} relative to $\mathfrak{A}$.

(2) If the subspace M is $\mathfrak{A}$-invariant, then so is its orthogonal complement $H \ominus M$. In fact, let $x \in H \ominus M$, $A \in \mathfrak{A}$; then, for any $y \in M$,

$$(Ax, y) = (x, A^*y) = 0,$$

hence $Ax \in H \ominus M$.

(3) Let m be a family of closed linear subspaces of H, satisfying the following conditions.

(i) Every M in m is cyclic relative to $\mathfrak{A}$.

(ii) If $M, N \in m$ and $M \neq N$, then $M \perp N$.

Let $\mathfrak{F}$ denote the totality of such families m. It follows from part (1)

[14] *Translator's note:* Scrutiny of the proof reveals that, in general, the localizability requirement cannot be satisfied (at least, not via the author's method of proof) unless μ is defined on a certain σ-algebra $\mathfrak{B}'$ which may be much larger than $\mathfrak{B}$. However, if H is separable, then one can take $\mathfrak{B}' = \mathfrak{B}$, and in this case μ is, in fact, σ-finite. See footnote 16.

that $\mathfrak{F}$ is nonempty; in fact, $m = \{H_{\xi_1}\}$ is one such family. $\mathfrak{F}$ is partially ordered by inclusion, that is, $m \prec m'$ provided that $m \subset m'$. Clearly, every totally ordered subset $\mathfrak{F}'$ of $\mathfrak{F}$ has a supremum

$$m_1 = \bigcup_{m \in \mathfrak{F}'} m \in \mathfrak{F}.$$

Therefore, by Zorn's lemma, $\mathfrak{F}$ has a maximal element m_0.

(4) Let $H' = \sum_{M \in m_0} \oplus M$. We shall prove that $H' = H$. Obviously, the subspace H' is $\mathfrak{A}$-invariant. If $H' \neq H$, then $H \ominus H' \neq (0)$. By part (2), $H \ominus H'$ is also $\mathfrak{A}$-invariant. Choose any element $\xi_0 \in H \ominus H'$, $\xi_0 \neq 0$; as in part (1), we obtain a cyclic subspace H_{ξ_0}. Since $H_{\xi_0} \perp M$ for every $M \in m_0$, it follows that the family $m_0 \cup \{H_{\xi_0}\}$ also belongs to $\mathfrak{F}$, and is strictly greater than m_0. But this contradicts the maximality of m_0.]

Proof of Theorem 2.4.5. Let $\{H_\xi, \xi \in \Xi\}$ be a family of subspaces of H satisfying the conditions of Lemma 2.4.6, and let P_ξ denote the projection operator from H onto H_ξ. Since H_ξ is invariant under $\mathfrak{A}$, it follows by Lemmas 2.3.7 and 2.4.2 that $P_\xi \in \mathfrak{A}' = \mathfrak{A}$. For each $\xi \in \Xi$, let

$$\Gamma_\xi = \{\gamma \mid P_\xi(\gamma) = 1\}.$$

If $\xi \neq \xi'$, then $H_\xi \perp H_{\xi'}$, that is, $P_\xi P_{\xi'} = 0$, hence $P_\xi(\gamma) \cdot P_{\xi'}(\gamma) = 0$ for every $\gamma \in \Gamma$, so that $\Gamma_\xi \cap \Gamma_{\xi'}$ is empty.

For each $\xi \in \Xi$, define a positive functional F_ξ on $\mathfrak{A}$, as follows:

$$F_\xi(A) = (A\xi, \xi), \qquad A \in \mathfrak{A}.$$

Now, by Theorem 2.2.6, the correspondence $A \to A(\gamma)$, $\gamma \in \Gamma$, is a symmetric isometric isomorphism from $\mathfrak{A}$ onto $C(\Gamma)$. Thus, F_ξ may be regarded as a positive functional on $C(\Gamma)$, hence, by a theorem of Riesz (see, e.g., Halmos [1], §56), there exists a unique regular finite Borel measure μ_ξ on $(\Gamma, \mathfrak{B})$ such that

$$F_\xi(A) = \int_\Gamma A(\gamma)\, d\mu_\xi(\gamma).$$

Since $P_\xi \xi = \xi$, we have $F_\xi(P_\xi A) = F_\xi(A)$ for all $A \in \mathfrak{A}$, hence

$$F_\xi(A) = \int_\Gamma P_\xi(\gamma)\, A(\gamma)\, d\mu_\xi(\gamma) = \int_{\Gamma_\xi} A(\gamma)\, d\mu_\xi(\gamma). \tag{2.4.5}$$

Since $C(\Gamma_\xi)$ is dense in $L^1(\Gamma, \mathfrak{B}, \mu_\xi)$, it follows easily that $\mu_\xi(\Gamma - \Gamma_\xi) = 0$. Let $\mathfrak{B}_\xi = \{B \mid B \in \mathfrak{B}, B \subset \Gamma_\xi\}$. We thus obtain a family of finite measure spaces $\Omega_\xi = (\Gamma_\xi, \mathfrak{B}_\xi, \mu_\xi)$, $\xi \in \Xi$.

Consider the dense linear subspace $M_\xi = \{A\xi \mid A \in \mathfrak{A}\}$ of H_ξ, and the space $C(\Gamma_\xi)$ of continuous complex-valued functions on Γ_ξ. We define a mapping φ_ξ from M_ξ onto[15] $C(\Gamma_\xi)$, as follows: if $A \in \mathfrak{A}$, then

$$\varphi_\xi(A\xi) = A(\gamma). \tag{2.4.6}$$

By (2.4.5),

$$(A\xi, B\xi) = F_\xi(B^*A) = \int_{\Gamma_\xi} A(\gamma)\,\overline{B(\gamma)}\,d\mu_\xi(\gamma) \tag{2.4.7}$$

for any $A, B \in \mathfrak{A}$. If we regard $C(\Gamma_\xi)$ as a subspace of $L^2(\Omega_\xi)$, then relation (2.4.7) means that φ_ξ is isometric. Since M_ξ is dense in H_ξ and $C(\Gamma_\xi)$ is dense in $L^2(\Omega_\xi)$ (see Halmos [1]), it follows that φ_ξ can be uniquely extended to a unitary operator from H_ξ onto $L^2(\Omega_\xi)$.

Let $A_\xi = \varphi_\xi A \varphi_\xi^{-1}$. Then

$$(A_\xi f)(\gamma) = A(\gamma) f(\gamma), \qquad f \in L^2(\Omega_\xi). \tag{2.4.8}$$

In fact, from (2.4.6) and the relation $(AB)(\gamma) = A(\gamma)\,B(\gamma)$, we see at once that (2.4.8) holds when $f(\gamma) = B(\gamma)$, $B \in \mathfrak{A}$. By virtue of the density of $C(\Gamma_\xi)$ in $L^2(\Omega_\xi)$, this implies that (2.4.8) holds for all $f \in L^2(\Omega_\xi)$.

Let

$$\Gamma_0 = \Gamma - \bigcup_{\xi \in \Xi} \Gamma_\xi, \qquad \mathfrak{B}_0 = \{B \mid B \in \mathfrak{B},\ B \subset \Gamma_0\},$$

$$\mu_0 = 0, \qquad \Omega_0 = (\Gamma_0, \mathfrak{B}_0, \mu_0).$$

Form the direct sum $\Omega = (\Gamma, \mathfrak{B}, \mu)$ of the measure spaces[16] $\{\Omega_\xi, \xi \in \Xi\}$ and Ω_0. By Example 1.2.1 and Theorem 1.2.6, Ω is localizable. Moreover, by Lemma 1.1.10,

$$L^2(\Omega) = \sum_{\xi \in \Xi} \oplus L^2(\Omega_\xi). \tag{2.4.9}$$

[15] Since $\Gamma - \Gamma_\xi = \{\gamma \mid P_\xi(\gamma) = 0\}$, it follows that Γ_ξ is both open and closed in Γ. Hence, any function in $C(\Gamma_\xi)$ can be extended to a continuous function on Γ by simply assigning the value zero to all points of $\Gamma - \Gamma_\xi$. By Theorem 2.2.6, $\{A(\gamma) \mid A \in \mathfrak{A}\} = C(\Gamma)$, therefore the image of φ_ξ does cover all of $C(\Gamma_\xi)$.

[16] *Translator's note:* There is a difficulty here. If the indexing set Ξ is uncountable, then the σ-algebra obtained by taking the direct sum of these measure spaces will, in general, be larger than $\mathfrak{B}$. If one regards each μ_ξ as defined on $\mathfrak{B}$, and simply defines $\mu = \sum_{\xi \in \Xi} \mu_\xi$, then one obtains the desired result, except that μ may not be localizable. Of course, if H is separable, then Ξ is countable, and μ is, in fact, σ-finite.

We now combine the family of mappings $\{\varphi_\xi, \xi \in \Xi\}$ to form a single mapping φ, as follows. If $x \in H$, define

$$\varphi x = \sum_{\xi \in \Xi} \varphi_\xi P_\xi x. \tag{2.4.10}$$

It is clear from (2.4.3) and (2.4.9) that φ is a unitary operator from H to $L^2(\Omega)$.

When $f \in L^2(\Omega_\xi)$, we have $\varphi^{-1} f = \varphi_\xi^{-1} f$, hence, by (2.4.8),

$$(\varphi A \varphi^{-1}) f(\gamma) = \varphi \varphi_\xi^{-1} A_\xi f(\gamma) = A(\gamma) f(\gamma). \tag{2.4.11}$$

But since the operators appearing on both sides of (2.4.11) are bounded, it follows that (2.4.11) holds for all $f \in L^2(\Omega)$.

Let $C = \{\varphi A \varphi^{-1} \mid A \in \mathfrak{A}\}$; then $C \subset \mathfrak{M}(\Omega)$. Since (2.4.10) is a unitary operator, it follows from the remarks in part 4° of §2.3 that C is a maximal commutative weakly closed operator ring over $L^2(\Omega)$. Obviously, $\mathfrak{M}(\Omega)$ is symmetric and commutative, hence, by Lemma 2.4.1, the weakly closed ring $R(\mathfrak{M}(\Omega))$ generated by $\mathfrak{M}(\Omega)$ is also commutative. Since $R(\mathfrak{M}(\Omega)) \supset \mathfrak{M}(\Omega) \supset C$, the maximality of C implies that $R(\mathfrak{M}(\Omega)) = \mathfrak{M}(\Omega) = C$.]

While continuing to use the notation of Theorem 2.4.5, we now consider the case of uniform multiplicity k.

Corollary 2.4.7. Let $\mathfrak{A}$ be a commutative weakly closed operator ring having uniform multiplicity k over H. Then, there exists a measure μ on $(\Gamma, \mathfrak{B})$ such that $\Omega = (\Gamma, \mathfrak{B}, \mu)$ is a localizable measure space, and there is a unitary operator φ from H onto $\mathfrak{L}_k^2(\Omega)$ such that, for every $A \in \mathfrak{A}$,

$$(\varphi A \varphi^{-1}) f(\gamma) = A(\gamma) f(\gamma), \qquad f \in \mathfrak{L}_k^2(\Omega),$$

moreover, the correspondence

$$A \to \varphi A \varphi^{-1}$$

is an isomorphism from $\mathfrak{A}$ onto $\mathfrak{M}_k(\Omega)$.

The above corollary may be easily deduced from Definition 2.4.2, Theorem 2.4.5 and Theorem 1.1.4; the details are left to the reader.

Corollary 2.4.8. Let $\mathfrak{A}$ be a commutative[17] weakly closed operator ring over the Hilbert space H. If $\mathfrak{A}$ has a cyclic element in H, then there

[17] Notice here that we do not hypothesize the maximality of $\mathfrak{A}$.

exists a finite measure μ on $(\Gamma, \mathfrak{B})$ and a unitary operator φ from H onto $L^2(\Gamma, \mathfrak{B}, \mu)$ such that

$$\varphi A \varphi^{-1} f(\gamma) = A(\gamma) f(\gamma)$$

for every $A \in \mathfrak{A}$ and $f \in L^2(\Gamma, \mathfrak{B}, \mu)$, moreover, the correspondence $A \to \varphi A \varphi^{-1}$ is an isomorphism from $\mathfrak{A}$ onto $\mathfrak{M}(\Gamma, \mathfrak{B}, \mu)$.

PROOF. By assumption, $H = H_\xi$, where ξ is a cyclic element of H relative to $\mathfrak{A}$. Thus, in the proof of Theorem 2.4.5, $P_\xi = I$, and we obtain a finite regular measure space $\Omega = (\Gamma, \mathfrak{B}, \mu)$ and an isomorphism of $\mathfrak{A}$ onto $C(\Gamma)$, regarded as a subring of the multiplication algebra $\mathfrak{M}(\Omega)$. Using the density of $C(\Gamma)$ (*qua* function space) in $L^2(\Omega)$, and the fact that $C(\Gamma)$ (*qua* operator ring) is weakly closed in $\mathfrak{B}(L^2(\Omega))$, it is easily verified that $C(\Gamma) = \mathfrak{M}(\Omega)$.]

Corollary 2.4.9. Let $\mathfrak{A}$ be a maximal commutative weakly closed operator ring over a separable Hilbert space H. Then $\mathfrak{A}$ has a cyclic element in H.

PROOF. Since H is separable, the indexing set Ξ in the proof of Theorem 2.4.5 is, in this case, finite or countable. Consequently, Ω is the direct sum of finitely many or countably many finite measure spaces $(\Gamma_k, \mathfrak{B}_k, \mu_k)$, $k = 0, 1, 2, \ldots$, and hence is σ-finite. Thus, $\{\varphi A \varphi^{-1} \mid A \in \mathfrak{A}\}$ has a cyclic element in $L^2(\Omega)$, namely,

$$\sum_{\mu(\Gamma_k) > 0} \frac{C_{\Gamma_k}(\gamma)}{(2^k \mu(\Gamma_k))^{1/2}},$$

where $C_{\Gamma_k}(\gamma)$ denotes the characteristic function of Γ_k. It follows immediately that $\mathfrak{A}$ has a cyclic element in H.]

Lemma 2.4.10. If $\Omega = (G, \mathfrak{B}, \mu)$ is a finite measure space, then the multiplication algebra $\mathfrak{M}(\Omega)$ over $L^2(\Omega)$ is maximal commutative weakly closed.

PROOF. Let $u \in (\mathfrak{M}(\Omega))'$; then $u1 \in L^2(\Omega)$. Let $u1$ be represented by the measurable function $u(g)$ on Ω. For any bounded measurable function φ on Ω, we have

$$u\varphi = (u\varphi)1 = (\varphi u)1 = \varphi(g)\, u(g), \tag{2.4.12}$$

hence $\int_G | u(g)\varphi(g)|^2 \, d\mu(g) \leqslant \| u \|^2 \| \varphi \|^2$. Given any positive number ϵ, let $\varphi(g)$ be the characteristic function of the set $\{g \mid | u(g)|^2 \geqslant \| u \|^2 + \epsilon\} = E$. Applying the preceding inequality, we have $\mu(E)(\| u \|^2 + \epsilon) \leqslant \| u \|^2 \cdot \mu(E)$, hence $\mu(E) = 0$. Thus, $| u(g)| \leqslant \| u \|$ almost everywhere. Consequently for any $\varphi \in L^2(\Omega)$, both sides of (2.4.12) belong to $L^2(\Omega)$. Since the totality of bounded measurable functions φ is dense in $L^2(\Omega)$, it follows that $u\varphi(g) = u(g)\varphi(g)$ for every $\varphi \in L^2(\Omega)$, that is, $u \in \mathfrak{M}(\Omega)$.]

We can now state a sufficient condition for the maximality of a commutative weakly closed operator ring.

Corollary 2.4.11. Let $\mathfrak{A}$ be a commutative weakly closed operator ring over a Hilbert space H. If $\mathfrak{A}$ has a cyclic element in H, then $\mathfrak{A}$ is maximal commutative weakly closed.

PROOF. By Corollary 2.4.8, $\mathfrak{A}$ is unitarily equivalent to a multiplication algebra $\mathfrak{M}(\Omega)$, where Ω is a finite measure space. By Lemma 2.4.10, $\mathfrak{M}(\Omega)$ is maximal commutative weakly closed, therefore $\mathfrak{A}$ is also maximal commutative weakly closed.]

We now consider a more general situation.

Theorem 2.4.12. Let $\Omega = (G, \mathfrak{B}, \mu)$ be a localizable measure space. Then the multiplication algebra $\mathfrak{M}(\Omega)$ over $L^2(\Omega)$ is a maximal commutative weakly closed operator ring.

PROOF. Obviously, $\mathfrak{M}(\Omega)$ is a symmetric operator algebra, and is closed in the uniform topology. We need only prove that $(\mathfrak{M}(\Omega)')^P \subset \mathfrak{M}(\Omega)$. For then, given any self-adjoint operator $A \in \mathfrak{M}(\Omega)'$, we know by Corollary 2.3.4 that all the projections P_λ associated with A belong to $(\mathfrak{M}(\Omega)')^P$, and hence to $\mathfrak{M}(\Omega)$, so again, by Corollary 2.3.4, $A \in \mathfrak{M}(\Omega)$. But since $\mathfrak{M}(\Omega)$ is symmetric, this implies that $\mathfrak{M}(\Omega)' \subset \mathfrak{M}(\Omega)$; obviously, since $\mathfrak{M}(\Omega)$ is commutative, $\mathfrak{M}(\Omega) \subset \mathfrak{M}(\Omega)'$. Thus, we obtain $\mathfrak{M}(\Omega) = \mathfrak{M}(\Omega)'$, whence it follows by Lemma 2.4.2 that $\mathfrak{M}(\Omega)$ is a maximal commutative weakly closed operator algebra.

Let $P \in (\mathfrak{M}(\Omega)')^P$, and consider the collection of sets

$$\mathfrak{F} = \{E \mid E \in \mathfrak{B}, \mu(E) < \infty\}.$$

For each $E \in \mathfrak{F}$, let C_E denote the characteristic function of E, and P_E the multiplication operator corresponding to C_E:

$$(P_E f)(g) = C_E(g) f(g), \qquad g \in G, \quad f \in L^2(\Omega).$$

Denote the restrictions of $\mathfrak{B}$, μ to E by $\mathfrak{B}_E$, μ_E, respectively. Since $\mu(E) < \infty$, $\Omega_E = (E, \mathfrak{B}_E, \mu_E)$ is a finite measure space. Regarding the functions in $L^2(\Omega_E)$ as functions in $L^2(\Omega)$ which vanish outside of E, we have $L^2(\Omega_E) = P_E L^2(\Omega)$, and it is clear that $\mathfrak{M}(\Omega_E)$ is then just the restriction of $\mathfrak{M}(\Omega)$ to $L^2(\Omega_E)$. Since $P \in (\mathfrak{M}(\Omega)')^P$, the projections P and P_E commute, hence PP_E is a projection operator on $L^2(\Omega_E)$, moreover, since $P_E \in \mathfrak{M}(\Omega_E)$ and $P \in \mathfrak{M}(\Omega)'$, we have $PP_E \in \mathfrak{M}(\Omega_E)'$. However, by

Lemma 2.4.10, $\mathfrak{M}(\Omega_E) = \mathfrak{M}(\Omega_E)'$, hence $PP_E \in \mathfrak{M}(\Omega_E)$, that is, there exists a set $\lambda_E \in B_E$ such that

$$(PP_E f)(g) = C_{\lambda_E}(g) f(g), \qquad g \in E, \quad f \in L^2(\Omega_E).$$

Obviously, the above formula is also valid for any $f \in L^2(\Omega)$, therefore PP_E is the projection operator corresponding to $\lambda_E \in \mathfrak{B}$. Now, $G \sim \bigvee_{A \in \mathfrak{F}} A$, and so, by Lemma 1.2.5, $I = \bigvee_{E \in \mathfrak{F}} P_E$, hence, $P = \bigvee_{E \in \mathfrak{F}} PP_E$; moreover, by the localizability of Ω, there exists $Q \in \mathfrak{B}$ such that

$$Q = \bigvee_{E \in \mathfrak{F}} \lambda_E .$$

Again, by Lemma 1.2.5, P is the projection operator P_Q corresponding to the set Q, hence $P \in \mathfrak{M}(\Omega)$. **]**

Corollary 2.4.13. Let $\Omega = (G, \mathfrak{B}, \mu)$ be a localizable measure space, k a cardinal number. Then the multiplication algebra $\mathfrak{M}_k(\Omega)$ over $\mathfrak{L}_k{}^2(\Omega)$ is a commutative weakly closed operator ring, and has uniform multiplicity k.

This corollary is a direct consequence of Theorem 2.4.12; the details of the proof are left to the reader.

3° Some Properties of Localizable Measure Spaces

We shall now make use of Theorem 2.4.12 to study in further detail the properties of localizable measure spaces.

Theorem 2.4.14. Let $\Omega = (G, \mathfrak{B}, \mu)$ be a localizable measure space, and let F be any continuous linear functional on $L^1(\Omega)$. Then, there exists an essentially bounded measurable function $F(g)$, $g \in G$, such that

$$F(\varphi) = \int_G \varphi(g) F(g) \, d\mu(g), \qquad \varphi \in L^1(\Omega), \tag{2.4.13}$$

moreover, $\| F \| = \| F \|_\infty$.

Proof. Define a bilinear functional on $L^2(\Omega)$, as follows:

$$F(\varphi, \psi) = F(\varphi\bar{\psi}), \qquad \varphi, \psi \in L^2(\Omega). \tag{2.4.14}$$

If $\varphi, \psi \in L^2(\Omega)$, then $\varphi\bar{\psi} \in L^1(\Omega)$, hence $F(\varphi, \psi)$ is well defined. Moreover,

$$| F(\varphi, \psi)| \leqslant \| F \| \, \| \varphi \|_2 \, \| \psi \|_2 ,$$

hence $F(\varphi, \psi)$ is continuous. Therefore, by a well known theorem, there exists a bounded linear operator A on $L^2(\Omega)$ such that $\| A \| \leqslant \| F \|$ and

$$F(\varphi, \psi) = (A\varphi, \psi), \qquad \varphi, \psi \in L^2(\Omega).$$

Let $B \in \mathfrak{M}(\Omega)$, and suppose that B corresponds to the bounded measurable function $b(g)$. Then, for any φ, $\psi \in L^2(\Omega)$,

$$(AB\varphi, \psi) = F(b\varphi\bar{\psi}) = (A\varphi, \bar{b}\psi) = (BA\varphi, \psi),$$

that is, A and B commute. Thus, $A \in \mathfrak{M}(\Omega)'$. By Theorem 2.4.12, $\mathfrak{M}(\Omega)$ is maximal commutative weakly closed, hence $A \in \mathfrak{M}(\Omega)$. Therefore, there exists an essentially bounded measurable function $F(g)$, $g \in G$ such that $A\varphi(g) = F(g)\varphi(g)$, whence

$$F(\varphi\bar{\psi}) = \int_G F(g)\, \varphi(g)\, \bar{\psi}(g)\, d\mu(g) \tag{2.4.15}$$

for any φ, $\psi \in L^2(\Omega)$. But, for any $f \in L^1(\Omega)$, the functions $\psi = | f |^{1/2}$ and $\varphi = f \cdot | f |^{-1/2}$ belong to $L^2(\Omega)$. Substituting these functions into (2.4.15), we obtain formula (2.4.13).

From (2.4.13), it follows at once that $\| F \| \leqslant \| F \|_\infty$, and since $\| F \|_\infty = \| A \| \leqslant \| F \|$, we get $\| F \| = \| F \|_\infty$.]

The reader may also prove for himself that, if Ω is a measure space and F is a continuous linear functional on $L^p(\Omega)$, $1 < p < \infty$,[18] then there exists a function $F(\cdot) \in L^q(\Omega)$, $(1/q) + (1/p) = 1$, such that

$$F(\varphi) = \int_G \varphi(g) F(g)\, d\mu(g), \qquad \varphi \in L^p(\Omega),$$

and $\| F \| = \| F \|_q$.

The following theorem extends the scope of the classical Radon–Nikodym theorem to localizable measure spaces.

Theorem 2.4.15. Let $\Omega_k = (G, \mathfrak{B}, \mu_k)$, $k = 1, 2$, be localizable measure spaces, with $\mu_1 \ll \mu_2$. Then, there exists a measurable function, which we denote by $d\mu_1(g)/d\mu_2(g)$, having the following properties:

(i) $0 \leqslant d\mu_1(g)/d\mu_2(g) < \infty$,

(ii) for any $\varphi \in L^1(\Omega_1)$, we have $(d\mu_1(g)/d\mu_2(g))\, \varphi \in L^1(\Omega_2)$, and

$$\int_G \varphi(g)\, d\mu_1(g) = \int_G \varphi(g)\, \frac{d\mu_1(g)}{d\mu_2(g)}\, d\mu_2(g). \tag{2.4.16}$$

We call $d\mu_1(g)/d\mu_2(g)$ the *Radon–Nikodym derivative* of μ_1 with respect to μ_2.

[18] The result stated here cannot be extended to the case $p = \infty$.

PROOF. (1) Consider the measure $\mu = \mu_1 + \mu_2$ on $(G, \mathfrak{B})$. It is easily verified that the measure space $\Omega = (G, \mathfrak{B}, \mu)$ is also localizable, and since $\mu_k \leqslant \mu$, $k = 1, 2$, we have $L^1(\Omega_k) \supset L^1(\Omega)$, $k = 1, 2$. Define two linear functionals F_1, F_2 on $L^1(\Omega)$, as follows:

$$F_k(\varphi) = \int_G \varphi(g)\, d\mu_k(g), \qquad \varphi \in L^1(\Omega), \quad k = 1, 2. \tag{2.4.17}$$

Obviously, $\| F_k \| \leqslant 1$. By Theorem 2.4.14, there exist bounded measurable functions on Ω, which we denote by $d\mu_k(g)/d\mu(g)$, $k = 1, 2$, such that

$$F_k(\varphi) = \int_G \varphi(g) \frac{d\mu_k(g)}{d\mu(g)}\, d\mu(g), \qquad k = 1, 2. \tag{2.4.18}$$

(2) We assert that $d\mu_2(g)/d\mu(g)$ is positive almost everywhere. For otherwise, there would exist a set $E \in \mathfrak{B}$, $0 < \mu(E) < \infty$, such that $d\mu_2(g)/d\mu(g) \leqslant 0$ on E. If, in formulas (2.4.17) and (2.4.18), we let φ be the characteristic function of E, then, for $k = 2$, we obtain $\mu_2(E) \leqslant 0$, hence also $\mu_1(E) = 0$; but this contradicts $\mu(E) > 0$. Thus, we may assume that $d\mu_2(g)/d\mu(g) > 0$ for all $g \in G$. A similar argument shows that we may assume $d\mu_1(g)/d\mu(g) \geqslant 0$ for all $g \in G$. We now form the nonnegative finite-valued measurable function

$$\frac{d\mu_1(g)}{d\mu_2(g)} = \frac{d\mu_1(g)}{d\mu(g)} \Big/ \frac{d\mu_2(g)}{d\mu(g)}\,.$$

(3) Let ψ be a nonnegative finite-valued $\mathfrak{B}$-measurable function on G such that the set $\{g \mid \psi(g) \neq 0\}$ is σ-finite with respect to the measure μ. Then there is a monotonically increasing sequence of functions $\{\psi_n(g)\}$ in $L^1(\Omega)$ such that $\lim_{n\to\infty} \psi_n(g) = \psi(g)$ for all $g \in G$. Substituting $\varphi = \psi_n$ in formulas (2.4.17) and (2.4.18), then letting $n \to \infty$ and applying the Levi lemma, we get

$$\int_G \psi(g)\, d\mu_k(g) = \int_G \psi(g) \frac{d\mu_k(g)}{d\mu(g)}\, d\mu(g), \qquad k = 1, 2. \tag{2.4.19}$$

Let

$$G_0 = \left\{ g \,\middle|\, \frac{d\mu_1(g)}{d\mu(g)} = 0 \right\}, \qquad G_n = \left\{ g \,\middle|\, \frac{d\mu_1(g)}{d\mu(g)} \geqslant \frac{1}{n} \right\}, \qquad n = 1, 2, \ldots .$$

Using an argument similar to that given in part (2) above, one may show that $\mu_1(G_0) = 0$, and that, if $E \subset G_n$, $\mu(E) < \infty$, then $\mu_1(E) \geqslant \mu(E)/n$. Consequently, any measurable subset of $G - G_0 = \bigcup_{n=1}^{\infty} G_n$ which is σ-finite relative to the measure μ_1 is also σ-finite relative to the measure μ.

Now, consider any nonnegative function $\varphi \in L^1(\Omega_1)$. Since the set $\{g \mid \varphi(g) \neq 0\}$ is σ-finite relative to μ_1, it follows from the foregoing remark that the set $\{g \mid \varphi(g) \neq 0\} \cap (G - G_0)$ is σ-finite relative to μ. Form the function

$$\psi(g) = \begin{cases} \varphi(g), & g \bar{\in} G_0, \\ 0, & g \in G_0. \end{cases} \tag{2.4.20}$$

Then, the set $\{g \mid \psi(g) \neq 0\}$ is σ-finite relative to μ. Hence, applying (2.4.19) with $k = 1$, we obtain

$$\int_{G-G_0} \varphi(g)\, d\mu_1(g) = \int_{G-G_0} \varphi(g) \frac{d\mu_1(g)}{d\mu(g)}\, d\mu(g). \tag{2.4.21}$$

Since $d\mu_1(g)/d\mu(g) = 0$ on G_0, it follows from (2.4.21) that

$$\int_G \varphi(g)\, d\mu_1(g) = \int_G \varphi(g) \frac{d\mu_1(g)}{d\mu(g)}\, d\mu(g). \tag{2.4.22}$$

On the other hand, if we choose $\psi(g) = \varphi(g)\, d\mu_1(g)/d\mu_2(g)$, then the set

$$\{g \mid \psi(g) \neq 0\} = \{g \mid \varphi(g) \neq 0\} \cap (G - G_0)$$

is also σ-finite relative to μ, hence, applying (2.4.19) with $k = 2$, we get

$$\int_G \varphi(g) \frac{d\mu_1(g)}{d\mu_2(g)}\, d\mu_2(g) = \int_G \varphi(g) \frac{d\mu_1(g)}{d\mu(g)}\, d\mu(g). \tag{2.4.23}$$

Combining (2.4.22) and (2.4.23), we conclude that whenever $\varphi \in L^1(\Omega_1)$ and $\varphi \geqslant 0$, relation (2.4.16) holds and so $\varphi(g)(d\mu_1(g)/d\mu_2(g)) \in L^1(\Omega_2)$. The proof is completed by observing that any function φ in $L^1(\Omega_1)$ can be expressed as the difference of two nonnegative functions in $L^1(\Omega_1)$.]

Corollary 2.4.16. Under the hypotheses of Theorem 2.4.15,

$$\mu_1(E) = \int_E \frac{d\mu_1(g)}{d\mu_2(g)}\, d\mu_2(g) \tag{2.4.24}$$

for any $E \in \tilde{\mathfrak{B}}$. Moreover, if E is σ-finite relative to μ_2, then E is also σ-finite relative to μ_1.[19]

PROOF. If $E \in \tilde{\mathfrak{B}}$, $\mu_1(E) < \infty$, then, in (2.4.16), we let φ be the characteristic function of E, and so deduce that (2.4.24) holds in this case.

[19] *Translator's note:* The statement in the original text was (perhaps unintentionally) the converse of this, that is, that σ-finiteness relative to μ_1 implies σ-finiteness relative to μ_2; the reader should have no difficulty in constructing both trivial and nontrivial counterexamples.

On the other hand, if $\mu_1(E) = \infty$, then there exists a sequence of measurable sets $E_n \subset E$, $E_1 \subset E_2 \subset E_3 \subset \cdots$ such that $\mu_1(E_n) < \infty$ and $\mu_1(E_n) \to \infty$. Setting $E = E_n$ in both sides of (2.4.24) and letting $n \to \infty$, we get

$$\int_{\bigcup_{n=1}^{\infty} E_n} \frac{d\mu_1(g)}{d\mu_2(g)}\, d\mu_2(g) = \infty,$$

whence, *a fortiori*

$$\int_E \frac{d\mu_1(g)}{d\mu_2(g)}\, d\mu_2(g) = \infty.$$

Next, suppose that E is σ-finite with respect to μ_2. Then, there exist measurable sets E_i, $i = 1, 2,\ldots$, such that $E = \bigcup_{i=1}^{\infty} E_i$ and $\mu_2(E_i) < \infty$. Let

$$F_n = \left\{ g \,\middle|\, \frac{d\mu_1(g)}{d\mu_2(g)} < n \right\}, \qquad n = 1, 2,\ldots .$$

Then, by (2.4.24),

$$\mu_1(E_i \cap F_n) = \int_{E_i \cap F_n} \frac{d\mu_1(g)}{d\mu_2(g)}\, d\mu_2(g) \leqslant n\mu_2(E_i \cap F_n) \leqslant n\mu_2(E_i) < \infty.$$

Since E is the union of the countable collection of sets

$$E_i \cap F_n, \qquad i, n = 1, 2,\ldots,$$

this shows that E is also σ-finite with respect to μ_1.]

Theorem 2.4.17. Let $\Omega_k = (G, \mathfrak{B}, \mu_k)$, $k = 1, 2, 3$, be localizable measure spaces such that $\mu_1 \ll \mu_2$, $\mu_2 \ll \mu_3$, and let $d\mu_1(g)/d\mu_2(g)$, $d\mu_1(g)/d\mu_3(g)$, $d\mu_2(g)/d\mu_3(g)$ denote the appropriate Radon–Nikodym derivatives. Then,

$$\frac{d\mu_1(g)}{d\mu_3(g)} = \frac{d\mu_1(g)}{d\mu_2(g)} \cdot \frac{d\mu_2(g)}{d\mu_3(g)}. \tag{2.4.25}$$

Proof. If $\varphi \in L^1(\Omega_1)$, then, by Theorem 2.4.15,

$$\frac{d\mu_1(g)}{d\mu_2(g)}\, \varphi \in L^1(\Omega_2),$$

whence, applying Theorem 2.4.15 to the measures μ_2, μ_3, we obtain

$$\int_G \varphi(g) \frac{d\mu_1(g)}{d\mu_2(g)}\, d\mu_2(g) = \int_G \varphi(g) \frac{d\mu_1(g)}{d\mu_2(g)} \cdot \frac{d\mu_2(g)}{d\mu_3(g)}\, d\mu_3(g). \tag{2.4.26}$$

Again, applying Theorem 2.4.15 to the measures μ_1, μ_3, we know that, if $\varphi \in L^1(\Omega_1)$, then $\varphi(d\mu_1(g)/d\mu_3(g)) \in L^1(\Omega_3)$, and

$$\int_G \varphi(g)\, d\mu_1(g) = \int_G \varphi(g) \frac{d\mu_1(g)}{d\mu_3(g)}\, d\mu_3(g). \tag{2.4.27}$$

Combining (2.4.16), (2.4.26), and (2.4.27), we see that

$$\int_G \varphi(g) \frac{d\mu_1(g)}{d\mu_3(g)}\, d\mu_3(g) = \int_G \varphi(g) \frac{d\mu_1(g)}{d\mu_2(g)} \frac{d\mu_2(g)}{d\mu_3(g)}\, d\mu_3(g) \tag{2.4.28}$$

for any $\varphi \in L^1(\Omega_1)$. Now, let ϵ be an arbitrary positive number, and let

$$E = \left\{ g \;\middle|\; \frac{d\mu_1(g)}{d\mu_3(g)} > \frac{d\mu_1(g)}{d\mu_2(g)} \frac{d\mu_2(g)}{d\mu_3(g)} + \epsilon \right\}.$$

Suppose that $\mu_3(E) > 0$; this implies the existence of a measurable set $A \subset E$ such that $\mu_3(A) > 0$ and $\mu_1(A) < \infty$. Let C_A denote the characteristic function of A; since $C_A \in L^1(\Omega_1)$, we may substitute $\varphi = C_A$ in (2.4.28), and thus obtain a contradiction. Therefore, $\mu_3(E) = 0$; since $\epsilon > 0$ is arbitrary, we conclude that

$$\frac{d\mu_1(g)}{d\mu_3(g)} \leqslant \frac{d\mu_1(g)}{d\mu_2(g)} \frac{d\mu_2(g)}{d\mu_3(g)} \tag{2.4.29}$$

for almost all $g \in G$ (relative to μ_3). An entirely similar argument shows that the reverse inequality also holds for almost all g. Hence, (2.4.25) is valid for almost all $g \in G$ (relative to the measure μ_3).]

Corollary 2.4.18. Let $\Omega_k = (G, \mathfrak{B}, \mu_k)$, $k = 1, 2$, be equivalent localizable measure spaces. Then the Radon–Nikodym derivative $d\mu_1(g)/d\mu_2(g)$ of μ_1 with respect to μ_2 may be chosen so that

$$0 < \frac{d\mu_1(g)}{d\mu_2(g)} < \infty.$$

PROOF. Taking $\mu_3 = \mu_1$ in Theorem 2.4.17, we may choose $d\mu_1(g)/d\mu_3(g) = 1$, hence

$$\frac{d\mu_1(g)}{d\mu_2(g)} \frac{d\mu_2(g)}{d\mu_3(g)} = 1 \tag{2.4.30}$$

for almost all g. Since $d\mu_2(g)/d\mu_3(g) < \infty$, it follows from (2.4.30) that

$d\mu_1(g)/d\mu_2(g) > 0$ almost everywhere. Thus, we need only alter the values of $d\mu_1(g)/d\mu_2(g)$ on a null set and so obtain the required function.]

Corollary 2.4.19. Under the hypotheses of Theorem 2.4.15, the mapping

$$T: \quad \varphi \to \varphi(d\mu_1/d\mu_2)^{1/2}$$

is an isometric operator from $L^2(\Omega_1)$ into $L^2(\Omega_2)$.

PROOF. Let $\varphi, \psi \in L^2(\Omega_1)$; then

$$(T\varphi)\overline{(T\psi)} = \varphi(g)\,\overline{\psi(g)}\,\frac{d\mu_1}{d\mu_2}.$$

However, $\varphi\bar{\psi} \in L^1(\Omega_1)$, hence, applying (2.4.16), we obtain

$$\begin{aligned}(\varphi, \psi) &= \int_G \varphi(g)\,\overline{\psi(g)}\,d\mu_1(g) = \int_G \varphi(g)\,\overline{\psi(g)}\,\frac{d\mu_1(g)}{d\mu_2(g)}\,d\mu_2(g)\\ &= (T\varphi, T\psi),\end{aligned}$$

that is, T is isometric.]

We shall now establish the general form of an operator in $[\mathfrak{M}_k(\Omega)']^U$, where Ω is a localizable measure space. For convenience, we shall explicitly discuss only the case $k \leqslant \aleph_0$.

Let $\{e_\lambda, \lambda \in \Lambda\}$ be an orthonormal basis for the k-dimensional Hilbert space H_k, let $H_\lambda = \{\varphi e_\lambda \mid \varphi \in L^2(\Omega)\}$, and let P_λ be the projection operator from $\mathfrak{L}_k^2(\Omega)$ onto H_λ. It is easily verified that $P_\lambda \in [\mathfrak{M}_k(\Omega)]'$, moreover, by (1.1.7), we have

$$\sum_{\lambda \in \Lambda} P_\lambda = I$$

in the strong topology.

Let $U \in [\mathfrak{M}_k(\Omega)']^U$, and write $U_{\lambda\lambda'} = P_\lambda U P_{\lambda'}$. We assert that there exists a bounded measurable function $u_{\lambda\lambda'}(g)$ on Ω such that

$$U_{\lambda\lambda'}(\varphi e_{\lambda'}) = u_{\lambda\lambda'}\varphi e_\lambda \tag{2.4.31}$$

for every $\varphi \in L^2(\Omega)$. In fact, let ψ be an arbitrary bounded measurable function on Ω; since $U_{\lambda\lambda'} \in [\mathfrak{M}_k(\Omega)]'$, we have

$$U_{\lambda\lambda'}(\psi\varphi e_\lambda) = \psi(U_{\lambda\lambda'}(\varphi e_\lambda)). \tag{2.4.32}$$

Define an operator $u_{\lambda\lambda'}$ from $L^2(\Omega)$ to $L^2(\Omega)$, as follows:

$$u_{\lambda\lambda'}\varphi = (U_{\lambda\lambda'}\varphi e_{\lambda'}, e_\lambda), \qquad \varphi \in L^2(\Omega).$$

Using (2.4.32), it is easily seen that $u_{\lambda\lambda'} \in [\mathfrak{M}(\Omega)]'$, hence, by Theorem 2.4.12, $u_{\lambda\lambda'} \in \mathfrak{M}(\Omega)$, that is, there exists a bounded measurable function $u_{\lambda\lambda'}(g)$, $g \in G$ such that

$$(u_{\lambda\lambda'}\varphi)(g) = u_{\lambda\lambda'}(g)\,\varphi(g).$$

Thus, we obtain (2.4.31).

Now, any $\xi \in H_k$ can be expressed in the form $\xi = \sum \xi_\lambda e_\lambda$ (where $\xi_\lambda \neq 0$ for at most countably many indices λ), and if $\varphi \in L^2(\Omega)$, then $\varphi\xi \in \mathfrak{L}_k{}^2(\Omega)$, hence

$$U\varphi\xi = \sum_\lambda \Big(\sum_{\lambda'} u_{\lambda\lambda'}\xi_{\lambda'}\varphi\Big)\, e_\lambda\,. \tag{2.4.33}$$

Therefore, since U is unitary,

$$\begin{aligned}\sum_{\lambda'} |\,\xi_{\lambda'}\,|^2 \int_G |\,\varphi(g)|^2\, d\mu(g) &= \|\,\varphi\xi\,\|^2 = \|\,U\varphi\xi\,\|^2\\ &= \int_G |\,\varphi(g)|^2 \sum_\lambda \Big|\sum_{\lambda'} u_{\lambda\lambda'}(g)\xi_{\lambda'}\Big|^2\, d\mu(g).\end{aligned} \tag{2.4.34}$$

Consider the subspace

$$\mathfrak{D} = \Big\{\xi\,\Big|\,\xi = \sum \xi_\lambda e_\lambda \in H_k\,,\ \xi_\lambda \text{ rational},\ \xi_\lambda \neq 0 \text{ for only finitely many } \lambda\Big\}.$$

Clearly, $\mathfrak{D}$ is countable and dense in H_k. Using (2.4.34) and the countability of $\mathfrak{D}$, we see that there exists a μ-null set E in G such that

$$\sum_\lambda \Big|\sum_{\lambda'} u_{\lambda\lambda'}(g)\xi_{\lambda'}\Big|^2 = \sum_{\lambda'} |\,\xi_{\lambda'}\,|^2 \tag{2.4.35}$$

for all $g \in G - E$, $\xi \in \mathfrak{D}$. Then, using the density of $\mathfrak{D}$, it follows easily that (2.4.35) holds for all $\xi \in H_k$, $g \in G - E$. Now, for each $g \in G - E$, define an operator $U(g)$ as follows: if $\xi = \{\xi_\lambda\} \in H_k$, then $U(g)\xi = \sum_\lambda \xi_\lambda' e_\lambda$, where

$$\xi_\lambda' = \sum_{\lambda'} u_{\lambda\lambda'}(g)\xi_{\lambda'}\,.$$

We see from (2.4.35) that $U(g)$ is an isometric linear operator from H_k into H_k. Moreover, using (2.4.33), one easily verifies that, if $\xi(\cdot) \in \mathfrak{L}_k{}^2(\Omega)$, then

$$(U\xi)(g) = U(g)\,\xi(g)$$

for almost all $g \in G$. Applying the foregoing argument to the unitary operator U^{-1}, we obtain a μ-null set $E' \subset G$ and, for each $g \in G - E'$, an isometric linear operator $U^{-1}(g)$ from H_k into H_k, such that, given any $\xi(\cdot) \in \mathfrak{L}_k^2(\Omega)$, the relation

$$(U^{-1}\xi)(g) = U^{-1}(g)\, \xi(g)$$

holds for almost all $g \in G$. Using the relation $U^{-1}U = UU^{-1} = I$ and the separability of H_k, it is easily proved that

$$U(g)\, U^{-1}(g) = U^{-1}(g)\, U(g) = I,$$

that is, that $U(g)$ is a unitary operator from H_k onto H_k, for almost all $g \in G$. Moreover, by supplementing or altering the definition of $U(g)$ on a μ-null set, if necessary, we may thus ensure that $U(g)$ is defined and unitary for all $g \in G$. Furthermore, for every $\xi(\cdot) \in M(H_k, \Omega)$, we have $U(\cdot)\xi(\cdot) \in M(H_k, \Omega)$.

Definition 2.4.4. Let $\Omega = (G, \mathfrak{B}, \mu)$ be a measure space, and H_k a k-dimensional Hilbert space. Suppose that, to every $g \in G$, there corresponds a unitary operator $U(g)$ from H_k to H_k, and that $U(\cdot)\xi(\cdot) \in M(H_k, \Omega)$ whenever $\xi(\cdot) \in M(H_k, \Omega)$. Then, $U(\cdot)$ is said to be a measurable k-dimensional unitary operator-valued function on Ω.

The foregoing result may then be stated as follows.

Lemma 2.4.20. Let Ω be a localizable[20] measure space, $k \leqslant \aleph_0$, and let $U \in [\mathfrak{M}_k(\Omega)']^U$. Then, there exists a measurable k-dimensional unitary operator-valued function $U(\cdot)$ on Ω, such that

$$(U\xi)(g) = U(g)\, \xi(g)$$

for every $\xi \in \mathfrak{L}_k^2(\Omega)$.

Next, we consider a condition for the equivalence of two localizable measure spaces (see Definition 1.2.4).

Theorem 2.4.21. Let $\Omega = (G, \mathfrak{B}, \mu)$ and $\Omega' = (G', \mathfrak{B}', \mu')$ be localizable measure spaces. Then Ω and Ω' are equivalent if and only if there exists a unitary mapping Q from $\mathfrak{L}_k^2(\Omega)$ onto $\mathfrak{L}_k^2(\Omega')$ such that the correspondence

$$A \to QAQ^{-1}$$

defines an isomorphism of $\mathfrak{M}_k(\Omega)$ onto $\mathfrak{M}_k(\Omega')$.

[20] *Translator's note:* In the original statement, Ω is assumed to be σ-finite. However, the above proof appears to be valid for any localizable measure space.

PROOF. In the sequel, we shall only use the sufficiency of this condition, therefore, we omit the proof of necessity.

Let (M, μ) and (M', μ') be the measure rings corresponding to Ω and Ω', respectively. Given any $E \in \tilde{\mathfrak{B}}$, denote the characteristic function of E by $C_E(g)$, $g \in G$, and let $P_E = T_{C_E}$. Since $QP_EQ^{-1} \in \mathfrak{M}_k(\Omega')$, there exists a function $f \in L^\infty(\Omega')$ such that $T_f = QP_EQ^{-1}$. Moreover, since Q is a unitary mapping, the relations $P_E{}^* = P_E$, $P_E{}^2 = P_E$ imply that

$$\overline{f(g)} = f(g), \qquad f(g)^2 = f(g).$$

Consequently, there exists a set $E' \in \tilde{\mathfrak{B}}'$ such that $QP_EQ^{-1} = P_{E'}$. Define a mapping φ from M to M', as follows:

$$\varphi([E]) = [E'].$$

Using the fact that Q is unitary, one may easily verify that φ is a one-to-one mapping of M onto M' and satisfies conditions (1.2.9). Furthermore, $\mu([E]) = 0$ is equivalent to $P_E = 0$, which is equivalent to $P_{E'} = 0$, which is in turn equivalent to $\mu(\varphi[E]) = 0$. Thus, Ω and Ω' are equivalent.]

4° Spectral Measure Spaces

Definition 2.4.5. Let $(\Gamma, \mathfrak{B})$ be a measurable space, where $\mathfrak{B}$ is a σ-algebra in Γ, and let H be a Hilbert space. Suppose that, for every $E \in \mathfrak{B}$, there corresponds a projection operator $P(E)$ on H, and that the following conditions are satisfied.

(i) $P(0) = 0$, $P(\Gamma) = I$.

(ii) If $E = \bigcup_{\nu=1}^{\infty} E_\nu$, $E_\nu \in \mathfrak{B}$, and $E_\mu \cap E_\nu = 0$ when $\mu \neq \nu$, then, for any $\xi, \eta \in H$,

$$(P(E)\xi, \eta) = \sum_{\nu=1}^{\infty} (P(E_\nu)\xi, \eta).$$

The abstract set function $E \to P(E)$ [simply written as $P(\cdot)$] is then said to be a *spectral measure* on $(\Gamma, \mathfrak{B})$ over the Hilbert space H, or, alternatively, $(\Gamma, \mathfrak{B}, P)$ is said to be a *spectral measure space* over H.

Let us record some simple properties of spectral measures:

(iii) If the sets $E_1, E_2, \ldots, E_n \in \mathfrak{B}$ are pairwise disjoint, then

$$P(E_1 + \cdots + E_n) = P(E_1) + \cdots + P(E_n).$$

(iv) For any $E_1, E_2 \in \mathfrak{B}$, we have $P(E_1)P(E_2) = P(E_1 \cap E_2)$.

In fact, from (iii), one easily deduces that, if $F_1, F_2 \in \mathfrak{B}$, $F_1 \cap F_2 = 0$, then

$$P(F_1)\,P(F_2) = 0.$$

Hence, for any $E_1, E_2 \in \mathfrak{B}$, we have

$$\begin{aligned} &P(E_1)\,P(E_2) \\ &\quad = (P(E_1 - E_1 \cap E_2) + P(E_1 \cap E_2))(P(E_2 - E_1 \cap E_2) + P(E_1 \cap E_2)) \\ &\quad = P(E_1 \cap E_2). \end{aligned}$$

In particular, (iv) shows that $\mathfrak{P} = \{P(E) \mid E \in \mathfrak{B}\}$ is a commutative family of operators. If $P(E)H$ cannot be decomposed into an orthogonal sum of uncountably many $\mathfrak{P}$-invariant nonzero closed linear subspaces, then $P(E)$ is said to be *countably decomposable*. If, for every $F \in \mathfrak{B}$ such that $P(F) \neq 0$, there exists a set $E \in \mathfrak{B}$, $E \subset F$ such that $P(E)$ is nonzero and countably decomposable, then the spectral measure $P(\cdot)$ is said to be *normal*.

Theorem 2.4.22. Let $(\Gamma, \mathfrak{B}, P)$ be a spectral measure space over the Hilbert space H, and let $\mathfrak{A}$ be the weakly closed operator algebra generated by $\{P(E) \mid E \in \mathfrak{B}\}$. If $\mathfrak{A}$ is maximal commutative weakly closed, then there exists a measure μ on $(\Gamma, \mathfrak{B})$, and a unitary mapping U from H onto $L^2(\Gamma, \mathfrak{B}, \mu)$, such that every projection $Q(E) = UP(E)\,U^{-1}$, $E \in \mathfrak{B}$, takes the form

$$(Q(E)\varphi)(\gamma) = C_E(\gamma)\,\varphi(\gamma), \qquad \varphi \in L^2(\Gamma, \mathfrak{B}, \mu), \tag{2.4.36}$$

where C_E is the characteristic function of the set E.

Proof. In accordance with Lemma 2.4.6, we decompose H into an orthogonal sum of closed linear subspaces $\{H_\xi, \xi \in \Xi\}$, where ξ is a cyclic element of $\mathfrak{A}$ in H_ξ. For each $\xi \in \Xi$, define a set function μ_ξ on $\mathfrak{B}$, as follows:

$$\mu_\xi(E) = (P(E)\xi, \xi), \qquad E \in \mathfrak{B}.$$

Since $P(\cdot)$ is a spectral measure, it follows easily that $\Omega_\xi = (\Gamma, \mathfrak{B}, \mu_\xi)$ is a finite measure space. Next, we construct a mapping U_ξ from H to $L^2(\Omega_\xi)$, in the following manner. For any vector in H_ξ of the form

$$\eta = \sum_{k=1}^{n} \lambda_k P(E_k)\xi \tag{2.4.37}$$

(where the E_k belong to $\mathfrak{B}$ and the λ_k are numbers), we define

$$U_\xi \eta = \sum_{k=1}^{n} \lambda_k C_{E_k}(\gamma), \qquad \gamma \in \Gamma. \tag{2.4.38}$$

Obviously, $U_\xi \eta \in L^2(\Omega_\xi)$, and it is easily verified that the correspondence $\eta \to U_\xi \eta$ is linear and isometric; since the totality of vectors of the form (2.4.37) is dense in H_ξ, and the totality of vectors of the form (2.4.38) is dense in $L^2(\Omega_\xi)$, it follows that this correspondence extends uniquely to a unitary operator U_ξ from H_ξ onto $L^2(\Omega_\xi)$.

Let $Q_\xi(E) = U_\xi P(E) U_\xi^{-1}$, $E \in \mathfrak{B}$. It is easily seen that

$$(Q_\xi(E)\varphi)(\gamma) = C_E(\gamma)\, \varphi(\gamma), \qquad \gamma \in \Gamma$$

for every $\varphi \in L^2(\Omega_\xi)$.

We now proceed to prove that if $\xi, \xi' \in \Xi$, $\xi \neq \xi'$, then μ_ξ and $\mu_{\xi'}$ are mutually singular. Suppose this were not the case. Then, there would exist a set $A \in \mathfrak{B}$ such that the restrictions of μ_ξ and $\mu_{\xi'}$ to A are equivalent (and not identically zero). Let $M \subset L^2(\Omega_\xi)$ and $M' \subset L^2(\Omega_{\xi'})$ be the closed linear subspaces defined by those functions in $L^2(\Omega_\xi)$ and $L^2(\Omega_{\xi'})$, respectively, which vanish outside of A. For every function $f \in M$, let

$$(Vf)(\gamma) = \begin{cases} \left(\dfrac{d\mu_\xi(\gamma)}{d\mu_{\xi'}(\gamma)}\right)^{1/2} f(\gamma), & \gamma \in A, \\ 0, & \gamma \,\overline{\in}\, A. \end{cases} \tag{2.4.39}$$

Since μ_ξ and $\mu_{\xi'}$ are equivalent on A, the correspondence $f \to Vf$ is clearly a unitary operator from M onto M'. Let P_ξ denote the projection of H onto H_ξ; by Lemmas 2.3.7 and 2.4.2, we know that $P_\xi \in \mathfrak{A}' = \mathfrak{A}$. Write $U = U_{\xi'}^{-1} V Q_\xi(A)\, U_\xi P_\xi$, and let E be any set in $\mathfrak{B}$; from (2.4.39), we obtain

$$\begin{aligned} UP(E) &= U_{\xi'}^{-1} V Q_\xi(A)\, Q_\xi(E)\, U_\xi P_\xi \\ &= U_{\xi'}^{-1} Q_{\xi'}(E)\, V Q_\xi(A)\, U_\xi P_\xi = P(E) U. \end{aligned} \tag{2.4.40}$$

Thus, $U \in \{P(E) \mid E \in \mathfrak{B}\}' = \mathfrak{A}'$, hence $U = P_{\xi'} U = U P_{\xi'} = 0$. But since $\mu_\xi(A) \neq 0$, $\mu_{\xi'}(A) \neq 0$, we have

$$U\xi = U_{\xi'}^{-1} \left(\frac{d\mu_\xi(\gamma)}{d\mu_{\xi'}(\gamma)}\right)^{1/2} C_A(\cdot) \neq 0.$$

This is a contradiction, hence, we conclude that the measures $\{\mu_\xi\,,\ \xi \in \Xi\}$ are mutually singular.

Let $\mathfrak{B}_0$ denote the totality of sets $E \in \mathfrak{B}$ such that $P(E)$ is countably decomposable. Clearly, if $E \in \mathfrak{B}_0$, then there are at most countably many indices ξ such that $\mu_\xi(E) > 0$. We define a measure μ on $\mathfrak{B}_0$, as follows:

$$\mu(E) = \sum_{\mu_\xi(E) \neq 0} \mu_\xi(E).$$

It is easily verified that $(\Gamma, \mathfrak{B}_0, \mu)$ is a measure space. Since $\mathfrak{A}$ is maximal commutative weakly closed, the spectral measure $P(\cdot)$ is normal[21]. Moreover, it is easily proved that[21] μ can be extended to $\mathfrak{B}$ and that[21] $L^2(\Gamma, \mathfrak{B}, \mu) = \sum_\xi \oplus L^2(\Omega_\xi)$. Let $U = \sum_{\xi \in \Xi} U_\xi P_\xi$; then U is a unitary operator from H onto $L^2(\Gamma, \mathfrak{B}, \mu)$, and (2.4.40) follows easily from (2.4.36).]

Corollary 2.4.23. Let H be a Hilbert space, let $(\Gamma, \mathfrak{B}, P)$ be a normal[22] spectral measure space over H, and let $\mathfrak{A}$ be the weakly closed operator algebra generated by $\{P(E) \mid E \in \mathfrak{B}\}$. Suppose that $\mathfrak{A}$ has uniform multiplicity k. Then, there exists a measure μ on $(\Gamma, \mathfrak{B})$ and a unitary operator U from H onto $\mathfrak{L}_k^2(\Gamma, \mathfrak{B}, \mu)$ such that every operator $Q(E) = UP(E)U^{-1}$, $E \in \mathfrak{B}$, assumes the form

$$(Q(E)\varphi)(\gamma) = C_E(\gamma)\,\varphi(\gamma), \qquad \varphi \in \mathfrak{L}_k^2(\Gamma, \mathfrak{B}, \mu),$$

where C_E denotes the characteristic function of the set E.

PROOF. Since $\mathfrak{A}$ has uniform multiplicity k, the Hilbert space H can be decomposed into an orthogonal sum of k closed linear subspaces H_λ, $\lambda = 1, 2,..., k$,

$$H = \sum_{\lambda=1}^{k} \oplus H_\lambda ,$$

such that each H_λ is $\mathfrak{A}$-invariant, and all the restrictions

$$\mathfrak{A} \mid H_\lambda , \qquad \lambda = 1, 2,..., k,$$

are maximal commutative weakly closed and unitarily equivalent. We arbitrarily choose an index $\lambda_0 \leqslant k$, say, $\lambda_0 = 1$. By Theorem 2.4.22, there exists a measure μ on $(\Gamma, \mathfrak{B})$ and a unitary operator U_1 from H_1 onto $L^2(\Gamma, \mathfrak{B}, \mu)$ such that every operator $Q(E) = U_1 P(E) U_1^{-1}$, $E \in \mathfrak{B}$, takes the form

$$(Q_1(E))\,\varphi(\gamma) = C_E(\gamma)\,\varphi(\gamma), \qquad \varphi \in L^2(\Gamma, \mathfrak{B}, \mu).$$

[21] *Translator's note:* More detailed proofs of these assertions would be welcome. Of course, if H is separable, then, by Corollary 2.4.9, the set Ξ reduces to a single index ξ, and the proof of the theorem offers no difficulty.

[22] *Translator's note:* If one accepts the validity of Theorem 2.4.22, then the assumption that $P(\cdot)$ is normal appears to be redundant.

For each $\lambda \leqslant k$, let $U_{\lambda 1}: H_\lambda \to H_1$ be a unitary operator which effects an equivalence between $\mathfrak{A} \mid H_1$ and $\mathfrak{A} \mid H_\lambda$, and let $U_\lambda = U_1 U_{\lambda 1}$. Then, a trivial calculation shows that every operator $Q_\lambda(E) = U_\lambda P(E)\, U_\lambda^{-1}$, $E \in \mathfrak{B}$ takes the form

$$(Q_\lambda(E))\,\varphi(\gamma) = C_E(\gamma)\,\varphi(\gamma), \qquad \varphi \in L^2(\Gamma, \mathfrak{B}, \mu).$$

Let $\mathfrak{H}_k$ be a k-dimensional Hilbert space, let $\{e_\lambda, \lambda = 1, 2, \ldots, k\}$ be a complete orthonormal set in $\mathfrak{H}_k$, and let $\tilde{H}_\lambda = \{f(\cdot)\, e_\lambda \mid f \in L^2(\Gamma, \mathfrak{B}, \mu)\}$. According to Theorem 1.1.4, we have

$$\mathfrak{L}_k{}^2(\Gamma, \mathfrak{B}, \mu) = \sum_{\lambda=1}^{k} \oplus \tilde{H}_\lambda .$$

Using the family of operators $\{U_\lambda, \lambda = 1, 2, \ldots, k\}$, we construct a unitary operator U from H onto $\mathfrak{L}_k{}^2(\Gamma, \mathfrak{B}, \mu)$, as follows. If $h \in H$, $h = \sum h_\lambda$, $h_\lambda \in H_\lambda$, then

$$(Uh)(\gamma) = \sum_{\lambda=1}^{k} (U_\lambda h_\lambda)(\gamma) e_\lambda, \qquad \gamma \in \Gamma.$$

It is easily verified that U is a unitary operator having the required property.]

CHAPTER

III

HARMONIC ANALYSIS ON GROUPS WITH QUASI-INVARIANT MEASURES

As is well known, every regular measure on n-dimensional Euclidean space which is invariant under all translations is just Lebesgue measure multiplied by some constant factor, and this translation invariance forms the basis of ordinary harmonic analysis (e.g., the theory of Fourier transforms). The theory of invariant measure and harmonic analysis on n-dimensional Euclidean space has long since been extended to locally compact topological groups (especially locally compact topological Abelian groups), and abundant results have been obtained.

It is natural to ask whether such a theory of translation invariant measure, and corresponding results in harmonic analysis, can be established on infinite-dimensional spaces. Actually, in dealing with the so-called continual integrals of quantum field theory, physicists have freely manipulated Fourier transforms on infinite-dimensional spaces.[1] Unfor-

[1] Brush [1] gives an extensive bibliography on the use of infinite-dimensional integrals in quantum field theory.

tunately, however, there can be no hope of extending the theory of invariant measure from finite to infinite-dimensional spaces. For example, one may easily prove that no reasonably well-behaved translation invariant measure exists on any infinite-dimensional Hilbert space; more specifically, for any translation invariant measure on an infinite-dimensional Hilbert space such that all balls are measurable sets, there must exist many balls whose measure is either zero or ∞. Consequently, we are compelled to relax the requirement of translation invariance. We note that if a σ-finite measure on n-dimensional Euclidean space is such that any translate of a null set is also a null set (this property is called quasi-invariance), then this measure must be equivalent to Lebesgue measure (see Theorem 3.1.5), and mutually equivalent (and, e.g., localizable) measures can be related by means of the Radon–Nikodym derivative. Thus, we may consider generalizing this quasi-invariance property of Lebesgue measure. But again, for the most useful infinite-dimensional spaces (e.g., infinite-dimensional Hilbert spaces), σ-finite quasi-invariant (i.e., all translates of any null set are null sets) measures do not exist. Therefore, we must relax our requirements still further; we shall consider measures such that null sets are transformed into null sets, not by all translations, but only by translations corresponding to the vectors in some fixed linear subspace. The notion of quasi-invariance in this form was introduced by Gel'fand [1] in 1959; he used measures on rigged Hilbert spaces to obtain preliminary results concerning representations of the commutation relations in quantum field theory (see Gel'fand and Vilenkin [1]). In this connection, Segal [5] had somewhat earlier (1958) introduced the notion of a quasi-invariant weak distribution on a Hilbert space; this notion is essentially equivalent to that of a quasi-invariant measure.

However, the above-mentioned authors did not establish a theory of harmonic analysis for quasi-invariant measures. Clearly, in order to give a thorough treatment of quasi-invariant measures, one should first discuss quasi-invariant measures on groups. Accordingly, we shall, in this chapter, develop a theory of harmonic analysis for quasi-invariant measures on groups.

Quasi-invariant measures constitute a very broad class, with general properties much weaker than those of invariant measures. Consequently, one cannot expect to create a general theory of harmonic analysis for quasi-invariant measures which is as orderly and complete as the theory of harmonic analysis on locally compact groups. Nevertheless, some of the problems to be discussed below are closely related with the corresponding problems in harmonic analysis on locally compact groups. However, the methods to be used here are radically different from those ordinarily used in harmonic analysis on locally compact groups, that is, we lean heavily toward measure-theoretic techniques and the theory of weakly closed commutative operator rings over Hilbert spaces. Some of the theory and methods developed here are intended to form a basis for the further study of the many problems associated with the "continual integrals" of quantum field theory.

In §3.1, we consider some of the basic properties of quasi-invariant measures. In §3.2, we introduce the notion of quasi-characters, which is essential for the study of quasi-invariant measures on groups, and proceed to discuss quasi-character groups in considerable detail. In §3.3, the integral representation theorem for positive definite functions on locally compact groups will be generalized to groups with quasi-invariant measures. In §3.4, we present the basic theory of Fourier transforms.

§3.1. Basic Properties of Quasi-Invariant Measures

1° The Concept of a Quasi-Invariant Measure

We first generalize the idea of a translation invariant measure.

Definition 3.1.1. Let $\Omega = (G, \mathfrak{B}, \mu)$ be a measure space, and let h be a mapping from a measurable subset $\mathfrak{D}(h)$ of G into G, where $G - \mathfrak{D}(h)$ is a null set. Also, suppose that, for every $A \in \mathfrak{B}$,

$$h^{-1}(A) = \{g \mid hg \in A;\ g \in \mathfrak{D}(h)\} \in \mathfrak{B}.$$

Then, we say that h is a *measurable transformation* in Ω. In this case, we write

$$\mu_h(A) = \mu(h^{-1}A), \qquad A \in \mathfrak{B}. \tag{3.1.1}$$

Clearly, μ_h is a measure on $(G, \mathfrak{B})$.

Let h be a measurable transformation[2] of Ω which has the following properties: (i) the range $\mathfrak{R}(h)$ of h is measurable and $G - \mathfrak{R}(h)$ is a μ-null set; (ii) h is one-to-one from $\mathfrak{D}(h)$ onto $\mathfrak{R}(h)$; (iii) the measures μ_h and μ are equivalent; (iv) h^{-1} is a measurable transformation. We then say that Ω is *quasi-invariant* under h, or *h-quasi-invariant*; the totality of such transformations h will be denoted by $\mathfrak{G}(\Omega)$.

If $h, h' \in \mathfrak{G}(\Omega)$, we may define a mapping hh' as follows: the domain of hh' is

$$\mathfrak{D}(hh') = \{g \mid g \in \mathfrak{D}(h'),\ h'g \in \mathfrak{D}(h)\},$$

[2] The notion of a measurable transformation, as defined here, differs somewhat from that used by Halmos [1].

and, when $g \in \mathfrak{D}(hh')$,

$$(hh')g = h(h'g).$$

Lemma 3.1.1. If $h, h' \in \mathfrak{G}(\Omega)$, then $hh' \in \mathfrak{G}(\Omega)$.

PROOF. Obviously, $\mathfrak{D}(h') - \mathfrak{D}(hh') = h'^{-1}(G - \mathfrak{D}(h))$, but $G - \mathfrak{D}(h)$ is a μ-null set, hence also a $\mu_{h'}$-null set, and therefore $\mathfrak{D}(h') - \mathfrak{D}(hh')$ is a μ-null set. The range $\mathfrak{R}(hh')$ of hh' is the set $h(\mathfrak{R}(h') \cap \mathfrak{D}(h))$ and

$$h(\mathfrak{R}(h') \cap \mathfrak{D}(h)) \supset \mathfrak{R}(h) - h(\mathfrak{D}(h) - \mathfrak{R}(h')).$$

Since $\mathfrak{D}(h) - \mathfrak{R}(h')$ is a μ-null set, and since μ, μ_h are equivalent, $h(\mathfrak{D}(h) - \mathfrak{R}(h'))$ is a μ-null set, therefore $G - \mathfrak{R}(hh')$ is a μ-null set. Obviously, hh' is one-to-one, and hh' is a measurable transformation, moreover, since $\mu_{hh'} = (\mu_{h'})_h$, it is clear that $\mu_{hh'}$ and μ_h are equivalent. Thus, we have $hh' \in \mathfrak{G}(\Omega)$.]

Let $h, h' \in \mathfrak{G}(\Omega)$, and suppose there exists a measurable set $\mathfrak{D} \subset \mathfrak{D}(h) \cap \mathfrak{D}(h')$, such that $G - \mathfrak{D}$ is a null set, and such that $hg = h'g$ for all $g \in \mathfrak{D}$. We then say that h and h' are equivalent, and write $h \equiv h'$; this is clearly an equivalence relation in the usual sense. Moreover, it is easily seen that, if $h_j \in \mathfrak{G}(\Omega)$, $j = 1, 2, 3, 4$, and $h_1 \equiv h_2$, $h_3 \equiv h_4$, then $h_1h_3 \equiv h_2h_4$. Therefore, if we identify equivalent elements of $\mathfrak{G}(\Omega)$, then, using Lemma 3.1.1, we easily deduce the following result.

Corollary 3.1.2. $\mathfrak{G}(\Omega)$ forms a group with respect to composition.

We call $\mathfrak{G}(\Omega)$ the *quasi-invariant transformation group* of Ω. Let $h \in \mathfrak{G}(\Omega)$, and suppose that both h and h^{-1} transform sets of finite measure into sets of finite measure; we then say that Ω is *strongly quasi-invariant* under h, or *strongly h-quasi-invariant.* If h and h^{-1} transform σ-finite sets into σ-finite sets, we say that Ω is *weakly quasi-invariant* under h, or *weakly h-quasi-invariant.* If $\mathfrak{G}$ is a subgroup of $\mathfrak{G}(\Omega)$, then we say that Ω is *quasi-invariant* under $\mathfrak{G}$, or *$\mathfrak{G}$-quasi-invariant*; if Ω is strongly (weakly) quasi-invariant under every $h \in \mathfrak{G}$, we say that Ω is *strongly* (*weakly*) *quasi-invariant* under $\mathfrak{G}$, or *strongly* (*weakly*) *$\mathfrak{G}$-quasi-invariant.* In particular, if $\mu(hB) = \mu(B)$ for all $h \in \mathfrak{G}$ and all $B \in \mathfrak{B}$ such that $B \subset \mathfrak{D}(h)$, then Ω is said to be *invariant* under $\mathfrak{G}$, or *$\mathfrak{G}$-invariant.*

In what follows, we shall be primarily interested in measures defined on groups. Let G be a group, $\mathfrak{G}$ a subgroup of G, and $(G, \mathfrak{B})$ a measurable space. If

$$hB = \{hg \mid g \in B\} \in \mathfrak{B}$$

for every $B \in \mathfrak{B}$, $h \in \mathfrak{G}$, then we say that the measurable space $(G, \mathfrak{B})$ is *left quasi-invariant* under $\mathfrak{G}$, or *left $\mathfrak{G}$-translation quasi-invariant.* If

μ is a measure on $(G, \mathfrak{B})$ such that $\mu(E) = 0$ and $\mu(hE) = 0$ are equivalent for all $B \in \mathfrak{B}$, $h \in \mathfrak{G}$, then we say that the measure space $(G, \mathfrak{B}, \mu)$ is *left quasi-invariant* under $\mathfrak{G}$, or *left $\mathfrak{G}$-translation quasi-invariant.*

Concisely stated, $(G, \mathfrak{B}, \mu)$ is (strongly, weakly) left quasi-invariant under the subgroup $\mathfrak{G}$, provided that every translation $g \to hg$, $g \in G$, $h \in \mathfrak{G}$, transforms measurable sets and null sets (sets of finite measure, σ-finite sets) into measurable sets and null sets (sets of finite measure, σ-finite sets), respectively.

In particular, if $\mu(hE) = \mu(E)$ for all $h \in \mathfrak{G}$ and $E \in \mathfrak{B}$, then $(G, \mathfrak{B}, \mu)$ is said to be *left invariant* under $\mathfrak{G}$, or *left-$\mathfrak{G}$-translation invariant.* For example, in every locally compact group G, there exists a so-called *Haar measure* which is left invariant under the entire group G (see Halmos [1]).

Let $(G, \mathfrak{B}, \mu)$ be a measure space which is quasi-invariant under the transformation group $\mathfrak{G}$, and let ν be a measure on $(G, \mathfrak{B})$ which is equivalent to μ; then $(G, \mathfrak{B}, \nu)$ is also quasi-invariant under $\mathfrak{G}$. Let f be a measurable function on $(G, \mathfrak{B})$ such that $0 < f(g) < \infty$ for all $g \in G$; if $(G, \mathfrak{B}, \mu)$ is weakly quasi-invariant under the transformation group $\mathfrak{G}$, and if ν is the measure on $(G, \mathfrak{B})$ defined by

$$\nu(B) = \int_B f(g)\, d\mu(g), \qquad B \in \mathfrak{B},$$

then $(G, \mathfrak{B}, \nu)$ is also weakly quasi-invariant under $\mathfrak{G}$. Of course, any measure space $\Omega = (G, \mathfrak{B}, \mu)$ is, trivially, invariant under the group $\mathfrak{G}$ which contains only the identity transformation of G.

Let G be a group, $\Omega = (G, \mathfrak{B}, \mu)$ a measure space, and $h \in G$; if the left translations $\tau_h\colon g \to hg$ and $\tau_{h^{-1}}$ are both measurable transformations of $(G, \mathfrak{B}, \mu)$, and if the measures μ_h (see (3.1.1)) and μ are equivalent, then we say that h is a *left quasi-invariant element* of Ω. Clearly, the totality of left quasi-invariant elements of Ω form a subgroup $\mathfrak{G}$ of G; we call $\mathfrak{G}$ the *maximal quasi-invariant left translation group* of Ω.

Naturally, all the above definitions are also applicable if the word "left" is replaced by "right" throughout, and of course, in the case of commutative groups, the words "left" and "right" may be omitted.

If $(G, \mathfrak{B}, \mu)$ is a σ-finite measure space which is quasi-invariant under the transformation group $\mathfrak{G}$, then there exists a finite measure ν on $(G, \mathfrak{B})$ which is equivalent to μ and is, therefore, also quasi-invariant under $\mathfrak{G}$. In fact, let B_k, $k = 1, 2,\ldots$, be a sequence of pairwise disjoint sets such that

$$B_k \in \mathfrak{B}, \qquad 0 < \mu(B_k) < \infty, \qquad G = \bigcup_{k=1}^{\infty} B_k\,.$$

Construct the measure ν as follows:

$$\nu(E) = \sum_{k=1}^{\infty} \frac{1}{2^k \mu(B_k)} \mu(E \cap B_k), \qquad E \in \mathfrak{B}.$$

It is easily seen that $\nu(G) = 1$, and that ν, μ are equivalent.

In the remainder of this book, the emphasis will be on localizable quasi-invariant measures, and, in particular, on finite quasi-invariant measures.

2° Measures Quasi-Invariant under the Entire Group G

Throughout this subsection, we shall assume, without further explicit mention, that G is a group, that $(G, \mathfrak{B})$ is a measurable space, and that the correspondence $(g, h) \to gh^{-1}$ is a measurable mapping from the product measurable space $(G \times G, \mathfrak{B} \times \mathfrak{B})$ to the measurable space $(G, \mathfrak{B})$.

Lemma 3.1.3. Let $(G, \mathfrak{B}, \mu_k)$, $k = 1, 2$, be finite measure spaces which are right and left quasi-invariant, respectively, under G. If μ_1 is not identically zero, then μ_2 is absolutely continuous with respect to μ_1.

PROOF. Form the product measure space $(G \times G, \mathfrak{B} \times \mathfrak{B}, \mu_1 \times \mu_2)$. For each $E \in \mathfrak{B} \times \mathfrak{B}$, let $\chi_E(x, y)$ denote the characteristic function of E. It is easily seen that, for any $E \in \mathfrak{B} \times \mathfrak{B}$,

$$f(x, y) = \chi_E(x, x^{-1}y)$$

is a measurable function on $(G \times G, \mathfrak{B} \times \mathfrak{B})$. We define a finite measure ν on $(G \times G, \mathfrak{B} \times \mathfrak{B})$, as follows:

$$\nu(E) = \int_G \int_G \chi_E(x, x^{-1}y)\, d\mu_1(x)\, d\mu_2(y). \tag{3.1.2}$$

We shall now prove that ν is absolutely continuous with respect to $\mu_1 \times \mu_2$. Take any $E \in \mathfrak{B} \times \mathfrak{B}$ such that $\mu_1 \times \mu_2(E) = 0$, and let $E^x = \{y \mid (x, y) \in E\}$. Then,

$$\int_G \mu_2(E^x)\, d\mu_1(x) = 0,$$

hence, there is a set $\Delta \in \mathfrak{B}$ such that $\mu_1(\Delta) = 0$ and $\mu_2(E^x) = 0$ whenever

$x \bar{\in} \Delta$. By the left quasi-invariance of μ_2, it follows that $\mu_2(xE^x) = 0$ whenever $x \bar{\in} \Delta$, hence, using Fubini's theorem, we obtain

$$\nu(E) = \int_G \Big(\int_G \chi_E(x, x^{-1}y)\, d\mu_2(y) \Big)\, d\mu_1(x)$$

$$= \int_G \mu_2(xE^x)\, d\mu_1(x) = 0.$$

Having proved that ν is absolutely continuous with respect to $\mu_1 \times \mu_2$, we know that there exists a nonnegative measurable function $g(x, y)$ on $(G \times G, \mathfrak{B} \times \mathfrak{B}, \mu_1 \times \mu_2)$, such that

$$\nu(E) = \iint \chi_E(x, y)\, g(x, y)\, d\mu_1(x)\, d\mu_2(x). \tag{3.1.3}$$

On the other hand, given any $A \in \mathfrak{B}$, since $(x, y) \to xy$ is a measurable mapping, we have

$$E = \{(x, y) \mid xy \in A\} \in \mathfrak{B} \times \mathfrak{B}.$$

Evaluating (3.1.2) for this set E, we obtain

$$\nu(E) = \iint \chi_A(y)\, d\mu_1(x)\, d\mu_2(y) = \mu_1(G)\, \mu_2(A),$$

but from (3.1.3), using Fubini's theorem, we get

$$\nu(E) = \int_G \Big(\int_{Ay^{-1}} g(x, y)\, d\mu_1(x) \Big)\, d\mu_2(y). \tag{3.1.4}$$

Consequently, if $\mu_1(A) = 0$, then, by the right quasi-invariance of μ_1, we have $\mu_1(Ay^{-1}) = 0$, whence, by (3.1.4), we get $\nu(E) = 0$, and so $\mu_2(A) = 0$. Thus, μ_2 is absolutely continuous with respect to μ_1.]

Corollary 3.1.4. Let $(G, \mathfrak{B}, \mu)$ be a finite measure space which is right G-translation quasi-invariant, and let μ_{-1} be the measure on $(G, \mathfrak{B})$ defined by $\mu_{-1}(E) = \mu(E^{-1})$, $E \in \mathfrak{B}$. Then μ and μ_{-1} are equivalent.

Proof. It is easily seen that the measure μ_{-1} is left G-translation quasi-invariant. Therefore, by Lemma 3.1.3, μ_{-1} is absolutely continuous with respect to μ, that is, if $E \in \mathfrak{B}$ and $\mu(E) = 0$, then $\mu(E^{-1}) = 0$. Hence, if $E \in \mathfrak{B}$ and $\mu_{-1}(E) = \mu(E^{-1}) = 0$, then $\mu(E) = \mu_{-1}(E^{-1}) = 0$, that is, μ is also absolutely continuous with respect to μ_{-1}. Thus, μ and μ_{-1} are equivalent.]

Theorem 3.1.5. Let $(G, \mathfrak{B}, \mu)$ be a σ-finite measure space which is left (right) quasi-invariant under G; then $(G, \mathfrak{B}, \mu)$ is also right (left) quasi-invariant under G. If $(G, \mathfrak{B}, \mu_k)$, $k = 1, 2$, are σ-finite nontrivial[3] measure spaces which are both (right or left) quasi-invariant under G, then μ_1 and μ_2 are equivalent.

PROOF. By virtue of the remarks in the preceding subsection, the σ-finite measures under consideration may be replaced by equivalent finite measures. Assume that $(G, \mathfrak{B}, \mu)$ is right G-translation quasi-invariant; then μ_{-1} is left G-translation quasi-invariant. But, by Lemma 3.1.4, μ and μ_{-1} are equivalent, hence μ is also left G-translation quasi-invariant. Conversely, assuming that $(G, \mathfrak{B}, \mu)$ is left G-translation quasi-invariant, then μ_{-1} is right G-translation quasi-invariant. Hence, by what has already been proved, μ_{-1} is also left G-translation quasi-invariant, therefore μ is right G-translation quasi-invariant.

Let $(G, \mathfrak{B}, \mu_k)$, $k = 1, 2$, be finite nontrivial measure spaces which are both G-translation quasi-invariant. By Lemma 3.1.3, μ_1 is absolutely continuous with respect to μ_2, and likewise μ_2 is absolutely continuous with respect to μ_1.]

Remark. Theorem 3.1.5 can be extended to pseudo-σ-finite measures (see Definition 1.1.9). For example, if $(G, \mathfrak{B}, \mu)$ is a pseudo-σ-finite measure space which is left G-translation quasi-invariant, then $(G, \mathfrak{B}, \mu)$ is also right G-translation quasi-invariant. In fact, let $E \in \mathfrak{B}$, $\mu(E) = 0$, $h \in G$; then there exists a subgroup G_1 of G, containing E and h, such that G_1 belongs to $\mathfrak{B}$ and is σ-finite.[4] Let $\mathfrak{B}_1$ and μ_1 denote the restrictions to G_1 of $\mathfrak{B}$ and μ, respectively. It is easily seen that $(G_1, \mathfrak{B}_1, \mu_1)$ is a left G_1-translation quasi-invariant σ-finite measure space. By Theorem 3.1.5, μ_1 is also right G_1-translation quasi-invariant. Consequently $\mu(E) = 0$ implies that $\mu(Eh) = 0$, that is, $(G, \mathfrak{B}, \mu)$ is right G-translation quasi-invariant.

Similarly, one can prove that if $(G, \mathfrak{B}, \mu_k)$, $k = 1, 2$, are pseudo-σ-finite measure spaces which are both G-translation quasi-invariant, then μ_1 and μ_2 are equivalent.[5]

Using the preceding remark and the results of §1.1, 6°, we immediately obtain the following corollary.

[3] That is, the measures are not identically zero.

[4] *Translator's note:* Such a subgroup certainly exists if the subgroup G_0 in Definition 1.1.9 happens to be normal in G, for in that case $G_1 = \bigcup_{n=-\infty}^{\infty} h^n G_0$ obviously satisfies the required conditions. If G_0 is not normal, then it is not clear how G_1 is to be constructed.

[5] *Translator's note:* This is easily proved if the same group G_0 (see Definition 1.1.9) works for both μ_1 and μ_2 (fortunately, this is the case in Corollary 3.1.6). Otherwise, the proof is not obvious.

Corollary 3.1.6. Let G be a locally compact group, and let $\mathfrak{B}$ be the σ-ring generated by the totality of compact subsets of G. Suppose that $(G, \mathfrak{B}, \mu)$ is a (left or right) G-translation quasi-invariant locally σ-finite[6] measure space. Then μ is equivalent to any Haar measure on G.

3° A Fundamental Property of Quasi-Invariant Measures

Definition 3.1.2. Let G be a topological space, and let $\mathfrak{G}$ be a family of continuous mappings of G into G. Suppose that $\mathfrak{G}$ also has a topology, and that the following condition is satisfied: given any $h_0 \in \mathfrak{G}$, any compact subset K of G, and any open set U containing h_0K, there exists a neighborhood V of h_0 such that $hK \subset U$ for every $h \in V$. Then, we say that the topology of $\mathfrak{G}$ is *admissible*.

We now consider a particular case which occurs frequently in the sequel.

Lemma 3.1.7. Let $\mathfrak{G}$ be a subgroup of a topological group G, and let each $h \in \mathfrak{G}$ be identified with the corresponding left translation $g \rightarrow hg$, $g \in G$. Then, the relative topology on $\mathfrak{G}$, induced by the topology of G, is admissible.

Proof. Let $h_0 \in \mathfrak{G}$, let K be any compact subset of G, and let U be any open set containing h_0K. For each $g \in h_0K$, choose neighborhoods U_g and V_g of the identity in G such that $U_g g \subset U$ and $V_g V_g \subset U_g$; then

$$h_0K \subset \bigcup_{g \in h_0K} V_g g \subset U.$$

Since h_0K is compact, there exist elements $g_1, \ldots, g_n \in h_0K$, such that

$$h_0K \subset \bigcup_{\nu=1}^{n} V_{g_\nu} g_\nu .$$

Obviously,

$$V = \bigcap_{\nu=1}^{n} V_{g_\nu}$$

is a neighborhood of the identity. We assert that $hK \subset U$ whenever h belongs to the neighborhood Vh_0 of h_0. In fact, for any $g \in hK$, $h_0h^{-1}g \in h_0K$, hence, there is an index ν such that $h_0h^{-1}g \in V_{g_\nu}g_\nu$, and so

$$g \in hh_0^{-1}V_{g_\nu}g_\nu \subset VV_{g_\nu}g_\nu \subset U_{g_\nu}g_\nu \subset U,$$

[6] *Translator's note:* The definition of a locally σ-finite measure (which the author has neglected to state explicitly) is an obvious modification of Definition 1.1.10, and it is clear that the proof of Lemma 1.1.11 goes through without any changes if the local finiteness hypothesis is replaced by local σ-finiteness.

which proves that $hK \subset U$. Thus, when $\mathfrak{G}$ is regarded as a group of transformations of G, the relative topology on $\mathfrak{G}$ is admissible.]

Lemma 3.1.8. Let G be a topological space, let $(G, \mathfrak{B}, \mu)$ be a regular measure space, and let $\mathfrak{G}$ be a topologized family of continuous $\mathfrak{B}$-measurable mappings of G into itself. If the topology of $\mathfrak{G}$ is admissable, then, for any fixed compact set $K \in \mathfrak{B}$, the value $\mu(hK)$, $h \in \mathfrak{G}$, is an upper semicontinuous[7] function on $\mathfrak{G}$.

PROOF. Choose any $h_0 \in \mathfrak{G}$. If $\mu(h_0K) = \infty$, then the function $\mu(hK)$ is obviously upper semicontinuous at h_0. Hence, suppose that $\mu(h_0K) < \infty$, and let ϵ be an arbitrary positive number. By the regularity of μ, there exists an open set $U \supset h_0K$ such that

$$\mu(U) < \mu(h_0K) + \epsilon. \tag{3.1.5}$$

Since the topology of $\mathfrak{G}$ is admissible, there exists a neighborhood V of h_0 such that $hK \subset U$ for all $h \in V$. Hence, since $hK \in \mathfrak{B}$, it follows from (3.1.5) that

$$\mu(hK) < \mu(h_0K) + \epsilon$$

for all $h \in V$. Consequently, the upper bound[7] $S(h_0, \mu(hK))$ of the function $\mu(hK)$ at h_0 satisfies

$$\mu(h_0K) \leqslant S(h_0, \mu(hK)) \leqslant S(\mu(hK), V) \leqslant \mu(h_0K) + \epsilon.$$

Letting $\epsilon \to 0$, we obtain $\mu(h_0K) = S(h_0 ; \mu(hK))$. Thus, the function $\mu(hK)$ is upper semicontinuous at the point $h = h_0$.]

As an immediate consequence of Lemmas 3.1.7 and 3.1.8, we obtain the following result.

Corollary 3.1.9. Let G be a topological group, let $(G, \mathfrak{B}, \mu)$ be a regular measure space, and let $\mathfrak{G}$ be a subgroup of G such that $hB \in \mathfrak{B}$ for every $h \in \mathfrak{G}$, $B \in \mathfrak{B}$. Then, for any fixed compact set $K \in \mathfrak{B}$, the function $\mu(hK)$, $h \in \mathfrak{G}$, is upper semicontinuous with respect to the relative topology of $\mathfrak{G}$.

In the following lemma, we establish a sort of continuity property of regular measures with respect to transformations; this, in turn, will be used to derive an important property of quasi-invariant measures.

Corollary 3.1.10. Let $(G, \mathfrak{B}, \mu)$ be a regular measure space, let $\mathfrak{G}$ be a topologized family of continuous $\mathfrak{B}$-measurable mappings of G

[7] For the definition of an upper semicontinuous function, along with the relevant notations and theorems, see Appendix I (§I.2).

into itself, and suppose that the topology of $\mathfrak{G}$ is admissable (see Definition 3.1.2). If the topological space $\mathfrak{G}$ is of the second category, then, for any fixed compact set $K \in \mathfrak{B}$, there exists a point $h_0 \in \mathfrak{G}$ such that the function $\mu(hK)$, $h \in \mathfrak{G}$, is continuous at h_0.

PROOF. By Lemma 3.1.8, the function $\mu(hK)$, $h \in \mathfrak{G}$, is upper semicontinuous on $\mathfrak{G}$. Therefore, since $\mu(hK)$ is nonnegative, and since $\mathfrak{G}$ is of the second category, it follows easily from Lemma I.2.2[8] of Appendix I that $\mathfrak{G}$ contains a continuity point h_0 of the function $\mu(hK)$.]

Applying Lemmas 3.1.7 and 3.1.10, we get the following result.

Corollary 3.1.11. Let G be a topological group, and let $\mathfrak{G}$ be a subgroup of G. Suppose that $\mathfrak{G}$ itself is a topological group of the second category with respect to a topology which is stronger than the topology induced by G. If $(G, \mathfrak{B})$ is a measurable space which is left (or right) quasi-invariant under $\mathfrak{G}$, and μ is a regular measure on $(G, \mathfrak{B})$, then, for any fixed compact set $K \in \mathfrak{B}$, there exists an $h_0 \in \mathfrak{G}$ such that the function $\mu(hK)$ [or $\mu(Kh)$], $h \in \mathfrak{G}$, is continuous at h_0.

Next, we use Lemma 3.1.10 to obtain a property of quasi-invariant measures which will be required in the ensuing discussion.

Lemma 3.1.12. Let $(G, \mathfrak{B}, \mu)$ be a regular measure space which is quasi-invariant under a group $\mathfrak{G}$ of continuous measurable transformations. Also, suppose that $\mathfrak{G}$ has an admissible topology which makes $\mathfrak{G}$ a topological group of the second category. Then, given any compact set $K \in \mathfrak{B}$, $\mu(K) > 0$, there exists a neighborhood V of the identity in $\mathfrak{G}$ such that

$$\mu(K \cap hK) > 0 \tag{3.1.6}$$

for every $h \in V$.[9]

PROOF. By Lemma 3.1.10, there exists an element $h_0 \in \mathfrak{G}$ such that the function $\mu(hK)$ is continuous at h_0. By quasi-invariance, we know that $\mu(h_0K) > 0$, and we may assume that[10] $\mu(h_0K) < \infty$. Since μ is regular, there is an open set $U \in \mathfrak{B}$ such that $U \supset h_0K$ and

$$\mu(U) < \tfrac{4}{3}\mu(h_0K). \tag{3.1.7}$$

[8] *Translator's note:* Lemma I.2.2 is valid only for finite-valued real functions. Therefore, in this lemma and the subsequent results which depend upon it, it seems necessary to assume that the measure of every compact set is finite, or some other suitable finiteness condition.

[9] *Translator's note:* The following simple counterexample shows that this conclusion is not valid unless one imposes some suitable finiteness condition on the measure μ. Let G be the additive group of reals with the usual topology, let $\mathfrak{B}$ be the totality of Borel sets in G, let $\mu(E) = \infty$ for all nonempty sets $E \in \mathfrak{B}$, let $\mathfrak{G} = G$, operating on G by translation, and let K be any singleton in G.

[10] *Translator's note:* The justification for this remark is obscure. See footnotes 8 and 9.

Since the topology of $\mathfrak{G}$ is admissible, there is a neighborhood V_{h_0} of h_0 such that $hK \subset U$ whenever $h \in V_{h_0}$. Moreover, V_{h_0} may be chosen sufficiently small so that

$$|\mu(hK) - \mu(h_0K)| < \tfrac{1}{3}\mu(h_0K) \tag{3.1.8}$$

for all $h \in V_{h_0}$. Since $hK \subset U$ and $h_0K \subset U$, we have

$$\mu(hK \cap h_0K) \geqslant \mu(hK) + \mu(h_0K) - \mu(U),$$

whence, using (3.1.7) and (3.1.8), we see that

$$\mu(hK \cap h_0K) \geqslant \tfrac{1}{3}\mu(h_0K) > 0 \tag{3.1.9}$$

for all $h \in V_{h_0}$. Again, using the quasi-invariance of μ, it follows from (3.1.9) that, if $g = h_0^{-1}h \in h_0^{-1}V_{h_0}$, then

$$\mu(gK \cap K) > 0.$$

Thus, $V = h_0^{-1}V_{h_0}$ is a neighborhood of the identity in $\mathfrak{G}$ which has the required property.]

Corollary 3.1.12′. Let G be a topological group, let $\mathfrak{G}$ be a subgroup of G, and suppose that $\mathfrak{G}$ itself is a topological group of the second category with respect to a topology which is stronger than that induced by G. If $(G, \mathfrak{B}, \mu)$ is a regular measure space which is left (or right) quasi-invariant under $\mathfrak{G}$, then, for any compact set $K \in \mathfrak{B}$, $\mu(K) > 0$, there exists a neighborhood V of the identity in $\mathfrak{G}$ such that $\mu(K \cap hK) > 0$ (or $\mu(K \cap Kh) > 0$) for all $h \in V$.

Using the above results, one may derive a simple necessary condition for the existence of quasi-invariant measures, as stated in the following theorem.

Theorem 3.1.13. Let G be a topological group, let $\mathfrak{G}$ be a subgroup of G, and suppose that $\mathfrak{G}$ itself is a topological group of the second category with respect to a topology which is stronger than the relative topology induced by G. If there exists a nontrivial[11] regular measure space $(G, \mathfrak{B}, \mu)$ which is quasi-invariant under $\mathfrak{G}$, then there is a neighborhood of the identity in $\mathfrak{G}$ which is contained in a compact subset of G.

PROOF. Since μ is regular, there exists a compact set $K \in \mathfrak{B}$ such that $\mu(K) > 0$. Moreover, by Corollary 3.1.12′, there is a neighborhood V of the identity in $\mathfrak{G}$ such that $K \cap hK \neq 0$ for all $h \in V$, that is,

[11] That is, μ is not identically zero.

$V \subset KK^{-1}$. We assert that KK^{-1} is a compact subset of G. In fact, by the Tychonoff theorem, $K \times K$ is a compact subset of $G \times G$; moreover, since G is a topological group, the correspondence

$$(x, y) \to xy^{-1}$$

is a continuous mapping from $G \times G$ to G. Hence, KK^{-1}, being the image of $K \times K$ under this mapping, is compact.]

Corollary 3.1.14. Let G be a topological group of the second category. Then, the local compactness of G is a necessary and sufficient condition for the existence of a regular measure space $(G, \mathfrak{B}, \mu)$ which is left quasi-invariant under G.

PROOF. The necessity follows at once from Theorem 3.1.13. Conversely, if G is locally compact, let $\mathfrak{B}$ be the σ-ring generated by the totality of compact subsets of G, and let μ be a left-invariant Haar measure on G. Then the measure space $(G, \mathfrak{B}, \mu)$ has the required properties.]

4° Functions on Topological Groups with Quasi-Invariant Measures

Throughout the present subsection, we shall assume that $(G, \mathfrak{B}, \mu)$ is a regular measure space which is quasi-invariant under a group of measurable continuous transformations $\mathfrak{G}$, and that $\mathfrak{G}$ is a topological group of the second category with respect to a certain admissible topology (see Definition 3.1.2); these assumptions will not be explicitly repeated.

Let $p(g)$ be a $\mathfrak{B}$-measurable real-valued function on G, and, for each $h \in \mathfrak{G}$, define

$$p^*(h) = \operatorname*{essential\ infimum}_{g \in G}(p(g) + p(h^{-1}g)). \tag{3.1.10}$$

Here, the term "essential infimum" means that one evaluates the infimum of the function on the complement of each μ-null set, and then takes the supremum of all these values.

Lemma 3.1.15. Let $p(x)$ be a nonnegative $\mathfrak{B}$-measurable function on G, and suppose that $p(x)$ is finite on some set $A \in \mathfrak{B}$, $0 < \mu(A)$. Then, there exists a neighborhood V of the identity in $\mathfrak{G}$ such that $p^*(h)$ is bounded on V.

PROOF. Since the set $\{x \mid p(x) < \infty\}$ contains A, there is a positive number $a < \infty$, such that the set $\{x \mid p(x) \leqslant a, x \in A\}$ has positive μ-measure, and since μ is regular, there is a compact set $K \in \mathfrak{B}$ such that $K \subset \{x \mid p(x) \leqslant a, x \in A\}$ and $\mu(K) > 0$. By Corollary 3.1.12′, there is a neighborhood V of the identity in $\mathfrak{G}$ such that $\mu(K \cap hK) > 0$ for all $h \in V$. Now, if $g \in K \cap hK$, then $p(g) \leqslant a$, $p(h^{-1}g) \leqslant a$. But when $\mu(K \cap hK) > 0$, the complement of every μ-null set E contains a point $g \in K \cap hK$, hence

$$\inf_{g \in G-E} (p(g) + p(h^{-1}g)) \leqslant 2a.$$

Consequently, $p^*(h) \leqslant 2a$ for all $h \in V$.]

With a view to the applications of Lemma 3.1.15, it is convenient to introduce the notion of a quasiconvex pair of functions, as follows.

Definition 3.1.3. Let $(G, \mathfrak{B}, \mu)$ be a measure space which is quasi-invariant under a group of transformations $\mathfrak{G}$. Let $p(g)$ be a $\mathfrak{B}$-measurable extended[12] real-valued function on G, let $\tilde{p}(h)$ be a real-valued function on $\mathfrak{G}$, and suppose that the following conditions are satisfied: (i) $0 \leqslant p(g) \leqslant \infty$; (ii) given any $h \in \mathfrak{G}$, the inequality

$$\tilde{p}(h) \leqslant p(g) + p(h^{-1}g)$$

holds for almost all $g \in G$; (iii) $\tilde{p}(h)$ is a convex function on $\mathfrak{G}$ (see §I.1 of Appendix I). Then, we say that $(p, \tilde{p})$ is a *quasiconvex pair of functions* over $((G, \mathfrak{B}, \mu), \mathfrak{G})$.

In particular, if p is a measurable convex function on G, then the conditions of Definition 3.1.3 are satisfied if one chooses $\tilde{p}(h) = p(h)$.

Corollary 3.1.16. Let $(p, \tilde{p})$ be a quasiconvex pair of functions over $((G, \mathfrak{B}, \mu), \mathfrak{G})$, and suppose that $p(g)$ is finite on some set $A \in \mathfrak{B}$, with $\mu(A) > 0$. Then $\tilde{p}(h)$ is locally bounded (i.e., every point in $\mathfrak{G}$ has a neighborhood upon which $\tilde{p}$ is bounded).

Proof. By Lemma 3.1.15, there exists a neighborhood V of the identity in $\mathfrak{G}$ such that $p^*(h) \leqslant 2a$, and hence $\tilde{p}(h) \leqslant 2a$, for all $h \in V$. Now, let h_0 be any element of $\mathfrak{G}$; if $h \in h_0 V^{-1}$, that is, if $h = h_0 h_1^{-1}$, $h_1 \in V$, then, by the convexity of $\tilde{p}$, we have

$$\tilde{p}(h) \leqslant \tilde{p}(h_0) + \tilde{p}(h_1) \leqslant \tilde{p}(h_0) + 2a.$$

Thus, $\tilde{p}$ is bounded on the neighborhood $h_0 V^{-1}$ of h_0.]

We may now sharpen Theorem 3.1.15, as follows.

Theorem 3.1.17. Suppose that the topological group $\mathfrak{G}$ satisfies the first axiom of countability, and let $A \in \mathfrak{B}$, with $\mu(A) > 0$. Then, there

[12] That is, infinite values are also allowed.

exists a neighborhood V of the identity in $\mathfrak{G}$, and a positive number c, such that the inequality

$$\sup_{h \in V} p^*(h) \leqslant c \int_A p(g)\, d\mu(g) \tag{3.1.11}$$

holds for every nonnegative $\mathfrak{B}$-measurable function p on $\mathfrak{G}$ [see (3.1.10) for the definition of p^*].

PROOF. If $\int_A p(g)\, d\mu(g) = 0$, then an argument similar to that used in the proof of Lemma 3.1.15 shows that $p^*(h) = 0$ on some neighborhood V of the identity in $\mathfrak{G}$. Hence, we may restrict our considerations to functions p such that $\int_A p(g)\, d\mu(g) > 0$.

Let $\{p_n\}$ be any sequence of nonnegative $\mathfrak{B}$-measurable functions on G such that

$$0 < a_n = \int_A p_n(g)\, d\mu(g) < \infty. \tag{3.1.12}$$

Let $\mathscr{L}$ denote the totality of real number sequences $\lambda = \{\lambda_1, \lambda_2, \ldots, \lambda_n, \ldots\}$ satisfying the condition

$$\| \lambda \| = \sum_{n=1}^{\infty} | \lambda_n |\, a_n < \infty. \tag{3.1.13}$$

$\mathscr{L}$ forms a Banach space with respect to the usual linear operations and the norm $\| \cdot \|$ defined by (3.1.13). For each $\lambda \in \mathscr{L}$, define a function $p(g; \lambda)$ on G, as follows:

$$p(g; \lambda) = \sum_{n=1}^{\infty} | \lambda_n |\, p_n(g).$$

Obviously, $p(g; \lambda)$ is nonnegative and $\mathfrak{B}$-measurable. Also, consider the function

$$F(h; \lambda) = \sum_{n=1}^{\infty} | \lambda_n |\, p_n{}^*(h), \qquad h \in \mathfrak{G}.$$

Obviously, $p^*(h; \lambda) \geqslant F(h; \lambda)$ for all $h \in \mathfrak{G}$. Now, since $\lambda \in \mathscr{L}$, it follows from (3.1.12) and the Levi lemma that

$$p(g; \lambda) < \infty \tag{3.1.14}$$

for almost all g in A. Thus, if we let A_λ denote the totality of elements $g \in A$ which satisfy (3.1.14), we have $\mu(A_\lambda) = \mu(A) > 0$. Hence, by

Lemma 3.1.15, the function $p^*(h; \lambda)$ is bounded on some neighborhood of the identity in $\mathfrak{G}$, therefore $F(h; \lambda)$ is also bounded on this neighborhood. Let $\{V_m\}$, $m = 1, 2,\ldots$, be a countable base for the neighborhoods of the identity in $\mathfrak{G}$; we may assume that $V_1 \supset V_2 \supset \cdots$. Thus, for every $\lambda \in \mathscr{L}$, there exists an integer m (possibly depending upon λ) such that $F(h; \lambda)$ is bounded on V_m. Notice that, for each fixed h, $F(h; \lambda)$ is the supremum of a sequence of continuous convex functionals on $\mathscr{L}$, namely, $\sum_{n=1}^{k} p_n^*(h)|\lambda_n|$, $k = 1, 2,\ldots$. Therefore, $F(h; \lambda)$ is a lower semicontinuous convex functional on $\mathscr{L}$, and hence

$$q_m(\lambda) = \sup_{h \in V_m} F(h; \lambda)$$

is also a lower semicontinuous convex functional on $\mathscr{L}$, moreover, for every λ, there is an m such that $q_m(\lambda) < \infty$. Furthermore, since $V_1 \supset V_2 \supset \cdots$, we have

$$q_1(\lambda) \geqslant q_2(\lambda) \geqslant \cdots \geqslant q_m(\lambda) \geqslant \cdots.$$

It follows by Theorem I.2.4 of Appendix I that there exists an integer m and a positive number a such that

$$q_m(\lambda) \leqslant a \, \| \lambda \| \tag{3.1.15}$$

for all λ in $\mathscr{L}$. From this, we deduce that

$$\sup_{h \in V_m} p_n^*(h) \leqslant a \int_A p_n(g) \, d\mu(g), \qquad n = 1, 2,\ldots. \tag{3.1.16}$$

In fact, one need only let λ in (3.1.15) be the sequence with nth term $\lambda_n = 1$ and all other terms zero.

Now, if the theorem were false, then, for every positive integer n, there would exist a nonnegative $\mathfrak{B}$-measurable function p_n on G such that

$$0 < \int_A p_n(g) \, d\mu(g) < \infty$$

and

$$\sup_{h \in V_n} p_n^*(h) > n \int_A p_n(g) \, d\mu(g),$$

which contradicts (3.1.16).]

Corollary 3.1.18. Suppose that G satisfies the first axiom of countability, let A be any set in $\mathfrak{B}$ such that $\mu(A) > 0$, and let h_0 be any element of $\mathfrak{G}$ such that[13] $\mu(A \cap h_0 A) > 0$. Then, there exists a neighborhood V of h_0 in $\mathfrak{G}$, and a positive number c, such that

$$\sup_{h \in V} \tilde{p}(h) \leqslant c \int_A p(g)\, d\mu(g) \tag{3.1.17}$$

for every quasiconvex pair of functions $(p, \tilde{p})$ over $((G, \mathfrak{B}, \mu), \mathfrak{G})$ satisfying the condition[13] $\int_A p(g)\, d\mu(g) > 0$.

Proof. We proceed as in the proof of Theorem 3.1.17, but in place of $F(h; \lambda)$ we consider the function

$$\tilde{p}(h; \lambda) = \sum_{n=1}^{\infty} |\lambda_n| \tilde{p}_n(h).$$

It is easily checked that the pair of functions $p(g; \lambda)$, $\tilde{p}(h; \lambda)$ has all the properties specified by Definition 3.1.3, except possibly that $\tilde{p}(h; \lambda)$ may not be everywhere finite; however, using (3.1.14), together with the inequality

$$\tilde{p}(h; \lambda) \leqslant p(g; \lambda) + p(h^{-1}g; \lambda)$$

and the condition $\mu(A \cap h_0 A) > 0$, it is easily shown that $\tilde{p}(h_0; \lambda) < \infty$, hence the proof of Corollary 3.1.16 shows that $\tilde{p}(h; \lambda)$ is bounded on some neighborhood V of h_0. Then, using this result instead of Corollary 3.1.15 in the argument following formula (3.1.14), the desired result follows without difficulty.]

The results obtained in the present subsection will find application in §3.2 and §4.2.

5° *k*-Pseudometrics on Transformation Groups

In this subsection, we restrict our considerations to finite measure spaces.

Definition 3.1.4. Let $\Omega = (G, \mathfrak{B}, \mu)$ be a finite measure space, and $\mathfrak{G}$ a group of measurable transformations[14] on Ω. Form the convex function

$$M_1(h) = d(\mu_h, \mu), \qquad h \in \mathfrak{G}, \tag{3.1.18}$$

[13] *Translator's note:* These two requirements were omitted in the original statement; however, it is easy to construct examples showing that the conclusion may be false in the absence of such conditions.

[14] Here, we assume that the transformations h in $\mathfrak{G}$ are all one-to-one mappings of the entire set G onto itself.

where d denotes the Kakutani distance (see §1.4) and μ_h is defined by (3.1.1). We call $M_1(h)$ the *k-pseudonorm* on $\mathfrak{G}$ defined by Ω. The corresponding left-invariant pseudometric,

$$d_1(h_1, h_2) = M_1(h_2^{-1}h_1), \qquad h_1, h_2 \in \mathfrak{G}, \tag{3.1.19}$$

will be called the *k-pseudometric* on $\mathfrak{G}$, and the topology on $\mathfrak{G}$ induced by the k-pseudometric (see Definition I.1.1) will be called the *k-topology*.

To verify that M_1 is indeed a convex function, pick any two elements $h, k \in \mathfrak{G}$, choose a finite measure ν on $(G, \mathfrak{B})$ such that the four measures $\mu, \mu_h, \mu_k, \mu_{kh^{-1}}$ are all absolutely continuous with respect to ν, and let $\|\cdot\|_2$ denote the norm in $L^2(G, \mathfrak{B}, \nu)$. Then, using the Minkowski inequality and the change of variable $g_1 = hk^{-1}g$, we have

$$\begin{aligned} M_1(kh^{-1}) &= \left\| \left(\frac{d\mu(hk^{-1}g)}{d\nu(g)}\right)^{1/2} - \left(\frac{d\mu(g)}{d\nu(g)}\right)^{1/2} \right\|_2 \\ &\leqslant \left\| \left(\frac{d\mu(hk^{-1}g)}{d\nu(g)}\right)^{1/2} - \left(\frac{d\mu(k^{-1}g)}{d\nu(g)}\right)^{1/2} \right\|_2 + \left\| \left(\frac{d\mu(k^{-1}g)}{d\nu(g)}\right)^{1/2} - \left(\frac{d\mu(g)}{d\nu(g)}\right)^{1/2} \right\|_2 \\ &= \left\| \left(\frac{d\mu(g_1)}{d\nu(g_1)}\right)^{1/2} - \left(\frac{d\mu(h^{-1}g_1)}{d\nu(g_1)}\right)^{1/2} \right\|_2 + \left\| \left(\frac{d\mu(k^{-1}g)}{d\nu(g)}\right)^{1/2} - \left(\frac{d\mu(g)}{d\nu(g)}\right)^{1/2} \right\|_2 \\ &= M_1(h) + M_1(k). \end{aligned} \tag{3.1.20}$$

It is easy to calculate that the pseudometric d_1 defined by (3.1.19) may also be expressed as

$$d_1(h_1, h_2) = d(\mu_{h_1}, \mu_{h_2}), \tag{3.1.21}$$

where d again denotes the Kakutani distance.

Lemma 3.1.19. Let $\Omega = (G, \mathfrak{B}, \mu)$ be a finite measure space, and $\mathfrak{G}$ a group of measurable transformations on Ω. Then, for each fixed $E \in \mathfrak{B}$, the function $\mu(h^{-1}E)$, $h \in \mathfrak{G}$, is continuous with respect to the k-topology on $\mathfrak{G}$.

PROOF. For any $h_1, h_2 \in \mathfrak{G}$, applying the Schwarz inequality, we get

$$\int_E (d\mu_{h_1}(g))^{1/2} (d\mu_{h_2}(g))^{1/2} \leqslant (\mu(h_1^{-1}E)\, \mu(h_2^{-1}E))^{1/2},$$

whence, using (3.1.21), we obtain

$$\begin{aligned} |(\mu(h_1^{-1}E))^{1/2} - (\mu(h_2^{-1}E))^{1/2}| &\leqslant \left(\int_E ((d\mu_{h_1}(g))^{1/2} - (d\mu_{h_2}(g))^{1/2})^2\right)^{1/2} \\ &= d_1(h_1, h_2). \quad] \end{aligned}$$

Lemma 3.1.20. Let $\Omega = (G, \mathfrak{B}, \mu)$ be a finite regular measure space, let $\mathfrak{G}$ be a group of continuous measurable transformations on Ω, and suppose that $\mathfrak{G}$ has an admissible topology $\mathfrak{T}$ (see Definition 3.1.2). Then, the convex function $M_1(h)$, $h \in \mathfrak{G}$, is lower semicontinuous on $(\mathfrak{G}, \mathfrak{T})$.

PROOF. Let $h \in \mathfrak{G}$, and let $\mathfrak{F}_0$ denote the totality of countable partitions of G (relative to μ, $\mu_{h^{-1}}$) consisting of compact sets in $\mathfrak{B}$; then, by Lemma 1.4.2,

$$\rho(\mu_{h^{-1}}, \mu) = \inf_{\{E_k\} \in \mathfrak{F}_0} \sum_k (\mu(hE_k)\, \mu(E_k))^{1/2}.$$

Hence, using (3.1.18) and (1.4.4), we obtain

$$\begin{aligned} M_1(h)^2 &= 2(\mu(G) - \rho(\mu_{h^{-1}}, \mu)) \\ &= 2\left(\mu(G) + \sup_{\{E_k\} \in \mathfrak{F}_0} \sum_k (-(\mu(hE_k)\, \mu(E_k))^{1/2})\right). \end{aligned} \tag{3.1.22}$$

In view of (3.1.22), it suffices to prove that, for any partition $\{E_k\}$, the quantity $\sum_k (\mu(hE_k)\, \mu(E_k))^{1/2}$ is an upper semicontinuous function on $(\mathfrak{G}, \mathfrak{T})$.

By the Schwarz inequality,

$$\sum_{k=n}^{\infty} (\mu(hE_k)\, \mu(E_k))^{1/2} \leqslant \left(\sum_{k=n}^{\infty} \mu(hE_k)\right)^{1/2} \left(\sum_{k=n}^{\infty} \mu(E_k)\right)^{1/2},$$

hence,

$$\sum_{k=n}^{\infty} (\mu(hE_k)\, \mu(E_k))^{1/2} \leqslant (\mu(G))^{1/2} \left(\sum_{k=n}^{\infty} \mu(E_k)\right)^{1/2}$$

for every n. Now, since

$$\sum_{k=1}^{\infty} \mu(E_k) = \mu(G) < \infty,$$

it follows that

$$\sum_{k=n}^{\infty} \mu(E_k) \to 0$$

as $n \to \infty$. Hence, given any $\epsilon > 0$, there exists an integer N such that

$$(\mu(G))^{1/2} \left(\sum_{k=n}^{\infty} \mu(E_k)\right)^{1/2} < \frac{\epsilon}{2}$$

for every $n > N$. On the other hand, by virtue of Lemma 3.1.8, we know that $\sum_{k=1}^{N} \mu(hE_k)\, \mu(E_k))^{1/2}$ is an upper semicontinuous function on

$(\mathfrak{G}, \mathfrak{T})$. Therefore, given any $h_0 \in \mathfrak{G}$, there exists a neighborhood V of h_0 such that

$$\sum_{k=1}^{N} (\mu(hE_k)\,\mu(E_k))^{1/2} < \frac{\epsilon}{2} + \sum_{k=1}^{N} (\mu(h_0E_k)\,\mu(E_k))^{1/2}$$

for all $h \in V$. Thus, when $h \in V$,

$$\sum_{k=1}^{\infty} (\mu(hE_k)\,\mu(E_k))^{1/2} < \frac{\epsilon}{2} + \sum_{k=1}^{\infty} (\mu(h_0E_k)\,\mu(E_k))^{1/2} + (\mu(G))^{1/2}\Big(\sum_{k=N+1}^{\infty} \mu(E_k)\Big)^{1/2}$$

$$< \epsilon + \sum_{k=1}^{\infty} (\mu(h_0E_k)\,\mu(E_k))^{1/2}. \quad]$$

In view of Lemma 3.1.7, we immediately deduce the following result.

Corollary 3.1.21. Let G be a topological group, let $(G, \mathfrak{B}, \mu)$ be a finite regular measure space, let $\mathfrak{G}$ be a subgroup of G, and let $\mathfrak{T}$ denote the relative topology on $\mathfrak{G}$, as a subset of G. If $\mathfrak{G}$ is regarded as operating on G by left translation, and if $hE \in \mathfrak{B}$ for all $E \in \mathfrak{B}$, $h \in \mathfrak{G}$, then the convex function $M_1(h)$, $h \in \mathfrak{G}$, is lower semicontinuous on $(\mathfrak{G}, \mathfrak{T})$.

Lemma 3.1.22. Let G be a topological group, $(G, \mathfrak{B}, \mu)$ a finite regular measure space, and $\mathfrak{G}$ a subgroup of G such that $hE \in \mathfrak{B}$ for all $E \in \mathfrak{B}$, $h \in \mathfrak{G}$. Then, given any $h_0 \in \mathfrak{G}$, there exists a neighborhood V of h_0, in the k-topology, and a compact subset K of G, such that $V \subset K$.

PROOF. Since μ is a finite regular measure, there is a compact set $E \in \mathfrak{B}$ such that

$$\mu(h_0^{-1}E) > \tfrac{2}{3}\mu(G).$$

By Lemma 3.1.19, there is a neighborhood V of h_0, in the k-topology, such that

$$|\,\mu(h^{-1}E) - \mu(h_0^{-1}E)| < \tfrac{1}{3}\mu(G)$$

for all $h \in V$. Therefore, if $h \in V$, then

$$\mu(h^{-1}E) > \tfrac{1}{3}\mu(G),$$

hence, $\mu(h^{-1}E \cap h_0^{-1}E) > 0$, consequently, there exists an element $g \in h^{-1}E \cap h_0^{-1}E$, and so

$$h \in Eg^{-1} \subset EE^{-1}h_0\,.$$

Thus, writing $K = EE^{-1}h_0$, we have

$$V \subset K.$$

But since $(x, y) \to xy^{-1}$ is a continuous mapping from $G \times G$ to G, the set EE^{-1} is compact, hence K is compact.]

Corollary 3.1.23.[15] Let G be a topological group, and let $(G, \mathfrak{B}, \mu)$ be a finite regular measure space which is left invariant under G. Then G is compact.

PROOF. Since μ is left invariant, $d(h_1, h_2) = 0$ for all $h_1, h_2 \in G$. Consequently, the k-topology on G has but a single nonempty open set, namely, G itself. Hence, by Lemma 3.1.22, G is contained in a compact subset of G, that is, G is compact.]

Note that the condition "invariant" in the hypothesis of Corollary 3.1.23 cannot be weakened to "quasi-invariant." For example, consider the additive group of real numbers G, with the Euclidean topology; every finite Borel measure on G which is equivalent to Lebesgue measure is also regular and quasi-invariant, and yet G is not compact.

Theorem 3.1.24. Let G be a topological group, let $(G, \mathfrak{B}, \mu)$ be a finite regular measure space, and suppose that $gE \in \mathfrak{B}$ for all $E \in \mathfrak{B}$, $g \in G$. If G is regarded as a group of left translations, then G is complete relative to the k-pseudometric $d_1(h_1, h_2)$.

PROOF. Let $\{h_n\} \subset G$ be a sequence which is fundamental relative to the pseudometric d_1. By Lemma 3.1.22, there exists a compact subset K of G such that K contains a set of the form $\{h \mid M_1(h) < \epsilon\}$, where ϵ is some positive number. By assumption, there is an integer N such that $M_1(h_N^{-1}h_n) < \epsilon$ when $n \geqslant N$. Hence, the elements h_n, $n = N, N+1, \ldots$, are contained in the compact set $h_N K$, and it follows that there exists an element $h_0 \in G$ such that every neighborhood U of h_0 contains a subsequence of $\{h_n\}$.

Now, for any fixed m, by Corollary 3.1.21, given any positive number δ, there exists a neighborhood V of $h_m^{-1}h_0$ such that

$$M_1(h_m^{-1}h_0) < M_1(g) + \delta$$

whenever $g \in V$. If we choose $g = h_m^{-1}h$, where h is any element of the subsequence of $\{h_n\}$ which is contained in the neighborhood $h_m V$ of h_0, we deduce that

$$M_1(h_m^{-1}h_0) \leqslant \overline{\lim_{n\to\infty}}\, M_1(h_m^{-1}h_n) + \delta.$$

[15] This theorem is, of course, well known, and may also be established by a direct proof.

But since $\{h_n\}$ is fundamental relative to d_1, there is an N such that $m > N$ implies

$$\overline{\lim_{n\to\infty}} M_1(h_m^{-1}h_n) < \delta.$$

Therefore, $d_1(h_m, h_0) < 2\delta$ whenever $m > N$. Thus, we have proved that $\{h_m\}$ converges to h_0 relative to the pseudometric d_1.]

Let $(G, \mathfrak{B}, \mu)$ be a finite measure space and $\mathfrak{G}$ a group of measurable transformations of $(G, \mathfrak{B}, \mu)$. Form the pseudometric

$$d_0(h_1, h_2) = \tfrac{1}{2}(d_1(h_1, h_2) + d_1(h_1^{-1}, h_2^{-1}))$$

on $\mathfrak{G}$. Notice that, if $\mathfrak{G}$ is commutative, then

$$d_1(h_1, h_2) = M_1(h_2^{-1}h_1) = M_1(h_1h_2^{-1}) = d_1(h_1^{-1}, h_2^{-1}),$$

and hence

$$d_0(h_1, h_2) = d_1(h_1, h_2). \tag{3.1.23}$$

Lemma 3.1.25. Let $\Omega = (G, \mathfrak{B}, \mu)$ be a finite measure space and $\mathfrak{G}$ a group of measurable transformations of Ω. Let T denote the totality of transformations $h \in \mathfrak{G}$ such that Ω is quasi-invariant under h. Then T is a closed subgroup of $\mathfrak{G}$ relative to the pseudometric d_0.

PROOF. It is easily seen that T is a subgroup of $\mathfrak{G}$. Thus, it remains to prove that T is closed. Suppose that $\{h_n\} \subset T$, $h_0 \in \mathfrak{G}$, $d_0(h_n, h_0) \to 0$, then

$$d_1(h_n, h_0) \to 0, \qquad d_1(h_n^{-1}, h_0^{-1}) \to 0.$$

For any $E \in \mathfrak{B}$, we know by Lemma 3.1.19 that

$$\lim_{n\to\infty} \mu(h_nE) = \mu(h_0E) \tag{3.1.24}$$

and

$$\lim_{n\to\infty} \mu(h_n^{-1}E) = \mu(h_0^{-1}E). \tag{3.1.25}$$

We shall now show that Ω is quasi-invariant under h_0. If $\mu(E) = 0$, then, since $h_n \in T$, $\mu(h_nE) = 0$, hence by (3.1.24) we have $\mu(h_0E) = 0$. Conversely, if $\mu(h_0E) = 0$, then

$$\mu(h_n^{-1}(h_0E)) = 0.$$

Hence, substituting h_0E for E in (3.1.25), we get $\mu(E) = 0$. Thus, Ω is quasi-invariant under h_0, that is, $h_0 \in T$.]

Theorem 3.1.26. Let G be a topological group, let $(G, \mathfrak{B}, \mu)$ be a finite regular measure space, and suppose that $gE \in \mathfrak{B}$ for all $E \in \mathfrak{B}$, $g \in G$. Then, the maximal quasi-invariant left translation group T of $(G, \mathfrak{B}, \mu)$ is complete relative to the pseudometric d_0.

PROOF. Let $\{h_n\} \subset G$ be a fundamental sequence relative to d_0. Then, $\{h_n\}$ is fundamental relative to both of the pseudometrics $d_1(h_1, h_2)$, $h_1, h_2 \in G$, and $d_1(h_1^{-1}, h_2^{-1})$, $h_1, h_2 \in G$. In the proof of Theorem 3.1.24, it is shown that $\{h_n\}$ converges to a certain element h_0, relative to the pseudometric $d_1(h_1, h_2)$, $h_1, h_2 \in G$, and a trivial modification of the proof shows that $\{h_n\}$ also converges to the same element h_0, relative to the pseudometric $d_1(h_1^{-1}, h_2^{-1})$, $h_1, h_2 \in G$. Hence, $\{h_n\}$ also converges to h_0 relative to d_0. Thus, G is complete relative to d_0. By Lemma 3.1.25, T is a closed subset of G relative to d_0, therefore T itself is complete relative to d_0.]

6° μ-Topologies on Transformation Groups

Throughout the present section, we shall assume that $\Omega = (G, \mathfrak{B}, \mu)$ is a localizable measure space which is quasi-invariant under the transformation group $\mathfrak{G}$. We let $L^2(\Omega)$ [or, sometimes, $L^2(\mu)$] denote the totality of $\mathfrak{B}$-measurable complex-valued functions on G which are quadratically integrable with respect to μ. If one identifies functions which coincide almost everywhere, then $L^2(\Omega)$ forms a Hilbert space relative to the usual linear operations and the inner product

$$(\varphi, \psi) = \int_G \varphi(g) \overline{\psi(g)} \, d\mu, \qquad \varphi, \psi \in L^2(\Omega).$$

For each $h \in \mathfrak{G}$, we form the measure μ_h on $(G, \mathfrak{B})$:

$$\mu_h(E) = \mu(h^{-1}E), \qquad E \in \mathfrak{B}.$$

Since Ω is quasi-invariant under $\mathfrak{G}$, the measures μ_h and μ are equivalent, and it is easily verified that μ_h is also localizable. Hence, by Corollary 2.4.18, there exists a Radon–Nikodym derivative $d\mu_h/d\mu$ such that $0 < d\mu_h/d\mu < \infty$. Define an operator $U(h)$ as follows: if $\varphi \in L^2(\Omega)$, then

$$(U(h)\varphi)(g) = \varphi(h^{-1}g) \left(\frac{d\mu_h(g)}{d\mu(g)}\right)^{1/2}.$$

Clearly, $U(h)\varphi$ is a measurable function on Ω, moreover,

$$\int_G | U(h)\varphi |^2 \, d\mu(g) = \int_G | \varphi(h^{-1}g)|^2 \, d\mu_h(g) = \int_G | \varphi(g)|^2 \, d\mu(g),$$

thus, $U(h)$ is an isometric linear operator from $L^2(\Omega)$ into $L^2(\Omega)$. Furthermore, by Theorem 2.4.17,

$$\frac{d\mu_{h_1h_2}(g)}{d\mu(g)} = \frac{d\mu_{h_1h_2}(g)}{d\mu_{h_1}(g)} \frac{d\mu_{h_2}(g)}{d\mu(g)} = \frac{d\mu_{h_2}(h_1^{-1}(g))}{d\mu(h_1^{-1}(g))} \frac{d\mu_{h_1}(g)}{d\mu(g)},$$

from which we deduce that, if $h_1, h_2 \in \mathfrak{G}$, then $U(h_1h_2) = U(h_1)\,U(h_2)$. Obviously, $U(e) = I$, hence it follows that $U(h)^{-1} = U(h^{-1})$, and therefore $U(h)$ is a unitary operator on $L^2(\Omega)$. Thus, the correspondence

$$h \to U(h), \qquad h \in \mathfrak{G} \tag{3.1.26}$$

is a unitary representation of $\mathfrak{G}$ in $L^2(\Omega)$. The group of unitary operators $\mathfrak{U} = \{U(h) \mid h \in \mathfrak{G}\}$ will be called the *transformation group of* $L^2(\Omega)$ *corresponding to* $\mathfrak{G}$. This group plays an important role in harmonic analysis on quasi-invariant measure spaces. We let $\mathfrak{A}(\Omega, \mathfrak{G})$ denote the smallest weakly closed operator algebra containing $\mathfrak{U}$, and call it the *algebra over* $L^2(\Omega)$ *corresponding to* $\mathfrak{G}$.

Definition 3.1.5. Let $\Omega = (G, \mathfrak{B}, \mu)$ be a localizable measure space which is quasi-invariant under a group of measurable transformations $\mathfrak{G}$. Let $\mathfrak{T}$ be the weakest topology on $\mathfrak{G}$ such that the unitary representation (3.1.26) is strongly continuous. We call $\mathfrak{T}$ the μ-*topology* on $\mathfrak{G}$.

Actually, the μ-topology is just the topology on $\mathfrak{G}$ induced by the family of convex functions $\{M_\varphi(h), \varphi \in L^2(\Omega)\}$, where

$$\begin{aligned} M_\varphi(h) &= \|(U(h) - I)\varphi\| \\ &= \left(\int |\varphi(hg)(d\mu_{h^{-1}}(g))^{1/2} - \varphi(g)(d\mu(g))^{1/2}|^2\right)^{1/2}. \end{aligned} \tag{3.1.27}$$

In fact, let $\mathfrak{T}_1$ denote the topology induced by the family of convex functions $\{M_\varphi(h), \varphi \in L^2(\Omega)\}$. Then, given any $\varphi \in L^2(\Omega)$, $h_0 \in \mathfrak{G}$, and positive number ϵ, there is a $\mathfrak{T}_1$-neighborhood of h_0, namely,

$$\{h \mid M_\varphi(h^{-1}h_0) < \epsilon\}, \tag{3.1.28}$$

such that, if h belongs to this neighborhood, then

$$\|(U(h) - U(h_0))\varphi\| = M_\varphi(h^{-1}h_0) < \epsilon;$$

thus, the unitary representation U is strongly continuous on $(\mathfrak{G}, \mathfrak{T}_1)$. Conversely, let $\mathfrak{T}_1'$ be any other topology on $\mathfrak{G}$ such that U is strongly continuous on $(\mathfrak{G}, \mathfrak{T}_1')$. Then, for each $\varphi \in L^2(\Omega)$, $h_0 \in \mathfrak{G}$, and $\epsilon > 0$, there exists a $\mathfrak{T}_1'$-neighborhood V of h_0 such that

$$M_\varphi(h^{-1}h_0) = \| (U(h) - U(h_0))\varphi \| < \epsilon$$

whenever $h_0 \in V$. This shows that $\mathfrak{T}_1' \supset \mathfrak{T}_1$. Thus, $\mathfrak{T}_1$ is just the μ-topology on $\mathfrak{G}$.

We now introduce still another family of convex functions on $\mathfrak{G}$. Let $\mathfrak{F}$ denote the totality of sets $A \in \mathfrak{B}$ such that $0 < \mu(A) < \infty$. Write

$$\mathscr{N}_A(h) = \left(2\mu(A) - 2\int_{A \cap hA} (d\mu_h(g)\, d\mu(g))^{1/2}\right)^{1/2}. \qquad (3.1.29)$$

We thus obtain a family of convex functions $\{\mathfrak{N}_A(h), A \in \mathfrak{F}\}$ on $\mathfrak{G}$. If we let C_A denote the characteristic function of the set A, then $C_A \in L^2(\Omega)$ for every $A \in \mathfrak{F}$, and one may easily calculate that

$$\mathscr{N}_A(h) = M_{C_A}(h). \qquad (3.1.29')$$

Lemma 3.1.27. Let $\Omega = (G, \mathfrak{B}, \mu)$ be a localizable measure space which is quasi-invariant under a transformation group $\mathfrak{G}$. Then the μ-topology on $\mathfrak{G}$ is induced by the family of convex functions $\{\mathfrak{N}_A(h), A \in \mathfrak{F}\}$.

PROOF. Let $\mathfrak{T}_1$ denote the topology on $\mathfrak{G}$ induced by $\{N_A, A \in \mathfrak{F}\}$. By virtue of (3.1.29′), we have $\{N_A ; A \in \mathfrak{F}\} \subset \{M_\varphi, \varphi \in L^2(\Omega)\}$, hence $\mathfrak{T}_1$ is weaker than the μ-topology. On the other hand, given any $\varphi \in L^2(\Omega)$ and any positive number ϵ, there exist sets $A_1, \ldots, A_n \in \mathfrak{F}$ and numbers $\lambda_1, \ldots, \lambda_n$ such that

$$\left\| \varphi - \sum_{k=1}^{n} \lambda_k C_{A_k} \right\| < \frac{\epsilon}{4}, \qquad (3.1.30)$$

(see Halmos [1]). For any $h, h_0 \in \mathfrak{G}$, it follows from (3.1.30) that

$$\|(U(h) - U(h_0))\varphi\| \leqslant \frac{\epsilon}{2} + \sum_{k=1}^{n} |\lambda_k| \, \|(U(h^{-1}h_0) - I)C_{A_k}\|. \qquad (3.1.31)$$

Choose

$$\delta < \epsilon \Big/ 2 \sum_{k=1}^{n} |\lambda_k|.$$

Then, we see from (3.1.27) and (3.1.31) that

$$\|(U(h) - U(h_0))\varphi\| < \epsilon,$$

provided that h belongs to the $\mathfrak{T}_1$-neighborhood

$$\{h \mid \mathcal{N}_{A_k}(h^{-1}h_0) < \delta, \quad k = 1, 2, \ldots, n\}$$

of h_0. This shows that the unitary representation $h \to U(h)$ is strongly continuous on $(\mathfrak{G}, \mathfrak{T}_1)$, hence, $\mathfrak{T}_1$ is stronger than the μ-topology. We conclude that the μ-topology is just $\mathfrak{T}_1$.]

Lemma 3.1.28. Let $\Omega_k = (G, \mathfrak{B}, \mu_k)$, $k = 1, 2$, be localizable measure spaces which are both quasi-invariant under the transformation group $\mathfrak{G}$. If μ_1 is absolutely continuous with respect to μ_2, then the μ_1-topology on $\mathfrak{G}$ is weaker than the μ_2-topology.

PROOF. Since μ_1 is absolutely continuous with respect to μ_2, there exists a Radon–Nikodym derivative $d\mu_1(g)/d\mu_2(g)$, moreover, by Corollary 2.4.19,

$$T: \quad \varphi \to \varphi \left(\frac{d\mu_1}{d\mu_2}\right)^{1/2}, \qquad \varphi \in L^2(\Omega_1)$$

is an isometric operator from $L^2(\Omega_1)$ to $L^2(\Omega_2)$. Let $U_k : k \to U_k(h)$ be the unitary representation of $\mathfrak{G}$ in $L^2(\Omega_k)$, $k = 1, 2$. It is easily verified that

$$U_2(h)\, T\varphi = TU_1(h)\varphi$$

for all $\varphi \in L^2(\Omega_1)$, $h \in \mathfrak{G}$. Hence, if $\varphi \in L^2(\Omega_1)$, $h, h_0 \in \mathfrak{G}$, then, writing $\psi = T\varphi \in L^2(\Omega_2)$, we have

$$\|(U_1(h) - U_1(h_0))\varphi\| = \|(U_2(h) - U_2(h_0))\psi\|.$$

This shows that the family of convex functions which defines the μ_1-topology is included in the family of convex functions which defines the μ_2-topology. Therefore, the μ_1-topology is weaker than the μ_2-topology.]

In particular, if the measure spaces Ω_1 and Ω_2 in Corollary 3.1.28 are equivalent, then the μ_1-topology and the μ_2-topology are identical.

Lemma 3.1.29. Let G be a locally compact topological group, $\mathfrak{B}$ the σ-ring generated by the compact subsets of G, and μ a left-invariant Haar measure on G. If G is regarded as a group of left translations of itself, then the μ-topology is weaker than the original topology of G.

PROOF. Notice that, since μ is invariant, $\mathfrak{N}_A(h) = (2\mu(A - hA))^{1/2}$, $A \in \mathfrak{F}$. Now, since every Haar measure is regular (see Halmos [1]), we know that for any $A \in \mathfrak{F}$, there is a compact set $C \subset A$ such that $\mu(A - C)$

is less than the prescribed positive number $\epsilon/2$, moreover, there is an open set $U \supset A$ such that $\mu(U - A) < \epsilon/2$. Furthermore, by Lemma 3.1.7, there is a neighborhood V of the identity in G such that $hC \subset U$ for all $h \in V$. Therefore, since $A - hA \subset U - hC$, it follows that if $h \in V$, then

$$\mu(A - hA) \leqslant \mu(U) - \mu(hC) = \mu(U) - \mu(C) < \epsilon.$$

This shows that $\mathfrak{N}_A(h)$ is a continuous function on G, which means that the μ-topology is weaker than the original topology of G.]

Lemma 3.1.30. Let G be a topological space, and let $(G, \mathfrak{B}, \mu)$ be a finite regular measure space which is quasi-invariant under a topological group of continuous measurable transformations $\mathfrak{G}$. If the topology $\mathfrak{T}$ of $\mathfrak{G}$ is admissible, then $\mathfrak{T}$ is stronger than the μ-topology if and only if the function

$$M_1(h) = \left(\int ((d\mu_h(g))^{1/2} - (d\mu(g))^{1/2})^2\right)^{1/2}, \qquad h \in \mathfrak{G}, \tag{3.1.32}$$

is continuous relative to $\mathfrak{T}$.

PROOF. Let $A \in \mathfrak{B}$. By the Cauchy inequality, we have

$$\begin{aligned} \mathcal{N}_A(h) &\leqslant \left(\int C_A(hg)((d\mu_{h^{-1}}(g))^{1/2} - (d\mu(g))^{1/2})^2\right)^{1/2} \\ &\quad + \left(\int (C_A(hg) - C_A(g))^2 \, d\mu(g)\right)^{1/2} \\ &\leqslant M_1(h) + (\mu(h^{-1}A) + \mu(A) - 2\mu(A \cap h^{-1}A))^{1/2}. \end{aligned} \tag{3.1.33}$$

Assume that $M_1(h)$ is continuous on $(\mathfrak{G}, \mathfrak{T})$. Then, for any $\epsilon > 0$, there is a $\mathfrak{T}$-neighborhood V of the identity in $\mathfrak{G}$ (we may suppose that $V = V^{-1}$) such that

$$M_1(h) < \epsilon \tag{3.1.34}$$

for all $h \in V$. From (3.1.34) and the last inequality in the proof of Lemma 3.1.19, it follows that

$$|(\mu(h^{-1}B))^{1/2} - (\mu(B))^{1/2}| \leqslant \epsilon$$

for all $B \in \mathfrak{B}$, $h \in V$. Multiplying both sides of the above inequality by $(\mu(h^{-1}B))^{1/2} + (\mu(B))^{1/2}$, we get

$$|\mu(h^{-1}B) - \mu(B)| < 2\epsilon(\mu(G))^{1/2}. \tag{3.1.35}$$

By the regularity of μ, there exists an open set $U \in \mathfrak{B}$ and a compact set $K \in \mathfrak{B}$ such that $U \supset A \supset K$ and

$$\mu(U - K) < \epsilon. \tag{3.1.36}$$

Since the topology of $\mathfrak{G}$ is admissible, we may assume that V is sufficiently small so that $h^{-1}K \subset U$ whenever $h \in V$. Now, from (3.1.35) and (3.1.36), we see that

$$\mu(h^{-1}(A - K)) + \mu(A - K) < 2\epsilon + 2\epsilon(\mu(G))^{1/2} \tag{3.1.37}$$

for all $h \in V$. Furthermore, if $h \in V$, then

$$h^{-1}K - (A \cap h^{-1}A) \subset h^{-1}K - (K \cap h^{-1}K) \subset U - K,$$

and hence

$$\mu(h^{-1}K) - \mu(A \cap h^{-1}A) < \epsilon. \tag{3.1.38}$$

Similarly, if $h \in V$, then, using (3.1.35) and (3.1.38), we obtain

$$\begin{aligned} \mu(K) - \mu(A \cap h^{-1}A) &= \mu(K) - \mu(h^{-1}K) + \mu(h^{-1}K) - \mu(A \cap h^{-1}A) \\ &< \epsilon + 2\epsilon(\mu(G))^{1/2}. \end{aligned} \tag{3.1.39}$$

Combining (3.1.33), (3.1.34), (3.1.37), (3.1.38), and (3.1.39), we deduce that

$$\mathscr{N}_A(h) < \epsilon + 2[\epsilon(1 + (\mu(G))^{1/2})]^{1/2}$$

for all $h \in V$. This shows that the μ-topology is weaker than $\mathfrak{T}$.

Conversely, if the μ-topology is weaker than $\mathfrak{T}$, then obviously $M_1(h) = \mathfrak{N}_G(h)$ is continuous on $(\mathfrak{G}, \mathfrak{T})$.]

7° Ergodic Measures

We now introduce the concepts of quasi-invariant sets and ergodicity.

Definition 3.1.6. Let $(G, \mathfrak{B}, \mu)$ be a measure space which is quasi-invariant under a transformation group $\mathfrak{G}$. If $B \in \mathfrak{B}$, and if $B - hB$ is a μ-null set for all $h \in \mathfrak{G}$, then we say that B is *quasi-invariant* under $\mathfrak{G}$, or $\mathfrak{G}$-*quasi-invariant*, or (if the group $\mathfrak{G}$ has been clearly specified) simply *quasi-invariant*.

We note the following simple facts. The totality of $\mathfrak{G}$-quasi-invariant sets of $(G, \mathfrak{B}, \mu)$ forms a σ-subring $\mathfrak{B}_1$ of $\mathfrak{B}$; if $\mathfrak{B}$ is an algebra, then $\mathfrak{B}_1$ is also an algebra. Any set which differs from a quasi-invariant set by a μ-null set is also quasi-invariant. Any quasi-invariant measure space

$(G, \mathfrak{B}, \mu)$ has at least one class of quasi-invariant sets, that is, the μ-null sets; if $\mathfrak{B}$ is an algebra, then there is at least one more class of quasi-invariant sets, namely, those sets which differ from G by a μ-null set. Any set which belongs to one of these two classes is said to be *trivially quasi-invariant*.

Definition 3.1.7. Let $(G, \mathfrak{B}, \mu)$ be a measure space which is quasi-invariant under a transformation group $\mathfrak{G}$. If there exist no nontrivially quasi-invariant sets, then $(G, \mathfrak{B}, \mu)$ is said to be *ergodic* with respect to $\mathfrak{G}$, or $\mathfrak{G}$-*ergodic*. If there exist no nontrivially quasi-invariant sets of finite measure (σ-finite measure), then $(G, \mathfrak{B}, \mu)$ is said to be *weakly* (*strongly*) *ergodic* with respect to $\mathfrak{G}$, or *weakly* (*strongly*) $\mathfrak{G}$-*ergodic*.[16]

Example 3.1.1. Let $(G, \mathfrak{B}, \mu_k)$, $k = 1, 2$, be mutually singular measure spaces which are both quasi-invariant under a transformation group $\mathfrak{G}$. Choose sets $B_k \in \mathfrak{B}$, $k = 1, 2$, such that μ_k is concentrated on B_k and $\mu_k(B_l) = 0$ for $k \neq l$. Then B_k is $\mathfrak{G}$-quasi-invariant relative to μ_l for any k, l. Let $\mu = (\mu_k + \mu_l)/2$. Obviously, the measure space $(G, \mathfrak{B}, \mu)$ is also quasi-invariant under $\mathfrak{G}$, and the sets B_k, $k = 1, 2$, are quasi-invariant relative to μ. If $\mu_k \not\equiv 0$, $k = 1, 2$, then it is easily seen that B_1 and B_2 are nontrivially quasi-invariant relative to μ, hence $(G, \mathfrak{B}, \mu)$ is not ergodic with respect to $\mathfrak{G}$.

Lemma 3.1.31. Let G be a group, and let $\mathfrak{B}$ be a σ-ring consisting of certain subsets of G, such that $gE^{-1} \in \mathfrak{B}$ whenever $g \in G$, $E \in \mathfrak{B}$. Let μ be a left-invariant σ-finite measure on $(G, \mathfrak{B})$, and suppose that there exists a right-invariant σ-finite measure ν on $(G, \mathfrak{B})$ such that ν is equivalent to μ. Then $(G, \mathfrak{B}, \mu)$ is ergodic with respect to the left translation group G.

Proof. Suppose, on the contrary, that μ were not ergodic. Then, there would exist disjoint quasi-invariant sets E, F, with $\mu(E) > 0$ and $\mu(F) > 0$.[17] Since $hE - E$ is a null set, we have

$$\mu(hE \cap F) = 0. \tag{3.1.40}$$

Moreover, since ν and μ are equivalent, $\nu(E) > 0$.

[16] *Translator's note:* Note the unfortunate fact that *strong* ergodicity, as defined here, is a *weaker* property than ergodicity.

[17] *Translator's note:* This inference seems to require that $\mathfrak{B}$ be a σ-algebra. Alternatively, one could modify the definition of ergodicity as follows: a measure space is ergodic provided that there do not exist two disjoint quasi-invariant sets which are both of positive measure.

Consider the measure μ_1 on $(G, \mathfrak{B})$, defined by

$$\mu_1(A) = \nu(A^{-1}), \qquad A \in \mathfrak{B}.$$

As usual, we denote the characteristic function of any set $\mathfrak{D}$ by $C_{\mathfrak{D}}$. One easily calculates that

$$\int_G C_E(h^{-1}g)\, d\mu_1(h) = \int_G C_E(h_1 g)\, d\nu(h_1) = \int_G C_E(h_1)\, d\nu(h_1) = \nu(E).$$

By Fubini's Theorem (see Halmos [1], §36, Theorem B),

$$\begin{aligned} \nu(E)\,\mu(F) &= \int_G \left(\int_G C_E(h^{-1}g)\, d\mu_1(h)\right) C_F(g)\, d\mu(g) \\ &= \int_G \left(\int_G C_E(h^{-1}g)\, C_F(g)\, d\mu(g)\right) d\mu_1(h) \\ &= \int_G \mu(hE \cap F)\, d\mu_1(h). \end{aligned} \tag{3.1.41}$$

But, in view of (3.1.40), the right-hand side of (3.1.41) is zero, which contradicts $\nu(E)\,\mu(F) > 0$. We conclude that μ is ergodic.]

Remark. As particular cases of Lemma 3.1.31, we have (i) any left-invariant Haar measure on a locally compact topological group, and (ii) any product $\mu_1 \times \cdots \times \mu_n$ of left-invariant Haar measures $\mu_1, \dots, \mu_n$ on locally compact topological groups $G_1, \dots, G_n$, respectively.

Next, we introduce a construction which will be used in the sequel.

Theorem 3.1.32. Let $\{G_\alpha, \alpha \in \mathfrak{A}\}$ be a family of locally compact topological groups, let $\mathfrak{B}_\alpha$ be the σ-ring generated by the totality of compact sets in G_α, and let $\Omega_\alpha = (G_\alpha, \mathfrak{B}_\alpha, \mu_\alpha)$ be a probability[18] measure space which is left quasi-invariant under G_α. Let $G = \mathsf{X}_{\alpha\in\mathfrak{A}}\, G_\alpha$; then G is a group with respect to the multiplication operation defined as follows:

$$\{g_\alpha, \alpha \in \mathfrak{A}\}\{h_\alpha, \alpha \in \mathfrak{A}\} = \{g_\alpha h_\alpha, \alpha \in \mathfrak{A}\}.$$

Let $\mathfrak{G}$ be the subgroup of G consisting of all elements $g = \{g_\alpha, \alpha \in \mathfrak{A}\}$ such that g_α differs from the identity element of G_α for at most a finite number of indices α. Let $\mathfrak{B} = \mathsf{X}_{\alpha\in\mathfrak{A}}\, \mathfrak{B}_\alpha$, $\mu = \mathsf{X}_{\alpha\in\mathfrak{A}}\, \mu_\alpha$; then, the measure space $\Omega = (G, \mathfrak{B}, \mu)$ is ergodic with respect to the group of left translations corresponding to $\mathfrak{G}$.

[18] In particular, this implies that $\mathfrak{B}_\alpha$ is a σ-algebra.

PROOF. Let $\lambda = \{\alpha_1, \ldots, \alpha_n\}$ be an arbitrary finite subset of $\mathfrak{A}$. Form the Cartesian product

$$G_\lambda = G_{\alpha_1} \times G_{\alpha_2} \times \cdots \times G_{\alpha_m},$$

and define multiplication in G_λ by the rule

$$\{g_{\alpha_1}, \ldots, g_{\alpha_n}\}\{h_{\alpha_1}, \ldots, h_{\alpha_n}\} = \{g_{\alpha_1}h_{\alpha_1}, \ldots, g_{\alpha_n}h_{\alpha_n}\}.$$

If G_λ is given the product topology, then it is easily shown that G_λ is also a locally compact topological group. Let $\mathfrak{B}_\lambda = \mathfrak{B}_{\alpha_1} \times \mathfrak{B}_{\alpha_2} \times \cdots \times \mathfrak{B}_{\alpha_n}$, $\mu_\lambda = \mu_{\alpha_1} \times \cdots \times \mu_{\alpha_n}$; then, using Fubini's theorem, it is easily proved that $\Omega_\lambda = (G_\lambda, \mathfrak{B}_\lambda, \mu_\lambda)$ is a probability measure space which is left quasi-invariant under G_λ. By Corollary 3.1.6, each μ_α is equivalent to a left-invariant Haar measure ν_α on G_α. Hence, μ_λ is equivalent to the measure $\nu_{\alpha_1} \times \cdots \times \nu_{\alpha_n}$ on $\mathfrak{B}_\lambda$. By the remark following Lemma 3.1.31, $\nu_{\alpha_1} \times \cdots \times \nu_{\alpha_n}$ is ergodic with respect to the left translations of G_λ, hence the same is true of μ_λ. If $\mathfrak{A}$ is finite, then $G_A = \mathfrak{G}$, hence, it only remains to consider the case where $\mathfrak{A}$ is infinite.

Suppose that E is any set in $\mathfrak{B}$ which is quasi-invariant under $\mathfrak{G}$. Now, there is a sequence $\{\alpha_n\} \subset \mathfrak{A}$, such that E is the direct product of $\mathsf{X}_{\alpha \bar{\in} \{\alpha_n\}} G_\alpha$ and some set in $\mathsf{X}_{n=1}^{\infty} \mathfrak{B}_{\alpha_n}$. Consequently, we may as well suppose that $\mathfrak{A}$ itself is countable, for example, $\mathfrak{A} = \{1, 2, \ldots, n, \ldots\}$. Let $\lambda_n = \{1, 2, \ldots, n\}$. If

$$f(g_{\lambda_n}) \in L^2(\Omega_{\lambda_n}),$$

then we may regard $f(\cdot)$ as a function in $L^2(\Omega)$, that is, $g \to f(g_{\lambda_n})$, $g \in G$, $\{g_{\lambda_n} = (g_1, g_2, \ldots, g_n)\}$. Thus, $L^2(\Omega_{\lambda_n})$ is imbedded as a closed linear subspace of $L^2(\Omega)$; let $P^{(n)}$ denote the corresponding projection operator. If $S \in L^2(\Omega)$, $X \in L^2(\Omega_{\lambda_n})$, then, since $P^{(n)}$ is self-adjoint, we have

$$\begin{aligned}\int_{G_{\lambda_n}} P^{(n)}S(g_{\lambda_n})\,\overline{X(g_{\lambda_n})}\,d\mu_{\lambda_n}(g_{\lambda_n}) &= \int_G P^{(n)}S(g_{\lambda_n})\,\overline{X(g_{\lambda_n})}\,d\mu(g)\\ &= \int_G S(g)\,\overline{P^{(n)}X(g_{\lambda_n})}\,d\mu(g)\\ &= \int_G S(g)\,\overline{X(g_{\lambda_n})}\,d\mu(g). \qquad (3.1.42)\end{aligned}$$

In particular, when the function S is nonnegative, formula (3.1.42) holds for any nonnegative measurable function X on Ω_{λ_n}, provided that $(P^{(n)}S)\,X \in L^1(\Omega_{\lambda_n})$. In fact, for such an X, we have

$$X_m = \min(X, m) \in L^2(\Omega_{\lambda_n}),$$

hence, setting $X = X_m$ in (3.1.42), and letting $m \to \infty$, we obtain the desired result.

Now, let S be the characteristic function of the set E; obviously, $S \in L^2(\Omega)$. Write $S^{(n)} = P^{(n)}(S)$. Choose an arbitrary element $h = \{h_1, \ldots, h_n\} \in G_{\lambda_n}$, let Y be any nonnegative measurable function on Ω_{λ_n} such that $S^{(n)}(hg_{\lambda_n})\, Y(g_{\lambda_n}) \in L^1(\Omega_{\lambda_n})$, and define

$$X(g_{\lambda_n}) = Y(h^{-1}g_{\lambda_n}) \frac{d\mu_{\lambda_n h}(g_{\lambda_n})}{d\mu_{\lambda_n}(g_{\lambda_n})}.$$

Since

$$\int_{G_{\lambda_n}} S^{(n)}(hg_{\lambda_n})\, Y(g_{\lambda_n})\, d\mu_{\lambda_n}(g_{\lambda_n})$$

$$= \int_{G_{\lambda_n}} S^{(n)}(g_{\lambda_n})\, Y(h^{-1}g_{\lambda_n}) \frac{d\mu_{\lambda_n h}(g_{\lambda_n})}{d\mu_{\lambda_n}(g_{\lambda_n})}\, d\mu_{\lambda_n}(g_{\lambda_n}), \qquad (3.1.43)$$

we have $S^{(n)}X \in L^1(\Omega_{\lambda_n})$. Hence, using (3.1.42) and (3.1.43), we get

$$\int_{G_{\lambda_n}} S^{(n)}(hg_{\lambda_n})\, Y(g_{\lambda_n})\, d\mu_{\lambda_n}(g_{\lambda_n})$$

$$= \int_G S(g)\, Y(h^{-1}g_{\lambda_n}) \frac{d\mu_{\lambda_n h}(g_{\lambda_n})}{d\mu_{\lambda_n}(g_{\lambda_n})}\, d\mu(g).$$

Let $\tilde{h} = \{\tilde{h}_\alpha, \alpha \in \mathfrak{A}\}$ be the element of $\mathfrak{G}$ whose coordinates are $\tilde{h}_\alpha = h_\alpha$ for $\alpha = 1, \ldots, n$ and $\tilde{h}_\alpha = e_\alpha$ (the unit element of G_α) for $\alpha > n$. Then, by virtue of (1.1.13), the above relation may be written as

$$\int_{G_{\lambda_n}} S^{(n)}(hg_{\lambda_n})\, Y(g_{\lambda_n})\, d\mu_{\lambda_n}(g_{\lambda_n})$$

$$= \int_G S(g)\, Y(h^{-1}g_{\lambda_n}) \frac{d\mu_{\tilde{h}}(g)}{d\mu(g)}\, d\mu(g). \qquad (3.1.44)$$

Since E is a quasi-invariant set, the functions $S(g)$ and $S(\tilde{h}g)$ are equal almost everywhere. Therefore, applying the change of variable $g \to \tilde{h}g$ in (3.1.44), we obtain

$$\int_{G_{\lambda_n}} S^{(n)}(hg_{\lambda_n})\, Y(g_{\lambda_n})\, d\mu_{\lambda_n}(g_{\lambda_n})$$

$$= \int_G S(g)\, Y(g_{\lambda_n})\, d\mu(g)$$

$$= \int_{G_{\lambda_n}} S^{(n)}(g_{\lambda_n})\, Y(g_{\lambda_n})\, d\mu_{\lambda_n}(g_{\lambda_n}),$$

whence it follows easily that $S^{(n)}(hg_{\lambda_n})$ and $S^{(n)}(g_{\lambda_n})$ are equal almost everywhere. Since h is an arbitrary element of G_{λ_n}, and since Ω_{λ_n} is ergodic with respect to the left translations of G_{λ_n}, it is not difficult to deduce that $S^{(n)}(g_{\lambda_n})$ is almost everywhere equal to some constant M_n. But according to Lemma 1.1.7, the sequence of projections $\{P^{(n)}\}$ converges strongly to I, hence

$$S = \lim_{n\to\infty} P^{(n)}S = \lim_{n\to\infty} M_n ,$$

which is obviously also equal to some constant almost everywhere. This means that C_E is either equal to zero almost everywhere or equal to 1 almost everywhere, that is, $\mu(E) = 0$ or $\mu(G - E) = 0$. Thus, Ω is ergodic with respect to the group of left translations $\mathfrak{G}$.]

§3.2. Characters and Quasi-Characters

1° Definition and Basic Properties of Characters

The class of functions known as characters is of basic importance for harmonic analysis on groups. Their definition is as follows.

Definition 3.2.1. A complex-valued function α on a group G is said to be a *character*, provided (i) $|\alpha(g)| = 1$ for all $g \in G$, and (ii) $\alpha(gh) = \alpha(g)\,\alpha(h)$ for all $g, h \in G$.

If we let e denote the unit element of G, then any character α of G satisfies the condition (iii) $\alpha(e) = 1$. In fact, by (ii), $\alpha(e)^2 = \alpha(e^2) = \alpha(e)$, and by (i), $\alpha(e) \neq 0$, hence $\alpha(e) = 1$. In the present section, we shall use C to denote the multiplicative group of complex numbers of unit modulus; thus, a character α of a group G is just a homomorphism of G into C:

$$\alpha : \quad g \to \alpha(g), \qquad g \in G.$$

Let α and β be any two characters of the group G. We define the product $\alpha\beta$ by the ordinary rule for multiplication of functions:

$$\alpha\beta(g) = \alpha(g)\,\beta(g), \qquad g \in G.$$

Obviously, $\alpha\beta$ is also a character of G. The totality of characters of G clearly forms a commutative group with respect to multiplication. This group is known as the *character group* of G (or the *algebraic dual* of G); we denote it by G'. The unit element of G' is the constant function 1.

Example 3.2.1. Let I be the additive group of all integers. For each c in C, define a character β_c of I as follows:

$$\beta_c(n) = c^n, \qquad n \in I.$$

The correspondence $c \to \beta_c$ is obviously an injective homomorphism from C into I'. Moreover, given any $\alpha \in I'$, we have $\alpha = \beta_c$, where $c = \alpha(1)$. Thus, the correspondence $c \to \beta_c$ is an isomorphism from C onto I'. Using this isomorphism, we may also consider C and I' as identical: $I' = C$.

Theorem 3.2.1. (Character Extension Theorem). Let G be a commutative group, G_1 a subgroup of G, and α_1 a character of G_1. Then, there exists a character α of G such that $\alpha(g) = \alpha_1(g)$ for all $g \in G_1$.

PROOF. If $G \neq G_1$, choose any $g_1 \in G - G_1$, and let G_2 be the smallest subgroup of G containing g_1 and G_1. We distinguish two cases. (i) For every natural number n, $g_1^n \notin G_1$. In this case, every element of G_2 is uniquely expressible in the form

$$gg_1^n, \qquad g \in G_1, \quad n = 0, \pm 1, \pm 2, \dots .$$

Choose any $c_1 \in C$, and define a character α_2 on G_2 as follows:

$$\alpha_2(gg_1^n) = \alpha_1(g)c_1^n. \tag{3.2.1}$$

It is easily seen that α_2 is an extension of α_1. (ii) There is a natural number n such that $g_1^n \in G_1$. Let n_1 be the smallest such number. Then, every element of G_2 is uniquely expressible in the form

$$gg_1^n, \qquad g \in G_1, \quad n = 0, 1, 2, \dots, n_1 - 1.$$

Choose a complex number c_1 such that $c_1^{n_1} = \alpha_1(g_1^{n_1})$, and define a character α_2 on G_2 as in formula (3.2.1). Again, it is easily verified that α_2 is an extension of α_1. Thus, in either of the two possible cases, α_1 can be extended to G_2. Using Zorn's lemma, we can extend α_1 to a character α on G.]

Example 3.2.2. Let $\mathscr{R}$ be the additive group of all rational numbers, let t be an arbitrary real number, and consider the character of $\mathscr{R}$ defined by

$$\alpha_t(g) = e^{itg}, \qquad g \in \mathscr{R}.$$

The multiplicative group of all characters of this form is obviously isomorphic with the additive group of real numbers R. However, this group is not the whole of $\mathscr{R}'$. In fact, choose any sequence of natural numbers $\{q_n\}$ such that q_n divides q_{n+1}, $2q_n < q_{n+1}$, and such that every natural number divides at least one of the q_n. Clearly, such a sequence does exist. Define another sequence of natural numbers $\{p_n\}$ by the recursion formula

$$p_0 = 0; \qquad p_n = p_{n-1} + 2q_{n-1}, \quad n = 1, 2, \ldots .$$

Obviously, $p_n \to \infty$. Now, form a character α on $\mathscr{R}$ as follows: for each integer k,

$$\alpha\left(\frac{k}{q_n}\right) = \exp\left(i\frac{p_n k}{q_n} 2\pi\right).$$

We must first show that α is well defined. If $k/q_n = k'/q_{n'}$, and, say, $n < n'$, then $p_{n'} - p_n$ is an integral multiple of q_n, hence

$$\frac{p_{n'}}{q_{n'}} k' = \frac{p_{n'}}{q_n} k = \frac{p_n}{q_n} k + r,$$

where r is an integer. Thus, α is well defined, and now one may easily verify that α is a character. Now, we assert that this α cannot be equal to any α_t. For suppose, on the contrary, that $\alpha = \alpha_t$ for some t. Then, since $\alpha(1/q_n) = \alpha_t(1/q_n)$, it follows that

$$\left(p_n - \frac{t}{2\pi}\right)\Big/ q_n$$

is an integer. However, when n is sufficiently large, we have $| t/2\pi | < \frac{1}{2}q_n$, and hence $| p_n - (t/2\pi)|/q_n < 1$, which is possible only if $p_n - (t/2\pi) = 0$. But this cannot hold for two different values of n. We conclude that $\alpha \neq \alpha_t$ for any t, and thus $\mathscr{R}' \neq R$.

Example 3.2.3. Let R be the additive group of all real numbers. For each fixed real number t, we have the character

$$\alpha_t(x) = e^{itx}, \qquad x \in R.$$

Denote the totality of such characters by R^*. We assert that $R^* \neq R'$. In fact, in Example 3.2.2 we constructed a character α on the subgroup $\mathscr{R}$ of R, such that α is not of the form α_t for any $t \in R$; using the character extension theorem, we may extend this α to a character of R, and of course this extension is also different from every α_t. Actually, one can even extend a character of the form α_t on $\mathscr{R}$ to a character on R which is not of the form α_t. For example, choose any irrational number x_0, and let G_1 be the subgroup of R consisting of all numbers of the form

$$mx_0 + x, \qquad m \in I, \quad x \in \mathscr{R}.$$

Let β be the character of G_1 defined by

$$\beta(mx_0 + x) = (-1)^m.$$

Then, any extension of β to a character of R is obviously not of the form α_t.

In a similar manner, one can prove that $C' \neq I$, that is, there exist characters of C which are not of the form

$$\alpha_n(a) = a^n, \qquad n \in I.$$

When G is a topological group, we shall be primarily interested in continuous characters of G, that is, characters which are continuous functions on G. The totality of continuous characters of G is called the *dual* of G, and will be denoted by G^*. Obviously, G^* is a subgroup of G'. Of course, if G is a discrete topological group, then every character of G is continuous, so that $G^* = G'$. However, for an arbitrary topological group G, in general, G^* and G' do not coincide.

Lemma 3.2.2. Let β be a character of the topological group G. If β is continuous at the identity e of G, then $\beta \in G^*$.

PROOF. Let h_0 be any element of G, and let ϵ be an arbitrary positive number. By hypothesis, there exists a neighborhood V of e such that $|\beta(g) - \beta(e)| < \epsilon$ for all $g \in V$. Hence, if h belongs to the neighborhood Vh_0 of h_0, then

$$|\beta(h) - \beta(h_0)| = |\beta(hh_0^{-1}) - \beta(e)| < \epsilon.$$

Thus, h_0 is also a continuity point of β.]

Let $\mathfrak{B}$ be a σ-algebra consisting of certain subsets of a group G, and let $G^{\mathfrak{B}}$ denote the totality of $\mathfrak{B}$-measurable characters of G. Obviously, $G^{\mathfrak{B}}$ is also a subgroup of G'. If $\mathfrak{B}$ consists of all the subsets of G, then of course $G^{\mathfrak{B}} = G'$. On the other hand, if $\mathfrak{B}$ contains only the null set and G, then $G^{\mathfrak{B}}$ consists of just the single character 1. If G is a topological group and if $\mathfrak{B}$ contains all the closed subsets of G, then every continuous character of G is $\mathfrak{B}$-measurable, that is, $G^* \subset G^{\mathfrak{B}}$. In the following example, we actually have $G^* = G^{\mathfrak{B}}$.

Example 3.2.4. Let R be the additive group of real numbers with the Euclidean topology. Then, every continuous character on the topological group R is of the form $\alpha_t(x) = e^{itx}$, $x \in R$. Moreover, if $\mathfrak{B}$

is the σ-algebra generated by the totality of closed subsets of R, then $R^{\mathfrak{B}} = R^*$.

In fact, let $\alpha \in R^{\mathfrak{B}}$, and consider the Lebesgue integral

$$\lambda = \int_0^\infty \alpha(x)e^{-x}\,dx.$$

Since

$$\alpha(h)\lambda = \int_0^\infty \alpha(h + x)e^{-x}\,dx = e^h \int_h^\infty \alpha(x)e^{-x}\,dx, \qquad (3.2.2)$$

it follows that $\alpha(h)$ is continuous. Thus, $R^{\mathfrak{B}} = R^*$. Furthermore, using (3.2.2), we obtain

$$\frac{e^h - \alpha(h)}{h}\lambda = e^h \frac{1}{h}\int_0^h \alpha(x)e^{-x}\,dx \to 1, \qquad h \to 0,$$

hence, as $h \to 0$,

$$\frac{\alpha(h) - 1}{h} \to 1 - \frac{1}{\lambda},$$

therefore $\alpha'(0)$ exists. Moreover, since

$$\frac{\alpha(g + h) - \alpha(g)}{h} = \alpha(g)\frac{\alpha(h) - 1}{h},$$

$\alpha'(g)$ exists and is equal to $\alpha'(0)\,\alpha(g)$. Integrating, we obtain

$$\alpha(g) = e^{\alpha'(0)g}.$$

Therefore, since $|\alpha(g)| = 1$, it follows that $\alpha'(0)$ is a pure imaginary number, which we call it, $t \in R$. Thus, every measurable character of R is of the form α_t.

The preceding example may also be regarded as a special case of Lemma II.3.2 of Appendix II.

2° Quasi-characters

In what follows, we shall require a class of functions more general than measurable characters, namely, quasi-characters. We shall not describe the concept of a quasi-character in its most complete generality; for our purposes, it suffices to restrict our considerations to groups.

Definition 3.2.2. Let $(G, \mathfrak{B}, \mu)$ be a measure space which is quasi-invariant under a group of measurable transformations $\mathfrak{G}$. Let α be a measurable function on $(G, \mathfrak{B})$, satisfying the following conditions. (i) $|\alpha(g)| = 1$ for all $g \in G$; (ii) for every fixed $h \in \mathfrak{G}$, the function

$$\alpha(hg)/\alpha(g), \qquad g \in G$$

is equal to a certain constant $\tilde{\alpha}(h)$ almost everywhere (relative to the measure μ). Then, we say that α is a *quasi-character*[19] of $(G, \mathfrak{B}, \mu)$ relative to $\mathfrak{G}$.

First, we observe that the function $\tilde{\alpha}(h)$, $h \in \mathfrak{G}$, is a character of $\mathfrak{G}$. In fact, we obviously have $|\tilde{\alpha}(h)| = 1$, moreover, for any $h, x \in \mathfrak{G}$,

$$\frac{\alpha(hxg)}{\alpha(g)} = \frac{\alpha(hxg)}{\alpha(xg)} \frac{\alpha(xg)}{\alpha(g)},$$

hence, it follows from the quasi-invariance of μ that $\tilde{\alpha}(hx) = \tilde{\alpha}(h)\,\tilde{\alpha}(x)$.[20]

We call $\tilde{\alpha}$ the character of $\mathfrak{G}$ *induced* by the quasi-character α; the totality of such induced characters will be denoted by $\tilde{\mathfrak{G}}^{\mu}$. We also refer to α as a quasi-character corresponding to $\tilde{\alpha}$.

We shall regard two quasi-characters which are equal almost everywhere (relative to μ) as identical. If two quasi-characters differ only by a constant factor (i.e., if the ratio of these two functions is equal almost everywhere to some constant), then we say that they are *similar*.

We proceed to state some simple properties of quasi-characters.

I. Similar quasi-characters induce the same character.

II. If α is a quasi-character and c is a complex number with $|c| = 1$, then $c\alpha$ is also a quasi-character.

III. If the measure μ is ergodic with respect to $\mathfrak{G}$, then quasi-characters which correspond to the same character are similar.

In fact, suppose that the quasi-characters α_1 and α_2 correspond to the same character of $\mathfrak{G}$, and consider the function $\alpha = \alpha_1\alpha_2^{-1}$. Then, for any $h \in \mathfrak{G}$, we have $\alpha(hg) = \alpha(g)$ for almost all $g \in G$. Consequently,

[19] A quasi-character can be regarded as an eigenfunction of a certain operator. In fact, consider the totality $U(G)$ of $\mathfrak{B}$-measurable complex-valued functions α on G such that $|\alpha(g)| = 1$ for all g; functions in $U(G)$ which are equal almost everywhere (μ) are to be regarded as identical. For each element h of $\mathfrak{G}$, define an operator T_h on $U(G)$ as follows:

$$T_h\alpha(g) = \alpha(h^{-1}g), \qquad \alpha \in U(G), \quad h \in \mathfrak{G}.$$

Then, the correspondence $h \to T_h$ is a representation of $\mathfrak{G}$ in $U(G)$ [i.e., a homomorphism of $\mathfrak{G}$ into a group of operators on $U(G)$]. A quasi-character α is just a common eigenfunction of all the operators T_h, $h \in \mathfrak{G}$, and $\tilde{\alpha}(h)$ is the eigenvalue of the operator $T_{h^{-1}}$, corresponding to the eigenfunction α.

[20] *Translator's note:* To draw this conclusion, one must assume that the measure μ is not identically zero.

if γ is any arc on the unit circle of the complex plane, then $\{g \mid \alpha(g) \in \gamma\}$ is a quasi-invariant set. But if $\alpha(g)$ were not equal to some constant almost everywhere, then there would exist two disjoint circular arcs γ_1, γ_2 such that both of the (disjoint) sets $\{g \mid \alpha(g) \in \gamma_1\}$ and $\{g \mid \alpha(g) \in \gamma_2\}$ are of positive measure. This contradicts the ergodicity of μ.

IV. If all the quasi characters corresponding to any given character of $\mathfrak{G}$ are similar, then μ is ergodic with respect to $\mathfrak{G}$.

In fact, if μ is not ergodic, then let A be a quasi-invariant set in G such that $\mu(A) > 0$ and $\mu(G - A) > 0$.[21] Define a function α as follows:

$$\alpha(g) = -1 \qquad \text{for} \quad g \in A,$$

$$\alpha(g) = 1 \qquad \text{for} \quad g \in G - A.$$

It is easily seen that α is a quasi-character which induces the character 1 on $\mathfrak{G}$. But α is not similar to the quasi-character 1 of G.

We shall denote the totality of quasi-characters by G^μ. Clearly, G^μ forms a group with respect to ordinary multiplication of functions. The totality of quasi-characters which correspond to the character 1 of $\mathfrak{G}$ will be denoted by $\mathfrak{N}$. It is easily verified that $\mathfrak{N}$ is a subgroup of G^μ. Moreover, it follows from properties I–IV that $\mathfrak{N}$ is the totality of constant functions c, $c \in C$, if and only if μ is ergodic with respect to G.

Consider the factor group $G_0{}^\mu = G^\mu/\mathfrak{N}$. For each $\eta \in G_0{}^\mu$, choose a representative element α in η. The mapping

$$S: \quad \eta \to \tilde{\alpha} \tag{3.2.3}$$

is clearly an isomorphism from $G_0{}^\mu$ onto $\tilde{\mathfrak{G}}^\mu$.

Example 3.2.5. Let G be a group, let $\mathfrak{G}$ be the totality of left translations of G, let $(G, \mathfrak{B}, \mu)$ be a finite measure space which is quasi-invariant under $\mathfrak{G}$, and suppose that the correspondence $(g, h) \to gh$ is a measurable mapping from $(G \times G, \mathfrak{B} \times \mathfrak{B})$ to $(G, \mathfrak{B})$. Then, the group of quasi-characters G^μ is just the totality of functions of the form

$$c\tilde{\alpha}, \qquad c \in C, \quad \tilde{\alpha} \in G^{\mathfrak{B}}.$$

Proof. For any $\alpha \in G^\mu$, the functions $\alpha(hg)$ and $\alpha(g)$, $(g, h) \in G \times G$, are both measurable with respect to $\mathfrak{B} \times \mathfrak{B}$, hence, the same is true

[21] *Translator's note:* Again, see footnote 17 concerning the proof of Lemma 3.1.31.

of their quotient $\alpha(hg)/\alpha(g)$. Since μ is finite, we may assume that[22] $\mu(G) = 1$. By Fubini's theorem, we know that

$$\tilde{\alpha}(h) = \int \frac{\alpha(hg)}{\alpha(g)}\, d\mu(g)$$

is a measurable function on $(G, \mathfrak{B})$, that is, $\tilde{\alpha} \in G^{\mathfrak{B}}$. In particular, $\tilde{\alpha}(h)$, regarded as a function of the two variables (g, h), is also measurable, hence

$$E = \left\{(g, h) \,\middle|\, \frac{\alpha(hg)}{\alpha(g)} \neq \tilde{\alpha}(h)\right\}$$

is a measurable set. Since

$$\mu\left(\left\{g \,\middle|\, \frac{\alpha(hg)}{\alpha(g)} \neq \tilde{\alpha}(h)\right\}\right) = 0$$

for every $h \in G$, it follows from Fubini's theorem that $(\mu \times \mu)(E) = 0$, and therefore

$$\mu\left(\left\{h \,\middle|\, \frac{\alpha(hg)}{\alpha(g)} \neq \tilde{\alpha}(h)\right\}\right) = 0$$

for almost every fixed value of g. We arbitrarily choose an element $g = g_0$ such that the above equation holds. Then

$$\alpha(hg_0) = \alpha(g_0)\, \tilde{\alpha}(h)$$

for almost all h in G. Writing $hg_0 = h_1$, it follows by the quasi-invariance of μ that

$$\alpha(h_1) = \frac{\alpha(g_0)}{\tilde{\alpha}(g_0)}\, \tilde{\alpha}(h_1)$$

for almost every h_1, that is, $\alpha = c\tilde{\alpha}$, where $c = \alpha(g_0)/\tilde{\alpha}(g_0)$.]

Example 3.2.6. Let G be a locally compact group, $\mathfrak{G}$ the group of all left-translations of G, and $(G, \mathfrak{B}, \nu)$ a left-invariant Haar measure space. Then, given any quasi-character α of $(G, \mathfrak{B}, \nu)$ relative to $\mathfrak{G}$, there exists a complex constant $c \in C$ such that $\alpha(g) = c\tilde{\alpha}(g)$ for almost every $g \in G$.

PROOF. Let G_0 be a subgroup of G constructed as in the proof of Lemma 1.1.11, and let $\mathfrak{B}_0$, ν_0 denote the restrictions to G_0 of $\mathfrak{B}$, ν, respectively. Then, the measure space $(G_0, \mathfrak{B}_0, \nu_0)$ is σ-finite, hence,

[22] *Translator's note:* Again, it is implicitly assumed that the measure μ is not identically zero.

there exists a finite measure μ on $(G_0, \mathfrak{B}_0)$, such that μ is equivalent to ν_0. Now, let α be any quasi-character of $(G, \mathfrak{B}, \nu)$, relative to $\mathfrak{G}$. Then, the restriction of α to G_0 (which we also denote by α) is a quasi-character of $(G_0, \mathfrak{B}_0, \nu_0)$, hence also a quasi-character of $(G_0, \mathfrak{B}_0, \mu)$, relative to the group $\mathfrak{G}_0$ of all left-translations of G_0. According to Example 3.2.5,[23] there exists a constant c, $|c| = 1$, such that

$$\alpha(g) = c\tilde{\alpha}(g) \tag{3.2.4}$$

for almost all $g \in G_0$; here, $\tilde{\alpha}(g)$ is the character of $\mathfrak{G}_0$ induced by α, which is just the restriction to G_0 of the character of G induced by α. Consequently, if h is an arbitrary element of G, then, for almost every g in the coset $h^{-1}G_0$, since $hg \in G_0$, we have

$$\alpha(g) = \tilde{\alpha}(h^{-1})\,\alpha(hg) = c\tilde{\alpha}(h^{-1})\,\tilde{\alpha}(hg) = c\tilde{\alpha}(g).$$

It follows that (3.2.4) holds for almost every $g \in G$.]

Examples 3.2.5 and 3.2.6 show that the idea of a quasi-character generalizes the notion of a measurable character.

We observe that, if $(G, \mathfrak{B}, \mu)$ is a localizable measure space which is quasi-invariant under the transformation group $\mathfrak{G}$, then any $\tilde{x} \in \tilde{G}^{\mu}$ is continuous relative to the μ-topology on $\mathfrak{G}$. In fact, given any $h \in \mathfrak{G}$, consider the operator $U(h)$ defined in §3.1, 6°. Then, for any appropriate choice of the vectors $\xi, \eta \in L^2(G, \mathfrak{B}, \mu)$, we have[24]

$$\tilde{x}(h) = \frac{(xU(h)\xi, \eta)}{(U(h)(x\xi), \eta)},$$

whence it is clear that $\tilde{x}(h)$ is continuous relative to the μ-topology.

We shall now make use of Lemma 3.1.12 to investigate the continuity of characters which are induced by quasi-characters.

Theorem 3.2.3. Let $(G, \mathfrak{B}, \mu)$ be a regular measure space[25] which is quasi-invariant under a group of continuous measurable transformations $\mathfrak{G}$, and suppose that $\mathfrak{G}$ is a topological group of the second category relative to an admissible topology. Then any quasi-character α of $(G, \mathfrak{B}, \mu)$ (relative to $\mathfrak{G}$) induces a continuous character on $\mathfrak{G}$.

[23] *Translator's note:* To apply Example 3.2.5, one must make some additional assumption to ensure the measurability of the mapping $(g, h) \to gh$.

[24] *Translator's note:* Given any $h_0 \in \mathfrak{G}$, and any nonzero vector $\xi \in L^2(G, \mathfrak{B}, \mu)$, we may choose $\eta = U(h_0)(x\xi)$, thus ensuring that the denominator is nonzero in some sufficiently small μ-neighborhood of h_0.

[25] *Translator's note:* Again, we must require that μ is not identically zero.

PROOF. Given any positive number ϵ, we partition the unit circle of the complex plane into a finite collection of disjoint Borel sets $A_1, \ldots, A_n$, such that the diameter of each A_k is less than ϵ. Accordingly, G is the union of the disjoint sets

$$B_k = \{g \mid \alpha(g) \in A_k\} \in \mathfrak{B}, \qquad k = 1, 2, \ldots, n. \tag{3.2.5}$$

Therefore, at least one of the sets (3.2.5), say B_{k_0}, has positive measure. By the regularity of μ, there exists a compact set $K \in \mathfrak{B}$ such that $K \subset \mathfrak{B}_{k_0}$ and $\mu(K) > 0$; in particular, if $g_1, g_2 \in K$, then

$$|\alpha(g_1) - \alpha(g_2)| < \epsilon. \tag{3.2.6}$$

Now, $\alpha(h^{-1}g)$ is almost everywhere equal to $\alpha(g)\,\tilde{\alpha}(h^{-1})$, where $\tilde{\alpha}$ is the character of $\mathfrak{G}$ induced by α. Hence, by virtue of Lemma 3.1.12,[26] there is a neighborhood V of the identity in $\mathfrak{G}$ such that, for any $h \in V$, there exists an element $g \in K \cap hK$ satisfying

$$\alpha(h^{-1}g) = \alpha(g)\,\tilde{\alpha}(h^{-1}). \tag{3.2.7}$$

Thus, if $h \in V$, then it follows from (3.2.6) and (3.2.7) that

$$|\tilde{\alpha}(h) - 1| = |1 - \tilde{\alpha}(h^{-1})| = |\alpha(g) - \alpha(h^{-1}g)| < \epsilon.$$

This shows that $\tilde{\alpha}$ is continuous at the identity of $\mathfrak{G}$. Hence, by Lemma 3.2.2, $\tilde{\alpha}$ is continuous at all points of $\mathfrak{G}$.]

Corollary 3.2.4. Let G be a topological group, let $\mathfrak{G}$ be a subgroup of G, and suppose that $\mathfrak{G}$ itself is a topological group of the second category with respect to a topology which is stronger than that induced by G. Let $(G, \mathfrak{B}, \mu)$ be a regular measure space which is left (or right) quasi-invariant under $\mathfrak{G}$. Then, every quasi-character α of $(G, \mathfrak{B}, \mu)$ (relative to $\mathfrak{G}$) induces a continuous character on $\mathfrak{G}$. In particular, every $\mathfrak{B}$-measurable character α of G, when restricted to $\mathfrak{G}$, defines a continuous character on $\mathfrak{G}$.

Remark. As a particular case of Corollary 3.2.4, if G is a locally compact group, $\mathfrak{B}$ is the σ-ring generated by the compact subsets of G, and μ is a left-invariant Haar measure, then every measurable character of G is continuous, and every quasi-character of $(G, \mathfrak{B}, \mu)$ (relative to the left translation group G) induces a continuous character on G. In fact, G is then a topological group of the second category (see Guan

[26] *Translator's note:* Recall, however, the additional finiteness restrictions on μ which were necessary to prove Lemma 3.1.12.

Zhao-zhi [1]), and $(G, \mathfrak{B}, \mu)$ is invariant (hence *a fortiori* quasi-invariant) under left translations. Applying Corollary 3.2.4, with $\mathfrak{G} = G$, the conclusion follows at once.

3° Topologies on Character Groups and Quasi-Character Groups

In the subsequent discussion, we shall require several kinds of topologies on character groups and quasi-character groups. We now proceed to describe these topologies.

I. Let G be a group, and let H be a group consisting of certain characters of G. Let $\alpha_0 \in H$, let $g_1, \ldots, g_n$ be any finite set of elements of G, and let ϵ be any positive number. Form the set

$$U(\alpha_0 ; g_1, \ldots, g_n, \epsilon) = \{\alpha \mid | \alpha(g_k) - \alpha_0(g_k)| < \epsilon, \quad k = 1, 2, \ldots, n\}.$$

For each $\alpha_0 \in H$, we take the totality of sets of the above form as a neighborhood basis at α_0. Thus, we obtain a topology on H, which we call the *weak topology*. This is, in fact, the weakest topology such that each $g \in G$, when considered as a function $\alpha(g)$, $\alpha \in H$, is continuous on H. It is easily seen that the weak topology makes H a topological group.

II. Let G be a topological group,[27] and let H be a group consisting of certain characters of G. Let $\alpha_0 \in H$, let Q be any compact subset of G, and let ϵ be any positive number. Form the set

$$V(\alpha_0, Q, \epsilon) = \{\alpha \mid \sup_{g \in Q} | \alpha(g) - \alpha_0(g)| < \epsilon\}.$$

For each $\alpha_0 \in H$, we take the totality of sets of the above form as a neighborhood basis at α_0. Thus, we obtain a topology on H, which we call the *strong topology*; it may be described as the topology of uniform convergence on compact subsets of G. It is easily verified that the strong topology makes H a topological group. Since every finite subset of G is compact, it follows that every weak neighborhood[28] in H is also a strong neighborhood, and consequently the strong topology is, in fact, stronger than the weak topology. On the other hand, there do exist topological groups G and corresponding groups of characters H such that the weak and strong topologies on H are actually different.

III. Again, let G be a topological group and let H be some group of

[27] *Translator's note:* Actually, the definition given here is applicable even if the group operations are not continuous relative to the topology of G.

[28] By a weak (strong) neighborhood, we mean, of course, a neighborhood in the weak (strong) topology.

characters on G. Let $\mathfrak{U}$ be a fixed open set in G. Then, for each $\alpha_0 \in H$ and positive number ϵ, we form the set

$$W(\alpha_0 ; \mathfrak{U}, \epsilon) = \{\alpha \mid \sup_{g \in \mathfrak{U}} |\alpha(g) - \alpha_0(g)| < \epsilon\}.$$

We take the totality of sets $W(\alpha_0 ; \mathfrak{U}, \epsilon)$ for all $\epsilon > 0$, as a neighborhood basis at α_0. In this manner, we obtain a topology T_U on H; we call T_U the $\mathfrak{U}$-*topology*. Again, H is a topological group relative to this topology.

Let Q be any compact subset of G. Then, there exist elements $g_1, \ldots, g_n \in G$, such that $Q \subset \sum_{k=1}^{n} g_k \mathfrak{U}$. Hence, if α satisfies

$$\sup_{g \in \mathfrak{U}} |\alpha(g) - \alpha_0(g)| < \epsilon, \qquad \max_{1 \leqslant k \leqslant n} |\alpha(g_k) - \alpha_0(g_k)| < \epsilon,$$

then

$$\sup_{g \in g_k \mathfrak{U}} |\alpha(g) - \alpha_0(g)| \leqslant \sup_{h \in \mathfrak{U}} |\alpha(h) - \alpha_0(h)| + |\alpha(g_k) - \alpha_0(g_k)| < 2\epsilon,$$

where $g = g_k h$. Thus, we have

$$V(\alpha_0 ; Q, 2\epsilon) \supset W(\alpha_0 ; \mathfrak{U}, \epsilon) \cap U(\alpha_0 ; g_1, \ldots, g_n, \epsilon),$$

whence it is clear that, if the $\mathfrak{U}$-topology is stronger than the weak topology, then it is also stronger than the strong topology.

Next, we shall consider certain topologies on groups of quasi-characters, or groups of measurable characters.

IV. Let $\Omega = (G, \mathfrak{B}, \mu)$ be a measure space, and let $L(\Omega)$ denote the Banach space formed by the totality of integrable functions on Ω. Let H be a family of bounded measurable functions on Ω, functions which are equal almost everywhere being regarded as identical. (Two cases occur most frequently in the subsequent applications: either Ω is quasi-invariant under a transformation group $\mathfrak{G}$, and H is a group of quasi-characters on Ω, relative to $\mathfrak{G}$, or G is a group and H is a group of $\mathfrak{B}$-measurable characters on G.) Each element α of H defines a continuous linear functional on $L(\Omega)$, as follows:

$$(f, \alpha) = \int_G f(g)\, \alpha(g)\, d\mu(g).$$

Thus, H may be imbedded[29] in the conjugate space of $L(\Omega)$. In this way,

[29] *Translator's note:* In order that the mapping $\alpha \to (\cdot, \alpha)$ be injective, it is necessary and sufficient that Ω satisfy condition (i) of Definition 1.2.3. But in any case, we may still define the μ-weak topology as the weakest topology on H such that the mapping $\alpha \to (\cdot, \alpha)$ is continuous [taking the weak topology on the conjugate space of $L(\Omega)$].

the weak topology on the conjugate space of $L(\Omega)$ induces a topology on H; a neighborhood basis for this topology may be constructed as follows. Let α_0 be an arbitrary point of H, choose any finite set of functions $f_1, \ldots, f_n$ in $L(\Omega)$, take any positive number ϵ, and form the set

$$X(\alpha_0 ; f_1, \ldots, f_n, \epsilon) = \{\alpha \mid |(f_k, \alpha) - (f_k, \alpha_0)| < \epsilon, \quad k = 1, \ldots, n\}.$$

We take the totality of sets of the above form as a neighborhood basis at the point α_0. The topology obtained in this manner will be called the *μ-weak topology*. If H is a group, then, in general, it does not necessarily follow that the μ-weak topology makes H a topological group.

Further on, however, we shall prove that, under certain circumstances, H does form a topological group relative to the μ-weak topology, and we shall compare the strengths of this and other topologies.

V. Let $(G, \mathfrak{B}, \mu)$ be a measure space which is quasi-invariant under a transformation group $\mathfrak{G}$, and let H be some group of quasi-characters[30] of $(G, \mathfrak{B}, \mu)$ relative to $\mathfrak{G}$ (in particular, if G is a group, then H may be just a group of $\mathfrak{B}$-measurable characters on G). Take any set $A \in \mathfrak{B}$, $\mu(A) < \infty$, and consider the function

$$N_A(\alpha) = \left(\int_A |\alpha(g) - 1|^2 \, d\mu(g)\right)^{1/2}$$

on H. Since

$$|\alpha\beta - 1| \leqslant |\alpha\beta - \beta| + |\beta - 1| = |\alpha - 1| + |\beta - 1|$$

for all $\alpha, \beta \in H$, it follows that $N_A(\cdot)$ is a pseudonorm on H. We use the family of pseudonorms $\{N_A(\cdot)\}$ to define a topology on H, as follows. Given any element $\alpha_0 \in H$, we choose an arbitrary set $A \in \mathfrak{B}$, $\mu(A) < \infty$, and an arbitrary positive number ϵ, and form the set

$$Y(\alpha_0 ; A, \epsilon) = \{\alpha \mid N_A(\alpha\alpha_0^{-1}) < \epsilon\}.$$

We then take the totality of such sets as a subbasis for the neighborhood system of α_0. The topology so obtained will be called the *μ-topology*. Using the fact that $N_A(\cdot)$ is a pseudonorm, it is easily verified that the μ-topology makes H a topological group. If μ is a finite measure, then the μ-topology is equivalent to the topology induced by the metric $N_G(\alpha\beta^{-1})$. To show this, it suffices to prove that every set of the form $Y(\alpha_0 ; A, \epsilon)$ is open in the topology induced by $N_G(\alpha\beta^{-1})$. Take any

[30] In particular, if $\mathfrak{G}$ contains only the identity transformation of G, then H is simply a multiplicative group of $\mathfrak{B}$-measurable functions α which satisfy the condition $|\alpha(g)| = 1$.

$\alpha_1 \in Y(\alpha_0 ; A, \epsilon)$, and let $\epsilon_1 = \epsilon - N_A(\alpha_1\alpha_0^{-1})$; then, since $N_A(\alpha) \leqslant N_G(\alpha)$, it is easily seen that

$$Y(\alpha_1 ; G, \epsilon_1) \subset Y(\alpha_0 ; A, \epsilon).$$

This shows that $Y(\alpha_0 ; A, \epsilon)$ is indeed an open set in the topology induced by the metric $N_G(\alpha\beta^{-1})$.

The pseudonorm $N_A(\cdot)$ may be used to introduce a pseudonorm on G_0^μ. Choose any $A \in \mathfrak{B}$ such that $0 < \mu(A) < \infty$, and define a function $N_A(\eta)$, $\eta \in G_0^\mu$, as follows:

$$N_A(\eta) = \inf_{\alpha\in\eta} N_A(\alpha).$$

It is easily verified that this is a pseudonorm on G_0^μ. If F is a subgroup of G_0^μ, and $\eta_0 \in F$, we may take the totality of sets of the form

$$Z(\eta_0 ; A, \epsilon) = \{\eta \mid N_A(\eta\eta_0^{-1}) < \epsilon\}$$

as a neighborhood basis at η_0. In this manner, we obtain a topology on F, which we also call the μ-topology. Relative to this topology, F forms a topological group. If μ is a finite measure, we write $N_G(\eta) = N(\eta)$, and in this case the μ-topology is equivalent to the topology induced by the metric[31] $N(\xi\eta^{-1})$, $\xi, \eta \in G_0^\mu$.

We now proceed to consider some properties of the topologies introduced above, and to give some examples.

Theorem 3.2.5. Let $\Omega = (G, \mathfrak{B}, \mu)$ be a measure space which is quasi-invariant under a transformation group $\mathfrak{G}$, and let H be some group of quasi-characters of Ω relative to $\mathfrak{G}$. Then, the μ-topology on H is stronger than the μ-weak topology.

PROOF. Given any $\alpha_0 \in H$ and any μ-weak neighborhood

$$X(\alpha_0 ; f_1 ,..., f_n , \epsilon),$$

we need only prove that there exists a set $A \in \mathfrak{B}$, $0 < \mu(A) < \infty$, and a positive number ϵ_1 such that

$$Y(\alpha_0 ; A, \epsilon_1) \subset X(\alpha_0 ; f_1 ,..., f_n , \epsilon).$$

Now, since $f_k \in L(G, \mathfrak{B}, \mu)$, there exists a set $A \in \mathfrak{B}$, $0 < \mu(A) < \infty$, such that

$$\int_{G-A} | f_k(g)| \, d\mu(g) < \frac{\epsilon}{4}, \qquad k = 1, 2,..., n,$$

[31] Since the subgroup $\mathfrak{N}$ is closed with respect to the metric $N(\alpha\beta^{-1})$, $\alpha, \beta \in G^\mu$, it follows that $N(\xi\eta^{-1})$, $\xi, \eta \in G_0^\mu$, is indeed a metric on G_0^μ. In fact, G_0^μ, with the metric $N(\xi\eta^{-1})$, $\xi, \eta \in G_0^\mu$, is just a quotient metric space of G_0^μ (relative to the metric $N_G(\alpha\beta^{-1})$, $\alpha, \beta \in G^\mu$).

and such that

$$\sup_{g \in A} |f_k(g)| < \infty, \qquad k = 1, 2, \ldots, n.$$

Choose

$$\epsilon_1 < \frac{\epsilon}{2(\mu(A))^{1/2} \sup_{g\in A} |f_k(g)|}, \qquad k = 1, 2, \ldots, n.$$

Then, if $\alpha \in Y(\alpha_0 ; A, \epsilon_1)$, we have

$$\begin{aligned} &\left| \int_G f_k(g)(\alpha(g) - \alpha_0(g))\, d\mu(g) \right| \\ &\qquad \leqslant \int_A |f_k(g)|\, |\alpha(g) - \alpha_0(g)|\, d\mu(g) \\ &\qquad\quad + \int_{G-A} |f_k(g)|\, |\alpha(g) - \alpha_0(g)|\, d\mu(g) \\ &\qquad < \sup_{g \in A} |f_k(g)|\, N_A(\alpha\alpha_0^{-1})(\mu(A))^{1/2} + 2 \cdot \frac{\epsilon}{4} < \epsilon. \end{aligned}$$

This shows that $Y(\alpha_0 ; Q, \epsilon_1) \subset X(\alpha_0 ; f_1, \ldots, f_n, \epsilon)$.]

Theorem 3.2.6. Let G be a topological space, and let $\Omega = (G, \mathfrak{B}, \mu)$ be a regular measure space which is quasi-invariant under a transformation group $\mathfrak{G}$. Let H be a group of quasi-characters of Ω (relative to $\mathfrak{G}$). Then, the μ-topology on H is weaker than the strong topology.

Proof. Let $\alpha_0 \in H$, and let $Y(\alpha_0 ; A, \epsilon)$ be a μ-neighborhood of α_0, where $\epsilon > 0$ and $A \in \mathfrak{B}$, $0 < \mu(A) < \infty$. By the regularity of μ, there is a compact set $Q \in \mathfrak{B}$ such that $Q \subset A$ and

$$\mu(A - Q) < \frac{\epsilon^2}{8}.$$

Choose

$$\epsilon_1 = \frac{\epsilon}{2(\mu(Q))^{1/2}}.$$

Then, if $\alpha \in V(\alpha_0 ; Q, \epsilon_1)$, we have

$$\begin{aligned} &\int_A |\alpha(g) - \alpha_0(g)|^2\, d\mu(g) \\ &\qquad \leqslant \int_{A-Q} |\alpha(g) - \alpha(g_0)|^2\, d\mu(g) + \mu(Q) \sup_{g \in Q} |\alpha(g) - \alpha_0(g)|^2 \\ &\qquad < \frac{\epsilon^2}{2} + \frac{\epsilon^2}{2} = \epsilon^2, \end{aligned}$$

that is, $N_A(\alpha\alpha_0^{-1}) < \epsilon$. Thus, $V(\alpha_0 ; Q, \epsilon_1) \subset Y(\alpha_0 ; A, \epsilon)$.]

Theorem 3.2.7. Let $\Omega = (G, \mathfrak{B}, \mu)$ be a finite measure space which is quasi-invariant under a transformation group $\mathfrak{G}$. Then, the group G^μ of all quasi-characters of Ω (relative to $\mathfrak{G}$) is complete relative to the metric $N_G(\alpha\beta^{-1})$, $\alpha, \beta \in G^\mu$.

PROOF. Let $\{\alpha_n\}$ be a fundamental sequence in G^μ, that is,

$$N_G(\alpha_n\alpha_m^{-1})^2 = \int_G |\alpha_n(g) - \alpha_m(g)|^2 \, d\mu(g) \to 0$$

as $m, n \to \infty$. Since the space $L^2(\Omega)$ is complete, there exists a quadratically integrable function α such that

$$\lim_{n\to\infty} \int_G |\alpha_n(g) - \alpha(g)|^2 \, d\mu(g) = 0, \tag{3.2.8}$$

furthermore, we can select a subsequence $\{\alpha_{n_k}(g)\}$ which converges to α almost everywhere. Therefore, since $|\alpha_{n_k}(g)| = 1$, we may choose α such that $|\alpha(g)| = 1$. Moreover, for any $h \in \mathfrak{G}$, it follows from the quasi-invariance of μ that

$$\frac{\alpha(hg)}{\alpha(g)} = \lim_{k\to\infty} \frac{\alpha_{n_k}(hg)}{\alpha_{n_k}(g)} = \lim_{k\to\infty} \tilde{\alpha}_{n_k}(h)$$

holds for almost all $g \in G$, that is, α is a quasi-character. And, by (3.2.8), $\lim_{n\to\infty} N_G(\alpha_n\alpha^{-1}) = 0$. Hence, G^μ is complete.]

Remark. If μ is not finite, but is localizable, then it can be proved that the multiply pseudometric space (for the definition, see Guan Zhao-zhi [1])[32] defined by the family of pseudometrics $\{N_A(\alpha\beta^{-1})\}$ is sequentially complete.

Corollary 3.2.8. Under the hypotheses of Theorem 3.2.7, the group $G_0{}^\mu$ is complete relative to the metric $N(\xi\eta^{-1})$, $\xi, \eta \in G_0{}^\mu$.

PROOF. Let $\{\xi_n\}$ be a fundamental sequence in $G_0{}^\mu$, that is, $\lim_{m,n\to\infty} N(\xi_m^{-1}\xi_n) = 0$. Then, we can select a subsequence $\{\xi_{n_k}\}$, $k = 1, 2, \ldots$, such that $N(\xi_{n_{k-1}}^{-1}\xi_{n_k}) < 1/2^k$. For each $k \geqslant 2$, choose an element $\alpha_k \in \xi_{n_{k-1}}^{-1}\xi_{n_k}$ such that $N(\alpha_k) < 1/2^k$, choose any element $\alpha_1 \in \xi_{n_1}$, and form the product $\beta_k = \prod_{l=1}^k \alpha_l$. Then, obviously $\beta_k \in \xi_{n_k}$, and when $l > k$, we have $\beta_k^{-1}\beta_l = \prod_{\nu=k+1}^l \alpha_\nu$, hence

$$N(\beta_k^{-1}\beta_l) < \sum_{\nu=k+1}^{l} N(\alpha_\nu) < \sum_{\nu=k+1}^{l} \frac{1}{2^\nu}.$$

[32] *Translator's note:* Unfortunately, the translator does not have access to the reference cited here, and has no idea whether the definition in question has been given or discussed elsewhere in the literature. Consequently, the word "multiply" in the English rendering can only be regarded as tentative.

Therefore, $\{\beta_k\}$ is a fundamental sequence in G^μ. By Theorem 3.2.7, there is an element $\beta \in G^\mu$ such that $\lim_{k\to\infty} N_G(\beta_k^{-1}\beta) = 0$. Let ξ be the residue class containing β:

$$\xi = \{\alpha\beta \mid \alpha \in \mathfrak{N}\}.$$

Then $\beta_k^{-1}\beta \in \xi_{n_k}^{-1}\xi$, hence

$$\lim_{k\to\infty} N(\xi_{n_k}^{-1}\xi) = 0,$$

and since $\lim_{m,k\to 0} N(\xi_{n_k}^{-1}\xi_m) = 0$, it follows easily that

$$\lim_{m\to\infty} N(\xi^{-1}\xi_m) = 0. \quad]$$

Next, we turn our attention to the mapping (3.2.3).

Theorem 3.2.9. Let $(G, \mathfrak{B}, \mu)$ be a regular measure space which is quasi-invariant under a group $\mathfrak{G}$ of continuous measurable transformations. Suppose that $\mathfrak{G}$ has an admissible topology (see Definition 3.1.2) which makes $\mathfrak{G}$ a topological group of the second category satisfying the first axiom of countability. Then, for any $A \in \mathfrak{B}$, with $\mu(A) > 0$, there exists a positive number c and a neighborhood V of the identity in $\mathfrak{G}$ such that

$$\sup_{h\in V} |\tilde{\alpha}(h) - 1| \leqslant c \int_A |\alpha(g) - 1| \, d\mu(g) \tag{3.2.9}$$

holds for all $\alpha \in G^\mu$.

Furthermore, if μ is a finite measure, then there exists an open set V in $\mathfrak{G}$, such that, if $\tilde{\mathfrak{G}}^\mu$ is given the V-topology, and $G_0{}^\mu$ is given the μ-topology, then the mapping S from $G_0{}^\mu$ onto $\tilde{\mathfrak{G}}^\mu$ is continuous.

Proof. Let $A \in \mathfrak{B}$, $\mu(A) > 0$. By Lemma 3.1.12,[33] there exists a neighborhood V_0 of the identity e in $\mathfrak{G}$ such that $\mu(A \cap h^{-1}A) > 0$ for every $h \in V_0$. Furthermore, by Corollary 3.1.18, there exists a neighborhood V_1 of e and a positive number c, such that (3.1.17) holds for every quasi-convex pair $(p, \tilde{p})$ for which $\int_A p(g)\, d\mu(g) > 0$, and every neighborhood V of e with $V \subset V_1$. We choose $V = V_0 \cap V_1$. Now, let $\alpha \in G^\mu$, and let $\tilde{\alpha}$ be the character induced by α. Consider the pair of functions

$$p(g) = |\alpha(g) - 1|, \qquad g \in G, \tag{3.2.10}$$

$$\tilde{p}(h) = |\tilde{\alpha}(h) - 1|, \qquad h \in \mathfrak{G}. \tag{3.2.11}$$

[33] *Translator's note:* See, however, footnote 9.

For almost every $g \in G$, we have

$$p(g) + p(gh^{-1}) = |\, \alpha(g) - 1 \,| + |\, \alpha(gh^{-1}) - 1 \,| \geqslant |\, \alpha(g) - \alpha(gh^{-1})|$$
$$= |\, \tilde{\alpha}(h^{-1}) - 1 \,| = \tilde{p}(h),$$

hence, $(p, \tilde{p})$ is a quasi-convex pair of functions over $((G, \mathfrak{B}, \mu), \mathfrak{G})$. First, suppose that $\int_A p(g)\, d\mu(g) = 0$. Then, $\alpha(g) = 1$ for almost all $g \in A$, hence $\alpha(hg) = \alpha(g) = 1$ for almost all $g \in A \cap h^{-1}A$. Therefore, if $h \in V$, so that $\mu(A \cap h^{-1}A) > 0$, it follows that $\tilde{\alpha}(h) = 1$; consequently, both sides of (3.2.9) vanish. On the other hand, if $\int_A p(g)\, d\mu(g) > 0$, then, since $V \subset V_1$, (3.1.17) holds. Thus, in either case, we have

$$\sup_{h \in V} |\, \tilde{\alpha}(h) - 1 \,| \leqslant c \int_A |\, \alpha(g) - 1 \,|\, d\mu(g) \leqslant c' N_A(\alpha), \qquad (3.2.12)$$

where $c' = c(\mu(A))^{1/2}$. Let $\eta = \alpha\mathfrak{N}$ be the residue class containing α. Taking the supremum of the right-hand side of (3.2.12) over all α in η, we get

$$\sup_{h \in V} |\, \tilde{\alpha}(h) - 1 \,| \leqslant c' N_A(\eta).$$

Hence, given any point $\eta_0 \in G_0{}^{\mu}$, and any V-neighborhood

$$\{\tilde{\alpha} \mid \sup_{h \in V} |\, \tilde{\alpha}(h) - \tilde{\alpha}_0(h)| < \epsilon\}$$

of the point $\tilde{\alpha}_0 = S\eta_0$, there exists a μ-neighborhood $Z(\eta_0\,; A, \epsilon/c')$ of η_0 such that

$$SZ(\eta_0\,; A, \epsilon/c') \subset \{\tilde{\alpha} \mid \sup_{h \in V} |\, \tilde{\alpha}(h) - \tilde{\alpha}_0(h)| < \epsilon\},$$

which shows that S is continuous.]

Theorem 3.2.10. Let $(G, \mathfrak{B}, \mu)$ be a regular measure space which is strongly quasi-invariant and ergodic with respect to a group $\mathfrak{G}$ of continuous measurable transformations. Suppose that $\mathfrak{G}$ has an admissible topology which makes $\mathfrak{G}$ a connected topological group of the second category. Choose any nonzero element ψ of $L(G, \mathfrak{B}, \mu)$ such that $\psi(g) \geqslant 0$, $g \in G$. Form the convex function

$$N_\psi(\alpha) = \left(\int_G |\, \alpha(g) - 1 \,|^2\, \psi(g)\, d\mu(g)\right)^{1/2}, \qquad \alpha \in G^{\mu}.$$

Then, the topology on G^{μ} induced by N_ψ is independent of the choice of ψ, and is, in fact, equivalent to the μ-topology.

PROOF. We shall first prove that, given any sequence $\{\alpha_n\} \subset G^\mu$ such that $\lim_{n\to\infty} N_\psi(\alpha_n) = 0$, there exists a subsequence $\{\alpha_{n'}\}$ which converges to 1 almost everywhere on G. Choose a positive number a, sufficiently small so that the set $A = \{g \mid \psi(g) > a\}$ has finite positive measure. Then, by virtue of the inequality

$$\mu\{g \mid | \alpha_n(g) - 1 | \geqslant \epsilon, g \in A\} \leqslant \frac{1}{\epsilon^2 a} N_\psi{}^2(\alpha_n),$$

we know that $\{\alpha_n\}$ converges in measure to 1 on the set A, hence, there exists a subsequence $\{\alpha_{n'}\}$ which converges to 1 almost everywhere on A. Consider the set

$$E = \{g \mid \lim_{n'\to\infty} \alpha_{n'}(g) = 1\}.$$

Since A is contained in E (modulo a null set), we have $\mu(E) > 0$. We shall prove that $\mu(G - E) = 0$. In view of the ergodicity of μ with respect to $\mathfrak{G}$, it suffices to prove that E is a quasi-ergodic set. Now, by the regularity of μ, there is a compact set $K \in \mathfrak{B}$, $0 < \mu(K)$, such that $K \subset E$. According to Lemma 3.1.12,[34] there is a neighborhood V of the identity in $\mathfrak{G}$, such that

$$\mu(K \cap hK) > 0$$

for all $h \in V$. Therefore, for each $h \in V$, there is an element $g \in K \cap hK$, such that

$$\alpha_{n'}(h^{-1}g) = \tilde{\alpha}_{n'}(h^{-1})\, \alpha_{n'}(g) \tag{3.2.13}$$

holds for all n'. But $g \in K \cap hK$ implies that $g \in E$, $h^{-1}g \in E$, hence, it follows that

$$\lim_{n'\to\infty} \tilde{\alpha}_{n'}(h) = 1 \tag{3.2.14}$$

for every $h \in V$. Since $\tilde{\alpha}$ is a character of $\mathfrak{G}$, the set

$$\mathfrak{G}_1 = \{h \mid \lim_{n'\to\infty} \tilde{\alpha}_{n'}(h) = 1\}$$

is a subgroup of $\mathfrak{G}$, and, by (3.2.14), we have $\mathfrak{G}_1 \supset V$. By virtue of the connectedness of $\mathfrak{G}$, it follows that $\mathfrak{G}_1 = \mathfrak{G}$, that is, $\lim_{n'\to\infty} \tilde{\alpha}_{n'}(h) = 1$ for all $h \in \mathfrak{G}$. On the other hand, for each $h \in \mathfrak{G}$, there is a null set G_h in G such that (3.2.13) holds for all n', provided $g \,\overline{\in}\, G_h$. Consequently,

[34] *Translator's note:* Again, recall the qualification concerning the validity of Lemma 3.1.12.

$g \in hE - G_h$ implies that $g \in E$; thus, $hE - E$ is a null set. This shows that E is indeed a quasi-invariant set, and therefore

$$\lim_{n' \to \infty} \alpha_{n'}(g) = 1 \tag{3.2.15}$$

holds for almost all $g \in G$. Hence, for any set $A \in \mathfrak{B}$, with $\mu(A) < \infty$, it follows by the Lebesgue dominated convergence theorem that

$$\lim_{n' \to \infty} N_A(\alpha_{n'}) = 0.$$

From this, we easily deduce that the topology induced by N_ψ is stronger than the μ-topology.

Conversely, given any positive number ϵ, we choose a set $A \in \mathfrak{B}$, $0 < \mu(A) < \infty$, such that

$$\int_{G-A} \psi(g)\, d\mu(g) < \frac{\epsilon^2}{16}$$

and

$$c = \sup_{g \in A} | \psi(g)| < \infty.$$

Then, for any $\alpha \in \{\alpha \mid N_A(\alpha) < \epsilon/2\sqrt{c}\}$, we have

$$N_\psi(\alpha) \leqslant 2\left(\int_{G-A} \psi(g)\, d\mu(g)\right)^{1/2} + \sqrt{c}\, N_A(\alpha) < \epsilon,$$

that is, $\{\alpha \mid N_A(\alpha) < \epsilon/2\sqrt{c}\} \subset \{\alpha \mid N_\psi(\alpha) < \epsilon\}$. Therefore, the μ-topology is stronger than the topology induced by N_ψ. Thus, the topology induced by N_ψ is identical with the μ-topology, and is therefore independent of the choice of ψ.]

The above theorem may be used as a criterion for the equivalence of two measures. In this connection, see Theorem 4.2.14 (which is an analog of Theorem 3.2.10) and Corollary 4.2.15.

Theorem 3.2.11. Let $\Omega = (G, \mathfrak{B}, \mu)$ be a localizable measure space which is quasi-invariant under a transformation group $\mathfrak{G}$. Suppose that $\mathfrak{G}$ has been given a topology which is stronger than the μ-topology. Let $\tilde{\mathfrak{G}}^\mu$ be given the strong topology, and let G^μ be given the μ-weak topology. Then, the mapping $\alpha \to \tilde{\alpha}$, from G^μ to $\tilde{\mathfrak{G}}^\mu$, is continuous.

PROOF. First, we shall prove that, for any fixed $f \in L(\Omega)$, $G^\mu \times \mathfrak{G}$, the function

$$\varphi(\alpha, h) = \int \alpha(hg) f(g)\, d\mu(g), \qquad (\alpha, h) \in G^\mu \times \mathfrak{G},$$

is jointly continuous in the variables α, h. Choose any two functions ξ, $\eta \in L^2(\Omega)$, such that $\xi(g)\,\overline{\eta(g)} = f(g)$; then, we have

$$\varphi(\alpha, h) = \int \alpha(g) f(h^{-1}g) \frac{d\mu_h(g)}{d\mu(g)}\, d\mu(g) = (\alpha U(h)\xi, U(h)\eta),$$

where $U(h)$ is the operator introduced in §3.1. Hence,

$$\begin{aligned} |\varphi(\alpha, h) - \varphi(\alpha_1, h_1)| &\leqslant \left| \int (\alpha(g) - \alpha_1(g)) f(h^{-1}g) \frac{d\mu_h(g)}{d\mu(g)}\, d\mu(g) \right| \\ &\quad + |(\alpha_1(U(h) - U(h_1))\xi, U(h)\eta)| \\ &\quad + |(\alpha_1 U(h_1)\xi, (U(h) - U(h_1))\eta| \\ &\leqslant \left| \int (\alpha(g) - \alpha_1(g)) f_1(g)\, d\mu(g) \right| \\ &\quad + \|(U(h) - U(h_1))\xi\| \, \|\eta\| \\ &\quad + \|(U(h) - U(h_1))\eta\| \, \|\xi\|, \end{aligned}$$

where $f_1(g) \in L(\Omega)$. Therefore, given any positive number ϵ, if h belongs to the neighborhood

$$\left\{ h \,\middle|\, \|(U(h) - U(h_1))\xi\| < \frac{\epsilon}{\|\xi\| + \|\eta\|},\ \|(U(h) - U(h_1))\eta\| < \frac{\epsilon}{\|\xi\| + \|\eta\|} \right\}$$

of h_1, and if α belongs to the neighborhood

$$\left\{ \alpha \,\middle|\, \left| \int (\alpha(g) - \alpha_1(g)) f_1(g)\, d\mu(g) \right| < \epsilon \right\}$$

of α_1, then $|\varphi(\alpha, h) - \varphi(\alpha_1, h_1)| < 2\epsilon$. This proves the continuity of $\varphi(\alpha, h)$. Now, given any $\alpha_1 \in G^{\mu}$, we choose an $f \in L(\Omega)$ such that

$$\int \alpha_1(g) f(g)\, d\mu(g) \neq 0.$$

Also, we choose a sufficiently small neighborhood V of α_1, such that

$$\int \alpha(g) f(g)\, d\mu(g) \neq 0$$

for all $\alpha \in V$. Thus, when $\alpha \in V$, the character

$$\tilde{\alpha}(h) = \frac{\int \alpha(hg) f(g)\, d\mu(g)}{\int \alpha(g) f(g)\, d\mu(g)}$$

is a continuous function of (α, h). It follows that, for any positive number ϵ and any compact set Q in $\mathfrak{G}$, the inverse image (under the mapping $\alpha \to \tilde{\alpha}$) of the strong neighborhood

$$\{\tilde{\alpha} \mid \sup_{h \in Q} |\tilde{\alpha}(h) - \tilde{\alpha}_1(h)| < \epsilon\}$$

(of $\tilde{\alpha}_1 \in \tilde{\mathfrak{G}}^\mu$) is open in G^μ. From this, one easily deduces that the mapping $\alpha \to \tilde{\alpha}$ is continuous.]

Corollary 3.2.12. Let G be a locally compact topological group, and $(G, \mathfrak{B}, \mu)$ a left-invariant Haar measure space. Then, the μ-weak topology and the strong topology on G^* are identical.

Proof. By Lemma 3.1.29, the original topology of G is stronger than the μ-topology. Following the argument used in the proof of Theorem 3.2.11, it is easily seen that the μ-weak topology on G^* is stronger than the strong topology. On the other hand, it follows from Theorems 3.2.5 and 3.2.6 that the strong topology on G^* is stronger than the μ-weak topology. Hence, these two topologies are identical.]

The next corollary is an immediate consequence of Theorems 3.2.5 and 3.2.11.

Corollary 3.2.13. Under the hypotheses of Theorem 3.2.11, if G^μ is given the μ-topology, then the mapping $\alpha \to \tilde{\alpha}$ is also continuous.

Note, however, that the conclusion of Corollary 3.2.13 is weaker than that of Theorem 3.2.9.

Theorem 3.2.14. Let $\Omega = (G, \mathfrak{B}, \mu)$ be a localizable measure space which is quasi-invariant and ergodic with respect to a transformation group $\mathfrak{G}$. Let $\mathfrak{H} = \{c\alpha \mid \alpha \in G^\mu, 0 \leqslant c \leqslant 1\}$. Then $\mathfrak{H}$ is compact with respect to the μ-weak topology.

Proof. By Theorem 2.4.14, the conjugate space of $L(\Omega)$ is isomorphic to $L^\infty(\Omega)$, the totality of essentially bounded measurable functions on Ω. Moreover, since the unit ball in the conjugate space of $L(\Omega)$ is compact in the weak topology, the same is true of the unit ball B in $L^\infty(\Omega)$. Hence, we need only prove that $\mathfrak{H}$ is a weakly closed subset of B, and it will follow that $\mathfrak{H}$ itself is weakly compact.

Let $\{\alpha_\lambda, \lambda \in \Lambda\}$ be a generalized sequence in $\mathfrak{H}$ which converges weakly to an element $\alpha \in B$. Since $0 \in \mathfrak{H}$, we need only consider the case $\alpha \neq 0$, and so we may as well suppose that $\alpha_\lambda \neq 0$ for all $\lambda \in \Lambda$. If $\alpha_\lambda = c_\lambda \beta_\lambda$, where $0 < c_\lambda \leqslant 1$ and $\beta_\lambda \in G^\mu$, then we write $\tilde{\alpha}_\lambda = \tilde{\beta}_\lambda$. We shall now prove that, for every fixed $h \in \mathfrak{G}$, the generalized sequence of complex numbers $\{\tilde{\alpha}_\lambda(h), \lambda \in \Lambda\}$ is fundamental.

Since Ω is localizable and quasi-invariant under the transformation h, it follows that the measure space $(G, \mathfrak{B}, \mu_h)$ [see (3.1.1)] is also localizable. Let $f \in L(\Omega)$. Then, by Theorem 2.4.15, we also have

$$f(h^{-1}g) \frac{d\mu_h(g)}{d\mu(g)} \in L(\Omega).$$

Therefore, given any $\epsilon > 0$, since $\lim_{\lambda \in \Lambda} \alpha_\lambda = \alpha$, there is an index $\lambda_0 \in \Lambda$ such that both of the inequalities

$$\left| \int (\alpha(g) - \alpha_\lambda(g)) f(g) \, d\mu(g) \right| < \epsilon \tag{3.2.16}$$

and

$$\left| \int (\alpha(g) - \alpha_\lambda(g)) f(h^{-1}g) \frac{d\mu_h(g)}{d\mu(g)} \, d\mu(g) \right| < \epsilon \tag{3.2.17}$$

hold whenever $\lambda_0 \prec \lambda$. Replacing g by hg in (3.2.17), we obtain

$$\left| \int (\alpha(hg) - \alpha_\lambda(hg)) f(g) \, d\mu(g) \right| < \epsilon. \tag{3.2.18}$$

Since α is not almost everywhere zero, we may choose f so that

$$b = \int \alpha(g) f(g) \, d\mu(g) \neq 0. \tag{3.2.19}$$

Using (3.2.16), (3.2.19), and the relation $\alpha_\lambda(hg) = \tilde{\alpha}_\lambda(h) \, \alpha_\lambda(g)$, we easily calculate that, if ϵ is sufficiently small, and $\lambda_0 \prec \lambda, \lambda'$, then

$$| \tilde{\alpha}_\lambda(h) - \tilde{\alpha}_{\lambda'}(h) | < \frac{4\epsilon}{| b |}.$$

Consequently, there is a complex number $\tilde{\alpha}(h)$, of modulus 1, such that

$$\lim_{\lambda \in \Lambda} \tilde{\alpha}_\lambda(h) = \tilde{\alpha}(h). \tag{3.2.20}$$

Furthermore, we know from (3.2.16) and (3.2.18) that, if $\lambda_0 \prec \lambda$, then

$$\left| \int (\alpha(hg) - \tilde{\alpha}_\lambda(h) \, \alpha(g)) f(g) \, d\mu(g) \right| \leqslant 2\epsilon.$$

Hence, using (3.2.20), we obtain

$$\int (\alpha(hg) - \tilde{\alpha}(h) \, \alpha(g)) f(g) \, d\mu(g) = 0, \tag{3.2.21}$$

which holds for all functions f in $L(\Omega)$ which satisfy (3.2.19). It follows easily that

$$\alpha(hg) = \tilde{\alpha}(h)\,\alpha(g) \tag{3.2.22}$$

for almost all $g \in G$. Since $|\,\tilde{\alpha}(h)| = 1$, it is obvious from (3.2.22) that, for any two positive numbers $c_1 < c_2$, the set

$$\{g \mid c_1 < |\,\alpha(g)| < c_2\}$$

is quasi-invariant. Consequently, there must exist a positive number c such that $|\,\alpha(g)|$ is equal to c almost everywhere, and so we may, in fact, assume that $|\,\alpha(g)| = c$ for all $g \in G$. Thus, setting $\beta = \alpha/c$, we see from (3.2.22) that $\beta \in G^{\mu}$, that is, $\alpha \in \mathfrak{H}$.]

Corollary 3.2.15. Let G be a locally compact group, and let $(G, \mathfrak{B}, \mu)$ be a left-invariant Haar measure space. Then G^*, the totality of continuous characters of G, is locally compact with respect to the μ-weak topology.

PROOF. Let $\mathfrak{H}_1 = G^* \cup \{0\}$. We shall prove that $\mathfrak{H}_1$ is a closed subset of B (see the proof of Theorem 3.2.14). Let $\{\alpha_\lambda\,,\ \lambda \in \Lambda\}$ be any μ-weakly convergent generalized sequence in $\mathfrak{H}_1$, and write $\alpha = \lim_{\lambda\in\Lambda} \alpha_\lambda$. If $\alpha \neq 0$, then, proceeding as in the proof of Theorem 3.2.14, one can show that there is an element $\beta \in G^{\mu}$ such that $\alpha = c\beta$, $0 < c \leqslant 1$. Furthermore, by Example 3.2.6 and the remark following Corollary 3.2.4, there is an $\tilde{\alpha} \in G^*$ such that $\beta = e^{i\theta}\tilde{\alpha}$. Thus, there is a complex number $\kappa = e^{i\theta}c$, with $0 < |\,\kappa\,| \leqslant 1$, such that

$$\alpha = \kappa\tilde{\alpha}, \qquad \tilde{\alpha} \in G^*.$$

To prove that $\alpha \in G^*$, we only show that $\kappa = 1$. Choose a continuous function φ on G, vanishing outside of some compact set, and such that

$$\int \tilde{\alpha}(g)\,\varphi(g)\,d\mu(g) \neq 0.$$

Then, $\varphi \in L(G, \mathfrak{B}, \mu)$, and also

$$f(g) = \int \varphi(h^{-1}g)\,\varphi(h)\,d\mu(h) \in L(G, \mathfrak{B}, \mu).$$

Moreover, using Fubini's Theorem, one may easily calculate that, for every $\gamma \in G^*$,

$$\int \gamma(g) f(g)\,d\mu(g) = \left(\int \gamma(h)\,\varphi(h)\,d\mu(h)\right)^2.$$

Obviously, we may assume that $\alpha_\lambda \in G^*$, hence, applying the preceding formula, we get

$$\kappa^2 \left(\int \tilde{\alpha}(g)\, \varphi(g)\, d\mu(g)\right)^2$$
$$= \lim_{\lambda \in \Lambda} \left(\int \alpha_\lambda(g)\, \varphi(g)\, d\mu(g)\right)^2 = \lim_{\lambda \in \Lambda} \int \alpha_\lambda(g) f(g)\, d\mu(g)$$
$$= \kappa \int \tilde{\alpha}(g) f(g)\, d\mu(g) = \kappa \left(\int \tilde{\alpha}(g)\, \varphi(g)\, d\mu(g)\right)^2.$$

This shows that $\kappa = 0$ or $\kappa = 1$, but since we have assumed that $\alpha \neq 0$, we must have $\kappa = 1$. Therefore, $\mathfrak{H}_1$ is closed in B, and hence compact. Moreover, it is easily seen that $\mathfrak{H}_1$ is Hausdorff, therefore, G^* is locally compact with respect to the μ-weak topology. **]**

Using Corollaries 3.2.12 and 3.2.15, we immediately obtain the following result.

Corollary 3.2.16. Let G be a locally compact group, and G^* the group of all continuous characters of G. Then, when given the strong topology, G^* becomes a locally compact topological group.

4° Characteristic and Quasi-Characteristic Functions

We now proceed to describe some commonly used σ-algebras on groups.

Definition 3.2.3. Let G be a group, and $\mathfrak{H}$ a subgroup of G'. Any set in $\mathfrak{H}$ of the form

$$\{\alpha \mid (\alpha(g_1), \ldots, \alpha(g_n)) \in E\},$$

where n is an arbitrary natural number, E is an arbitrary Borel set in complex n-dimensional space, and $g_1, \ldots, g_n$ are arbitrary elements of G, is called a *Borel cylinder* with *base* E (corresponding to $\{g_1, \ldots, g_n\}$). Let $\mathfrak{B}$ be the smallest σ-algebra containing all the Borel cylinders in $\mathfrak{H}$. The members of $\mathfrak{B}$ are called *weak Borel sets* in $\mathfrak{H}$ (corresponding to G).

Similarly, any set in G of the form

$$\{g \mid (\alpha_1(g), \ldots, \alpha_n(g)) \in E\},$$

where n is an arbitrary natural number, E is an arbitrary Borel set in complex n-dimensional space, and $\alpha_1, \ldots, \alpha_n$ are arbitrary elements of $\mathfrak{H}$, is called a *Borel cylinder* with *base* E (corresponding to $\{\alpha_1, \ldots, \alpha_n\}$). Let $\mathfrak{F}$ be the smallest σ-algebra containing all Borel cylinders in G. The members of $\mathfrak{F}$ are called *weak Borel sets* in G (corresponding to $\mathfrak{H}$).

Definition 3.2.4. Let G be a group, let $(G, \mathfrak{B}, \mu)$ be a finite measure space, and let $\mathfrak{H}$ be a subgroup of $G^{\mathfrak{B}}$. Then, the function

$$\varphi(\beta) = \int_G \beta(g)\, d\mu(g), \qquad \beta \in \mathfrak{H}$$

is called the *characteristic function* of $(G, \mathfrak{B}, \mu)$ on $\mathfrak{H}$.

In what follows, we shall usually assume, for convenience, that $\varphi(1) = \mu(G) = 1$. With reference to Definition 3.2.4, we shall frequently deal with the following type of situation. Namely, $\mathfrak{H}$ is a group, G is a subgroup of $\mathfrak{H}'$, and $\mathfrak{B}$ is the totality of weak Borel sets in G. To each $h \in \mathfrak{H}$, there corresponds an element $\alpha \in G'$, defined as follows:

$$\alpha(g) = g(h), \qquad g \in G.$$

Clearly, $\alpha \in G^{\mathfrak{B}}$. In this fashion, $\mathfrak{H}$ is imbedded[35] in $G^{\mathfrak{B}}$. Identifying each h with the corresponding α, we obtain the characteristic function of $(G, \mathfrak{B}, \mu)$ on $\mathfrak{H}$, defined by

$$\varphi(h) = \int_G g(h)\, d\mu(g), \qquad h \in \mathfrak{H}.$$

Example 3.2.7. Let $(R, \mathfrak{B})$ be the measurable space discussed in Example 3.2.4. In this case, $R^{\mathfrak{B}}$ is precisely the totality of functions of the form $\alpha_t(x) = e^{itx}$, $x \in R$, $t \in R$. If we identify α_t with t, then the characteristic function of a probability measure space $(R, \mathfrak{B}, \mu)$ is just the Fourier–Stieltjes transform:

$$\varphi(t) = \int_R e^{itx}\, d\mu(x).$$

Definition 3.2.5. Let $(G, \mathfrak{B}, \mu)$ be a finite measure space which is quasi-invariant under a transformation group $\mathfrak{G}$. Form a complex-valued function φ, on the quasi-character group G^{μ}, as follows:

$$\varphi(\alpha) = \int_G \alpha(g)\, d\mu(g), \qquad \alpha \in G^{\mu}.$$

We call φ the *quasi-characteristic function* corresponding to $(G, \mathfrak{B}, \mu)$ and the group $\mathfrak{G}$.

Again, for convenience, we shall assume that $\varphi(1) = \mu(G) = 1$.

[35] *Translator's note*: Of course, for the correspondence $h \to \alpha$ to be injective, certain additional assumptions must be made. In any case, the characteristic function $\varphi(h)$ can still be defined in the indicated manner.

Since $G^{\mathfrak{B}} \subset G^{\mu}$, the quasi-characteristic function is an extension of the characteristic function.

If $\mathfrak{N}$ and C are isomorphic, that is, if μ is ergodic, then the absolute value $|\varphi(\alpha)|$ depends only upon the coset of $\mathfrak{N}$ containing α. In fact, if α, $\alpha' \in \eta \in G_0^{\mu}$, then there exists a number $c \in C$, such that $\alpha(g) = c\alpha'(g)$, and hence $\varphi(\alpha) = c\varphi(\alpha')$, that is, $|\varphi(\alpha)| = |\varphi(\alpha')|$. Thus, we may regard $|\varphi(\alpha)|$ as a function on G_0^{μ}, and we write $\psi(\eta) = |\varphi(\alpha)|$, $\alpha \in \eta$. Alternatively, we may regard $|\varphi(\alpha)|$ as a function on $\tilde{\mathfrak{G}}^{\mu}$, that is,

$$\tilde{\varphi}(\tilde{\alpha}) = |\varphi(\alpha)|, \qquad \tilde{\alpha} \in \tilde{\mathfrak{G}}^{\mu},$$

where α is any quasi-character corresponding to the character $\tilde{\alpha}$. When referring to the functions ψ, $\tilde{\varphi}$, we shall also use the term *quasi-characteristic function*, since they differ from $|\varphi(\alpha)|$ only in having different domains of definition.

Observe that the quasi-characteristic function and the norm $N_G(\alpha)$, introduced in 3°, are related by the formula

$$N_G(\alpha)^2 = 2(1 - \mathfrak{R}\varphi(\alpha)), \qquad \alpha \in G^{\mu}. \tag{3.2.23}$$

In fact, if $\alpha \in G^{\mu}$, then

$$|1 - \alpha(g)|^2 = 1 - 2\mathfrak{R}\alpha(g) + |\alpha(g)|^2 = 2(1 - \mathfrak{R}\alpha(g)),$$

hence,

$$N_G(\alpha)^2 = \int_G |\alpha(g) - 1|^2 \, d\mu(g) = 2\int_G (1 - \mathfrak{R}\alpha(g)) \, d\mu(g),$$

and using $\varphi(1) = 1$, we obtain (3.2.23).

In case $\mathfrak{N}$ and C are isomorphic, we have

$$\sup_{\alpha\in\eta} \mathfrak{R}\varphi(\alpha) = \sup_{c\in C} \mathfrak{R}c\varphi(\alpha) = |\varphi(\alpha)|, \qquad \alpha \in \eta,$$

hence, from (3.2.23), we obtain

$$N(\eta)^2 = 2(1 - \psi(\eta)), \qquad \eta \in G^{\mu}. \tag{3.2.24}$$

Lemma 3.2.17. Let μ be a finite measure space which is quasi-invariant under a transformation group $\mathfrak{G}$. Let G_0^{μ} be given the topology induced by the metric $N(\xi\eta^{-1})$, and let $\mathfrak{T}$ be any topology which makes $\tilde{\mathfrak{G}}^{\mu}$ a topological group. Then, the inverse S^{-1} of the mapping defined by (3.2.3) is continuous if and only if the quasi-characteristic function $\tilde{\varphi}(\tilde{\alpha})$ on $(\tilde{\mathfrak{G}}^{\mu}, \mathfrak{T})$ is continuous at the point $\tilde{\alpha} = 1$.

PROOF. Suppose that $\tilde{\varphi}$ is continuous at the point 1. Then, given any positive number ϵ, there is a $\mathfrak{T}$-neighborhood V of 1 in $\mathfrak{G}^\mu$ such that $2(1 - \tilde{\varphi}(\tilde{\alpha})) < \epsilon^2$ whenever $\tilde{\alpha} \in V$ [since $\tilde{\varphi}(1) = 1$]. Hence, if $\tilde{\alpha} \in V$, it then follows by (3.2.24) that $N(\eta) < \epsilon$, where $\eta = S^{-1}\tilde{\alpha}$. Therefore,

$$S^{-1}V \subset Z(1; G, \epsilon). \tag{3.2.25}$$

Thus, S^{-1} is continuous at the point 1. Since S^{-1} is an algebraic isomorphism between the topological groups $\mathfrak{G}^\mu$ and $G_0{}^\mu$, it follows that S^{-1} is continuous throughout $\mathfrak{G}^\mu$.

Conversely, assume that the inverse mapping S^{-1} is continuous at the point 1. Then, given any positive number ϵ, there is a neighborhood V of 1 such that (3.2.25) holds. Hence, if $\tilde{\alpha} \in V$, then $0 \leqslant 1 - \tilde{\varphi}(\tilde{\alpha}) < \epsilon^2/2$. Thus, $\tilde{\varphi}$ is continuous at 1.]

We also observe that the continuity of $\tilde{\varphi}(\tilde{\alpha})$ at the point $\tilde{\alpha} = 1$ implies continuity throughout $\mathfrak{G}^\mu$. In fact, it is easily shown [see the proof of (3.3.3)] that

$$|\varphi(\alpha) - \varphi(\alpha_0)| \leqslant 2(1 - \Re\varphi(\alpha\alpha_0^{-1}))$$

for any $\alpha, \alpha_0 \in G^\mu$. Hence, for any $\tilde{\alpha}, \tilde{\alpha}_0 \in \mathfrak{G}^\mu$, and any $c \in C$, we have

$$\begin{aligned} |\tilde{\varphi}(\tilde{\alpha}) - \tilde{\varphi}(\tilde{\alpha}_0)| &= |\,|\varphi(\alpha)| - |\varphi(\alpha_0)|\,| \\ &\leqslant |c\varphi(\alpha) - \varphi(\alpha_0)| \leqslant 2(1 - \Re c\varphi(\alpha\alpha_0^{-1})), \end{aligned}$$

whence one easily deduces that

$$|\tilde{\varphi}(\tilde{\alpha}) - \tilde{\varphi}(\tilde{\alpha}_0)| \leqslant 2(1 - \tilde{\varphi}(\tilde{\alpha}\tilde{\alpha}_0^{-1})).$$

From this inequality, it is obvious that $\tilde{\varphi}$ is continuous at $\tilde{\alpha}_0$ provided it is continuous at 1.

Definition 3.2.6. Let G be a group, let $(G, \mathfrak{B}, \mu)$ be a finite measure space, let $\mathfrak{H}$ be a subgroup of $G^{\mathfrak{B}}$, and let $\mathfrak{T}$ be a topology on $\mathfrak{H}$, stronger than the weak topology, such that $(\mathfrak{H}, \mathfrak{T})$ is a topological group. Then, μ is said to be *continuous* with respect to the topology $\mathfrak{T}$, provided that, given any two positive numbers ϵ, η, there exists a $\mathfrak{T}$-neighborhood V of the identity 1 in $\mathfrak{H}$ such that

$$\mu(\{g \mid |1 - \beta(g)| \geqslant \eta\}) < \epsilon \tag{3.2.26}$$

for all $\beta \in V$.

Lemma 3.2.18. Under the hypotheses of Definition 3.2.6, μ is continuous with respect to $\mathfrak{T}$ if and only if the characteristic function of $(G, \mathfrak{B}, \mu)$ on $(\mathfrak{H}, \mathfrak{T})$ (see Definition 3.2.4) is continuous.

Proof. Suppose that the characteristic function φ is continuous. Then, given any two positive numbers ϵ and η, there is a $\mathfrak{T}$-neighborhood V of the identity 1 in $\mathfrak{H}$, such that

$$1 - \Re\varphi(\beta) < \frac{\eta^2}{2}\epsilon \tag{3.2.27}$$

for all $\beta \in V$. Also, for any $\beta \in \mathfrak{H}$,

$$\begin{aligned} 1 - \Re\varphi(\beta) &= \int_G (1 - \Re\beta(g))\, d\mu(g) = \tfrac{1}{2}\int_G |1 - \beta(g)|^2\, d\mu(g) \\ &\geqslant \frac{\eta^2}{2}\mu(\{g \mid |1 - \beta(g)| \geqslant \eta\}). \end{aligned} \tag{3.2.28}$$

Combining (3.2.27) and (3.2.28), we see that (3.2.26) holds for all $\beta \in V$.

Conversely, suppose that μ is continuous with respect to $\mathfrak{T}$. Then, given any $\epsilon > 0$, we choose a $\mathfrak{T}$-neighborhood V of 1, such that

$$\mu(\{g \mid |1 - \beta(g)| \geqslant \sqrt{\epsilon}\}) < \tfrac{1}{4}\epsilon$$

for all $\beta \in V$. Hence, when $\beta \in V$, we have [taking $\mu(G) = 1$]

$$\begin{aligned} 1 - \Re\varphi(\beta) &= \tfrac{1}{2}\int_G |1 - \beta(g)|^2\, d\mu(g) \\ &\leqslant 2\mu(\{g \mid |1 - \beta(g)| \geqslant \sqrt{\epsilon}\}) + \tfrac{1}{2}(\sqrt{\epsilon})^2 < \epsilon. \end{aligned}$$

This shows that φ is continuous at the identity of $\mathfrak{H}$, whence it follows easily that φ is continuous everywhere in $\mathfrak{H}$.]

Lemma 3.2.19. Under the hypotheses of Definition 3.2.6, if $\mathfrak{T}$ satisfies the first axiom of countability, then μ is continuous with respect to $\mathfrak{T}$.

Proof. Let $\{\beta_n\}$ be a sequence in $\mathfrak{H}$ which converges in the topology $\mathfrak{T}$ to a point $\beta \in \mathfrak{H}$. Since $\mathfrak{T}$ is stronger than the weak topology, we have $\lim_{n\to\infty} \beta_n(g) = \beta(g)$ for every $g \in G$. By the Lebesgue dominated convergence theorem, it follows that $\varphi(\beta_n) \to \varphi(\beta)$. Since $\mathfrak{T}$ satisfies the first axiom of countability, this shows that φ is a continuous function on $(\mathfrak{H}, \mathfrak{T})$. Therefore, by Lemma 3.2.18, μ is continuous with respect to $\mathfrak{T}$.]

§3.3. Integral Representation of Positive Definite Functions on Groups

1° Positive Definite Functions on Groups

Definition 3.3.1. Let G be a group, and let $f(g)$, $g \in G$, be a function on G having the following property: for any natural number n, any n complex numbers $z_1, \ldots, z_n$, and any n elements $g_1, \ldots, g_n$ in G,

$$\sum_{k,l=1}^{n} f(g_k^{-1} g_l) z_k \bar{z}_l \geqslant 0. \tag{3.3.1}$$

Then, f is said to be a *positive definite* function on G.

In particular, choosing $n = 2$, $g_1 = e$ (the identity of G) and $g_2 = g$ (an arbitrary element of G), it follows easily from (3.3.1) that

$$|f(g)| \leqslant f(e), \qquad f(g^{-1}) = \overline{f(g)} \tag{3.3.2}$$

for every $g \in G$.

Again, in (3.3.1), choosing $n = 3$, $g_1 = e$, $g_2 = g$, $g_3 = g'$, $z_1 = \overline{(f(g') - f(g))}/f(e)$, $z_2 = 1$, $z_3 = -1$, and using (3.3.2), we obtain[36]

$$|f(g) - f(g')|^2 \leqslant 2(f(e) - \Re f(g'^{-1}g)) f(e). \tag{3.3.3}$$

From (3.3.3), we immediately deduce the following fact: if f is a positive definite function on a topological group G, then f is continuous on G if and only if f is continuous at the identity of G.

It is easy to check that any character of a group is positive definite. As we shall see below, characters may be regarded as the most basic of the positive definite functions.

We note, for future reference, that if H is an inner product space, and $U: g \to U(g)$ ($g \in G$) is a unitary representation of G in H, then, for any fixed $\xi \in H$, the function

$$(U(h)\xi, \xi), \qquad h \in G$$

is positive definite on G. This fact follows from Definition 3.3.1 by a straightforward calculation.

Definition 3.3.2. Let $\mathfrak{G}$ be a topological group, $\mathfrak{G}^*$ the dual of $\mathfrak{G}$. Let Γ be a set such that $\mathfrak{G}^* \subset \Gamma$, and let $\mathfrak{B}$ be a σ-algebra consisting of certain subsets of Γ. Also, suppose that, to each $h \in \mathfrak{G}$, there corre-

[36] If $f(e) = 0$, then, by (3.3.2), f is identically zero, so that (3.3.3) is trivial.

sponds a $\mathfrak{B}$-measurable function $h(\gamma)$, $\gamma \in \Gamma$, satisfying the following conditions: (i) $|h(\gamma)| = 1$; (ii) the family of functions $\mathfrak{E} = \{h(\cdot) \mid h \in \mathfrak{G}\}$ is a determining set on $(\Gamma, \mathfrak{B})$; (iii) $\mathfrak{E}$ forms a group with respect to multiplication, and the correspondence $h \rightarrow h(\cdot)$ is an isomorphism; (iv) when $\gamma \in \mathfrak{G}^*$, then $h(\gamma)$ is the value of the character γ at the point h. If, for every positive definite continuous function f on $\mathfrak{G}$, there exists a unique measure μ on $(\Gamma, \mathfrak{B})$ such that

$$f(g) = \int g(\gamma)\, d\mu(\gamma), \qquad g \in \mathfrak{G}, \tag{3.3.4}$$

then we say that $(\Gamma, \mathfrak{B})$ is a *spectral space* for $\mathfrak{G}$.

In particular, when Γ is a group consisting of certain characters of $\mathfrak{G}$ (i.e., $\Gamma \subset \mathfrak{G}'$), and, for each $g \in \mathfrak{G}$, the value of $g(\cdot)$ at the point γ is just the value of $\gamma(\cdot)$ at the point g [that is, $g(\gamma) = \gamma(g)$], then we call Γ a *spectral group* for $\mathfrak{G}$. In this case, setting $g = e$ in (3.3.4), we see at once that, for any f, the corresponding measure μ is finite.

The problem of finding so-called integral representations of positive functionals is just the problem of finding appropriate spectral groups. It is not at all difficult to show the existence of spectral groups (see, e.g., Theorem 3.3.4 below); however, our ultimate objective in this direction is to find spectral groups which are as "small" as possible. Before pursuing this problem any further, we shall first explain the use of spectral groups in the study of unitary representations.[37] This will demonstrate the connection between unitary representations of groups and the theory of positive definite functions.

Theorem 3.3.1. Let $\mathfrak{G}$ be a topological group, let $(\Gamma, \mathfrak{B})$ be a spectral group[38] for $\mathfrak{G}$, and let $U: h \rightarrow U(h)$ be a weakly continuous unitary representation of $\mathfrak{G}$ in the Hilbert space H. Then, there is a spectral measure $\{P(E), E \in \mathfrak{B}\}$ over H such that

$$U(h) = \int_{\Gamma} h(\gamma)\, dP(\gamma)$$

for every $h \in \mathfrak{G}$.

Proof. Take any $\xi \in H$, and form the function

$$\varphi_{\xi}(h) = (U(h)\xi, \xi), \qquad h \in \mathfrak{G}. \tag{3.3.5}$$

[37] Unitary representations of groups are defined and discussed briefly in Appendix II, §II.3.

[38] *Translator's note:* In the original, it was specified that $\mathfrak{B}$ is the σ-algebra of all weak Borel sets in Γ. However, the proof seems to be valid without this restriction.

φ_ξ is a positive definite continuous function on $\mathfrak{G}$. Hence, by the definition of a spectral group, there is a unique finite measure μ_ξ on $(\Gamma, \mathfrak{B})$ such that

$$\varphi_\xi(h) = \int_\Gamma h(\gamma)\, d\mu_\xi(\gamma); \tag{3.3.6}$$

in particular, setting $h = 1$, we get

$$\mu_\xi(\Gamma) = \| \xi \|^2.$$

Now, for each pair of vectors $\xi, \eta \in H$ and every set $E \in \mathfrak{B}$, we define

$$\mu(E; \xi, \eta) = \tfrac{1}{4}(\mu_{\xi+\eta}(E) - \mu_{\xi-\eta}(E) + i\mu_{\xi+i\eta}(E) - i\mu_{\xi-i\eta}(E)). \tag{3.3.7}$$

Then, from (3.3.5), (3.3.6), and (3.3.7), we calculate that

$$(U(h)\xi, \eta) = \int_\Gamma h(\gamma)\, d\mu(\gamma; \xi, \eta). \tag{3.3.8}$$

Since $\{h(\cdot) \mid h \in \mathfrak{G}\}$ is a determining set of functions on $(\Gamma, \mathfrak{B})$, the complex measure $\mu(E; \xi, \eta)$ is uniquely determined by relation (3.3.8). Therefore, since the left-hand side of (3.3.8) is a bilinear functional on $H \times H$, it follows that $\mu(E; \xi, \eta)$ is also bilinear in ξ, η. Similarly, it follows easily from (3.3.8) that $\mu(E; \xi, \eta)$ is Hermitian in ξ, η, and that $0 \leqslant \mu(E; \xi, \xi) = \mu_\xi(E) \leqslant \| \xi \|^2$. Consequently, $\mu(E; \xi, \eta)$ is a positive definite continuous bilinear functional on $H \times H$, and so, by a well-known theorem, there is a bounded positive self-adjoint operator $P(E)$ on H such that

$$(P(E)\xi, \eta) = \mu(E; \xi, \eta), \qquad \xi, \eta \in H. \tag{3.3.9}$$

By (3.3.7), we know that $\mu(E; \xi, \eta)$, $E \in \mathfrak{B}$, is a countably additive set function on $\mathfrak{B}$; obviously, $P(0) = 0$, and since

$$(P(\Gamma)\xi, \eta) = \mu(\Gamma, \xi, \eta) = (\xi, \eta),$$

we have $P(\Gamma) = I$. Thus, to show that P is a spectral measure, it only remains to prove that each $P(E)$ is a projection operator.

Fix any element $h_1 \in \mathfrak{G}$, and consider the following complex measure on $(\Gamma, \mathfrak{B})$:

$$\mu_1(E; \xi, \eta) = \int_E h_1(\gamma)\, d(P(\gamma)\xi, \eta). \tag{3.3.10}$$

Since

$$\int_\Gamma h(\gamma)\,d\mu_1(\gamma;\,\xi,\eta) = \int_\Gamma h(\gamma)\,h_1(\gamma)\,d(P(\gamma)\xi,\eta)$$
$$= (U(hh_1)\xi,\eta)$$
$$= \int_\Gamma h(\gamma)\,d\mu(\gamma,\,U(h_1)\xi,\eta)$$

for every $h \in \mathfrak{G}$, it follows by the uniqueness of the complex measure μ in (3.3.7) that

$$\mu_1(E;\,\xi,\eta) = \mu(E;\,U(h_1)\xi,\eta), \qquad E \in \mathfrak{B}. \tag{3.3.11}$$

Using (3.3.9), (3.3.10), and (3.3.11), we obtain

$$\int_\Gamma h_1(\gamma)\,C_E(\gamma)\,d(P(\gamma)\xi,\eta) = (P(E)\,U(h_1)\xi,\eta)$$
$$= \int_\Gamma h_1(\gamma)\,d(P(\gamma)\xi,\,P(E)\eta), \tag{3.3.12}$$

where C_E denotes the characteristic function of the set E. Again, since $\{h_1(\cdot) \mid h_1 \in \mathfrak{G}\}$ is a determining set of functions on $(\Gamma, \mathfrak{B})$, it follows from (3.3.12) that, for any bounded $\mathfrak{B}$-measurable function q on Γ,

$$\int_\Gamma q(\gamma)\,C_E(\gamma)\,d(P(\gamma)\xi,\eta) = \int_\Gamma q(\gamma)\,d(P(\gamma)\xi,\,P(E)\eta).$$

In particular, choosing $q = C_E$, we get

$$(P(E)\xi,\eta) = \int_\Gamma C_E(\gamma)\,d(P(\gamma)\xi,\eta) = \int_E d(P(\gamma)\xi,\,P(E)\eta)$$
$$= (P(E)\xi,\,P(E)\eta).$$

Thus, $P(E)^*\,P(E) = P(E)$, and since $P(E)$ is self-adjoint, we conclude that $P(E)$ is a projection operator.]

From Theorem 3.3.1 and Corollary 2.4.23,[39] we immediately deduce the following result.

Corollary 3.3.2. Under the hypotheses of Theorem 3.3.1, let $\mathfrak{A}$ be the commutative weakly closed operator algebra over H generated by the family of operators $\{U(h) \mid h \in \mathfrak{G}\}$. If $\mathfrak{A}$ has uniform multiplicity k,

[39] *Translator's note:* Recall that the proof of Corollary 2.4.23 depended upon Theorem 2.4.22, and the proof of the latter, in the general case of nonseparable H, was open to question.

then there is a measure μ on $(\Gamma, \mathfrak{B})$ and a unitary mapping Q from H onto $\mathfrak{L}_k^2(\Gamma, \mathfrak{B}, \mu)$ such that the relation

$$(QU(h)Q^{-1})\,\varphi(\gamma) = h(\gamma)\,\varphi(\gamma), \qquad \gamma \in \Gamma$$

holds for every $h \in \mathfrak{G}$ and every $\varphi \in \mathfrak{L}_k^2(\Gamma, \mathfrak{B}, \mu)$.

2° Positive Definite Functions on Discrete Groups

The following discussion will be restricted to commutative groups. First, we merely consider abstract groups, that is, we take all the subsets of a group G as open sets, thereby obtaining a topology on G, known as the discrete topology. All functions on G are then continuous.

Let G be a commutative group, G' the totality of characters of G. We construct the *group ring* $R(G)$ of G, as follows. Let φ be a complex-valued function on G, such that the set $\{g \mid \varphi(g) \neq 0\}$ is finite or countable, and

$$\| \varphi \| = \sum_{g \in G} | \varphi(g)| < \infty. \tag{3.3.13}$$

The totality $R(G)$ of such functions φ clearly forms a Banach space with respect to the ordinary linear operations and the norm $\| \varphi \|$ defined in (3.3.13). If $\varphi, \psi \in R(G)$, the function

$$(\varphi * \psi)(g) = \sum_{g_1 g_2 = g} \varphi(g_1)\,\psi(g_2)$$

is called the *convolution* of φ and ψ. Taking convolution as a multiplication operation, $R(G)$ becomes a commutative Banach algebra. Also, if $\varphi \in R(G)$, we define

$$\varphi^*(g) = \overline{\varphi(g^{-1})}.$$

It is easily checked that $\varphi^* \in R(G)$, and that the correspondence $\varphi \to \varphi^*$ is an involution in $R(G)$. Obviously, the norm (3.3.13) is symmetric with respect to this involution.

We now consider the space $\mathfrak{M}$ of symmetric multiplicative linear functionals on $R(G)$. Take any $\alpha \in G'$, and define a linear functional F_α on $R(G)$, as follows:

$$F_\alpha(\varphi) = \sum \varphi(g)\,\alpha(g). \tag{3.3.14}$$

It is easily verified that $F_\alpha \in \mathfrak{M}$. Conversely, let $F \in \mathfrak{M}$; for each $h \in G$, form the function

$$\varphi_h(g) = \begin{cases} 1 & \text{when } g = h, \\ 0 & \text{when } g \neq h. \end{cases}$$

Let

$$\alpha(h) = F(\varphi_h).$$

Since φ_e (where e denotes the identity of G) is the unit element of $R(G)$, we have $\alpha(e) = F(\varphi_e) = 1$. Also, since $\varphi_{h_1} * \varphi_{h_2} = \varphi_{h_1 h_2}$, we have

$$\alpha(h_1 h_2) = F(\varphi_{h_1 h_2}) = F(\varphi_{h_1}) F(\varphi_{h_2}) = \alpha(h_1)\, \alpha(h_2).$$

Moreover, $|\alpha(h)| \leqslant \|F\| \|\varphi_h\| = 1$, hence $|\alpha(h)^{-1}| = |\alpha(h^{-1})| \leqslant 1$, and therefore $|\alpha(h)| = 1$. This shows that $\alpha \in G'$. Finally, using Lemma 2.2.1, it is easily seen that $F = F_\alpha$. Thus, we have proved the following result.

Lemma 3.3.3. Let G be a commutative group, G' the character group of G, and $\mathfrak{M}$ the space of symmetric multiplicative linear functionals on the group ring $R(G)$. Then, the correspondence T: $\alpha \to F_\alpha$ [where F_α is defined by (3.3.14)] is a one-to-one mapping from G' onto $\mathfrak{M}$.

Let $\mathfrak{F}'$ be the smallest σ-algebra containing all Borel cylinders in G', and let $\mathfrak{B}$ be the smallest σ-algebra containing all Borel cylinders in $\mathfrak{M}$. Then, the correspondence T maps $\mathfrak{F}'$ onto $\mathfrak{B}$, that is, $\mathfrak{B} = \{T(E) \mid E \in \mathfrak{F}'\}$. In fact, let E be any Borel cylinder in G', say,

$$E = \{\alpha \mid (\alpha(g_1),..., \alpha(g_n)) \in B\}.$$

Then,

$$TE = \{F \mid (F(\varphi_{g_1}),..., F(\varphi_{g_n})) \in B\}.$$

Therefore, $\mathfrak{B} \supset \{T(E) \mid E \in \mathfrak{F}'\} = \mathfrak{B}'$. Conversely, given any $\varphi \in R(G)$, there is a sequence $\{g_k\} \subset G$ such that $\varphi = \lim_{m\to\infty} \sum_{k=1}^{m} \varphi(g_k)\, \varphi_{g_k}$. Hence, $F(\varphi) = \sum_{k=1}^{\infty} \varphi(g_k)(T^{-1}F)(g_k)$. This shows that the function $F \to F(\varphi)$ is $\mathfrak{B}'$-measurable, whence it follows that $\mathfrak{B} \subset \mathfrak{B}'$. Thus, $\mathfrak{B} = \mathfrak{B}'$, as asserted.

Theorem 3.3.4. Let G be a group, and let f be any positive definite function on G, with $f(e) = 1$. Then, there exists a unique probability measure P' on $(G', \mathfrak{F}')$ such that

$$f(g) = \int_{G'} \alpha(g)\, dP'(\alpha) \tag{3.3.15}$$

holds for all $g \in G$.

PROOF. Using the given positive definite function f, we construct a linear functional F on $R(G)$, as follows:

$$F(\varphi) = \sum_{g\in G} f(g)\, \varphi(g), \qquad \varphi \in R(G).$$

Since $|f(g)| \leqslant 1$, the series $\sum_{g \in G} f(g)\, \varphi(g)$ converges for every $\varphi \in R(G)$. Furthermore, by the positive definiteness of f, we have

$$F(\varphi * \varphi^*) = \sum_{g,h \in G} f(gh^{-1})\, \varphi(g)\, \overline{\varphi(h)} \geqslant 0, \qquad \varphi \in R(G),$$

hence F is a positive functional on $R(G)$. Since $R(G)$ is a symmetric commutative Banach algebra with unit, it follows (see the proof of Theorem 2.2.4) that there exists a unique probability measure μ on $(\mathfrak{M}, \mathfrak{B})$ such that

$$F(\varphi) = \int_{\mathfrak{M}} \varphi(M)\, d\mu(M). \tag{3.3.16}$$

Now, by Lemma 3.3.3 and the remark following it, there is a one-to-one mapping T from G' onto $\mathfrak{M}$ which induces a mapping from $\mathfrak{F}'$ onto $\mathfrak{B}$. Accordingly, we may define a set function P' on $\mathfrak{F}'$ by letting

$$P'(E) = \mu(TE), \qquad E \in \mathfrak{F}', \tag{3.3.17}$$

and it is clear that P' is a probability measure on $(G', \mathfrak{F}')$. Moreover, from (3.3.16) and (3.3.17), we obtain

$$F(\varphi) = \int_{G'} F_\alpha(\varphi)\, dP'(\alpha). \tag{3.3.18}$$

In particular, choosing $\varphi = \varphi_h$ in (3.3.14) and (3.3.18), we obtain (3.3.15). The uniqueness of P' may be deduced from the uniqueness of μ.]

Theorem 3.3.4 may also be stated in the following form.

Theorem 3.3.4′. Let G be a commutative topological group, and let G' be the algebraic dual of G. Then G' is a spectral group for G.

By following a procedure analogous to the proof of Theorem 3.3.4, one can also show that the smallest spectral group for a locally compact group G is just the dual G^*. However, such a proof would by itself be fairly lengthy, and since the method requires the use of either the group ring or invariant measure, it does not lend itself to further generalization. Therefore, we shall not use such an approach; instead, we shall continue our researches by measure-theoretic methods.

Proceeding from Theorem 3.3.4, we see that, to find smaller spectral groups, it suffices to prove that the measure P' in (3.3.15) can be concentrated on certain subgroups of G'. For this purpose, we first require some preparatory work.

3° Positive Definite Functions on Groups with Measures

We now introduce a type of topological space which will be useful in the subsequent discussion. Let G be a group, and for each $g \in G$, let C_g be a copy of the unit circle in the complex plane, with the topology induced by the usual metric

$$\rho(z_1, z_2) = | z_1 - z_2 |, \qquad z_1, z_2 \in C_g .$$

Let C^G be the topological product of all the spaces C_g, $g \in G$; the elements of C^G may be regarded as complex-valued functions f on G, satisfying the condition $|f(g)| = 1$. Take any finite set of elements $g_1, \ldots, g_n \in G$, and any Borel set B in real n-dimensional space. The set

$$\{f \mid (f(g_1), \ldots, f(g_n)) \in B\}$$

in C^G is called the *Borel cylinder* with *base* B corresponding to the elements $g_1, \ldots, g_n$. Let $\mathfrak{L}$ be the smallest σ-algebra containing all Borel cylinders in C^G.

Observe that $G' \subset C^G$, that the topology on G' induced by C^G is just the weak topology, and that the σ-algebra in G' induced by $\mathfrak{L}$ is just $\mathfrak{F}'$.

Henceforth, we shall further assume that $(G, \mathfrak{B}, \mu)$ is a complete localizable measure space. Let $\{E_\sigma, \sigma \in \Sigma\}$ be a partition[40] of G (see Definition 1.1.8), and suppose that, for each $\sigma \in \Sigma$, there exists a sequence of mappings $\varphi_n^{(\sigma)} : E_\sigma \to E_\sigma$, $n = 1, 2, \ldots$, satisfying the following conditions:

(i) The values of $\varphi_n^{(\sigma)}(g)$ form an (at most) countable set $\{g_{k,n}^{(\sigma)}, k = 1, 2, \ldots\}$.
(ii) For each k, the set $\{g \mid \varphi_n^{(\sigma)}(g) = g_{k,n}^{(\sigma)}\}$ belongs to $\mathfrak{B}$.
(iii) If $m \geqslant n$, then $\varphi_m^{(\sigma)}(g_{k,n}^{(\sigma)}) = g_{k,n}^{(\sigma)}$, $k = 1, 2, \ldots$.

We let $\mathfrak{F}^{\mathfrak{B}}$ denote the σ-algebra generated by the totality of Borel cylinders in $G^{\mathfrak{B}}$ (see §3.2).

Lemma 3.3.5. Let G be a commutative group, and let f be a $\mathfrak{B}$-measurable positive definite function on G, with $f(e) = 1$. Also,

[40] *Translator's note:* The intended meaning of the term *partition*, as used in this subsection, is, presumably, as follows. Let $E_\sigma \in \mathfrak{B}$, $\sigma \in \Sigma$, with $\mu(E_\sigma) < \infty$, let $\mathfrak{B}_\sigma$, μ_σ denote the restrictions of $\mathfrak{B}$, μ, respectively, to E_σ, and let $(G, \mathfrak{B}', \mu')$ denote the direct sum of all the measure spaces $(E_\sigma, \mathfrak{B}_\sigma, \mu_0)$. Then $\tilde{\mathfrak{B}}' = \mathfrak{B}$ and $\tilde{\mu}' = \mu$ (see Definitions 1.1.1 and 1.1.2).

suppose that, for every σ, the sequence of measurable[41] functions

$$\{\Re f(\varphi_n^{(\sigma)}(g)^{-1}g)\}$$

on E_σ converges to 1 almost everywhere (relative to μ). Then, there exists a unique probability measure $P^{\mathfrak{B}}$ on $(G^{\mathfrak{B}}, \mathfrak{F}^{\mathfrak{B}})$ such that

$$f(g) = \int_{G^{\mathfrak{B}}} \alpha(g)\, dP^{\mathfrak{B}}(\alpha) \tag{3.3.19}$$

for all $g \in G$.

Proof. We divide the proof into several steps, as follows.

(1) By Theorem 3.3.4, there exists a probability measure space $(G', \mathfrak{F}', P')$ such that (3.3.15) holds. We first extend P' to a measure on $(C^G, \mathfrak{L})$. If $A \in \mathfrak{L}$, then, since $A \cap G' \in \mathfrak{F}'$, we may define

$$Q(A) = P'(A \cap G'). \tag{3.3.20}$$

Clearly, Q is a probability measure on $(C^G, \mathfrak{L})$. From (3.3.15) and (3.3.20), we obtain

$$f(g) = \int_{C^G} \alpha(g)\, dQ(\alpha), \qquad g \in G;$$

moreover, for any two elements $g, g' \in G$, we have

$$\int_{G'} \alpha(g)\, \overline{\alpha(g')}\, dP'(\alpha) = \int_{C^G} \alpha(g)\, \overline{\alpha(g')}\, dQ(\alpha). \tag{3.3.21}$$

(2) Every $\alpha \in G'$ is a character of G, hence, using (3.3.21), we get

$$\begin{aligned} 2(1 - \Re f(\varphi_n^{(\sigma)}(g)^{-1}g)) &= \int_{G'} 2(1 - \Re \overline{\alpha(\varphi_n^{(\sigma)}(g))}\, \alpha(g))\, dP'(\alpha) \\ &= \int_{C^G} 2(1 - \Re \overline{\alpha(\varphi_n^{(\sigma)}(g))}\, \alpha(g))\, dQ(\alpha) \\ &= \int_{C^G} |\alpha(g) - \alpha(\varphi_n^{(\sigma)}(g))|^2\, dQ(\alpha). \end{aligned} \tag{3.3.22}$$

By hypothesis, the sequence of functions $1 - \Re f(\varphi_n^{(\sigma)}(g)^{-1}g)$ converges

[41] *Translator's note:* Apparently, the author has tacitly assumed that the group operation $(g, h) \to g^{-1}h$ is a measurable mapping from $(G \times G, \mathfrak{B} \times \mathfrak{B})$ to $(G, \mathfrak{B})$.

to zero almost everywhere (relative to the measure μ), and is uniformly bounded [see (3.3.2)], therefore, using (3.3.22), we obtain

$$\lim_{n\to\infty} \int_{E_\sigma} \left(\int_{C^G} | \alpha(\varphi_n^{(\sigma)}(g)) - \alpha(g)|^2 \, dQ(\alpha) \right) d\mu(g)$$

$$= \lim_{n\to\infty} \int_{E_\sigma} 2(1 - \Re f(\varphi_n^{(\sigma)}(g)^{-1}g)) \, d\mu(g) = 0. \qquad (3.3.23)$$

Consider the product measure space $(E_\sigma \times C^G, \mathfrak{B}_\sigma \times \mathfrak{L}, \mu \times Q)$ (here, $\mathfrak{B}_\sigma$ denotes the restriction of $\mathfrak{B}$ to E_σ). By virtue of the conditions (i) and (ii) satisfied by $\varphi_n^{(\sigma)}(g)$, the quantity $\alpha(\varphi_n^{(\sigma)}(g))$, regarded as a function of $(g, \alpha) \in E_\sigma \times C^G$, is measurable with respect to $\mathfrak{B}_\sigma \times \mathfrak{L}$. Furthermore, using Fubini's theorem and (3.3.23), we obtain

$$\lim_{m,n\to\infty} \int_{E_\sigma \times C^G} | \alpha(\varphi_n^{(\sigma)}(g)) - \alpha(\varphi_m^{(\sigma)}(g))|^2 \, d\mu \times Q$$

$$= \lim_{m,n\to\infty} \int_{E_\sigma} \left(\int_{C^G} | \alpha(\varphi_n^{(\sigma)}(g)) - \alpha(\varphi_m^{(\sigma)}(g))|^2 \, dQ(\alpha) \right) d\mu(g)$$

$$\leqslant \lim_{m,n\to\infty} 4 \int_{E_\sigma} [(1 - \Re f(\varphi_n^{(\sigma)}(g)^{-1}g)) + (1 - \Re f(\varphi_m^{(\sigma)}(g)^{-1}g))] \, d\mu(g)$$

$$= 0.$$

Consequently, there is a subsequence $\{\alpha(\varphi_{\beta_n}^{(\sigma)}(g))\}$[42] and a measurable function $U^{(\sigma)}(g, \alpha)$ on $(E_\sigma \times C^G, \mathfrak{B}_\sigma \times \mathfrak{L})$ such that

$$\lim_{n\to\infty} \alpha(\varphi_{\beta_n}^{(\sigma)}(g)) = U^{(\sigma)}(g, \alpha) \qquad (3.3.24)$$

holds for almost all (g, α) in $E_\sigma \times C^G$ (relative to the measure $\mu \times Q$). It follows that there is a set $F_\sigma \in \mathfrak{B}_\sigma$, with $\mu(F_\sigma) = 0$, such that, for each $g \in E_\sigma - F_\sigma$, $U^{(\sigma)}(g, \alpha)$ is a measurable function on $(C^G, \mathfrak{L})$, and

$$Q(\{\alpha \mid \lim_{n\to\infty} \alpha(\varphi_{\beta_n}^{(\sigma)}(g)) = U^{(\sigma)}(g, \alpha)\}) = 1.$$

From (3.3.23) and (3.3.24), we see that

$$\int_{E_\sigma} \left(\int_{C^G} | U^{(\sigma)}(g, \alpha) - \alpha(g)|^2 \, dQ(\alpha) \right) d\mu(g) = 0.$$

[42] Notice that the indices $\{\beta_n\}$ actually depend upon σ; however, to avoid unduly cumbersome notation, this has not been explicitly indicated.

Therefore, we can, if necessary, enlarge the null set F_σ, and so ensure that the equality

$$Q(\{\alpha \mid U^{(\sigma)}(g, \alpha) = \alpha(g)\}) = 1$$

holds for all $g \in E_\sigma - F_\sigma$. Hence, if $g \in E_\sigma - F_\sigma$, then

$$Q(\{\alpha \mid \lim_{n\to\infty} \alpha(\varphi_{\beta_n}^{(\sigma)}(g)) = \alpha(g)\}) = 1. \tag{3.3.25}$$

(3) Next, we define four functions $V_\nu(g; \alpha)$, $\nu = 1, 2, 3, 4$, on $G \times C^G$, as follows. Given any $(g, \alpha) \in G \times C^G$, let $\sigma \in \Sigma$ be the index such that $g \in E_\sigma$, and let

$$V_1(g; \alpha) = \varliminf_{n\to\infty} \mathfrak{R}\alpha(\varphi_{\beta_n}^{(\sigma)}(g)),$$

$$V_2(g; \alpha) = \varlimsup_{n\to\infty} \mathfrak{R}\alpha(\varphi_{\beta_n}^{(\sigma)}(g)),$$

$$V_3(g; \alpha) = \varliminf_{n\to\infty} \mathfrak{J}\alpha(\varphi_{\beta_n}^{(\sigma)}(g)),$$

$$V_4(g; \alpha) = \varlimsup_{n\to\infty} \mathfrak{J}\alpha(\varphi_{\beta_n}^{(\sigma)}(g)).$$

Since, for each $\sigma \in \Sigma$, the restriction of V_ν to $E_\sigma \times C^G$ is measurable with respect to $\mathfrak{B}_\sigma \times \mathfrak{L}$, it follows that V_ν is a measurable function on $(G \times C^G, \mathfrak{B} \times \mathfrak{L})$, $\nu = 1, 2, 3, 4$.

Let H be the set of all α in C^G such that the inequalities

$$\begin{aligned} V_1(g; \alpha) \leqslant \mathfrak{R}\alpha(g) \leqslant V_2(g; \alpha) \\ V_3(g; \alpha) \leqslant \mathfrak{J}\alpha(g) \leqslant V_4(g; \alpha) \end{aligned} \tag{3.3.26}$$

hold for almost all $g \in G$. We now proceed to prove that the outer measure $Q^*(H)$ is 1.

Let $\Gamma = \bigcup_{n=1}^{\infty} \Gamma_n$ be a union of Borel cylinders, with $H \subset \Gamma$; we may assume that Γ_n is a cylinder corresponding to the elements $g_1, \ldots, g_n \in G$, $n = 1, 2, \ldots$. For each n, there is an index $\sigma_n \in \Sigma$ such that $g_n \in E_{\sigma_n}$. Let $g_{n_1}, \ldots, g_{n_k}, \ldots$ denote the subsequence of $\{g_n\}$ consisting of all elements satisfying the condition[43] $g_n \in E_{\sigma_n} - F_{\sigma_n}$. Consider the set

$$\Lambda = \bigcap_k \{\alpha \mid \lim_{m\to\infty} \alpha(\varphi_{\beta_m}^{(\sigma_{n_k})}(g_{n_k})) = \alpha(g_{n_k})\}.$$

Using (3.3.25), we see that $Q(\Lambda) = 1$.

We shall show that $\Lambda \subset \Gamma$. Given any $\alpha_0 \in \Lambda$, define a function

[43] The subsequence $\{g_{n_k}\}$ may be empty, but in that case the intersection defining Λ is vacuous, that is, $\Lambda = C^G$, so that $Q(\Lambda) = 1$ is still valid.

$\alpha \in C^G$ as follows: for every g_ν, let

$$\alpha(g_\nu) = \alpha_0(g_\nu), \tag{3.3.27}$$

for every value $g_{k,\beta_m}^{(\sigma_n)}$ of $\varphi_{\beta_m}^{(\sigma_n)}(g)$, let

$$\alpha(g_{k,\beta_m}^{(\sigma_n)}) = \alpha_0(g_{k,\beta_m}^{(\sigma_n)}), \tag{3.3.28}$$

and, for every $g \in E_{\sigma_n} - (\{g_{k,\beta_m}^{(\sigma_n)}\} \cup \{g_\nu\})$, let $\alpha(g)$ be any complex number of modulus 1 which satisfies the conditions

$$V_1(g;\alpha_0) \leqslant \mathfrak{R}\alpha(g) \leqslant V_2(g;\alpha_0), \tag{3.3.29}$$

$$V_3(g;\alpha_0) \leqslant \mathfrak{J}\alpha(g) \leqslant V_4(g;\alpha_0); \tag{3.3.30}$$

it is easily seen that such a complex number does exist. Finally, for every $g \in G - \bigcup_{n=1}^\infty E_{\sigma_n}$, let

$$\alpha(g) = 1. \tag{3.3.31}$$

We assert that $\alpha \in H$. To prove this, it suffices to show that (3.3.26) holds for this α and any $g \in G - \bigcup_{n=1}^\infty F_{\sigma_n}$.

If $g \in G - \bigcup_{n=1}^\infty E_{\sigma_n}$, then, by (3.3.31), we have $V_\nu(g;\alpha) = 1 = \mathfrak{R}\alpha(g)$, $\nu = 1, 2$, and $V_\nu(g,\alpha) = 0 = \mathfrak{J}\alpha(g)$, $\nu = 3, 4$, hence (3.3.26) holds. Moreover, we see from (3.3.28) and the definition of $V_\nu(g,\alpha)$ that $V_\nu(g;\alpha) = V_\nu(g;\alpha_0)$ for all $g \in \bigcup_{n=1}^\infty E_{\sigma_n}$. Hence, it follows by (3.3.29) and (3.3.30) that (3.3.26) holds if $g \in E_{\sigma_n} - (\{g_{k,\beta_m}^{(\sigma_n)}\} \cup \{g_\nu\})$. Furthermore, since the $\varphi_n^{(\sigma)}(g)$ satisfy condition (iii), we have $\varphi_{\beta_l}^{(\sigma)}(g_{k,\beta_m}^{(\sigma)}) = g_{k,\beta_m}^{(\sigma)}$ whenever $l \geqslant m$, hence

$$\lim_{l\to\infty} \alpha(\varphi_{\beta_l}^{(\sigma_n)}(g_{k,\beta_m}^{(\sigma_n)})) = \alpha(g_{k,\beta_m}^{(\sigma_n)}).$$

Therefore, (3.3.26) also holds for $g = g_{k,\beta_m}^{(\sigma_n)}$. Finally, since $\alpha_0 \in \Lambda$, it follows from (3.3.27) and (3.3.28) that

$$\alpha(g_{n_k}) = \lim_{m\to\infty} \alpha(\varphi_{\beta_m}^{(\sigma_{n_k})}(g_{n_k})).$$

Hence, (3.3.26) also holds for $g = g_{n_k}$. Thus, we have shown that (3.3.26) holds for all $g \overline{\in} \bigcup_{n=1}^\infty F_{\sigma_n}$, so that $\alpha \in H$, as asserted.

Since $H \subset \Gamma$, we must have $\alpha \in \Gamma_k$ for some k. But Γ_k is a Borel cylinder corresponding to $g_1, \ldots, g_k$, hence, it follows by (3.3.27) that $\alpha_0 \in \Gamma_k$. This shows that $\Lambda \subset \Gamma$. Since $Q(\Lambda) = 1$, this implies that $Q(\Gamma) = 1$. We conclude that the outer measure of H is 1.

(4) Let $\mathfrak{L}_0 = \{A \cap H \mid A \in \mathfrak{L}\}$, and define a set function Q_0 on $\mathfrak{L}_0$ as follows:

$$Q_0(A \cap H) = Q(A), \qquad A \in \mathfrak{L}.$$

Since $Q^*(H) = 1$, it follows from the discussion in §1.1, 1°, that Q_0 is a probability measure on $(H, \mathfrak{L}_0)$.

Now, the functions $V_\nu(g, \alpha)$ on $G \times C^G$ are measurable with respect to $\mathfrak{B} \times \mathfrak{L}$, hence their restrictions to $G \times H$ are measurable with respect to $\mathfrak{B} \times \mathfrak{L}_0$. Let $F = \bigcup_{\sigma \in \Sigma} F_\sigma$. If $g \in F$, then, by (3.3.25),

$$Q(\{\alpha \mid V_1(g; \alpha) = V_2(g; \alpha); V_3(g; \alpha) = V_4(g; \alpha)\}) = 1.$$

Hence, if $g \in G - F$, then

$$Q_0(\{\alpha \mid V_1(g; \alpha) = V_2(g; \alpha); V_3(g; \alpha) = V_4(g; \alpha), \alpha \in H\}) = 1.$$

By Fubini's theorem,[44]

$$\mu \times Q_0(\{(g; \alpha) \mid V_1(g; \alpha) \neq V_2(g, \alpha) \text{ or } V_3(g; \alpha) \neq V_4(g; \alpha), \alpha \in H\}) = 0.$$

Consequently, there is a null set $H_0 \subset H$ such that

$$\mu(\{g \mid V_1(g; \alpha) \neq V_2(g; \alpha) \text{ or } V_3(g; \alpha) \neq V_4(g; \alpha)\}) = 0$$

for every $\alpha \in H - H_0$. But if $\alpha \in H$, then (3.3.26) holds for almost all g. Therefore, if $\alpha \in H - H_0$, then

$$\begin{aligned} V_1(g; \alpha) &= \mathfrak{R}\alpha(g) = V_2(g; \alpha) \\ V_3(g; \alpha) &= \mathfrak{J}\alpha(g) = V_4(g; \alpha) \end{aligned} \tag{3.3.32}$$

holds for almost all g. Since the functions $V_\nu(g; \alpha)$ are measurable in g for every fixed α, it follows from (3.3.32) and the completeness of the measure μ that $\alpha(g)$ is a measurable function of g for every $\alpha \in H - H_0 (Q_0(H_0) = 0)$.

(5) Let $H_1 = H - H_0$; then, $H_1 \cap G' \subset G^{\mathfrak{B}}$. Since $Q_0(H_1) = 1$, the outer measure $Q^*(H_1) = 1$. By (3.3.20), we know that[45]

$$Q^*(H_1) = P'^*(H_1 \cap G'),$$

[44] *Translator's note:* In, for example, the Halmos [1] version of Fubini's theorem, it is assumed that the measures concerned are σ-finite.

[45] *Translator's note:* The inequality $Q^*(H_1) \geqslant P'^*(H_1 \cap G')$ is obvious, but the reverse inequality is not.

hence $P'^*(H_1 \cap G') = 1$, and therefore $P'^*(G^{\mathfrak{B}}) = 1$. For every $A \in \mathfrak{F}'$, define

$$P^{\mathfrak{B}}(A \cap G^{\mathfrak{B}}) = P'(A). \tag{3.3.33}$$

Since $\mathfrak{F}^{\mathfrak{B}} = \{A \cap G^{\mathfrak{B}} \mid A \in \mathfrak{L}\}$ and $P'^*(G^{\mathfrak{B}}) = 1$, it follows again from §1.1, 1°, that $P^{\mathfrak{B}}$ is a probability measure on $(G^{\mathfrak{B}}, \mathfrak{F}^{\mathfrak{B}})$, moreover, it is easily seen that

$$\int_{G^{\mathfrak{B}}} \alpha(g)\, dP^{\mathfrak{B}}(\alpha) = \int_{G'} \alpha(g)\, dP'(\alpha)$$

holds for every $g \in G$. This, combined with (3.3.15), shows that (3.3.19) holds for all $g \in G$.

(6) It only remains to show the uniqueness of the measure $P^{\mathfrak{B}}$. Suppose that $P_1^{\mathfrak{B}}$ is another measure on $(G^{\mathfrak{B}}, \mathfrak{F}^{\mathfrak{B}})$ satisfying (3.3.19). Define a probability measure P_1' on $(G', \mathfrak{B}')$, as follows:

$$P_1'(A) = P_1^{\mathfrak{B}}(A \cap G^{\mathfrak{B}}), \qquad A \in \mathfrak{F}'. \tag{3.3.34}$$

It then follows from (3.3.19) and (3.3.34) that P_1' satisfies condition (3.3.15). But, by Theorem 3.3.4, this measure is unique, that is, $P_1' = P'$. Hence, from (3.3.33) and (3.3.34), we conclude that $P_1^{\mathfrak{B}} = P^{\mathfrak{B}}$.]

We point out that, in Theorem 3.3.5, the condition that $\{\Re f(\varphi_n^{(\sigma)}(g)^{-1}g)\}$ converges to 1 almost everywhere (relative to μ) cannot be dispensed with. This may be seen from the following example.

Example 3.3.1. Let R denote the additive group of all real numbers, $\mathfrak{B}$ the totality of Lebesgue measurable subsets of R, and μ any probability measure equivalent to Lebesgue measure. It was shown in Example 3.2.4 that $R^{\mathfrak{B}}$ is just the totality of functions of the form

$$\alpha_t(x) = e^{itx}, \qquad x \in R,$$

hence the right-hand side of (3.3.19), in this case, takes the form

$$\int_R e^{itx}\, dP(t), \qquad x \in R, \tag{3.3.35}$$

where P is a probability measure on $(R, \mathfrak{B})$. It is easily seen that, if R is given the ordinary Euclidean topology, then (3.3.35) must be a continuous function on R. Consider the function

$$f(x) = \begin{cases} 1, & x = 0, \\ 0, & x \neq 0. \end{cases}$$

Obviously, f is a $\mathfrak{B}$-measurable positive definite function on R. But f is not continuous, and hence cannot be expressed in the form (3.3.19).

Lemma 3.3.6. Let G be a topological group satisfying the first axiom of countability, let $\mathfrak{B}$ be a σ-algebra in G which contains all the open subsets of G, and let $(G, \mathfrak{B}, \mu)$ be a regular locally finite measure space. Then, there exists a partition $\{E_\sigma \mid \sigma \in \Sigma\}$ of G, and a corresponding family of mappings $\{\varphi_n^{(\sigma)} \mid \sigma \in \Sigma, n = 1, 2, 3,...\}$, satisfying the conditions (i), (ii), and (iii) (preceding Lemma 3.3.5) such that, for every $\sigma \in \Sigma$,

$$\lim_{n\to\infty} \varphi_n^{(\sigma)}(g) = g \tag{3.3.36}$$

holds for almost all $g \in E_\sigma$.

PROOF. Let $\mathfrak{U}$ be a family of pairwise disjoint subsets E of G, satisfying the following conditions: $E : E \in \mathfrak{B}$, $0 < \mu(E) < \infty$, and, for any open set V, either $E \cap V$ is empty or $\mu(V \cap E) > 0$. Let $\mathfrak{F}$ denote the totality of such families $\mathfrak{U}$, and let $\mathfrak{F}$ be partially ordered by inclusion, that is, $\mathfrak{U}_1 \prec \mathfrak{U}_2$ means that $\mathfrak{U}_1 \subset \mathfrak{U}_2$. Using Zorn's lemma, it is easily seen that $\mathfrak{F}$ has a maximal element,[46] which we denote by $\{E_\sigma \mid \sigma \in \Sigma\}$.

We assert that any compact set K intersects at most countably many E_σ. In fact, it follows easily from the local finiteness of μ that there exists an open set V, having finite measure, such that $V \supset K$. Therefore, since

$$\infty > \mu(V) \geqslant \sum \mu(E_\sigma \cap V),$$

there are at most countably many E_σ such that $\mu(E_\sigma \cap V) \neq 0$, that is, at most countably many E_σ such that $E_\sigma \cap V$ is nonempty. Hence, *a fortiori*, there are at most countably many E_σ such that $E_\sigma \cap K$ is nonempty.

Let $R = \bigcup_\sigma E_\sigma$. We shall prove that $G - R$ is a null set. To do this, it suffices to prove that, if K is an arbitrary totally bounded set, then $K \cap (G - R)$ is a null set.[47] Since K intersects at most countably many E_σ, we have $K \cap R \in \mathfrak{B}$, hence $K \cap (G - R) \in \mathfrak{B}$. Suppose that

[46] *Translator's note:* It is not immediately clear that $\mathfrak{F}$ is nonempty, unless one makes some additional assumption, for example, that every nonempty open set has positive measure.

[47] *Translator's note:* The justification for this assertion is obscure. The crucial point is proving that $R \in \mathfrak{B}$, and it is difficult to see how this follows from the argument given here. Of course, if μ is σ-finite, then there are at most countably many sets E_σ. In that case, R obviously belongs to $\mathfrak{B}$, and one can simply consider an arbitrary compact set $K \subset G - R$; the proof can then be completed in a manner similar to that indicated in the text.

the measure of $K \cap (G - R)$ were not zero; then $K \cap (G - R)$ must contain a totally bounded set E having positive measure. Obviously, the intersection $E \cap E_\sigma$ is empty for every $\sigma \in \Sigma$.

Let $U_1 \supset U_2 \supset \cdots \supset U_n \supset \cdots$ be a neighborhood basis at the identity e of G; we may assume that $\mu(U_1) < \infty$. Since E is totally bounded, there exists, for any n, a finite set of points $g_1^{(n)},\ldots, g_{m_n}^{(n)}$ in E such that $\bigcup_{\nu=1}^{m_n} g_\nu^{(n)} U_n \supset E$. It may happen that $E \cap (g_\nu^{(n)} U_n)$ is a null set for certain indices n, ν; let F be the union of all (at most countably many) such sets, and let $E_0 = E - F$. Then $E_0 \in \mathfrak{B}$, $E_0 \cap E_\sigma$ is empty for every $\sigma \in \Sigma$, and $0 < \mu(E_0) = \mu(E) < \infty$. We assert that, if V is any open set which intersects E_0, then $\mu(E_0 \cap V) > 0$. In fact, suppose $g \in E_0 \cap V$; since g is an interior point of V, there is an n such that $gU_n \subset V$, and, for any m, since $g \in E_0$, there is a ν such that $g \in g_\nu^{(m)} U_m$, that is, $g^{-1} g_\nu^{(m)} \in U_m^{-1}$. Hence, if m is sufficiently large, then $U_m^{-1} U_m \subset U_n$, and we have

$$g_\nu^{(m)} U_m \subset g U_m^{-1} U_m \subset V.$$

But $g \in g_\nu^{(m)} U_m$, $g \in E_0$, hence $E \cap (g_\nu^{(m)} U_m)$ is not a null set [since E_0 was obtained from E by removing all those sets $E \cap g_\nu^{(m)} U_m$ such that $\mu(E \cap g_\nu^{(m)} U_m) = 0$]. Since $E_0 \cap V \supset E_0 \cap (g_\nu^{(m)} U_m)$, this implies that $E_0 \cap V$ is not a null set. Thus, $\{E_\sigma, \sigma \in \Sigma\} \cup \{E_0\} \in \mathfrak{F}$. This contradicts the maximality of $\{E_\sigma \mid \sigma \in \Sigma\}$. Therefore, we conclude that $G - R$ is a null set. We may assume that $G = R$ (for otherwise, we could incorporate $G - R$ in one of the sets E_σ). Thus, $\{E_\sigma \mid \sigma \in \Sigma\}$ is a partition[48] of G.

Next, we proceed to construct the mappings $\{\phi_n^{(\sigma)}\}$. Clearly, we may assume that $U_n^{-1} U_n \subset U_{n-1}$, $n = 2, 3,\ldots$. Now, for each n, $n = 1, 2,\ldots$, and each $\sigma \in \Sigma$, one can choose a finite or countable set of elements $h_{\nu,n}^{(\sigma)} \in E_\sigma$, $\nu = 1, 2,\ldots$, such that

$$\mu\left(E_\sigma - \bigcup_\nu h_{\nu,n}^{(\sigma)} U_n\right) = 0. \tag{3.3.37}$$

In fact, by regularity, E_σ is the union of countably many compact sets and a null set, and each of these compact sets can be covered by finitely many sets of the form hU_n, where $h \in E_\sigma$, hence, there does exist a finite or countable set $\{h_{\nu,n}^{(\sigma)}\} \subset \mathfrak{B}$ such that (3.3.7) holds. Since $h_{\nu,n}^{(\sigma)} U_n \in \mathfrak{B}$, all the sets

$$E_{\nu,n}^{(\sigma)} = \left(h_{\nu,n}^{(\sigma)} U_n - \bigcup_{k=1}^{\nu-1} h_{k,n}^{(\sigma)} U_n\right) \cap E_\sigma$$

48 *Translator's note:* Again, the proof of this assertion is not obvious (unless μ is σ-finite).

also belong to $\mathfrak{B}$, and if $\nu \neq \nu'$, then $E^{(\sigma)}_{\nu,n}$ and $E^{(\sigma)}_{\nu',n}$ are disjoint. Moreover, if we let

$$E^{(\sigma)}_{0,n} = E_\sigma - \bigcup_\nu E^{(\sigma)}_{\nu,n},$$

then, by (3.3.37), we get

$$\mu(E^{(\sigma)}_{0,n}) = 0. \tag{3.3.38}$$

We shall define the mappings $\{\varphi^{(\sigma)}_n\}$ by induction. First, define $\varphi^{(\sigma)}_1$ as follows. For each nonempty set $E^{(\sigma)}_{\nu,1}$, choose any element $g^{(\sigma)}_{\nu,1} \in E^{(\sigma)}_{\nu,1}$, and let $\varphi^{(\sigma)}_1(g) = g^{(\sigma)}_{\nu,1}$ for all $g \in E^{(\sigma)}_{\nu,1}$. Clearly, $\varphi^{(\sigma)}_1$ satisfies conditions (i) and (ii). Now, assume that mappings $\varphi^{(\sigma)}_1, \ldots, \varphi^{(\sigma)}_{l-1}$ have already been defined, satisfying conditions (i), (ii) and (iii) (for $1 \leqslant n \leqslant m \leqslant l-1$). Given any ν, we use the letter s to denote all the indices such that $g^{(\sigma)}_{s,l-1} \in E^{(\sigma)}_{\nu,l}$. If $g \in E^{(\sigma)}_{\nu,l} \cap (\bigcup_s E^{(\sigma)}_{s,l-1})$, let $\varphi^{(\sigma)}_l(g) = \varphi^{(\sigma)}_{l-1}(g)$. If the set

$$F^{(\sigma)}_{\nu,l} = E^{(\sigma)}_{\nu,l} - \bigcup_s E^{(\sigma)}_{s,l-1}$$

is nonempty, we arbitrarily choose a point $g_0 \in F^{(\sigma)}_{\nu,l}$, and let $\varphi^{(\sigma)}_l(g) = g_0$ for all $g \in F^{(\sigma)}_{\nu,l}$. This completes the inductive definition of the mappings $\{\varphi^{(\sigma)}_n\}$, and one can easily verify that they satisfy conditions (i), (ii), and (iii). Notice also that these mappings have the following property:

$$\varphi^{(\sigma)}_n(E^{(\sigma)}_{\nu,n}) \subset E^{(\sigma)}_{\nu,n}.$$

Consequently, if $g \,\overline{\in}\, E^{(\sigma)}_{0,n}$, then $g^{-1}\varphi^{(\sigma)}_n(g) \in E^{(\sigma)-1}_{\nu,n} E^{(\sigma)}_{\nu,n} \subset U^{-1}_n U_n \subset U_{n-1}$. Therefore, on the set $E_\sigma - \bigcup_{n=1}^\infty E^{(\sigma)}_{0,n}$, we have

$$\lim_{n\to\infty} \varphi^{(\sigma)}_n(g) = g.$$

But, by (3.3.38), $\bigcup_{n=1}^\infty E^{(\sigma)}_{0,n}$ is a null set in E_σ.]

4° Positive Definite Continuous Functions on Commutative Topological Groups

Theorem 3.3.7. Let $(G, \mathfrak{T})$ be a commutative topological group, and let $\mathfrak{B}$ be a σ-algebra of sets in G, such that $\mathfrak{B}$ contains all the closed subsets of G. Suppose there exists a regular locally finite measure μ on $(G, \mathfrak{B})$. Then, given any positive definite continuous function f on G, with $f(e) = 1$, there exists a unique probability measure $P^{\mathfrak{B}}$ on $(G^{\mathfrak{B}}, \mathfrak{F}^{\mathfrak{B}})$, such that

$$f(g) = \int_{G^{\mathfrak{B}}} \alpha(g)\, dP^{\mathfrak{B}}(\alpha) \tag{3.3.39}$$

for all $g \in G$.

PROOF. Let f be a positive definite continuous function on G. Choose any neighborhood U of the identity e in G such that $\mu(U) < \infty$. Construct a sequence of neighborhoods U_n of e as follows:

$$U_n = \left\{ g \,\middle|\, 1 - \Re f(g) < \frac{1}{n} \right\} \cap U. \tag{3.3.40}$$

Then, taking $\{U_n\}$ as a neighborhood basis at e, we obtain a topology $\mathfrak{T}' \subset \mathfrak{T}$, such that $(G, \mathfrak{T}')$ is a commutative topological group[49] satisfying the first axiom of countability. Since $\mathfrak{B}$ contains all the closed sets of $(G, \mathfrak{T})$, we have $\mathfrak{T} \subset \mathfrak{B}$, hence $\mathfrak{T}' \subset \mathfrak{B}$. Moreover, $\mu(U_n) < \infty$. It is easily seen that the measure space $(G, \mathfrak{B}, \mu)$ is also locally finite and regular[50] with respect to the topology $\mathfrak{T}'$.

Using Lemma 3.3.6, we obtain a partition $\{E_\sigma \mid \sigma \in \Sigma\}$ of G and a collection of mappings $\{\varphi_n^{(\sigma)} \mid \sigma \in \Sigma, n = 1, 2, \ldots\}$, satisfying conditions (i), (ii), and (iii), such that (3.3.6) holds (for almost all $g \in E_\sigma$) relative to the topology $\mathfrak{T}'$. By virtue of (3.3.40), the function $\Re f$ is continuous at e relative to $\mathfrak{T}'$. Therefore, by (3.3.36), we deduce that

$$\lim_{n\to\infty} \Re(f(\varphi_n^{(\sigma)}(g)^{-1}g)) = 1$$

for almost all $g \in E_\sigma$. The conclusion of the theorem then follows by an application of Lemma 3.3.5. ∎

We now proceed to use Theorem 3.3.7, together with the properties of quasi-invariant measures, to establish a representation theorem for positive definite continuous functions on groups with quasi-invariant measures.

Theorem 3.3.8. Let G be a commutative topological group, let $\mathfrak{G}$ be a subgroup of G, and let $\mathfrak{T}$ be a topology on $\mathfrak{G}$, stronger than the relative topology induced by G, such that $(\mathfrak{G}, \mathfrak{T})$ is a topological group of the second category. Let $\mathfrak{B}$ be a σ-algebra of sets in G, containing all closed subsets of G. Suppose that there exists a locally finite, regular and strongly $\mathfrak{G}$-quasi-invariant measure μ on $(G, \mathfrak{B})$. Let $\mathfrak{G}^*$ denote the dual of $(\mathfrak{G}, \mathfrak{T})$, and $\mathfrak{F}^*$ the totality of weak Borel sets in $\mathfrak{G}^*$. Then, for each positive definite continuous function f on G, with $f(e) = 1$, there exists a unique probability measure P^* on $(\mathfrak{G}^*, \mathfrak{F}^*)$ such that

$$f(g) = \int_{\mathfrak{G}^*} \alpha(g)\, dP^*(\alpha), \qquad g \in \mathfrak{G}, \tag{3.3.41}$$

[49] *Translator's note:* There is a serious error here. For example, if f is identically equal to 1, then $U_n = U$ for all n, and $\{U_n\}$ cannot be a neighborhood basis for a topological group in the usual sense, unless $U = (e)$.

[50] *Translator's note:* The translator finds it difficult to verify these assertions.

moreover, P^* is continuous with respect to the topology induced by G on $\mathfrak{G}$.

PROOF. According to Corollary 3.2.4, if $\alpha \in G^{\mathfrak{B}}$, and if α' denotes the restriction of α to $\mathfrak{G}$, then $\alpha' \in \mathfrak{G}^*$. Thus, we have a homomorphic mapping from $G^{\mathfrak{B}}$ into $\mathfrak{G}^*$:

$$T: \quad \alpha \to \alpha'.$$

If $A \in \mathfrak{F}^*$, then $T^{-1}A = \{\alpha \mid T\alpha \in A\}$ belongs to $\mathfrak{F}^{\mathfrak{B}}$. In fact, let $\mathfrak{F}^{\dagger}$ denote the totality of sets A in $\mathfrak{F}^*$ such that $T^{-1}A \in \mathfrak{F}^{\mathfrak{B}}$; clearly, $\mathfrak{F}^{\dagger}$ is a σ-algebra. On the other hand, it is easily seen that, if A is any Borel cylinder in $\mathfrak{G}^*$, then $T^{-1}A \in \mathfrak{F}^{\mathfrak{B}}$. Therefore, since $\mathfrak{F}^*$ is the σ-algebra generated by the totality of Borel cylinders in $\mathfrak{G}^*$, it follows that $\mathfrak{F}^* \subset \mathfrak{F}^{\dagger}$.

Now, let f be any positive definite continuous function on G, with $f(e) = 1$. By Theorem 3.3.7, there is a unique probability measure $P^{\mathfrak{B}}$ on $(G^{\mathfrak{B}}, \mathfrak{F}^{\mathfrak{B}})$ such that f satisfies (3.3.39). We define a measure P^* on $\mathfrak{F}^*$, as follows:

$$P^*(A) = P^{\mathfrak{B}}(T^{-1}A).$$

It is easily seen that $(\mathfrak{G}^*, \mathfrak{F}^*, P^*)$ is a probability measure space, furthermore, if $g \in \mathfrak{G}$, then it follows from (3.3.39) that

$$f(g) = \int_{G^{\mathfrak{B}}} \alpha(g)\, dP^{\mathfrak{B}}(\alpha) = \int_{\mathfrak{G}^*} \alpha(g)\, dP^*(\alpha).$$

Next, we prove the uniqueness of P^*. Suppose that P^{**} is another probability measure on $(\mathfrak{G}^*, \mathfrak{F}^*)$, satisfying

$$f(g) = \int_{\mathfrak{G}^*} \alpha(g)\, dP^{**}(\alpha). \tag{3.3.42}$$

Let $\mathfrak{G}'$ be the algebraic dual of $\mathfrak{G}$, and let $\mathfrak{F}'$ be the σ-algebra generated by the totality of Borel cylinders in $\mathfrak{G}'$. Then, $\mathfrak{F}^*$ is precisely the totality of sets of the form $A \cap \mathfrak{G}^*$, where $A \in \mathfrak{F}'$, hence we may define set functions P' and P'' on $\mathfrak{F}'$, as follows:

$$P'(A) = P^*(A \cap \mathfrak{G}^*), \qquad P''(A) = P^{**}(A \cap \mathfrak{G}^*).$$

Clearly, P' and P'' are probability measures on $(\mathfrak{G}', \mathfrak{F}')$. Also, we have the relations

$$f(g) = \int_{\mathfrak{G}'} \alpha(g)\, dP'(\alpha), \qquad g \in \mathfrak{G},$$

$$f(g) = \int_{\mathfrak{G}'} \alpha(g)\, dP''(\alpha), \qquad g \in \mathfrak{G},$$

which follow from (3.3.41) and (3.3.42), respectively. Therefore, since the restriction $f \mid \mathfrak{G}$ is a positive definite function on the commutative group $\mathfrak{G}$, it follows by the uniqueness property in Theorem 3.3.4 that

$$P' = P'',$$

and hence $P^* = P^{**}$. Finally, the continuity of the measure P^* may be deduced by using Lemma 3.2.18.]

Note that Theorem 3.3.8 only provides an expression for the values of f on the subgroup $\mathfrak{G}$. Of course, if $\mathfrak{G}$ happens to be dense in G, then, by continuity, the values of f on $\mathfrak{G}$ completely determine the values of f on G.

Example 3.3.2. Let G be a locally compact commutative topological group, and let G^* be the dual of G. Then, given any positive definite continuous function f on G, there is a unique finite measure P^* on $(G^*, \mathfrak{F}^*)$ such that

$$f(g) = \int_{G^*} \alpha(g)\, dP^*(\alpha), \qquad g \in G. \tag{3.3.43}$$

In fact, we need only apply Theorem 3.3.8 to the Haar measure space[51] $\Omega = (G, \mathfrak{B}, \mu)$, with $\mathfrak{G} = G$; the local finiteness of μ follows at once from the local compactness of G.

Retaining the hypotheses and notation of Example 3.3.2, we consider any function $\xi \in L^1(\Omega)$, and form a corresponding function $\hat{\xi}$ on G^*, as follows:

$$\hat{\xi}(\alpha) = \int_G \overline{\alpha(g)}\, \xi(g)\, d\mu(g), \qquad \alpha \in G^*. \tag{3.3.44}$$

$\hat{\xi}$ is called the *L^1-Fourier transform* of ξ. We have the following result:

$$\text{if} \quad \xi \in L^1(\Omega), \qquad \hat{\xi}(\alpha) = 0, \qquad \text{then} \quad \xi = 0. \tag{3.3.45}$$

In fact, let η be any continuous function on G having a compact support. Consider the function

$$f(h) = \int \overline{\eta(gh)}\, \eta(g)\, d\mu(g). \tag{3.3.46}$$

[51] *Translator's note:* This must be understood as meaning the extended Haar measure space, constructed from the ordinary Haar measure space (in the sense of Halmos [1]) by the procedure described in Definitions 1.1.1 and 1.1.2.

It is easily shown that f is continuous on G, moreover, for any $g_1 \ldots, g_n \in G$, and any n complex numbers $z_1, \ldots, z_n$, we have

$$\sum_{k,l=1}^{n} f(g_k^{-1} g_l)\, z_k \bar{z}_l = \int \left| \sum_{k=1}^{n} \eta(g g_k) z_k \right|^2 d\mu(g) \geqslant 0.$$

Thus, f is a positive definite continuous function. Hence, there is a corresponding probability measure P^* on $(G^*, \mathfrak{F}^*)$ such that (3.3.43) holds. Now, multiply both sides of (3.3.43) by $\overline{\xi(hg)}\, \xi(h)$, and integrate over $\Omega \times \Omega$. Since

$$\begin{aligned} &\iint \alpha(g)\, \overline{\xi(hg)}\, \xi(h)\, d\mu(h)\, d\mu(g) \\ &\quad = \int \overline{\alpha(h)} \left(\int \alpha(hg)\, \overline{\xi(hg)}\, d\mu(g) \right) \xi(h)\, d\mu(h) \\ &\quad = |\, \hat{\xi}(\alpha)|^2, \end{aligned} \tag{3.3.47}$$

we get

$$\iint f(g)\, \overline{\xi(hg)}\, \xi(h)\, d\mu(h)\, d\mu(g) = 0.$$

Substituting (3.3.46) into the above relation, interchanging the order of integration, and performing a change of variable, we obtain

$$\int \left| \int \right| \eta(gh)\, \xi(h)\, d\mu(h)|^2\, d\mu(g) = 0.$$

Therefore,

$$\int \eta(gh)\, \xi(h)\, d\mu(h) = 0$$

for almost all $g \in G$. Since every open Borel set has positive measure, every neighborhood of the identity in G contains a point g satisfying the above equality. Using the fact that η vanishes outside a compact set, we deduce by a straightforward approximation argument that

$$\int \eta(h)\, \xi(h)\, d\mu(h) = 0.$$

Since the class of continuous functions with compact supports is dense in $L^1(\Omega)$, it follows by an obvious limiting process (using the Lebesgue dominated convergence theorem) that the last relation also holds when η is the characteristic function of an arbitrary compact subset of G. Using the regularity of μ, we deduce that ξ vanishes almost everywhere.]

We conclude this section by proving that the points of a locally compact group are distinguished by its continuous characters.

Lemma 3.3.9. Let G be a locally compact commutative topological group. Then, for any $g_0 \in G$, with $g_0 \neq e$, there exists a continuous character α_0 of G such that $\alpha_0(g_0) \neq 1$.

PROOF. It is easily seen that there exists a continuous function φ on G, with a compact support, such that the function

$$f(h) = \int_G \varphi(hg)\,\overline{\varphi(g)}\,d\mu(g)$$

has the property $f(g_0) \neq f(e) = 1$. Since f is a positive definite continuous function on G, there is a corresponding probability measure P^* on (G^*, F^*) such that (3.3.43) holds. Now, if we had $\alpha(g_0) = 1$ for all $\alpha \in G$, then, from (3.3.43), we would get $f(g_0) = 1$, which is impossible. Thus, there must be an $\alpha_0 \in G^*$, such that $\alpha_0(g_0) \neq 1$.]

§3.4. L^2-Fourier Transforms

Throughout this section, we assume that $\Omega = (G, \mathfrak{B}, \mu)$ is a localizable measure space which is quasi-invariant under a commutative group of measurable transformations $\mathfrak{G}$. We shall make use of the unitary representation U of $\mathfrak{G}$, defined in §3.1, 6°; in the present case, since $\mathfrak{G}$ is commutative, the group $\mathfrak{U} = \{U(h) \mid h \in \mathfrak{G}\}$ is commutative, and hence $\mathfrak{A}(\Omega, \mathfrak{G})$, the smallest weakly closed operator algebra over $L^2(\Omega)$ containing $\mathfrak{U}$ (denoted briefly by $\mathfrak{A}$), is also commutative (see Lemma 2.4.1).

1° kth Order Cyclic Measures

Definition 3.4.1. Let $\Omega = (G, \mathfrak{B}, \mu)$ be a nontrivial[52] localizable measure space which is quasi-invariant under a commutative group of measurable transformations $\mathfrak{G}$. If the operator algebra $\mathfrak{A}(\Omega, \mathfrak{G})$ has uniform multiplicity k, then we say that the measure space Ω (or the measure μ) is *cyclic of order k* relative to $\mathfrak{G}$.[53]

In particular, if $\mathfrak{A}(\Omega, \mathfrak{G})$ is maximal commutative weakly closed (that is, $k = 1$), then we simply say that Ω (or μ) is *cyclic* relative to $\mathfrak{G}$.

Example 3.4.1. If $\mathfrak{G}$ consists only of the identity element, and the dimension of $L^2(\Omega)$ is k, then the operator algebra $\mathfrak{A}(\Omega, \mathfrak{G}) = \{\lambda I\}$ (where λ ranges over all complex numbers) has uniform multiplicity k,

[52] That is, $\mu \not\equiv 0$.

[53] Obviously, k cannot exceed the dimension of $L^2(\Omega)$.

that is, Ω is cyclic of order k. Thus, for any cardinal number k, there exist cyclic measures of order k.

In the ensuing discussion, we shall decompose certain measures into sums of mutually singular measures having distinct cyclic orders k. For this purpose, we introduce the following concept.

Definition 3.4.2. Let $\Omega = (G, \mathfrak{B}, \mu)$ be a localizable measure space which is quasi-invariant under a commutative transformation group $\mathfrak{G}$. Let $\mathfrak{E}$ be some group of quasi-characters of Ω relative to $\mathfrak{G}$. If $\mathfrak{E}$ constitutes a determining set of functions on Ω, then we say that Ω is *normal* with respect to $\mathfrak{E}$. In particular, if G^{μ} is a determining set of functions on Ω, we say that Ω is normal with respect to $\mathfrak{G}$.

Notice that Ω can always be made into a normal measure space (with respect to $\mathfrak{E}$) by suitably cutting down the size of the σ-ring $\mathfrak{B}$, if necessary.[54]

Example 3.4.2. Let G be a commutative locally compact group. Then, the Haar measure space $\Omega = (G, \mathfrak{B}, \mu)$ is normal with respect to the translation group G.

PROOF. We shall prove that G^* constitutes a determining set of functions on Ω. Let $\mathfrak{B}_1$ denote the smallest σ-algebra which makes all the functions in G^* measurable. Given any set $E \in \mathfrak{B}$, it follows from the σ-finiteness and regularity of Haar measure that there is a sequence of compact sets

$$Q_1 \subset Q_2 \subset \cdots \subset Q_n \subset \cdots$$

in G such that $E - \bigcup_{n=1}^{\infty} Q_n$ is a μ-null set. Let R denote the set of all those linear combinations $\sum a_i\alpha_i$, $\alpha_i \in G^*$, such that $\sum a_i\alpha_i(g)$ is real for all $g \in G$. Consider any compact set $F \subset E$; obviously, we may assume that $F \subset Q_1$. By the regularity of μ, there is a sequence of open sets $\{O_n\}$, $O_n \supset F$, such that $\mu(O_n - F) \to 0$. For each n, since F and $Q_1 - O_n$ are disjoint closed subsets of the normal space Q_1, it follows from Urysohn's lemma that there exists a continuous real-valued function $f_n(q)$, $q \in Q_1$, $0 \leqslant f_n(q) \leqslant 1$, such that $f_n(q) = 1$ when $q \in F$, and $f_n(q) = 0$ when $q \in Q_1 - O_n$. Thus, the sequence of functions $\{f_n\}$ converges to C_F (the characteristic function of F) almost everywhere on Q_1. Now, by applying Lemma 3.3.9 and Theorem 2.1.2, one can

54 *Translator's note:* If $\mathfrak{B}$ is not a σ-algebra, then it is not $\mathfrak{B}$, but rather $\tilde{\mathfrak{B}}$ which must be cut down to size. In any case, one replaces $\tilde{\mathfrak{B}}$ by the σ-algebra generated by the totality of sets of the form (modulo a null set) $\alpha^{-1}(B)$, where $\alpha \in \mathfrak{E}$ and B is a Borel subset of the complex plane.

prove that R is dense in the totality of real continuous functions on Q_1. Consequently, for each n, there is a function $\varphi_n \in R$, such that

$$\max_{g \in Q_1} | \varphi_n(g) - f_n(g)| < \frac{1}{n}.$$

Therefore, the sequence $\{\varphi_n\}$ also converges to C_F almost everywhere on Q_1, whence it follows that if we define

$$F' = \bigcap_{n=1}^{\infty} \varliminf_{k\to\infty} \left\{ g \mid \varphi_k(g) \geqslant 1 - \frac{1}{n} \right\} \in \mathfrak{B}_1,$$

then F and $Q_1 \cap F'$ differ only by a μ-null set. Similarly, for each Q_n, there is a set $Q_n' \in \mathfrak{B}_1$ such that Q_n and $Q_{n+1} \cap Q_n'$ differ only by a μ-null set. This implies that F and $\bigcap_{n=1}^{\infty} Q_n' \cap F' \cap E$ differ only by a μ-null set. Again, by the regularity of μ, we deduce that, for any set $F \subset E$ with $\mu(F) < \infty$, there exists $F'' \in \mathfrak{B}_1$ such that $E \cap F''$ differs from F only by a μ-null set. Since E is σ-finite, it follows easily that the same conclusion applies to any Borel set $F \subset E$.]

Theorem 3.4.1. Let $\Omega = (G, \mathfrak{B}, \mu)$ be a localizable measure space which is quasi-invariant under a transformation group $\mathfrak{G}$, and let $\mathfrak{E}$ be a certain group of quasi-characters of Ω relative to $\mathfrak{G}$. If Ω is normal with respect to $\mathfrak{E}$, then one can assign a localizable measure space $(G, \mathfrak{B}, \mu_k)$ to each cardinal number[55] k, in a unique manner, such that the following conditions are satisfied: (i) the measures μ_k are mutually singular; (ii) if $\mu_k \neq 0$, then $(G, \mathfrak{B}, \mu_k)$ is quasi-invariant and cyclic of order k relative to $\mathfrak{G}$, and also normal with respect to $\mathfrak{E}$; (iii) $\mu = \sum_k \mu_k$.

In other words, every normal quasi-invariant localizable measure space can be uniquely decomposed into the sum of a family of mutually singular normal quasi-invariant localizable measure spaces having distinct cyclic orders k. Consequently, various problems concerning general quasi-invariant measures may henceforth be reduced to problems concerning kth order cyclic measures.

Before proving Theorem 3.4.1, we introduce a unitary representation V of $\mathfrak{E}$ in the Hilbert space $L^2(\Omega)$, as follows. If $x \in \mathfrak{E}$, let

$$(V(x)\varphi)(g) = x(g)\,\varphi(g), \qquad g \in G, \quad \varphi \in L^2(\Omega).$$

Let $\mathfrak{V}$ denote the totality of operators $V(x)$, $x \in \mathfrak{E}$; we call $\mathfrak{V}$ the *multiplication group* corresponding to $\mathfrak{E}$. Let C be the smallest weakly closed

[55] *Translator's note:* Or, more appropriately, to each "sufficiently small' cardinal number (to avoid logical difficulties).

operator algebra containing $\mathfrak{B}$; since $\mathfrak{E}$ is a determining set, it follows that C is maximal commutative weakly closed. Also, we note the following important "commutation relation" between the members of $\mathfrak{U}$ and $\mathfrak{B}$: for any $x \in \mathfrak{E}$, $h \in \mathfrak{G}$, we have[56]

$$\tilde{x}(h)\, U(h)\, V(x) = V(x)\, U(h). \tag{3.4.1}$$

In fact, (3.4.1) is a straightforward consequence of the definitions of U, V and the properties of x.

PROOF OF THEOREM 3.4.1. Since $\mathfrak{A}$ is a commutative weakly closed operator algebra, it follows from Theorem 2.4.3 that there is a unique system of mutually orthogonal projection operators $\{P_n\} \subset \mathfrak{A}'$, one for each cardinal number n, such that the restriction of $\mathfrak{A}$ to $P_n L^2(\Omega)$ has uniform multiplicity n, and $\sum P_n = I$.

We now proceed to prove that each P_n commutes with every $V(x)$, $x \in \mathfrak{E}$. For each $x \in \mathfrak{E}$, consider the correspondence

$$T(x)\colon \quad A \to V(x)\, A V(x)^{-1},$$

where A ranges over all bounded linear operators on $L^2(\Omega)$. We assert that $T(x)$ maps $\mathfrak{A}$ onto $\mathfrak{A}$. In fact, if $A = U(h) \in \mathfrak{U}$, then, by the commutation relation (3.4.1), we have $T(x)\, U(h) = \tilde{x}(h)\, U(h) \in \mathfrak{A}$, which shows that the commutative weakly closed operator algebra $T(x)^{-1}\mathfrak{A}$ contains $\mathfrak{U}$. But $\mathfrak{A}$ is generated by $\mathfrak{U}$, hence $T(x)^{-1}\mathfrak{A} \supset \mathfrak{A}$. Similarly, $T(x^{-1})^{-1}\, \mathfrak{A} \supset \mathfrak{A}$, but $T(x^{-1})^{-1} = T(x)$, therefore, $T(x)\, \mathfrak{A} = \mathfrak{A}$, and $T(x) \mid \mathfrak{A}$ is an automorphism of $\mathfrak{A}$. Let $Q_n = T(x)\, P_n$; then Q_n is a projection operator, and $Q_n \in \mathfrak{A}'$. Moreover, $V(x)$ maps $P_n L^2(\Omega)$ unitarily onto $Q_n L^2(\Omega)$, and, for any $A \in \mathfrak{A}$, the action of A on $P_n L^2(\Omega)$ is unitarily equivalent [under $V(x)$] to the action of $T(x)\, A$ on $Q_n L^2(\Omega)$. It follows that the restriction of $\mathfrak{A}$ to $Q_n L^2(\Omega)$ also has uniform multiplicity n. Furthermore, since the $\{P_n\}$ are mutually orthogonal, the $\{Q_n\}$ are also mutually orthogonal. Thus, it follows from the uniqueness of the correspondence $n \to P_n$ that $P_n = Q_n$, that is $P_n V(x) = V(x)\, P_n$. Hence, $P_n \in (\mathfrak{B})' = C'$, but since C is maximal commutative weakly closed, this means that $P_n \in C$, and since C is clearly just the multiplication algebra over $L^2(\Omega)$, there exists a bounded $\mathfrak{B}$-measurable function $P_n(g)$ on G, such that

$$(P_n\varphi)(g) = P_n(g)\, \varphi(g), \qquad \varphi \in L^2(\Omega).$$

Since $P_n = P_n^2$, we have $P_n(g) = P_n(g)^2$, which implies that $P_n(g)$ is the characteristic function of some set $E_n \in \mathfrak{B}$. If $n \neq n'$, then $P_n P_{n'} = 0$,

[56] Here, $\tilde{x}(h)$ denotes the character of $\mathfrak{G}$ induced by x (see §3.2).

that is, $P_n(g)\,P_{n'}(g) = 0$, hence $E_n \cap E_{n'}$ is a null set. Moreover, since $\sum_n P_n = I$, it follows by Lemma 1.2.5 that $\bigvee_n E_n$ differs from the whole of G by at most a null set.

We further assert that every E_n is quasi-invariant under $\mathfrak{G}$. In fact, if $h \in \mathfrak{G}$, then $U(h) \in \mathfrak{A}$, hence $U(h)\,P_n = P_n U(h)$. Therefore, since $d\mu_h/d\mu > 0$, we obtain

$$P_n(hg)\,\varphi(g) = P_n(g)\,\varphi(g)$$

for every $\varphi \in L^2(\Omega)$. Consequently, the equality $P_n(hg) = P_n(g)$ holds almost everywhere, that is, E_n and hE_n differ only by a null set. Thus, E_n is quasi-invariant under $\mathfrak{G}$.

For each E_n, we define a corresponding measure μ_n on $(G, \mathfrak{B})$, as follows:

$$\mu_n(E) = \mu(E \cap E_n), \qquad E \in \mathfrak{B}.$$

If $n \neq n'$, then, since $\mu(E_n \cap E_{n'}) = 0$, it is clear that μ_n and $\mu_{n'}$ are mutually singular. Since E_n is a quasi-invariant set, it is easily seen that, if $\mu_n \neq 0$, then μ_n is a quasi-invariant measure. Moreover, since G is a least upper bound for $\{E_n\}$, it clearly follows that E is a least upper bound for $\{E \cap E_n\}$, and by Lemma 1.2.4 we obtain

$$\mu(E) = \sum_n \mu_n(E).$$

The normality of $\Omega_n = (G, \mathfrak{B}, \mu_n)$ with respect to $\mathfrak{C}$ is obvious. Thus, a decomposition of the required kind does exist. Finally, the uniqueness of the decomposition can be deduced at once from the uniqueness of the system[57] $\{P_n\}$.]

Next, we use Theorem 3.4.1 to establish a certain property of ergodic measures.

Theorem 3.4.2. Let $\Omega = (G, \mathfrak{B}, \mu)$ be a localizable measure space which is quasi-invariant under a transformation group $\mathfrak{G}$. If Ω is ergodic and normal with respect to $\mathfrak{G}$, then there exists a cardinal number k such that Ω is cyclic of order k relative to $\mathfrak{G}$.

Proof. We decompose μ into the sum of a family of mutually singular quasi-invariant measures $\{\mu_k\}$, in accordance with Theorem 3.4.1. Suppose there existed two distinct indices k, k' such that μ_k and $\mu_{k'}$

[57] *Translator's note:* In order to deduce uniqueness in this manner, it seems that one must restrict one's considerations to systems of measures having the form $\mu_n'(E) = \mu(E \cap E_n')$, where $\{E_n'\}$ is some family of sets of the sort constructed in the foregoing proof. If the range of n is uncountable, then it is not immediately clear whether the conditions to be satisfied by $\{\mu_n'\}$ necessarily imply that it is of this form.

were both different from zero. Then, one easily deduces that both of the sets E_k (see the proof of Theorem 3.4.1) and $G - E_k$ have positive measure; since E_k and $G - E_k$ are quasi-invariant sets, this contradicts the ergodicity of μ. We conclude that there is an index k such that $\mu = \mu_k$.]

Problem. If $k \neq 1$, then, does there exist a localizable measure space which is quasi-invariant, ergodic, normal and cyclic of order k relative to some transformation group $\mathfrak{G}$?

This question may not be easy to answer. However, we can give an example of a nontrivial normal ergodic cyclic measure, as follows.

Example 3.4.3. Let G be a commutative locally compact group, and $\Omega = (G, \mathfrak{B}, \mu)$ a Haar measure space. Then, Ω is cyclic relative to the translation group G.

PROOF. According to Example 3.4.2, Ω is normal, hence, by Theorem 3.4.2, there is a cardinal number k such that Ω is cyclic of order k. Suppose that $k \geqslant 2$; then, there must exist two mutually orthogonal closed linear subspaces H_1 and H_2 in $L^2(\Omega)$, both invariant under $\mathfrak{A}(\Omega, \mathfrak{G})$, and a unitary mapping T from H_1 onto H_2, such that $TU(h) = U(h)\,T$. Choose any $a \in H_1$, $a \neq 0$, and let $b = Ta \in H_2$; then $b \neq 0$. Moreover,

$$(U(h)a, a) = (U(h)b, b), \qquad (U(h)a, b) = 0.$$

Now, consider any $\psi \in L^2(\Omega) \cap L^1(\Omega)$, and let $\varphi(h) = (\psi^* * U(q)\,\psi)(h)$. Multiplying the above relation by $\varphi(h)$, and integrating over Ω, we get

$$(U(q)\psi * a, \psi * a) = (U(q)\psi * b, \psi * b), \qquad (U(q)\psi * a, \psi * b) = 0. \quad (3.4.2)$$

Here, both $a_1 = \psi * a$ and $b_1 = \psi * b$ belong to[58] $L^1(\Omega) \cap L^2(\Omega)$. Multiplying (3.4.2) by $\alpha(q) (\alpha \in G^*)$, integrating over Ω, and using (3.3.47), we obtain

$$|\,\hat{a}_1(\alpha)|^2 = |\,\hat{b}_1(\alpha)|^2, \qquad \hat{a}_1(\alpha)\,\overline{\hat{b}_1(\alpha)} = 0, \quad \alpha \in G^*.$$

Therefore, $\hat{a}_1 \equiv 0$, $\hat{b}_1 \equiv 0$, and hence, by (3.3.45), we see that $a_1 = 0$, $b_1 = 0$. Thus, it follows by Corollary 1.1.15 that both a and b vanish almost everywhere, which contradicts our assumption. We conclude that $k = 1$, that is, Ω is cyclic.]

[58] *Translator's note:* It is certainly not obvious that a_1 and b_1 belong to $L^1(\Omega)$. Of course, if $\mu(G)$ is finite, then $L^2(\Omega) \subset L^1(\Omega)$, so that the example is at least valid for compact groups.

2° Dual Measure Spaces

We have already discussed the notion of the L^1-Fourier transform on a locally compact group. We now proceed to generalize this notion, as follows.

Definition 3.4.3. Let $\Omega = (G, \mathfrak{B}, \mu)$ be a localizable measure space which is quasi-invariant under and cyclic of order k relative to a transformation group $\mathfrak{G}$. Let $\hat{\Omega} = (\hat{G}, \hat{\mathfrak{B}}, \hat{\mu})$ be a measure space which is quasi-invariant under a transformation group $\hat{\mathfrak{G}}$, and such that the following conditions are fulfilled. (i) There exists an isomorphism

$$\tilde{\alpha} \to \hat{\alpha} \tag{3.4.3}$$

from $\tilde{\mathfrak{G}}^{\mu}$ onto $\hat{\mathfrak{G}}$ and an isomorphism

$$h \to h(\hat{g}), \qquad \hat{g} \in \hat{G} \tag{3.4.4}$$

from $\mathfrak{G}$ into $\hat{G}^{\hat{\mu}}$ such that $\hat{\mathfrak{E}} = \{h(\cdot) \mid h \in \mathfrak{G}\}$ constitutes a determining set of functions on $\hat{\Omega}$, moreover, for each $h \in \mathfrak{G}$, the character $\tilde{h}$ of $\hat{\mathfrak{G}}$ induced by $h(\cdot)$ satisfies the relation

$$\tilde{h}(\hat{\alpha}) = \tilde{\alpha}(h), \qquad \hat{\alpha} \in \hat{\mathfrak{G}}. \tag{3.4.5}$$

(ii) There exists a unitary mapping F from $L^2(\Omega)$ onto $\mathfrak{L}_k^2(\hat{\Omega})$ (see §1.1, 2°), such that, for each $h \in \mathfrak{G}$, the unitary operator $\hat{U}(h) = FU(h)F^{-1}$ in $\mathfrak{L}_k^2(\hat{\Omega})$ takes the form

$$(\hat{U}(h)\varphi)(\hat{g}) = h(\hat{g})\,\varphi(\hat{g}), \qquad \hat{g} \in \hat{G}, \quad \varphi \in \mathfrak{L}_k^2(\hat{\Omega}). \tag{3.4.6}$$

Then, we say that $(\hat{\Omega}, \hat{\mathfrak{G}})$ is a *dual* of $(\Omega, \mathfrak{G})$ (or that $\hat{\Omega}$ is dual to Ω) and that F is an associated *Fourier transform* (from $L^2(\Omega)$ to $\mathfrak{L}_k^2(\hat{\Omega})$); we describe such a mapping F, in general, as an *L^2-Fourier transform.*

First, we prove the following existence theorem.

Theorem 3.4.3. Let $\Omega = (G, \mathfrak{B}, \mu)$ be a localizable measure space which is quasi-invariant under and cyclic of order k relative to a transformation group $\mathfrak{G}$. Then, there exists a dual $(\hat{\Omega}, \hat{\mathfrak{G}})$ of $(\Omega, \mathfrak{G})$ and an associated Fourier transform from $L^2(\Omega)$ to $\mathfrak{L}_k^2(\hat{\Omega})$.

Proof. Let $\hat{G}$ be the totality of symmetric multiplicative linear functionals on $\mathfrak{A}(\Omega, \mathfrak{G})$, let $\hat{\mathfrak{B}}$ be the totality of Borel sets in $\hat{G}$, and, for each $h \in \mathfrak{G}$, let $h(\hat{g}) = U(h)(\hat{g})$. Obviously, the function $h(\cdot)$ is $\hat{\mathfrak{B}}$-measurable, and $|h(\hat{g})| = 1$ for all $\hat{g} \in \hat{G}$. Since $\mathfrak{A}$ has uniform multiplicity k, we know by Corollary 2.4.7 that there exists a measure space $\hat{\Omega} = (\hat{G}, \hat{\mathfrak{B}}, \hat{\mu})$ and a unitary mapping F from $L^2(\Omega)$ onto $\mathfrak{L}_k^{\,2}(\hat{\Omega})$ such that, for every $A \in \mathfrak{A}$,

$$FAF^{-1}\xi(\hat{g}) = A(\hat{g})\, \xi(\hat{g}), \qquad g \in \hat{G}, \quad \xi \in \mathfrak{L}_k^{\,2}(\hat{\Omega}),$$

and, in particular, (3.4.6) holds. Let $\mathfrak{F}$ be the linear hull of the family of functions $\mathfrak{E} = \{h(\hat{g}) \mid h \in \mathfrak{G}\}$. If $\mathfrak{F}$ be regarded as a set of multiplication operators on $\mathfrak{L}_k^{\,2}(\Omega)$, then $F\mathfrak{A}F^{-1}$ is just the closure of $\mathfrak{F}$ in the strong topology. Now, take any $A_0 \in \mathfrak{A}$, and let E be any σ-finite set in $\hat{\Omega}$; let E be expressed as the union of a countable family of disjoint sets $\{E_i\} \subset \hat{\mathfrak{B}}$, with $\hat{\mu}(E_i) < \infty$, $i = 1, 2, \ldots$. If ξ_0 is any unit vector in k-dimensional Hilbert space, then the vector-valued function

$$\xi(\hat{g}) = \sum_j 2^{-j/2}\hat{\mu}(E_j)^{-1/2}\, C_{E_j}(\hat{g})\xi_0\,, \qquad \hat{g} \in \hat{G}$$

[the summation being extended over those indices j for which $\hat{\mu}(E_j) > 0$], belongs to $\mathfrak{L}_k(\hat{\Omega})$. Hence, for every natural number n, there is a function $f_n(\hat{g}) \in \mathfrak{F}$ such that

$$\frac{1}{n} > \|(f_n - FA_0F^{-1})\xi\,\|^2 = \int_{G} |\,f_n(\hat{g}) - A_0(\hat{g})|^2\, \|\,\xi(\hat{g})\|^2\, d\hat{\mu}(\hat{g})$$

$$= \sum_j \frac{1}{2^j\hat{\mu}(E_j)} \int_{E_j} |\,f_n(\hat{g}) - A_0(\hat{g})|^2\, d\hat{\mu}(\hat{g}).$$

Therefore, using a diagonal process, one may select a subsequence $\{f_{n'}\}$ of $\{f_n\}$ such that $f_{n'}(\hat{g}) \to A_0(\hat{g})$ for almost every $\hat{g} \in E$. Since $\{f_{n'}\} \subset \mathfrak{F}$, and since $\hat{\mathfrak{B}}$ is, by definition, determined by the totality of functions $A_0(\hat{g})$, $A_0 \in \mathfrak{A}$, it follows easily that $\mathfrak{F}$ is a determining set for $\hat{\Omega}$, whence $\mathfrak{E}$ is, *a fortiori*, a determining set for $\hat{\Omega}$.

Let $\mathscr{B}(L^2(\Omega))$ denote the totality of bounded linear operators on $L^2(\Omega)$, and, for each $x \in G^\mu$, define a transformation $T_x : B \to V(x)\, BV(x)^{-1}$, $B \in \mathscr{B}(L^2(\Omega))$. Obviously, T_x is an automorphism of $\mathscr{B}(L^2(\Omega))$, and $T_xB^* = (T_xB)^*$ for every $B \in \mathscr{B}(L^2(\Omega))$. Moreover, if $h \in \mathfrak{G}$, then, by (3.4.1), we have $T_xU(h) = \tilde{x}(h)\, U(h)$. Consequently, T_x induces an automorphism of $\mathfrak{A}(\Omega, \mathfrak{G})$, and this automorphism depends only upon $\tilde{x}$. Define

$$(\hat{x}\hat{g})(A) = \hat{g}(T_xA), \qquad A \in \mathfrak{A}(\Omega, \mathfrak{G}), \quad \hat{g} \in \hat{G}.$$

It is easily verified that $\hat{x}\hat{g} \in \hat{G}$, that the mapping $\hat{g} \to \hat{x}\hat{g}$ is $\hat{\mathfrak{B}}$-measurable, that the correspondence $\tilde{x} \to \hat{x}$ is an isomorphism of $\tilde{\mathfrak{G}}^\mu$ onto $\hat{\mathfrak{G}} = \{\hat{x} \mid x \in G^\mu\}$, and that $\hat{h}(\hat{x}) = h(\hat{x}\hat{g})/h(\hat{g}) = \tilde{x}(h)$ for all $\hat{g} \in \hat{G}$. Thus, it only remains to prove that $\hat{\mu}$ is quasi-invariant under $\hat{\mathfrak{G}}$.

Write $\hat{V}(x) = FV(x)F^{-1}$. Obviously, $\hat{V}(x)$ is a unitary operator on $\mathfrak{L}_k^2(\hat{\Omega})$, and, by (3.4.1),

$$\tilde{x}(h)\, \hat{U}(h)\, \hat{V}(x) = \hat{V}(x)\, \hat{U}(h) \tag{3.4.7}$$

for every $x \in G^\mu$, $h \in \mathfrak{G}$. If $\xi \in \mathfrak{L}_k^2(\hat{\Omega})$, then, using (3.4.6) and (3.4.7), we get

$$\hat{V}(x)\, h(\hat{g})\, \xi(\hat{g}) = \tilde{x}(h)\, h(\hat{g})\, \hat{V}(x)\, \xi(\hat{g}) = h(\hat{x}\hat{g})\, \hat{V}(x)\, \xi(\hat{g}),$$

therefore, for any $f \in \mathfrak{F}$,

$$\hat{V}(x) f(\hat{g})\, \xi(\hat{g}) = f(\hat{x}\hat{g})\, \hat{V}(x)\, \xi(\hat{g}). \tag{3.4.8}$$

Write $\xi'(\hat{g})$ for $\hat{V}(x)\, \xi(\hat{g})$; since $\hat{V}(x)$ is unitary, it follows from (3.4.8) that

$$\int |f(\hat{g})|^2\, \|\xi(\hat{g})\|^2\, d\hat{\mu}(\hat{g}) = \int |f(\hat{x}\hat{g})|^2\, \|\xi'(\hat{g})\|^2\, d\hat{\mu}(\hat{g})$$

$$= \int |f(\hat{g})|^2\, \|\xi'(\hat{x}^{-1}\hat{g})\|^2\, d\hat{\mu}_{\hat{x}}(\hat{g}). \tag{3.4.9}$$

Now, the two set functions

$$\mu_1(E) = \int_E \|\xi(\hat{g})\|^2\, d\hat{\mu}(\hat{g}), \qquad \mu_2(E) = \int_E \|\xi'(\hat{x}^{-1}\hat{g})\|^2\, d\hat{\mu}_{\hat{x}}(\hat{g})$$

are finite measures on $(\hat{G}, \hat{\mathfrak{B}})$. Since $\hat{\mathfrak{E}}$ is a determining set for $\hat{\Omega}$, it is clear by Lemma 1.1.6[59] that $\mathfrak{F}$ is dense in $L^2(\hat{G}, \hat{\mathfrak{B}}, \mu_1 + \mu_2)$. Hence, any bounded measurable function on $\hat{G}$ may be simultaneously approximated in $L^2(\hat{G}, \hat{\mathfrak{B}}, \mu_1)$ and $L^2(\hat{G}, \hat{\mathfrak{B}}, \mu_2)$ by a sequence of functions $\{f_n\} \subset \mathfrak{F}$. Therefore, it follows from (3.4.9) that

$$\int_{\hat{G}} |f(\hat{g})|^2\, d\mu_1(\hat{g}) = \int_{\hat{G}} |f(\hat{g})|^2\, d\mu_2(\hat{g})$$

for any bounded measurable function f. In particular, choosing $f = C_E$, the characteristic function of an arbitrary set $E \in \hat{\mathfrak{B}}$, we get $\mu_2(E) = \mu_1(E)$, that is,

$$\int_E \|\xi(\hat{g})\|^2\, d\hat{\mu}(\hat{g}) = \int_E \|\xi'(\hat{x}^{-1}\hat{g})\|^2\, d\hat{\mu}_{\hat{x}}(\hat{g}) \tag{3.4.10}$$

[59] *Translator's note:* The author's proof of Lemma 1.1.6 is valid only for the real case. However, it is a simple exercise to show that the result is also valid for the complex case, provided that the algebra $\mathfrak{D}$ is invariant under complex conjugation, which condition is obviously satisfied by $\mathfrak{F}$ in the present situation.

for any $E \in \hat{\mathfrak{B}}$. If $\hat{\mu}(\hat{x}^{-1}E) = 0$, then the right-hand side of (3.4.10) is zero, hence the left-hand side is zero for all $\xi \in \mathfrak{L}_k^2(\hat{\Omega})$, and it follows easily that $\hat{\mu}(E) = 0$. Since $\hat{x}$ is an arbitrary element of $\hat{\mathfrak{G}}$, this shows that $\hat{\Omega}$ is quasi-invariant under $\tilde{\mathfrak{G}}$.][60]

Using the proof of Theorem 3.4.3, together with Corollary 3.3.2, one can also obtain the following result.

Corollary 3.4.4. Let $\Omega = (G, \mathfrak{B}, \mu)$ be a localizable measure space which is quasi-invariant under and cyclic of order k relative to a transformation group $\mathfrak{G}$. Let $\mathfrak{T}$ denote the μ-topology on $\mathfrak{G}$, and let $\hat{G}$ be a spectral group[61] for $(\mathfrak{G}, \mathfrak{T})$ such that $\tilde{\mathfrak{G}}^\mu \subset \hat{G}$. Let $\hat{\mathfrak{B}}$ be the totality of weak Borel sets in $\hat{G}$, and take $\hat{\mathfrak{G}} = \tilde{\mathfrak{G}}^\mu$. Then, there exists a $\hat{\mathfrak{G}}$-quasi-invariant measure $\hat{\mu}$ on $(\hat{G}, \hat{\mathfrak{B}})$, such that $((\hat{G}, \hat{\mathfrak{B}}, \hat{\mu}), \hat{\mathfrak{G}})$ is a dual of $(\Omega, \mathfrak{G})$ with respect to an appropriate Fourier transform from $L^2(\Omega)$ to $\mathfrak{L}_k^2(\hat{G}, \hat{\mathfrak{B}}, \hat{\mu})$.

Next, we proceed to find the general form of all L^2-Fourier transforms associated with a fixed dual.

Let $\mathfrak{U}_1 = \{U(\hat{x}) \mid \hat{x} \in \hat{\mathfrak{G}}\}$ be the transformation group in $L^2(\hat{\Omega})$ corresponding to $\hat{\mathfrak{G}}$. We may also regard U_1 as a group of transformations of $\mathfrak{L}_k^2(\hat{\Omega})$, namely, if $\xi \in \mathfrak{L}_k^2(\hat{\Omega})$, $\hat{x} \in \hat{\mathfrak{G}}$, we define

$$(U(\hat{x})\xi)(\hat{g}) = \xi(\hat{x}^{-1}\hat{g}) \left(\frac{d\hat{\mu}_{\hat{x}}(\hat{g})}{d\hat{\mu}(\hat{g})}\right)^{1/2}. \tag{3.4.11}$$

Clearly, $U(\hat{x})$ is a unitary operator in $\mathfrak{L}_k^2(\hat{\Omega})$. We note that the group of multiplication operators on $\mathfrak{L}_k^2(\hat{\Omega})$, corresponding to $\hat{\mathfrak{C}}$, is just $\{\hat{U}(h) \mid h \in \mathfrak{G}\}$ [see (3.4.6)]. A trivial calculation yields the commutation relation

$$\tilde{\alpha}(h)\, U(\hat{\alpha})\, \hat{U}(h) = \hat{U}(h)\, U(\hat{\alpha}), \qquad h \in \mathfrak{G}, \quad \hat{\alpha} \in \hat{\mathfrak{G}}, \tag{3.4.12}$$

analogous to (3.4.1). For any $\alpha \in G^\mu$, let $W(\alpha) = \hat{V}(\alpha)\, U(\hat{\alpha})$, where $\hat{\alpha}$ denotes the image under the mapping (3.4.3) of the character $\tilde{\alpha}$ induced by α; thus,

$$\hat{V}(\alpha) = W(\alpha)\, U(\hat{\alpha}^{-1}), \qquad \alpha \in G^\mu. \tag{3.4.13}$$

[60] *Translator's note:* Since the mapping $h \to h(\hat{g})$ defined in this proof is the composition of the mappings $h \to U(h)$ and $U(h) \to U(h)(\hat{g})$, and since $h \to U(h)$ may not be injective, it is obviously impossible to prove that $h \to h(\cdot)$ is injective, unless further conditions are imposed. This suggests that the requirement of Definition 3.4.3 [that (3.4.4) be injective] is too strong, and that one should, instead, require the existence of an appropriate isomorphism $U(h) \to h(\cdot)$, from $\mathfrak{U}$ into $\hat{G}^{\hat{\mu}}$.

[61] *Translator's note:* The term "spectral group" was defined with reference to a topological group. In the absence of further assumptions, $(\mathfrak{G}, \mathfrak{T})$ may not be a topological group in the usual sense, since there is nothing to guarantee that the μ-topology is Hausdorff.

We see from (3.4.7) and (3.4.12) that $W(\alpha)$ and $\hat{U}(h)$ commute, that is,

$$W(\alpha)\,\hat{U}(h) = \hat{U}(h)\,W(\alpha), \qquad \alpha \in G^\mu. \tag{3.4.14}$$

Since $\{\hat{U}(h), h \in \mathfrak{G}\}$ generates the multiplication algebra $\mathfrak{M}_k$ over $\mathfrak{L}_k^2(\hat{\Omega})$, it follows from (3.4.14) that $W(\alpha) \in \{\hat{U}(h), h \in \mathfrak{G}\}' = \mathfrak{M}_k'$. If we assume that $k \leqslant \aleph_0$ and that $\hat{\mu}$ is localizable, then, by Lemma 2.4.20, we know that there exists a measurable k-dimensional unitary operator-valued function $z(\hat{g}; \alpha)$, $\hat{g} \in \hat{G}$, such that

$$(W(\alpha)\xi)(\hat{g}) = z(\hat{g}; \alpha)\,\xi(\hat{g}) \tag{3.4.15}$$

for every $\xi \in \mathfrak{L}_k^2(\hat{\Omega})$. Moreover, since

$$\begin{aligned} \hat{V}(\alpha_1\alpha_2) &= \hat{V}(\alpha_1)\,\hat{V}(\alpha_2), \qquad \alpha_1, \alpha_2 \in G^\mu, \\ \hat{V}(1) &= I, \end{aligned} \tag{3.4.16}$$

it follows by (3.4.15) that[62]

$$\begin{aligned} z(\hat{g}; \alpha_1\alpha_2) &= z(\hat{g}; \alpha_1)\,z(\hat{\alpha}_1\hat{g}; \alpha_2), \qquad \alpha_1, \alpha_2 \in G^\mu, \\ z(\hat{g}, 1) &= I. \end{aligned} \tag{3.4.17}$$

Thus, we obtain the following result.

Theorem 3.4.5. Let $\Omega = (G, \mathfrak{B}, \mu)$ be a localizable[63] measure space which is quasi-invariant under and cyclic of order k ($\leqslant \aleph_0$) relative to a transformation group $\mathfrak{G}$. Let $(\hat{\Omega}, \hat{\mathfrak{G}})$ be a dual of $(\Omega, \mathfrak{G})$ [with $\hat{\Omega} = (\hat{G}, \hat{\mathfrak{B}}, \hat{\mu})$ localizable[63]], let F be an associated Fourier transform, and, for each $\alpha \in G^\mu$, let $\hat{V}(\alpha) = FV(\alpha)F^{-1}$. Then, there exists a system $\{z(\cdot; \alpha), \alpha \in G^\mu\}$ of k-dimensional unitary operator-valued functions on $\hat{\Omega}$, satisfying relations (3.4.17), and such that

$$(\hat{V}(\alpha)\xi)(\hat{g}) = z(\hat{g}; \alpha)\,\xi(\hat{\alpha}\hat{g}) \left(\frac{d\hat{\mu}_{\hat{\alpha}^{-1}}(\hat{g})}{d\hat{\mu}(\hat{g})}\right)^{1/2}, \qquad \xi \in \mathfrak{L}_k^2(\hat{\Omega}) \tag{3.4.18}$$

holds[64] for each $\alpha \in G^\mu$.

Notice that when $k = 1$, that is, when Ω is cyclic relative to $\mathfrak{G}$, the conclusion of Theorem 3.4.5 may be further simplified, for in that case, $\mathfrak{L}_k^2(\hat{\Omega})$ is essentially just $L^2(\hat{\Omega})$, and each $z(\cdot; \alpha)$ is just a measurable

[62] Here, equality holds for almost all $\hat{g} \in \hat{G}$.

[63] As in the case of Lemma 2.4.20, the original statement of this theorem hypothesized σ-finiteness (in this case, for both Ω and $\hat{\Omega}$).

[64] The equality here holds for almost all $\hat{g} \in \hat{G}$.

complex-valued function whose modulus is almost everywhere equal to 1.

If the measure space Ω is ergodic with respect to $\mathfrak{G}$, then the totality $\mathfrak{N}$ of quasi-characters of Ω corresponding to the character 1 of $\mathfrak{G}$ is just the set of all complex constant functions c, $|c| = 1$. Hence, if $\tilde{\alpha}_1 = \tilde{\alpha}_2$, then $\alpha_1/\alpha_2 = c$, and so we have

$$z(\hat{g}; \alpha_1) = cz(\hat{g}; \alpha_2)$$

for almost all $\hat{g} \in \hat{G}$.

Using similar arguments, we may derive the general relationship between any two Fourier transforms associated with the same dual, namely,

Theorem 3.4.6. Let $\Omega = (G, \mathfrak{B}, \mu)$ be a localizable[65] measure space which is quasi-invariant under and cyclic of order k ($k \leqslant \aleph_0$) relative to a transformation group $\mathfrak{G}$. Let $(\hat{\Omega}, \hat{\mathfrak{G}})$ be a dual of $(\Omega, \mathfrak{G})$ [with $\hat{\Omega} = (\hat{G}, \hat{\mathfrak{B}}, \hat{\mu})$ localizable[65]], and let F, F' be any two Fourier transforms[66] associated with $(\hat{\Omega}, \hat{\mathfrak{G}})$. Then, there exists a measurable k-dimensional unitary operator-valued function $u(\hat{g})$, $\hat{g} \in \hat{G}$, such that

$$F' = u(\cdot)F. \tag{3.4.19}$$

Conversely, given any Fourier transform F from $L^2(\Omega)$ onto $\mathfrak{L}_k^2(\hat{\Omega})$, and any measurable k-dimensional unitary operator-valued function $u(\hat{g})$, $\hat{g} \in \hat{G}$, the operator F' defined by (3.4.19) is also a Fourier transform associated with the same dual [and the same mapping (3.4.4)].

PROOF. Given the two Fourier transforms F and F', form the unitary operator

$$u = F'F^{-1} \tag{3.4.20}$$

in $\mathfrak{L}_k^2(\hat{\Omega})$. Since

$$FU(h)F^{-1} = F'U(h)F'^{-1} = \hat{U}(h), \qquad h \in \mathfrak{G}, \tag{3.4.21}$$

we have

$$u\hat{U}(h) = \hat{U}(h)u, \qquad h \in \mathfrak{G}. \tag{3.4.22}$$

Hence, as in the derivation of (3.4.15), it follows by Lemma 2.4.20 that

[65] *Translator's note:* Again, in the original version, it was supposed that Ω and $\hat{\Omega}$ were σ-finite. Notice that the proof of the second (converse) part of the theorem does not make use of the localizability of $\hat{\Omega}$.

[66] It is assumed here that the mapping (3.4.4) is the same for both F and F'.

there is a measurable k-dimensional unitary operator-valued function $u(\hat{g})$, $\hat{g} \in \hat{G}$, such that, for all $\xi \in \mathfrak{L}_k^2(\hat{\Omega})$,

$$u\xi(\hat{g}) = u(\hat{g})\, \xi(\hat{g}), \qquad \hat{g} \in \hat{G}.$$

Combining this relation with (3.4.20), we get (3.4.19).

Conversely, given any measurable k-dimensional unitary operator-valued function $u(\cdot)$ on $\hat{\Omega}$, we form a unitary operator F', from $L^2(\Omega)$ to $\mathfrak{L}_k^2(\hat{\Omega})$, in accordance with (3.4.19). Since the operator

$$u: \quad \xi(\cdot) \to u(\cdot)\, \xi(\cdot), \qquad \xi \in \mathfrak{L}_k^2(\hat{\Omega})$$

satisfies (3.4.22), it follows that (3.4.21) also holds, which shows that F' is also a Fourier transform.]

Using Theorems 3.4.5 and 3.4.6, we see that the Fourier transforms associated with a given dual $(\hat{\Omega}, \hat{\mathfrak{G}})$ and a given mapping (3.4.4) have the general form[67]

$$(F'V(\alpha)F'^{-1}\xi)(\hat{g}) = u(\hat{g})\, z(\hat{g};\alpha)\, u(\hat{\alpha}\hat{g})^{-1}\, \xi(\hat{\alpha}\hat{g}) \left(\frac{d\hat{\mu}_{\hat{\alpha}^{-1}}(\hat{g})}{d\hat{\mu}(\hat{g})}\right)^{1/2},$$

$$\xi \in \mathfrak{L}_k^2(\hat{\Omega}), \tag{3.4.23}$$

where $z(\cdot;\alpha)$ and $u(\cdot)$ denote the operator-valued functions appearing in (3.4.18) and (3.4.19), respectively.

The following theorem shows that the dual measure space is, in a certain sense, unique.

Theorem 3.4.7. Let $\Omega = (G, \mathfrak{B}, \mu)$ be a localizable measure space which is quasi-invariant under and cyclic of order k relative to a transformation group $\mathfrak{G}$. Let $\hat{\Omega} = (\hat{G}, \hat{\mathfrak{B}}, \hat{\mu})$ and $\hat{\Omega}' = (\hat{G}', \hat{\mathfrak{B}}', \hat{\mu}')$ be localizable measure spaces which are quasi-invariant under the transformation groups $\hat{\mathfrak{G}}$ and $\hat{\mathfrak{G}}'$, respectively, moreover, suppose that $(\hat{\Omega}, \hat{\mathfrak{G}})$ and $(\hat{\Omega}', \hat{\mathfrak{G}}')$ are both duals of $(\Omega, \mathfrak{G})$. Then, the two measure spaces $\hat{\Omega}$ and $\hat{\Omega}'$ are equivalent.

Proof. Let $\hat{\alpha}$, $h(\cdot)$, F, $\hat{U}(h)$ and $\hat{\alpha}'$, $h'(\cdot)$, F', $\hat{U}'(h)$ denote the quantities in Definition 3.4.3 corresponding to the duals $(\hat{\Omega}, \hat{\mathfrak{G}})$ and $(\hat{\Omega}', \hat{\mathfrak{G}}')$, respectively. Form the unitary operator

$$Q = F'F^{-1}$$

from $\mathfrak{L}_k^2(\hat{\Omega})$ to $\mathfrak{L}_k^2(\hat{\Omega}')$. Obviously, $Q\hat{U}(h)Q^{-1} = \hat{U}'(h)$. Since the

[67] *Translator's note:* This is something of an overstatement, since (3.4.23) does not explicitly describe the operation of F' on the elements of $L^2(\Omega)$.

families of functions $\{h(\hat{g}) \mid h \in \mathfrak{G}\}$ and $\{h'(\hat{g}') \mid h \in \mathfrak{G}\}$ are determining sets on $\hat{\Omega}$ and $\hat{\Omega}'$, respectively, it follows that the families of operators $\{\hat{U}(h) \mid h \in \mathfrak{G}\}$ and $\{\hat{U}'(h) \mid h \in \mathfrak{G}\}$ generate the multiplication algebras $\hat{\mathfrak{M}}_k$ and $\hat{\mathfrak{M}}_k'$ over $\mathfrak{L}_k^2(\hat{\Omega})$ and $\mathfrak{L}_k^2(\hat{\Omega}')$, respectively. But clearly, $\{QAQ^{-1} \mid A \in \hat{\mathfrak{M}}_k\}$ is a commutative weakly closed operator algebra over $\mathfrak{L}_k^2(\hat{\Omega}')$, and contains $\{\hat{U}'(h) \mid h \in \mathfrak{G}\}$, hence $\{QAQ^{-1} \mid A \in \hat{\mathfrak{M}}_k\} \supset \hat{\mathfrak{M}}_k'$. Similarly, $\{QAQ^{-1} \mid A \in \hat{\mathfrak{M}}_k\} \subset \hat{\mathfrak{M}}_k'$. Thus, Q effects a unitary equivalence between $\hat{\mathfrak{M}}_k$ and $\hat{\mathfrak{M}}_k'$. Therefore, by Theorem 2.4.21, $\hat{\Omega}$ and $\hat{\Omega}'$ are equivalent.]

Similarly, one may prove the following result.

Theorem 3.4.8. Let $\mathfrak{G}$ be a group, and suppose that $\mathfrak{G}$ acts as a group of measurable transformations on each of the two localizable measure spaces $\Omega = (G, \mathfrak{B}, \mu)$ and $\Omega = (G', \mathfrak{B}', \mu')$, moreover, suppose that both Ω and Ω' are quasi-invariant under and cyclic of order k relative to $\mathfrak{G}$. If $(\Omega, \mathfrak{G})$ and $(\Omega', \mathfrak{G})$ are equivalent,[68] then they have a common dual.

Remark 1. Let us consider the "uniqueness" of the mapping (3.4.4) in Definition 3.4.3. Assume that $\hat{\Omega}$ is ergodic with respect to $\hat{\mathfrak{G}}$, and suppose that, besides (3.4.4), there is another homomorphism

$$h \to h'(\hat{g}), \qquad \hat{g} \in \hat{G}, \tag{3.4.24}$$

from $\mathfrak{G}$ into $\hat{G}^\mu$, satisfying condition (3.4.5), in other words, the quasi-characters $h(\cdot)$ and $h'(\cdot)$ induce the same character on $\hat{\mathfrak{G}}$, that is,

$$\hat{\alpha} \to \tilde{\alpha}(h), \qquad \hat{\alpha} \in \hat{\mathfrak{G}}.$$

Since $\hat{\Omega}$ is ergodic, it follows by property III of §3.2, 2°, that there exists a constant $c(h)$ (independent of $\hat{g}$), with $|c(h)| = 1$, such that

$$h'(\hat{g}) = c(h)\, h(\hat{g}), \qquad \hat{g} \in \hat{G}. \tag{3.4.25}$$

Moreover, since (3.4.4) and (3.4.24) are both homomorphisms, the function $c(h)$, $h \in \mathfrak{G}$, is a character of $\mathfrak{G}$.

Now, let F' be a Fourier transform associated with the homomorphism $h \to h'(\cdot)$, so that every operator $\hat{U}'(h) = F'U(h)F'^{-1}$, $h \in \mathfrak{G}$, takes the form

$$(\hat{U}'(h)\xi)(\hat{g}) = h'(\hat{g})\, \xi(\hat{g}), \qquad \hat{g} \in \hat{G}, \quad \xi \in \mathfrak{L}_k^2(\hat{\Omega}). \tag{3.4.26}$$

[68] That is, there exists an equivalence between Ω and Ω' which is consistent with the action of $\mathfrak{G}$.

Then, using (3.4.25), (3.4.26), and (3.4.6), one easily calculates that, if $\xi, \eta \in \mathfrak{L}_k^2(\hat{\Omega})$, then

$$c(h) = \frac{(U(h)F'^{-1}\xi, F'^{-1}\eta)}{(U(h)F^{-1}\xi, F^{-1}\eta)}.$$

From this formula,[69] we see that the character $c(h)$, $h \in \mathfrak{G}$, is continuous relative to the μ-topology on $\mathfrak{G}$. If we assume that every character of $\mathfrak{G}$ which is continuous relative to the μ-topology is induced by some quasi-character of Ω (relative to $\mathfrak{G}$), then there is an $\alpha_0 \in G^\mu$ such that

$$c(h) = \tilde{\alpha}_0(h), \qquad h \in \mathfrak{G}.$$

Hence, by (3.4.5), we get

$$h'(\hat{g}) = \tilde{\alpha}_0(h)\, h(\hat{g}) = h(\hat{\alpha}_0 \hat{g}). \tag{3.4.27}$$

Conversely, take any $\alpha_0 \in G^\mu$, and define a function $h'(\cdot)$ in accordance with formula (3.4.27). Then the correspondence (3.4.24) satisfies the conditions of Definition 3.4.3, and the operator $F' = QF$, where

$$(Q\xi)(\hat{g}) = \xi(\hat{\alpha}_0 \hat{g}) \left(\frac{d\hat{\mu}_{\hat{\alpha}_0^{-1}}(\hat{g})}{d\hat{\mu}(\hat{g})} \right)^{1/2}, \qquad \xi \in \mathfrak{L}_k^2(\Omega), \tag{3.4.28}$$

serves as an associated Fourier transform.

Also, consider the case where G is a commutative group, $\mathfrak{G}$ is the group of all translations of G, $\hat{G}$ is the group of all characters of G, $\hat{\mathfrak{B}}$ is the σ-algebra of all weak Borel sets in $\hat{G}$, and $\hat{\mathfrak{G}}$ is the group of all translations of $\hat{G}$. In this case, we need not assume that $\hat{\Omega}$ is ergodic with respect to $\hat{\mathfrak{G}}$, and still (3.4.25) holds, with $c \in \hat{G}$, moreover, if there exists a Fourier transform F' corresponding to (3.4.24), then, arguing as in the proof of Theorem 3.4.7, one can show that c is a quasi-invariant point[70] of $\hat{\Omega}$.

Remark 2. Dual measure spaces are not necessarily unique. In fact, let $(\hat{\Omega}, \hat{\mathfrak{G}})$ be any dual of $(\Omega, \mathfrak{G})$, and let F be an associated Fourier transform. Let $\hat{\mu}'$ be another measure on $(\hat{G}, \hat{\mathfrak{B}})$ such that $\hat{\Omega}$ and

[69] *Translator's note:* The author has overlooked the possibility that the denominator may vanish. However, in view of the *ad hoc* hypothesis in the next sentence, the question of whether or not $c(h)$ is continuous relative to the μ-topology hardly merits any special attention, since a still stronger *ad hoc* hypothesis could be made, if necessary.

[70] *Translator's note:* The translator is frankly puzzled by this assertion. Presumably, it has already been assumed that $\hat{\Omega}$ is quasi-invariant under the translation group $\hat{G}$, in accordance with the requirements of Definition 3.4.3.

$\hat{\Omega}' = (\hat{G}, \hat{\mathfrak{B}}, \hat{\mu}')$ are equivalent and have the same σ-finite sets. Then, leaving the mappings (3.4.4) (3.4.5), and (3.4.6) unchanged, we see that $(\hat{\Omega}', \mathfrak{G})$ is also a dual of $(\Omega, \mathfrak{G})$, with an associated Fourier transform

$$F' = \left(\frac{d\hat{\mu}'(\cdot)}{d\hat{\mu}(\cdot)}\right)^{1/2} F. \tag{3.4.29}$$

Remark 3. If we take the μ-topology on $\mathfrak{G}$ and the $\hat{\mu}$-topology on $\hat{\mathfrak{E}} = \{h(\hat{g}) \mid h \in \mathfrak{G}\}$, then the mapping (3.4.4) is a homeomorphism[71] from $\mathfrak{G}$ onto $\hat{\mathfrak{E}}$. In fact, for any $\xi \in L^2(\Omega)$,

$$\|(U(h) - I)\xi\| = \|(h(\cdot) - 1) F\xi\|,$$

but the left- and right-hand sides of this equality represent arbitrary members of the families of convex functions which define the μ-topology on $\mathfrak{G}$ and the $\hat{\mu}$-topology on $\hat{\mathfrak{E}}$, respectively.[72] Thus, (3.4.4) is a homeomorphism.

Example 3.4.4. Let G be a commutative locally compact topological group, and let $\Omega = (G, \mathfrak{B}, \mu)$ be a Haar measure space. Also, let G^* be the dual of G (given the strong topology, G^* becomes a locally compact group), and let $\Omega^* = (G^*, \mathfrak{B}^*, \mu^*)$ be a Haar measure space. If G and G^* are regarded as operating on themselves by translation, then (Ω^*, G^*) is a dual of (Ω, G), moreover, if the measure μ^* is suitably normalized, then there is a Fourier transform F_0 from $L^2(\Omega)$ to $L^2(\Omega^*)$ such that, when $f \in L^1(\Omega) \cap L^2(\Omega)$,

$$F_0(f) = \hat{f}, \tag{3.4.30}$$

where $\hat{f}$ is defined as in (3.3.44).

Proof. As was shown in Example 3.3.2, G^* is a spectral group for G. According to Example 3.4.3, Ω is cyclic relative to the translation group G, and by the remark following Corollary 3.2.4, $\tilde{G}^\mu = G^*$. Let $\hat{\mathfrak{B}}$ denote the totality of weak Borel sets in G^*. Using Corollary 3.4.4, we deduce the existence of a localizable measure $\hat{\mu}$ on $(G^*, \hat{\mathfrak{B}})$ such that $\hat{\mu}$ is quasi-invariant under the translation group G^*, and such

[71] *Translator's note:* Again, see footnote 60.

[72] For the left-hand side, this statement is obvious. As for the right-hand side, one can certainly choose ξ so that $\|F\xi(\hat{g})\|$ is the characteristic function of an arbitrary set $A \in \hat{\mathfrak{B}}$, with $\hat{\mu}(A) < \infty$, hence one obtains all the convex functions defining the $\hat{\mu}$-topology on $\hat{\mathfrak{E}}$; moreover, using the second part of the proof of Theorem 3.2.10, it is easily seen that $\|(h(\cdot) - 1)\eta\|$ is continuous relative to the $\hat{\mu}$-topology, for any $\eta \in \mathfrak{L}_k{}^2(\hat{\Omega})$.

that $(\hat{\Omega}, G^*) = ((G^*, \hat{\mathfrak{B}}, \hat{\mu}), G^*)$ is a dual of (Ω, G). In this case, the homomorphism (3.4.4) is

$$h \to \overline{\alpha(h)}, \qquad \alpha \in G^*. \tag{3.4.31}$$

Let F be any L^2-Fourier transform, from $L^2(\Omega)$ to $L^2(\hat{\Omega})$, corresponding to the homomorphism (3.4.31), and let $\xi, \eta \in L^2(\Omega)$. Then, for any $h \in G$,

$$(U(h)\xi, \eta) = \int_{G^*} \overline{\alpha(h)}\, F\xi(\alpha)\, \overline{F\eta(\alpha)}\, d\hat{\mu}(\alpha).$$

Now, take any $a \in L^1(\Omega) \cap L^2(\Omega)$, multiply both sides of the above formula by $a(h)$, and integrate with respect to h; using the notation (3.3.44), we get

$$(a * \xi, \eta) = \int_{G^*} \hat{a}(\alpha)\, F\xi(\alpha)\, \overline{F\eta(\alpha)}\, d\hat{\mu}(\alpha). \tag{3.4.32}$$

This shows that, if $a \in L^1(\Omega) \cap L^2(\Omega)$, then

$$F(a * \xi)(\alpha) = \hat{a}(\alpha)\, F\xi(\alpha). \tag{3.4.33}$$

Next, take any two elements $a, b \in L^1(\Omega) \cap L^2(\Omega)$; by Lemma 1.1.14, there is a sequence $\{\xi_n\} \subset L^2(\Omega)$ such that

$$\lim_{n\to\infty} \| a * \xi_n - a \|_2 = \lim_{n\to\infty} \| b * \xi_n - b \|_2 = 0.$$

Applying (3.4.33), it follows that

$$\| \hat{a}(\cdot) F(\xi_n) - F(a)\|_2 \to 0, \qquad \| \hat{b}(\cdot) F(\xi_n) - F(b)\|_2 \to 0. \tag{3.4.34}$$

From (3.4.34), we see that, if $a, b \in L^1(\Omega) \cap L^2(\Omega)$, then

$$Fa(\alpha)\, \hat{b}(\alpha) \doteq Fb(\alpha)\, \hat{a}(\alpha). \tag{3.4.35}$$

Note that if $\xi \in L^1(\Omega)$, then, according to Corollary 3.2.12, $\hat{\xi}(\alpha)$ is a continuous function on G^*. Consequently, if ξ is not almost everywhere zero, then by (3.3.45), $\hat{\xi}(\alpha)$ is not identically zero, and so $| \hat{\xi}(\alpha)|$ exceeds a certain positive constant ϵ on some open subset of G^*. Furthermore, note that, if $\xi \in L^1(\Omega)$ and $\chi, \alpha \in G^*$, then

$$\int_G \chi(g)\, \xi(g)\, \overline{\alpha(g)}\, d\mu(g) = \hat{\xi}(\alpha\chi^{-1}),$$

hence, for any $\chi \in G^*$, there is an element $\xi \in L^1(\Omega) \cap L^2(\Omega)$, such that the infimum of $| \hat{\xi}(\alpha)|$ on some neighborhood of χ is positive. Let $\mathfrak{B}'$

denote the totality of sets E in $\hat{\mathfrak{B}}$ such that there exists a sequence $\{a_n\} \subset L^1(\Omega) \cap L^2(\Omega)$ satisfying

$$E \subset \bigcup_{n=1}^{\infty} \{\alpha \mid \hat{a}_n(\alpha) \neq 0\}. \tag{3.4.36}$$

By virtue of the facts which have just been stated, every compact set in $\hat{\mathfrak{B}}$ belongs to $\mathfrak{B}'$, therefore, since $\mathfrak{B}'$ is obviously a σ-ring, and since $\mathfrak{B}^*$ is the σ-ring generated by the compact subsets of G^*, it follows[73] that $\mathfrak{B}^* \subset \mathfrak{B}'$.

Now, let $E \in \mathfrak{B}'$, and choose any sequence $\{a_n\} \subset L^1(\Omega) \cap L^2(\Omega)$ such that E satisfies (3.4.36). We construct a measurable function u_E on E, as follows. If $\hat{a}_n(\alpha) \neq 0$, let

$$u_E(\alpha) = \frac{Fa_n(\alpha)}{\hat{a}_n(\alpha)}. \tag{3.4.37}$$

In view of (3.4.35), it is clear that, apart from a set of measure zero, $u_E(\alpha)$ is well-defined and independent of the choice of the sequence $\{a_n\}$. Moreover, it is also clear from (3.4.35) and (3.4.37) that, if D is any other set in $\mathfrak{B}'$, then

$$u_D(\alpha) = u_E(\alpha) \tag{3.4.38}$$

holds for almost all $\alpha \in D \cap E$. In what follows, we need only consider sets E which belong to $\mathfrak{B}^*$. Define a set function μ' on $\mathfrak{B}^*$ as follows:

$$\mu'(E) = \int_E |u_E(\alpha)|^2 \, d\hat{\mu}(\alpha). \tag{3.4.39}$$

Using (3.4.38), it is easily verified that $(G^*, \mathfrak{B}^*, \mu')$ is a measure space. Furthermore, using (3.4.35), (3.4.37), and (3.4.39), one easily verifies that, if $\xi, \eta \in L^1(\Omega) \cap L^2(\Omega)$, then

$$\int_G \xi(g) \overline{\eta(g)} \, d\mu(g) = \int_{G^*} F\xi(\alpha) \overline{F\eta(\alpha)} \, d\hat{\mu}(\alpha)$$
$$= \int_{G^*} \hat{\xi}(\alpha) \overline{\hat{\eta}(\alpha)} \, d\mu'(\alpha). \tag{3.4.40}$$

Next, we assert that $H = \{\hat{\eta} \mid \eta \in L^1(\Omega) \cap L^2(\Omega)\}$ is dense in

[73] *Translator's note:* Obviously, in order to draw this conclusion, one must establish that *all* compact sets are in $\mathfrak{B}'$, and not just those which also happen to belong to $\hat{\mathfrak{B}}$. It is not clear how one does this.

$L^2(G^*, \mathfrak{B}^*, \mu')$. In fact, suppose that $\xi \in L^2(G^*, \mathfrak{B}^*, \mu')$, $\xi \perp H$; then, for any $\eta \in L^1(\Omega) \cap L^2(\Omega)$,

$$\int_{G^*} \xi(\alpha)\, \overline{\hat{\eta}(\alpha)}\, d\mu'(\alpha) = 0.$$

But since $\eta_h \in L^1(\Omega) \cap L^2(\Omega)$, where $\eta_h(g) = \eta(hg)$, we also have

$$\int_{G^*} \xi(\alpha)\, \overline{\alpha(h)\, \hat{\eta}(\alpha)}\, d\mu'(\alpha) = 0.$$

Since $\{\alpha(h), h \in G\}$ is a determining set of functions on $(G^*, \mathfrak{B}^*)$, it follows from Lemma 1.1.6 that $\xi(\alpha)\, \hat{\eta}(\alpha)$ is zero almost everywhere,[74] hence $\xi(\alpha)$ is zero almost everywhere.[74] Thus, H is indeed dense in $L^2(G^*, \mathfrak{B}', \mu')$. Therefore, in view of (3.4.40), the correspondence $\xi \to \hat{\xi}$ can be extended to a unitary mapping F_0 from $L^2(\Omega)$ onto $L^2(G^*, \mathfrak{B}^*, \mu')$. Moreover, when $\xi \in L^1(\Omega) \cap L^2(\Omega)$, we have

$$\begin{aligned} F_0 V(\chi)\, \xi(\alpha) &= \int \chi(g)\, \xi(g)\, \overline{\alpha(g)}\, d\mu(g) \\ &= \hat{\xi}(\alpha\chi^{-1}) = F_0 \xi(\alpha\chi^{-1}), \end{aligned} \tag{3.4.41}$$

hence, by (3.4.40),

$$\begin{aligned} \int \hat{\xi}(\alpha)\, \overline{\hat{\eta}(\alpha)}\, d\mu'(\alpha) &= \int_G V(\chi)\, \xi(g)\, \overline{V(\chi)\, \eta(g)}\, d\mu(g) \\ &= \int_{G^*} \hat{\xi}(\alpha\chi^{-1})\, \overline{\hat{\eta}(\alpha\chi^{-1})}\, d\mu'(\alpha). \end{aligned}$$

From this, one easily deduces that the measure μ' is invariant under translations. Furthermore, it is obvious from our earlier remarks that there exists an element $\xi \in L^1(\Omega) \cap L^2(\Omega)$ such that $|\hat{\xi}(\alpha)| > 1$ on some compact neighborhood N in G^*. Hence, by (3.4.40), we get

$$\mu'(N) \leqslant \int_{G^*} |\hat{\xi}(\alpha)|^2\, d\mu'(\alpha) = \int_G |\xi(g)|^2\, d\mu(g) < \infty.$$

Thus, μ' is locally finite, and we conclude that μ' is a Haar measure. Finally, using (3.4.40), it is easily shown that F_0 is the required L^2-Fourier transform.]

Example 3.4.5. Let $l = 1, 2, \ldots,$ be a finite or countable set of

[74] *Translator's note:* Although these assertions seem highly plausible, more detailed arguments would be welcome.

indices, and for each l, let $\Omega_l = (G_l, \mathfrak{B}_l, \mu_l)$ be a localizable measure space which is quasi-invariant under and cyclic of order k relative to a transformation group $\mathfrak{G}_l$. For any finite sequence $h = (h_1, ..., h_n)$, where $h_l \in \mathfrak{G}_l$, $l = 1, ..., n$, we define a transformation by

$$hg = \{h_1 g_1, ..., h_n g_n, g_{n+1}, ...\},$$

where $g = \{g_1, ..., g_n, g_{n+1}, ...\}$ is any element of $\times_l G_l$ such that $g_l \in \mathfrak{D}(h_l)$, $l = 1, 2, ..., n$. Let $\mathfrak{G}$ denote the totality of such transformations h, and suppose that Ω_l is a probability measure space for all but a finite number of indices l. Then, the product measure space $\Omega = \times_l \Omega_l$ is quasi-invariant under and cyclic of order $\prod_l k_l$ relative to $\mathfrak{G}$. Moreover, if, for each l, $(\hat{\Omega}_l, \hat{\mathfrak{G}}_l)$ is a dual of $(\Omega_l, \mathfrak{G}_l)$, if $\hat{\Omega}_l$ is a probability measure space for almost all l, and if we define a corresponding transformation group $\hat{\mathfrak{G}}$ in the same way that $\mathfrak{G}$ was defined above, then $(\times_l \hat{\Omega}_l, \hat{\mathfrak{G}})$ is a dual of $(\Omega, \mathfrak{G})$.

The proof is left to the reader.

3° Duals of Ergodic Measure Spaces

Lemma 3.4.9. Let $\Omega = (G, \mathfrak{B}, \mu)$ be a localizable measure space which is quasi-invariant under a transformation group $\mathfrak{G}$. Let $B \in \mathfrak{B}$, and let H_B be the closed linear subspace of $\mathfrak{L}_k^2(\Omega)$ formed by the totality of functions in $\mathfrak{L}_k^2(\Omega)$ which vanish almost everywhere outside of B. Then, B is quasi-invariant under $\mathfrak{G}$ if and only if H_B is invariant under the group of unitary operators $\mathfrak{U} = \{U(h) \mid h \in \mathfrak{G}\}$.

PROOF. Assume that B is quasi-invariant under $\mathfrak{G}$, and let $\varphi \in H_B$. Then, for any $h \in \mathfrak{G}$, it is clear that $\varphi(hg)$ is zero for almost all $g \in G - B$, hence

$$(U(h)\varphi)(g) = \varphi(hg) \left(\frac{d\mu_h(g)}{d\mu(g)}\right)^{1/2} = 0$$

for almost all $g \in G - B$, that is $U(h)\varphi \in H_B$. Thus, H_B is invariant under $\mathfrak{U}$.

Conversely, let $B \in \mathfrak{B}$, and assume that H_B is invariant under $\mathfrak{U}$. If B were not quasi-invariant, then there would exist an element $h \in \mathfrak{G}$ such that $hB - B$ is not a μ-null set, therefore, since Ω is localizable, one could find a set $E \in \mathfrak{B}$ such that $0 < \mu(E) < \infty$, $0 < \mu(hE) < \infty$, $E \subset B$, $hE \subset hB - B$. Then, if C_E denotes the characteristic function of E, and φ_0 is any fixed nonzero vector in k-dimensional Hilbert space, the function $\varphi_E(\cdot) = C_E(\cdot)\varphi_0$ would belong to H_B, and

$$(U(h)\varphi_E)(g) = \varphi_E(h^{-1}g) \left(\frac{d\mu_h(g)}{d\mu(g)}\right)^{1/2} = \varphi_{hE}(g) \left(\frac{d\mu_h(g)}{d\mu(g)}\right)^{1/2}.$$

Thus, $(U(h)\,\varphi_E)(g) \neq 0$ for all $g \in hE \subset G - B$, and hence $U(h)\,\varphi_E \,\overline{\in}\, H_B$. But this contradicts the invariance of H_B under $\mathfrak{U}$. We conclude that B is quasi-invariant under $\mathfrak{G}$.]

Corollary 3.4.10. Under the hypotheses of Lemma 3.4.9, let M be the smallest weakly closed operator ring containing both the multiplication algebra $\mathfrak{M}(\Omega)$ over $L^2(\Omega)$ and the group $\mathfrak{U} = \{U(h) \mid h \in \mathfrak{G}\}$. Then, Ω is ergodic with respect to $\mathfrak{G}$ if and only if M is a factor; Ω is weakly ergodic with respect to $\mathfrak{G}$ if and only if the center of M contains no countably decomposable projection operators.[75]

Theorem 3.4.11. Let Ω be a localizable measure space which is quasi-invariant under and cyclic of order k relative to a transformation group $\mathfrak{G}$. Suppose that Ω is normal with respect to $\mathfrak{G}$, and let $(\hat{\Omega}, \hat{\mathfrak{G}})$ be a dual of $(\Omega, \mathfrak{G})$. If Ω is ergodic with respect to $\mathfrak{G}$, then $\hat{\Omega}$ is ergodic with respect to $\hat{\mathfrak{G}}$.

PROOF. Assume that $\hat{\Omega}$ is not ergodic with respect to $\hat{\mathfrak{G}}$, and let $\hat{E}$ be a set in $\hat{\mathfrak{B}}$ which is quasi-invariant under $\hat{\mathfrak{G}}$, with $\hat{\mu}(\hat{E}) \neq 0$, $\hat{\mu}(\hat{G} - \hat{E}) \neq 0$. Let $H_{\hat{E}}$ denote the closed linear subspace of $\mathfrak{L}_k{}^2(\hat{\Omega})$ generated by the totality of vector-valued functions which vanish on $\hat{G} - \hat{E}$. Then,

$$(0) \neq H_{\hat{E}} \neq \mathfrak{L}_k{}^2(\hat{\Omega}). \tag{3.4.42}$$

Moreover, by Lemma 3.4.9, $H_{\hat{E}}$ is invariant under all the operators $U(\chi)$, $\chi \in \hat{\mathfrak{G}}$, and obviously $H_{\hat{E}}$ is also invariant under every operator of the form $\xi \to z(\cdot\,;\alpha)\,\xi(\cdot)$. Since, by (3.4.18), $\hat{V}(\alpha) = z(\cdot\,;\alpha)\,U(\hat{\alpha}^{-1})$, $\alpha \in G^\mu$, it follows that $H_{\hat{E}}$ is invariant under all the operators $\hat{V}(\alpha)$, $\alpha \in G^\mu$. It is also obvious that $H_{\hat{E}}$ is invariant under the operators $\hat{U}(h)$, $h \in \mathfrak{G}$.

Write $H = F^{-1}H_{\hat{E}}$. Then, H is a closed linear subspace of $L^2(\Omega)$ which is invariant under the groups $\mathfrak{U}$ and $\mathfrak{V}$. Let P denote the projection operator from $L^2(\Omega)$ onto H; we then have $P \in \mathfrak{V}'$. Since G^μ is a determining set of functions on Ω, the group $\mathfrak{V}$ generates the multiplication algebra $\mathfrak{M}$ over $L^2(\Omega)$, and since Ω is localizable, $\mathfrak{M}$ is maximal commutative weakly closed. Hence, $\mathfrak{V}' = \mathfrak{M}' = \mathfrak{M}$, so that $P \in \mathfrak{M}$. Let M be the weakly closed operator algebra generated by $\mathfrak{U} \cup \mathfrak{M}$; since $P \in \mathfrak{U}'$, it follows that $P \in M \cap M'$. But by Corollary 3.4.10, M is a factor, hence $P = 0$ or $P = I$, which contradicts (3.4.42). We conclude that $\hat{\Omega}$ is ergodic with respect to $\hat{\mathfrak{G}}$.]

In general, the converse of Theorem 3.4.11 is not valid, as is shown by the following example.

[75] *Translator's note:* This is clearly *not* a necessary condition for weak ergodicity, since it is obviously not satisfied when $L^2(\Omega)$ is separable, and examples of weakly ergodic σ-finite measure spaces Ω, such that $L^2(\Omega)$ is separable, are easily constructed.

Example 3.4.6. Let G be a group of just two elements $\{a, e\}$, where $a \neq e$, $a^2 = e$, let $\mathfrak{B}$ be the totality of subsets of G, and let μ be the measure on $\mathfrak{B}$ defined by $\mu(\{a, e\}) = 2$, $\mu(\{a\}) = \mu(\{e\}) = 1$, $\mu(\text{empty set}) = 0$. Let $\mathfrak{G}$ be the subgroup of G consisting of just the unit element e, and let $\mathfrak{E}$ be the totality of functions from G to the complex numbers of unit modulus. Then, $(G, \mathfrak{B}, \mu)$ is a finite measure space which is invariant under and cyclic of order 2 relative to the translation group $\mathfrak{G}$, and is obviously normal with respect to $\mathfrak{G}$. However, μ is not ergodic with respect to $\mathfrak{G}$, in fact, both of the singletons $\{e\}$ and $\{a\}$ are nontrivial quasi-invariant sets. Since the quasi-characters $\mathfrak{E}$ induce just the one character 1 on $\mathfrak{G}$, we see that $(G, \mathfrak{B}, \mu)$ has a dual measure space $(\hat{G}, \hat{\mathfrak{B}}, \hat{\mu})$ defined as follows: $\hat{G}$ is a group containing only a single element $\hat{e}$, $\hat{\mathfrak{B}}$ consists of $\{\hat{e}\}$ and the empty set, $\hat{\mu}(\{\hat{e}\}) = 1$, $\hat{\mu}(\text{empty set}) = 0$, and $\hat{\mathfrak{G}}$ consists of just the identity transformation. Obviously, $(\hat{G}, \hat{\mathfrak{B}}, \hat{\mu})$ is ergodic with respect to $\hat{\mathfrak{G}}$. This shows that a quasi-invariant measure space which is not ergodic may have an ergodic dual.

However, when $k = 1$, the converse of Theorem 3.4.11 is, in fact, valid.

Theorem 3.4.12. Let Ω be a localizable measure space which is quasi-invariant under and cyclic relative to a transformation group $\mathfrak{G}$. Suppose that Ω is normal with respect to $\mathfrak{G}$, and let $(\hat{\Omega}, \hat{\mathfrak{G}})$ be a dual of $(\Omega, \mathfrak{G})$. If $\hat{\Omega}$ is localizable and ergodic with respect to $\hat{\mathfrak{G}}$, then Ω is ergodic with respect to $\mathfrak{G}$.

Proof. We follow the argument used in proving Theorem 3.4.11. Assume that Ω is not ergodic, and let $B \in \mathfrak{B}$ be quasi-invariant under $\mathfrak{G}$, with $\mu(B) \neq 0$, $\mu(G - B) \neq 0$. Let H_B denote the totality of functions in $L^2(\Omega)$ which vanish almost everywhere in $G - B$. Then, H_B is a closed linear subspace of $L^2(\Omega)$ which is invariant under $\mathfrak{U}$ and $\mathfrak{V}$, moreover,

$$(0) \neq H_B \neq L^2(\Omega). \tag{3.4.43}$$

Let $\mathfrak{H} = FH_B$; then, $\mathfrak{H}$ is a closed linear subspace of $\mathfrak{L}_1{}^2(\hat{\Omega}) = L^2(\hat{\Omega})$ which is invariant under the groups $\hat{\mathfrak{U}} = \{\hat{U}(h) \mid h \in \mathfrak{G}\}$ and $\hat{\mathfrak{V}} = \{\hat{V}(\alpha) \mid \alpha \in G^\mu\}$. Let Q be the projection operator from $L^2(\hat{\Omega})$ onto $\mathfrak{H}$, and let $\hat{\mathfrak{M}}$ denote the multiplication algebra over $L^2(\hat{\Omega})$. Now, according to Definition 3.4.3, $\hat{\mathfrak{E}}$ is a determining set of functions on $\hat{\Omega}$, hence, it follows by the results of §2.4 that $\hat{\mathfrak{U}}$ generates $\hat{\mathfrak{M}}$. However, since $\hat{\Omega}$ is localizable, $\hat{\mathfrak{M}}$ is maximal commutative weakly closed. Thus, we have $Q \in \hat{\mathfrak{U}}' = \hat{\mathfrak{M}}' = \hat{\mathfrak{M}}$. If $\alpha \in G^\mu$, and $z(\hat{g}; \alpha)$ is the function appearing in (3.4.18) (since $k = 1$, this is presently a numerical-valued function), then the operator $\xi \to z(\cdot\,;\alpha)\,\xi(\cdot)$ belongs to $\hat{\mathfrak{M}}$. Therefore, since $U(\hat{\alpha}^{-1}) = z(\cdot\,;\alpha)^{-1}\,\hat{V}(\alpha)$, we see that $\mathfrak{H}$ is invariant under $\{U(\chi) \mid \chi \in \hat{\mathfrak{G}}\}$, and hence $Q \in \{U(\chi) \mid \chi \in \hat{\mathfrak{G}}\}'$. Let $\hat{M}$ denote the weakly closed operator algebra generated by $\hat{\mathfrak{M}}$ and $\{U(\chi) \mid \chi \in \hat{\mathfrak{G}}\}$. By virtue of Corollary 3.4.10, the ergodicity of $\hat{\Omega}$ implies that $\hat{M}$ is a factor. But $Q \in \hat{M}$ and $Q \in \hat{M}'$, hence Q is either 0 or I, which contradicts (3.4.43). We conclude that Ω is ergodic with respect to $\mathfrak{G}$.]

4° Strongly *k*th Order Cyclic Measures

We shall now consider, in particular, the case where the dual is a finite measure space.

Since every σ-finite measure is equivalent to a finite measure space, some of the following results may easily be extended to σ-finite measure spaces.

Definition 3.4.4. Let $\Omega = (G, \mathfrak{B}, \mu)$ be a localizable measure space which is quasi-invariant under a transformation group $\mathfrak{G}$. Let $\varphi \in L^2(\Omega)$, and let H_φ be the smallest closed linear subspace of $L^2(\Omega)$ which contains φ and is invariant under all the operators $U(h)$, $h \in \mathfrak{G}$. We call H_φ the subspace *generated by the cyclic element* φ (and the group $\mathfrak{U}$). Suppose that $\{\varphi_\lambda \mid \lambda \in \Lambda\}$ is a family of vectors in $L^2(\Omega)$ such that

$$L^2(\Omega) = \sum_{\lambda \in \Lambda} \oplus H_{\varphi_\lambda},$$

and that the function

$$\psi(h) = (U(h)\varphi_\lambda, \varphi_\lambda), \qquad h \in \mathfrak{G}, \tag{3.4.44}$$

is independent of λ. Then, we say that $\{\varphi_\lambda \mid \lambda \in \Lambda\}$ is a $\mathfrak{G}$-*cyclic family* of elements, that $\psi(h)$ is the corresponding *adjoint function*, and that $(G, \mathfrak{B}, \mu)$ is *strongly kth order cyclic* relative to $\mathfrak{G}$, where k is the cardinal number[76] of Λ.

Theorem 3.4.13. Let $\Omega = (G, \mathfrak{B}, \mu)$ be a localizable measure space which is quasi-invariant under a transformation group $\mathfrak{G}$. If Ω is strongly kth order cyclic relative to $\mathfrak{G}$, then Ω is cyclic of order k relative to $\mathfrak{G}$. Conversely, if $L^2(\Omega)$ is separable, and Ω is cyclic of order k relative to $\mathfrak{G}$, then $k \leqslant \aleph_0$, and Ω is strongly kth order cyclic relative to $\mathfrak{G}$.

PROOF. Suppose there exists a $\mathfrak{G}$-cyclic family of elements $\{\varphi_\lambda \mid \lambda \in \Lambda\}$, where the cardinal number of Λ is k. Let $\mathfrak{A}$ denote the operator algebra corresponding to the transformation group $\mathfrak{G}$. Then each H_{φ_λ} is an

[76] As before, when $k = 1$, we simply say *strongly cyclic*.

$\mathfrak{A}$-invariant subspace, and $\mathfrak{A}$ has the cyclic element φ_λ in H_{φ_λ}, hence, by Corollary 2.4.11, the restriction of $\mathfrak{A}$ to H_{φ_λ} is maximal commutative weakly closed. Furthermore, since the adjoint function $\psi(h)$ [see (3.4.44)] is independent of λ, one can deduce that, for any $\lambda, \lambda' \in \Lambda$, the restrictions of $\mathfrak{A}$ to H_{φ_λ} and $H_{\varphi_{\lambda'}}$ are unitarily equivalent. In fact, consider the mapping $U_{\lambda'\lambda}$, from H_{φ_λ} to $H_{\varphi_{\lambda'}}$, defined as follows: if $h_l \in \mathfrak{G}$, and z_l is a complex number, $l = 1, 2,\ldots, n$, let

$$U_{\lambda'\lambda} \sum_l z_l U(h_l)\varphi_\lambda = \sum z_l U(h_l)\varphi_{\lambda'}\,.$$

Then, using (3.4.44), we obtain

$$\Big\| \sum_l z_l U(h_l)\varphi_\lambda \Big\|^2 = \sum_{l,l'} \psi(h_l^{-1}h_{l'})\, z_{l'}\bar{z}_l = \Big\| \sum z_l U(h_l)\varphi_{\lambda'} \Big\|^2. \tag{3.4.45}$$

Hence, $U_{\lambda'\lambda}$ is an isometric linear mapping from the dense linear subspace $\{\sum z_l U(h_l)\, \varphi_\lambda \mid h_l \in \mathfrak{G}\}$ of H_{φ_λ} onto the dense linear subspace $\{\sum z_l U(h_l)\, \varphi_{\lambda'} \mid h_l \in \mathfrak{G}\}$ of $H_{\varphi_{\lambda'}}$, and can therefore be uniquely extended to a unitary operator from H_{φ_λ} onto $H_{\varphi_{\lambda'}}$. Moreover, it is easily verified that $U_{\lambda'\lambda} U(h)\, U_{\lambda'\lambda}^{-1} = U(h)$, and it follows that the restrictions of $\mathfrak{A}$ to H_{φ_λ} and $H_{\varphi_{\lambda'}}$ are unitarily equivalent. Thus, in accordance with Definition 2.4.2, $\mathfrak{A}$ has uniform multiplicity k over $L^2(\Omega)$.

On the other hand, suppose that $L^2(\Omega)$ is separable, and that Ω is cyclic of order k relative to $\mathfrak{G}$. Then there exists a family of closed linear subspaces $\{H_\lambda, \lambda \in \Lambda\}$ in $L^2(\Omega)$ (the cardinal number of Λ being k) such that each H_λ is $\mathfrak{A}$-invariant, the restrictions of $\mathfrak{A}$ to the H_λ are maximal commutative weakly closed and unitarily equivalent, and

$$L^2(\Omega) = \sum_{\lambda \in \Lambda} \oplus H_\lambda\,.$$

Since $L^2(\Omega)$ is separable, and each H_λ is nontrivial, it is obvious that $k \leqslant \aleph_0$, and that each H_λ is separable, whence it follows by Corollary 2.4.9 that $\mathfrak{A}$ has a cyclic element in each H_λ. We arbitrarily choose a fixed index $\lambda_0 \in \Lambda$, and in H_{λ_0} we choose any cyclic element φ_{λ_0} relative to $\mathfrak{A}$. For each $\lambda \in \Lambda$, $\lambda \neq \lambda_0$, let U_λ be a unitary operator from H_{λ_0} onto H_λ such that $U_\lambda^{-1} U(h)\, U_\lambda = U(h)$, and let $\varphi_\lambda = U_\lambda \varphi_{\lambda_0}$. Then φ_λ is clearly a cyclic element of H_λ relative to $\mathfrak{A}$, that is, $H_\lambda = H_{\varphi_\lambda}$, moreover,

$$(U(h)\varphi_\lambda, \varphi_\lambda) = (U(h)\, U_\lambda \varphi_0, U_\lambda \varphi_0) = (U(h)\varphi_{\lambda_0}, \varphi_{\lambda_0}),$$

that is, the function $\psi(h)$ in (3.4.44) is indeed independent of λ. Thus, $\{\varphi_\lambda \mid \lambda \in \Lambda\}$ is a $\mathfrak{G}$-cyclic family of elements.]

In particular, if there exists $\varphi \in L^2(\Omega)$ such that $L^2(\Omega) = H_\varphi$, then μ is cyclic relative to $\mathfrak{G}$.

Theorem 3.4.14. Let $\Omega = (G, \mathfrak{B}, \mu)$ be a measure space which is quasi-invariant under and strongly kth order cyclic relative to a transformation group $\mathfrak{G}$, and let ψ be the corresponding adjoint function. Then there exists a finite measure space $\hat{\Omega} = (\hat{G}, \hat{\mathfrak{B}}, \hat{\mu})$, dual to $\hat{\Omega}$, such that ψ coincides with the quasi-characteristic function corresponding to $\hat{\Omega}$, that is,

$$\psi(h) = \int_{\hat{G}} h(\hat{g})\, d\hat{\mu}(\hat{g}), \qquad h \in \mathfrak{G}. \tag{3.4.46}$$

Moreover, one may choose $\hat{G}$ to be a spectral group for $\mathfrak{G}$ (taking the μ-topology on $\mathfrak{G}$), and $\hat{\mathfrak{B}}$ to be the totality of weak Borel sets in $\hat{G}$.

Proof. Let $\{\varphi_\lambda \mid \lambda \in \Lambda\}$ be a $\mathfrak{G}$-cyclic family of elements in $L^2(\Omega)$, with corresponding adjoint function ψ. Clearly, ψ is a positive definite function[77] on $\mathfrak{G}$, moreover, since

$$\psi(0) - \Re\psi(h) = \tfrac{1}{2} \|(U(h) - I)\varphi_\lambda\|^2, \tag{3.4.47}$$

it follows from inequality (3.3.3) that ψ is continuous with respect to the μ-topology. Hence, if $\hat{G}$ is a spectral group for $\mathfrak{G}$ (the appropriate σ-algebra $\hat{\mathfrak{B}}$ being the totality of weak Borel sets in $\hat{G}$),[78] then there exists a finite measure $\hat{\mu}$ on $(\hat{G}, \hat{\mathfrak{B}})$ such that (3.4.46) holds; let $\hat{\Omega} = (\hat{G}, \hat{\mathfrak{B}}, \hat{\mu})$.

Choose a complete orthonormal system $\{e_\lambda \mid \lambda \in \Lambda\}$ in k-dimensional Hilbert space. We construct a unitary mapping F from $L^2(\Omega)$ onto $\mathfrak{L}_k{}^2(\hat{\Omega})$, as follows. For any $\lambda_1, \ldots, \lambda_n \in \Lambda$, $h_{\lambda_1}, \ldots, h_{\lambda_n} \in \mathfrak{G}$, and complex numbers $z_{\lambda_1}, \ldots, z_{\lambda_n}$, define

$$F\left(\sum_{l=1}^{n} z_{\lambda_l} U(h_{\lambda_l})\varphi_{\lambda_l}\right) = \sum_{l=1}^{n} z_{\lambda_l} h_{\lambda_l}(\hat{g}) e_{\lambda_l}. \tag{3.4.48}$$

F is an isometric mapping; in fact, using (3.4.46), we have

$$\begin{aligned}\left\| \sum_{l=1}^{n} z_{\lambda_l} U(h_{\lambda_l})\varphi_{\lambda_l} \right\|^2 &= \sum z_{\lambda_l} \bar{z}_{\lambda_{l'}} \psi(h_{\lambda_{l'}}^{-1} h_{\lambda_l})\, \delta_{\lambda_{l'}\lambda_l} \\ &= \int \left\| \sum z_{\lambda_l} h_{\lambda_l}(\hat{g}) e_{\lambda_l} \right\|^2 d\hat{\mu}(\hat{g}).\end{aligned}$$

Since $\{\varphi_\lambda \mid \lambda \in \Lambda\}$ is a $\mathfrak{G}$-cyclic family of elements, the totality of elements

[77] See (3.4.45).

[78] Theorem 3.3.4 ensures that such spectral groups do exist.

of the form $\sum_l z_{\lambda_l} U(h_{\lambda_l}) \varphi_{\lambda_l}$ is dense in $L^2(\Omega)$. Also, since $\{h(\cdot) \mid h \in \mathfrak{G}\}$ is a determining set of functions on $(\hat{G}, \hat{\mathfrak{B}})$, it follows by Lemma 1.1.6[79] that the totality of linear combinations of these functions is dense in $L^2(\hat{\Omega})$. Hence, we see from the discussion in §1.1, 2°, that the totality of elements of the form represented on the right-hand side of (3.4.48) is dense in $\mathfrak{L}_k^2(\hat{\Omega})$. Thus, F uniquely extends to a unitary mapping from $L^2(\Omega)$ onto $\mathfrak{L}_k^2(\hat{\Omega})$.

Let $\hat{U}(h) = FU(h)F^{-1}$. From (3.4.8), we see that, for any function of the form

$$\xi(\hat{g}) = \sum_{l=1}^{n} z_{\lambda_l} h_{\lambda_l}(\hat{g}) e_{\lambda_l},$$

the relation

$$(\hat{U}(h)\xi)(\hat{g}) = h(\hat{g})\, \xi(\hat{g})$$

holds, and again, since these functions are dense in $\mathfrak{L}_k^2(\hat{\Omega})$, this relation holds for all $\xi \in \mathfrak{L}_k^2(\hat{\Omega})$.

If $\tilde{\alpha} \in \tilde{\mathfrak{G}}^\mu$, then, according to the remark preceding Theorem 3.2.3, $\tilde{\alpha}$ is continuous with respect to the μ-topology, hence, by the definition of a spectral group, we have $\tilde{\alpha} \in \hat{G}$. We take $\mathfrak{G} = \tilde{\mathfrak{G}}^\mu$, operating on $\hat{G}$ by translation, and take $\hat{\alpha} = \tilde{\alpha}$. The proof may now be completed by following the reasoning used in the proof of Theorem 3.4.3.]

Next, we establish a converse to Theorem 3.4.14.

Theorem 3.4.15. Let $\Omega = (G, \mathfrak{B}, \mu)$ be a localizable measure space which is quasi-invariant under and cyclic of order k relative to a transformation group $\mathfrak{G}$. If Ω has a σ-finite dual $\hat{\Omega} = (\hat{G}, \hat{\mathfrak{B}}, \hat{\mu})$, then Ω is strongly kth order cyclic.

PROOF. Since any σ-finite measure is equivalent to a finite measure, we may assume that $\hat{\Omega}$ is a finite measure space dual to Ω. Let F be a Fourier transform from $L^2(\Omega)$ onto $\mathfrak{L}_k^2(\hat{\Omega})$, and let $\{e_\lambda \mid \lambda \in \Lambda\}$ be an orthonormal basis for the k-dimensional Hilbert space H_k. For each $\lambda \in \Lambda$, let $\varphi_\lambda = F^{-1}e_\lambda$; we assert that $\{\varphi_\lambda \mid \lambda \in \Lambda\}$ is a $\mathfrak{G}$-cyclic family of elements in $L^2(\Omega)$. In fact, if $\lambda \neq \lambda'$, h, $h' \in \mathfrak{G}$, then, since

$$U(h)\varphi_\lambda = F^{-1}(h(\cdot)e_\lambda), \qquad h \in \mathfrak{G}, \tag{3.4.49}$$

we have

$$(U(h)\varphi_\lambda, U(h')\varphi_{\lambda'}) = \int_{\hat{G}} h(\hat{g})\, h'(\hat{g})(e_\lambda, e_{\lambda'})\, d\hat{\mu}(\hat{g}) = 0.$$

[79] *Translator's note:* Again, Lemma 1.1.6 is valid here because the algebra under consideration is invariant under complex conjugation.

Hence, $H_{\varphi_\lambda} \perp H_{\varphi_{\lambda'}}$ for $\lambda \neq \lambda'$. Moreover, since the totality of elements of the form $\sum z_{\lambda_\nu} h_\nu(\cdot)\, e_{\lambda_\nu}$. is dense in $\mathfrak{L}_k{}^2(\hat{\Omega})$, it follows that $L^2(G, \mathfrak{B}, \mu) = \sum_{\lambda\in\Lambda} \oplus H_{\varphi_\lambda}$. Furthermore, by (3.4.49), we have

$$\psi(h) = (U(h)\varphi_\lambda\, , \varphi_\lambda) = (h(\cdot)e_\lambda\, , e_\lambda) = \int_{\hat{G}} h(\hat{g})\, d\hat{\mu}(\hat{g}),$$

which is independent of λ. Thus, Ω is strongly kth order cyclic relative to $\mathfrak{G}$.]

Next, we prove a lemma, concerning cyclic subspaces, which will be required in the subsequent discussion.

Lemma 3.4.16. Let $\Omega = (G, \mathfrak{B}, \mu)$ be a localizable measure space which is quasi-invariant under a transformation group $\mathfrak{G}$. Let φ_0 be any fixed vector in $L^2(\Omega)$, form the subspace H_{φ_0} (see Definition 3.4.4), and let $\{h_n\}$ be any sequence in $\mathfrak{G}$. Then $M_{\varphi_0}(h_n) \to 0$ if and only if the sequence of unitary operators $\{U(h_n)\}$ converges strongly to the identity operator on H_{φ_0}.

Proof. Sufficiency follows immediately from the relation

$$M_{\varphi_0}(h_n) = \|(U(h_n) - I)\varphi_0\|, \qquad \varphi_0 \in H_{\varphi_0}.$$

Conversely, suppose that $M_{\varphi_0}(h_n) \to 0$. Given any $\varphi \in H_{\varphi_0}$ and any positive number ϵ, there exist elements $g_k \in \mathfrak{G}$ and numbers λ_k, $k = 1, 2,..., n$, such that

$$\left\| \varphi - \sum_{\lambda=1}^{n} \lambda_k U(g_k)\varphi_0 \right\| < \frac{\epsilon}{4}. \tag{3.4.50}$$

Choose a positive integer N such that

$$M_{\varphi_0}(h_\nu) < \epsilon \Big/ 2 \sum_{k=1}^{n} |\lambda_k| \tag{3.4.51}$$

whenever $\nu \geqslant N$. Then, from (3.4.50) and (3.4.51), using the relation $U(h_\nu)\, U(g_k) = U(g_k)\, U(h_\nu)$, we get

$$\|(U(h_\nu) - I)\varphi\| \leqslant \|U(h_\nu) - I\| \left\| \varphi - \sum_{k=1}^{n} \lambda_k U(g_k)\varphi_0 \right\|$$

$$+ \sum_{k=1}^{n} |\lambda_k|\, \|(U(h_\nu) - I)\varphi_0\| < \epsilon.$$

Thus, $U(h_\nu)\,\varphi \to \varphi$.]

Corollary 3.4.17. Let $\Omega = (G, \mathfrak{B}, \mu)$ be a localizable measure space which is quasi-invariant under and strongly cyclic relative to a transformation group $\mathfrak{G}$, with φ_0 as $\mathfrak{G}$-cyclic element. Then the μ-topology on $\mathfrak{G}$ is equivalent to the topology induced by the convex function $M_{\varphi_0}(h)$, $h \in \mathfrak{G}$.

PROOF. Since $H_{\varphi_0} = L^2(\Omega)$, it follows from Lemma 3.4.16 that, for any $\varphi \in L^2(\Omega)$, $M_\varphi(h) = \|(U(h) - I)\varphi\|$ is continuous relative to M_{φ_0}. Hence, it is clear that the μ-topology on $\mathfrak{G}$ is weaker than the topology induced by M_{φ_0}. On the other hand, the topology induced by M_{φ_0} is obviously weaker than the μ-topology.]

Corollary 3.4.18. Let $\Omega_k = (G, \mathfrak{B}, \mu_k)$, $k = 1, 2$, be localizable measure spaces which are both quasi-invariant under and strongly cyclic relative to a transformation group $\mathfrak{G}$, and denote their respective $\mathfrak{G}$-cyclic elements by φ_k, $k = 1, 2$. Write

$$M^{(k)}(h) = \left(\int_G |\varphi_k(hg)(d\mu_k(hg))^{1/2} - \varphi_k(g)(d\mu_k(g))^{1/2}|^2\right)^{1/2}, \qquad h \in \mathfrak{G}.$$

If μ_1 is absolutely continuous with respect to μ_2, then the convex function $M^{(1)}(h)$ is continuous relative to $M^{(2)}(h)$. In particular, if μ_1 and μ_2 are equivalent measures, then $M^{(1)}(h)$ and $M^{(2)}(h)$ are topologically equivalent.

Corollary 3.4.18 is a direct consequence of Corollary 3.4.17 and Theorem 2.4.13. It can be used as a criterion for the inequivalence of two quasi-invariant measures.

CHAPTER

IV

QUASI-INVARIANT MEASURES AND HARMONIC ANALYSIS ON LINEAR TOPOLOGICAL SPACES

A linear topological space may be regarded as an Abelian group with respect to addition, hence, the theory of quasi-invariant measures on linear topological spaces may be regarded as a special case of the theory of quasi-invariant measures on groups. However, since linear topological spaces also possess the operation of scalar multiplication, certain new situations arise. In this chapter, we shall discuss various problems relating to the concepts and results of Chapter III in the special context of linear spaces; in many cases, fairly complete results are obtained. In this connection, it should be mentioned that linear topological spaces constitute the most important special case in the theory of quasi-invariant measures, or, at least, that most likely to find applications. For example, quantum field theory, stochastic processes, and the so-called generalized stochastic processes are all connected with the theory of quasi-invariant measures on linear topological spaces.

Section §4.1 is primarily concerned with the study of the topology induced by the s-pseudometric on a quasi-invariant linear subspace. In section §4.2, we discuss quasi-linear functionals, which are the analogs of the quasi-characters studied in Chapter III. Since the theory of L^2-Fourier transforms requires only a slight modification when passing from groups to linear topological spaces, this subject is touched upon only very briefly in §4.2. Section §4.3 is devoted mostly to the following three topics: (1) the countable additivity of cylinder measures on spaces of linear functionals (this topic is closely related to the concept of a commutative weak distribution, due to Segal [3, 4, 5]); (2) representation theorems for positive-definite continuous linear functionals, these being among the most important results of harmonic analysis on linear topological spaces; (3) measure-theoretic foundations of the theory of generalized stochastic processes.

§4.1. Quasi-Invariant Measures on Linear Topological Spaces

1° Pseudometrics on Quasi-Invariant Linear Spaces

We begin by introducing some new concepts.

Definition 4.1.1. Let $\Omega = (G, \mathfrak{B}, \mu)$ be a measure space, let $\mathfrak{G}$ be a commutative group[1] of measurable transformations of Ω, and suppose that a scalar[2] multiplication is also defined on $\mathfrak{G}$ in such a way that $\mathfrak{G}$ becomes a linear space. Then, $\mathfrak{G}$ is called a *linear space of measurable transformations* on Ω. If, moreover, the functional[1]

$$M_1(h) = \left(\int_G ((d\mu(g+h))^{1/2} - (d\mu(g))^{1/2})^2\right)^{1/2}, \qquad h \in \mathfrak{G} \qquad (4.1.1)$$

is well defined[3] and quasi-continuous on $\mathfrak{G}$, that is, for each fixed $h \in \mathfrak{G}$, the quantity $M_1(th)$ is a continuous function of t, $-\infty < t < \infty$, then we say that the measure space Ω is *quasi-continuous* with respect to $\mathfrak{G}$.

Lemma 4.1.1. Let $\Omega = (G, \mathfrak{B}, \mu)$ be a measure space, and let $\mathfrak{G}$ be a linear space of measurable transformations on Ω. Given any $\varphi \in L^2(G, \mathfrak{B}, \mu)$, we form the convex function

$$M_\varphi(h) = \left(\int_G |\varphi(g+h)(d\mu(g+h))^{1/2} - \varphi(g)(d\mu(g))^{1/2}|^2\right)^{1/2}, \quad h \in \mathfrak{G}. \qquad (4.1.2)$$

[1] Throughout the present chapter, when $h \in \mathfrak{G}$ and $g \in \Omega$, we shall write $h + g$ (or $g + h$) instead of hg, and $d\mu(g - h)$ instead of $d\mu_h(g)$.

[2] The researches in this chapter will be primarily concerned with real linear spaces.

[3] In particular, $M_1(h)$ is always well defined if Ω is a finite measure space.

Suppose that $M_\varphi(h)$ is quasi-continuous on $\mathfrak{G}$. Define

$$R_\varphi(h) = \left(\int_0^\infty e^{-t} M_\varphi(th)^2 \, dt\right)^{1/2}. \tag{4.1.3}$$

Then $\mathfrak{G}$ is a linear pseudometric space with respect to the invariant pseudometric $\rho_\varphi(h_1, h_2) = R_\varphi(h_1 - h_2)$, $h_1, h_2 \in \mathfrak{G}$. In particular, if Ω is quasi-continuous with respect to $\mathfrak{G}$, then $\mathfrak{G}$ is a linear pseudometric space with respect to $\rho_1(h_1, h_2)$.

PROOF. It is easily verified that the three conditions in Lemma I.1.2 of Appendix I are satisfied for $M(h) = M_\varphi(h)$, therefore, $\mathfrak{G}$ is a linear pseudometric space with respect to $\rho_\varphi(\cdot, \cdot)$.]

Henceforth, unless otherwise stated, $\mathfrak{G}$ will be regarded as a linear pseudometric space with respect to ρ_1 (that is, $\varphi \equiv 1$).

Definition 4.1.2. Let $\Omega = (G, \mathfrak{B}, \mu)$ be a measure space and $\mathfrak{G}$ a linear space of measurable transformations on Ω. Suppose that, for each $h_0 \in \mathfrak{G}$, $h_0 \neq 0$, there is a real measurable function f on Ω, such that, for almost all $g \in G$, the equality

$$f(g + th_0) = f(g) + t \tag{4.1.4}$$

holds identically in t, $-\infty < t < \infty$. We then say that Ω is *separated* with respect to $\mathfrak{G}$.

For example, suppose G is a linear space, $\mathfrak{G}$ is a linear subspace of G, and $\mathfrak{F}$ is a family of linear functionals on G, such that, if $h \in \mathfrak{G}$, $h \neq 0$, then $f(h) = 1$ for some $f \in \mathfrak{F}$. Let $\mathfrak{B}$ be the smallest σ-algebra in G with respect to which all the functions in $\mathfrak{F}$ are measurable, and let μ be any measure on $(G, \mathfrak{B})$. Then, $(G, \mathfrak{B}, \mu)$ is separated with respect to the space of translations corresponding to $\mathfrak{G}$.

Theorem 4.1.2. Let $\Omega = (G, \mathfrak{B}, \mu)$ be a nontrivial[4] finite measure space which is separated and quasi-continuous with respect to a linear space of measurable transformations $\mathfrak{G}$. Then $\mathfrak{G}$ is a linear metric space with respect to $\rho_1(h_1, h_2)$, $h_1, h_2 \in \mathfrak{G}$.

PROOF. By virtue of Lemma 4.1.1, it suffices to prove that ρ_1 is a metric, that is, that $R_1(h) > 0$ if $h \neq 0$.

Suppose that $h_0 \in \mathfrak{G}$, $h_0 \neq 0$, and that $R_1(h_0) = 0$. Then, by (4.1.3) and the continuity of $M_1(th_0)$ with respect to t, it follows that $M_1(th_0) = 0$ for all $t \geqslant 0$. But since $M_1(-th_0) = M_1(th_0)$, this implies that, for

[4] That is, $\mu(G) > 0$.

every real number t, $M_1(th_0) = 0$, whence one may easily deduce that $\mu_{th_0} \equiv \mu$, that is,

$$\mu(E + th_0) = \mu(E) \tag{4.1.5}$$

for all $-\infty < t < \infty$ and all $E \in \mathfrak{B}$. Now, since Ω is separated with respect to $\mathfrak{G}$, there is a real $\mathfrak{B}$-measurable function f on G such that (4.1.4) holds. Using this function, we construct a Borel measure on the real line as follows: for any Borel subset B of the real line, let

$$\mu_1(B) = \mu(f^{-1}(B)).$$

By (4.1.4), if B is any Borel set, then $f^{-1}(B) + th_0$ and $f^{-1}(B + t)$ differ by at most a μ-null set. Hence, using (4.1.5), we deduce that

$$\mu_1(B + t) = \mu_1(B)$$

for all $-\infty < t < \infty$, in other words, μ_1 is a translation invariant Borel measure on the real line. Therefore, μ_1 is just the ordinary Lebesgue measure multiplied by some constant factor. But since μ is finite (and nonzero), μ_1 is also finite (and nonzero); this is a contradiction. We conclude that $R_1(h_0) = 0$ is impossible if $h_0 \neq 0$, that is, ρ_1 is a metric.]

In Theorem 4.1.2, one cannot dispense with the assumption that Ω is separated with respect to $\mathfrak{G}$. This is shown by the following example.

Example 4.1.1. Let G be a nontrivial linear space, let $\mathfrak{B}$ consist of just two sets, namely, G and the empty set O, and let $\mu(G) = 1$, $\mu(O) = 0$. Then $(G, \mathfrak{B}, \mu)$ is not separated with respect to the group of all translations G. In this case, we have $R_1(h) = 0$ for all $h \in G$.

2° Quasi-Invariance under Translations

We proceed to specialize the situation by assuming that G is a linear space and that $\mathfrak{G}$ is a linear subspace of G, operating on G by translation.[5] Suppose that $\Omega = (G, \mathfrak{B}, \mu)$ is a measure space which is quasi-invariant (quasi-invariant and quasi-continuous) with respect to[6] $\mathfrak{G}$, and suppose that $\mathfrak{G}$ is not properly contained in any other linear subspace $\mathfrak{G}'$ of G such that Ω is quasi-invariant (quasi-invariant and quasi-continuous) with respect to $\mathfrak{G}'$. Then we say that $\mathfrak{G}$ is a *maximal quasi-invariant (quasi-invariant and quasi-continuous) linear subspace relative to* Ω.

[5] Henceforth, it will always be assumed, without explicit repetition, that linear subspaces are regarded as operating on their ambient spaces by translation.

[6] Or, in the phraseology of Chapter III, quasi-invariant *under* $\mathfrak{G}$.

Let N be a linear subspace of G such that

$$\mu(E + h) = \mu(E)$$

for every $h \in N$ and every $E \in \mathfrak{B}$. We then say that Ω is *invariant with respect to* N (or *invariant under* N), or, alternatively, that N is an *invariant linear subspace relative to* Ω. Clearly, there exists a greatest invariant linear subspace relative to Ω; we denote this subspace by $\mathfrak{G}_0$. It is obvious that the s-pseudometric is identically zero on $\mathfrak{G}_0$, moreover, if Ω is quasi-invariant and quasi-continuous with respect to the linear subspace $\mathfrak{G}$, then

$$\mathfrak{G} \cap \mathfrak{G}_0 = \{x \mid \rho_1(x, 0) = 0,\ x \in \mathfrak{G}\}.$$

Using Zorn's lemma, one may easily prove that, if G is a linear space and $\Omega = (G, \mathfrak{B}, \mu)$ is a measure space, then there exists a maximal[7] quasi-invariant quasi-continuous linear subspace $\mathfrak{G}$ relative to Ω, and obviously the greatest invariant linear subspace $\mathfrak{G}_0$ (relative to Ω) is contained in $\mathfrak{G}$.

Theorem 4.1.3. Let G be a linear topological space, $\Omega = (G, \mathfrak{B}, \mu)$ a regular finite measure space, and $\mathfrak{G}$ a maximal quasi-invariant quasi-continuous linear subspace relative to Ω. Then $\mathfrak{G}$ is a complete linear pseudometric space with respect to the pseudometric $\rho_1(h_1, h_2)$, $h_1, h_2 \in \mathfrak{G}$.

PROOF. We need only prove the completeness of $\mathfrak{G}$. Let $\{h_n\}$ be any fundamental sequence in $\mathfrak{G}$ relative to ρ_1. Choose a sequence of natural numbers $\{n_k\}$, $n_1 < n_2 < \cdots < n_k < \cdots$, such that

$$R_1(h_n - h_m) < \frac{1}{4^k} \tag{4.1.6}$$

when $m, n \geqslant n_k$. Let μ_1 denote the Lebesgue measure on the real line, and let

$$\Lambda_k = \left\{t \mid M_1((h_{n_{k+1}} - h_{n_k})\, t) \geqslant \frac{1}{2^k},\ t \in (0, 1]\right\}.$$

Then, using (4.1.6), we get

$$\mu_1(\Lambda_k) \leqslant e 4^k \int_0^1 e^{-t} M_1((h_{n_{k+1}} - h_{n_k})\, t)^2\, dt < \frac{e}{4^k}.$$

[7] *Translator's note:* The use of Zorn's lemma here is unnecessary. In fact, it is easily verified that the linear hull of all the linear subspaces of G with respect to which Ω is quasi-invariant and quasi-continuous is actually the *greatest* linear subspace having this property.

Hence, writing $\Lambda = \overline{\lim}_{k\to\infty} \Lambda_k$, we have $\mu_1(\Lambda) = 0$. Let $Q = (0, 1] - \Lambda$. Then, for each $t \in Q$, there is an integer k_t such that $t \in (0, 1] - \Lambda_k$, that is,

$$M_1((h_{n_{k+1}} - h_{n_k})\, t) < \frac{1}{2^k}$$

for all $k \geqslant k_t$. Therefore, if $t \in Q$, then

$$\sum_{k=1}^{\infty} M_1(t(h_{n_{k+1}} - h_{n_k})) < \infty,$$

which implies that the sequence $\{th_{n_k}\}$ is fundamental relative to the pseudometric $M_1(h_1 - h_2)$, $h_1, h_2 \in \mathfrak{G}$. To simplify notation, we shall write k in place of n_k throughout the remainder of the proof.

Now, given any $t_0 \in Q$, the argument used in the proof of Theorem 3.1.24 shows that there exists[8] an $h_0 \in G$ such that every neighborhood V of t_0h_0 (in the original topology of G) contains a subsequence of $\{t_0h_k\}$. Hence, for any $t \in Q$, the arbitrary neighborhood $(t/t_0)\, V$ of th_0 contains a subsequence of $\{th_k\}$. Again, following the proof of Theorem 3.1.24, we deduce that, if $t \in Q$, then

$$\lim_{k\to\infty} M_1(t(h_k - h_0)) = 0. \tag{4.1.7}$$

Let $\mathfrak{G}_1$ denote the totality of elements (i.e., translations) of G with respect to which Ω is quasi-invariant, and let Q_0 denote the totality of real numbers t such that $th_0 \in \mathfrak{G}_1$. Obviously, $\mathfrak{G}_1$ is a group, and therefore Q_0 is an additive subgroup of the real numbers. Moreover, since $th_k \in \mathfrak{G}_1$, it follows from (4.1.7) and Lemma 3.1.25 that[9] $th_0 \in \mathfrak{G}_1$ for every $t \in Q$, that is, $Q \subset Q_0$. Let Q_1 denote the set of all real numbers $t \in Q_0$ such that (4.1.7) holds. Then, we obviously have

$$Q \subset Q_1. \tag{4.1.8}$$

We assert that Q_1 is a group. In fact, note that M_1 satisfies the relations

$$\begin{gathered} M_1(h_1 + h_2) \leqslant M_1(h_1) + M_2(h_2), \qquad h_1, h_2 \in \mathfrak{G}_1, \\ M_1(h) = M_1(-h), \qquad h \in \mathfrak{G}. \end{gathered} \tag{4.1.9}$$

[8] *Translator's note:* At this stage, it has not yet been established that $h_0 \in \mathfrak{G}$. Therefore, it seems necessary to add the hypothesis that every translation of the space G is $\mathfrak{B}$-measurable. For otherwise, the quantities $M_1(h_0 - h_m)$, $M(t(h_k - h_0))$, and so on, would be undefined.

[9] *Translator's note:* Again, the hypotheses must ensure that the translation th_0 is $\mathfrak{B}$-measurable.

Therefore, if t_1, $t_2 \in Q_1$, then

$$\lim_{k\to\infty} M_1((t_1 + t_2)(h_k - h_0)) \leqslant \lim_{k\to\infty} M_1(t_1(h_k - h_0)) + \lim_{k\to\infty} M_1(t_2(h_k - h_0)),$$

hence, $t_1 + t_2 \in Q_1$. Similarly, if $t \in Q_1$, then $-t \in Q_1$. Thus, Q_1 is a group, and in view of (4.1.8), we have

$$Q - Q \subset Q_1. \tag{4.1.10}$$

Since the Lebesgue measure μ_1 is invariant, and $\mu_1(Q) > 0$, the reasoning used in the proof of Theorem 3.1.13 shows that $Q - Q$ contains a neighborhood of zero on the real line. Hence, by (4.1.10), we see that Q_1 contains some neighborhood of zero, for example, the open interval $(-\delta, \delta)$. If t is an arbitrary real number, we choose a sufficiently large positive integer n, so that $(t/n) \in (-\delta, \delta)$, whence $(t/n) \in Q_1$; since Q_1 is a group, this implies that $t = n(t/n) = (t/n) + \cdots + (t/n) \in Q_1$. Thus, Q_1 is the entire real line. Therefore, for every real number t, we have $th_0 \in \mathfrak{G}_1$, and (4.1.7) holds.

Let $\mathfrak{G}'$ be the linear hull of $\mathfrak{G}$ and the set $\{th_0 \mid -\infty < t < \infty\}$. We shall now prove that Ω is quasi-invariant and quasi-continuous with respect to $\mathfrak{G}'$. The elements of $\mathfrak{G}'$ are of the form $g + t_0 h_0$, $g \in \mathfrak{G}$, $-\infty < t_0 < \infty$. By (4.1.7), the sequence $\{t_0 h_k + g, k = 1, 2,\ldots\}$ converges to $t_0 h_0 + g$ in the pseudometric $M_1(h_1 - h_2)$, h_1, $h_2 \in G$. Since $t_0 h_k + g \in \mathfrak{G} \subset \mathfrak{G}_1$, it follows by Lemma 3.1.25 that $t_0 h_0 + g \in \mathfrak{G}_1$. Let H_1 be the smallest closed linear subspace of $L^2(\Omega)$ containing the constant function 1 and invariant under all the operators $\{U((t_0 h_k + g)\, t),$ $-\infty < t < \infty,\ k = 1, 2,\ldots\}$. It is easily verified that the totality of vectors $\varphi \in L^2(\Omega)$, such that

$$\lim_{k\to\infty} \|(U((t_0 h_k + g)\, t) - U((t_0 h_0 + g)\, t))\, \varphi\| = 0$$

for all real t, forms a closed linear subspace which is invariant under all the operators $\{U((t_0 h_k + g)\, t), -\infty < t < \infty, k = 1, 2,\ldots\}$, and by (4.1.7), this subspace contains 1, hence it contains H_1. This means that, for all real numbers t, the sequence of operators $\{U((t_0 h_k + g)\, t),\ k = 1, 2,\ldots\}$ converges strongly to $U((t_0 h_0 + g)\, t)$ on H_1. Now, since $M_1((t_0 h_k + g)\, t)$ is a continuous function of t, we see from Lemma 3.4.16 that, for each k, the one-parameter group of unitary operators

$$\{U((t_0 h_k + g)\, t) \mid -\infty < t < \infty\}$$

is strongly continuous on H_1. It follows that H_1 is separable, and moreover, for each pair of vectors ξ, $\eta \in H_1$, the function

$f(t) = (U((t_0h_0 + g)\,t)\,\xi, \eta)$, being the limit of the sequence of continuous functions $(U((t_0h_k + g)\,t)\,\xi, \eta)$, is Lebesgue measurable. By virtue of Lemma II.3.2, this implies that $\{U((t_0h_0 + g)\,t) \mid -\infty < t < \infty\}$ is a strongly continuous one-parameter group of unitary operators on H_1. Hence, the quantity

$$M_1((t_0h_0 + g)\,t) = \|(U((t_0h_0 + g)\,t) - I)\,1\,\|$$

is a continuous function of t, that is, M_1 is quasi-continuous on $\mathfrak{G}'$.

Thus, Ω is quasi-invariant and quasi-continuous with respect to the linear subspace $\mathfrak{G}'$ of G. Since $\mathfrak{G}' \supset \mathfrak{G}$, it follows from the maximality of $\mathfrak{G}$ that $\mathfrak{G}' = \mathfrak{G}$, so that $h_0 \in \mathfrak{G}$. Moreover, using (4.1.7) and the Lebesgue dominated convergence theorem, we get

$$\lim_{k\to\infty} \rho_1(h_k\,, h_0) = \lim_{k\to\infty} R_1(h_k - h_0) = 0.$$

Therefore, $\mathfrak{G}$ is complete.]

Theorem 4.1.4. Let G be a linear topological space, $\mathfrak{G}$ a linear subspace of G, and $(G, \mathfrak{B}, \mu)$ a regular finite measure space which is quasi-invariant and quasi-continuous with respect to $\mathfrak{G}$. If $\mathfrak{G}$ is assigned the topology induced by the pseudometric $\rho_1(h_1\,, h_2)$, $h_1\,, h_2 \in \mathfrak{G}$, then the functional $M_1(h)$ is continuous on $\mathfrak{G}$. Moreover, if $\{h_n\} \subset \mathfrak{G}$, $h_0 \in \mathfrak{G}$, then $\rho_1(h_n\,, h_0) \to 0$ if and only if

$$\lim_{n\to\infty} M_1(t(h_n - h_0)) = 0, \qquad -\infty < t < \infty. \tag{4.1.11}$$

Proof. We may assume that $\mathfrak{G}$ is a maximal quasi-invariant quasi-continuous linear subspace relative to Ω. For otherwise, we could replace $\mathfrak{G}$ by a maximal quasi-invariant quasi-continuous linear subspace containing $\mathfrak{G}$. Moreover, by Theorem 4.1.3 and Lemma I.2.3, in order to prove the continuity of $M_1(h)$, it suffices to prove that $M_1(h)$ is lower semicontinuous. Suppose that $\{h_n\} \subset \mathfrak{G}$, $h_0 \in \mathfrak{G}$, and $R_1(h_n - h_0) \to 0$. Choose a subsequence $\{h_{n'}\}$ of $\{h_n\}$ such that

$$\lim_{n'\to\infty} M_1(h_{n'}) = \varliminf_{n\to\infty} M_1(h_n).$$

As in the proof of Theorem 4.1.3, since $\{h_{n'}\}$ is fundamental relative to ρ_1, there exists a subsequence $\{h_{n''}\}$ of $\{h_{n'}\}$ and an element $h_0' \in \mathfrak{G}$, such that

$$\lim_{n''\to\infty} M_1(t(h_{n''} - h_0')) = 0 \tag{4.1.12}$$

for every real number t. Since $\rho_1(h_{n''}, h_0') \to 0$, we have $\rho_1(h_0, h_0') = 0$. Hence, by the quasi-continuity of M_1, it follows that

$$M_1(t(h_0 - h_0')) = 0 \tag{4.1.13}$$

for all t. From (4.1.12) and (4.1.13), we get

$$\lim_{n''\to\infty} M_1(t(h_{n''} - h_0)) = 0,$$

hence,

$$M_1(h_0) \leqslant \lim_{n''\to\infty} (M_1(h_{n''}) + M_1(h_{n''} - h_0)) = \lim_{n'\to\infty} M_1(h_{n'}) = \underline{\lim}_{n\to\infty} M_1(h_n).$$

This proves that M_1 is lower semicontinuous, and therefore continuous.

Obviously, (4.1.11) is a sufficient condition for $\rho_1(h_n, h_0) \to 0$. On the other hand, if $\rho_1(h_n, h_0) \to 0$, then, since $\mathfrak{G}$ is a linear pseudometric space with respect to ρ_1, it follows that $\rho_1(th_n, th_0) \to 0$ for every real number t. Hence, by the continuity of M_1 on $\mathfrak{G}$, we obtain (4.1.11).]

In particular, Theorem 4.1.4 shows that the topology $\mathfrak{T}_0$ on $\mathfrak{G}$ induced by the convex function $M_1(h)$ is weaker than the topology $\mathfrak{T}_1$ induced by the pseudometric ρ_1. The following theorem describes the relationship between $\mathfrak{T}_0$ and $\mathfrak{T}_1$ still more precisely.

Theorem 4.1.5. Let G be a linear topological space, $\mathfrak{G}$ a linear subspace of G, and $(G, \mathfrak{B}, \mu)$ a regular finite measure space which is quasi-invariant and quasi-continuous with respect to $\mathfrak{G}$. Then the topology $\mathfrak{T}_1$ on $\mathfrak{G}$ induced by the s-pseudometric ρ_1 is the weakest among all those topologies which are stronger than the topology $\mathfrak{T}_0$ [induced by $M_1(h)$, $h \in \mathfrak{G}$] *and* make $\mathfrak{G}$ a linear topological space.

PROOF. It suffices to prove that any topology which satisfies the above two conditions is stronger than $\mathfrak{T}_1$. Let $\mathfrak{T}$ be any such topology; then the functional $M_1(h)$, $h \in \mathfrak{G}$, is continuous with respect to $\mathfrak{T}$. Consider any generalized sequence $\{h_n\} \subset \mathfrak{G}$ converging to a limit h_0 in the topology $\mathfrak{T}$. Since $\mathfrak{G}$ is a linear topological space with respect to $\mathfrak{T}$, this implies that, for any real number t, the generalized sequence $\{th_n\}$ converges to th_0 in the topology $\mathfrak{T}$. Hence, since M_1 is continuous with respect to $\mathfrak{T}$, condition (4.1.11) holds, and therefore $\rho_1(h_n, h_0) \to 0$. This shows that $\mathfrak{T}$ is stronger than $\mathfrak{T}_1$.]

Next, we examine the relationship between the topology on $\mathfrak{G}$ induced by the s-pseudometric and the topology on $\mathfrak{G}$ induced by the original topology of G.

Theorem 4.1.6. Let G be a linear topological space, $\mathfrak{G}$ a linear subspace of G, and $(G, \mathfrak{B}, \mu)$ a regular finite measure space which is quasi-invariant and quasi-continuous with respect to $\mathfrak{G}$. Then the topology $\mathfrak{T}_1$ induced by the s-pseudometric ρ_1 on $\mathfrak{G}$ is stronger than that induced by the original topology of G. Moreover, there exists a compact subset of G which contains a $\mathfrak{T}_1$-neighborhood of the identity 0 in $\mathfrak{G}$.

PROOF. According to Lemma 3.1.22, there exists a compact subset K of G which contains a $\mathfrak{T}_0$-neighborhood of zero in $\mathfrak{G}$, say, the neighborhood $\{h \mid M_1(h) < \epsilon, h \in \mathfrak{G}\}$, $\epsilon > 0$. But then K must contain a $\mathfrak{T}_1$-neighborhood of zero in $\mathfrak{G}$, for example, $\{h \mid \rho_1(h, 0) < \delta, h \in \mathfrak{G}\}$, $\delta > 0$. For otherwise, one could find a sequence $\{h_n\} \subset \mathfrak{G}$ such that $\rho_1(h_n, 0) < 1/n$ and $h_n \bar{\in} K$. However, by Theorem 4.1.4, $\rho_1(h_n, 0) \to 0$ implies that $M_1(h_n) \to 0$, so that $M_1(h_n) < \epsilon$ for some n, and therefore $h_n \in K$, which is a contradiction. Consequently, K does contain a $\mathfrak{T}_1$-neighborhood of zero in $\mathfrak{G}$, and it follows easily that the inclusion mapping of $\mathfrak{G}$ (with the topology $\mathfrak{T}_1$) into G is completely continuous, hence, *a fortiori*, continuous. Thus, the topology $\mathfrak{T}_1$ is stronger than the topology on $\mathfrak{G}$ induced by G.]

Corollary 4.1.7. Let G be a linear topological space which satisfies some separation axiom, let $\mathfrak{G}$ be a linear subspace of G, and let $(G, \mathfrak{B}, \mu)$ be a regular finite measure space which is quasi-invariant and quasi-continuous with respect to $\mathfrak{G}$. Then the corresponding s-pseudometric ρ_1 is actually a metric on $\mathfrak{G}$.

PROOF. By Theorem 4.1.6, the topology $\mathfrak{T}_1$ induced by ρ_1 is stronger than the topology on $\mathfrak{G}$ induced by G; since the latter satisfies a separation axiom, $\mathfrak{T}_1$ certainly satisfies the same separation axiom. Therefore, ρ_1 is a metric.]

3° Topological Characterization of the s-Pseudometric

Theorem 4.1.8. Let G be a linear topological space, $\mathfrak{G}$ a linear subspace of G, and $(G, \mathfrak{B}, \mu)$ a regular finite measure space which is quasi-invariant and quasi-continuous with respect to $\mathfrak{G}$. Suppose that $\mathfrak{G}$ itself is a linear topological space of the second category with respect to a certain topology $\mathfrak{T}$ which is stronger than the topology induced by G. Then the functionals $M_1(h)$ and $R_1(h)$ are both continuous with respect to $\mathfrak{T}$; in other words, $\mathfrak{T}$ is stronger than the topology $\mathfrak{T}_1$ induced by the s-pseudometric ρ_1.

PROOF. Since $M_1(h)$ is lower semicontinuous with respect to the topology on $\mathfrak{G}$ induced by G (see Corollary 3.1.21), $M_1(h)$ is also lower semicontinuous with respect to $\mathfrak{T}$. Furthermore, by quasi-continuity, we have $\lim_{n\to\infty} M_1((1/n)h) = 0$. Hence, using Lemma I.2.3 of Appendix I, we see that $M_1(h)$ is continuous with respect to $\mathfrak{T}$, that is, the topology $\mathfrak{T}_0$ on $\mathfrak{G}$ induced by M_1 is weaker than $\mathfrak{T}$. Since $\mathfrak{G}$ is a linear topological space with respect to $\mathfrak{T}$, it follows immediately from Theorem 4.1.5 that $\mathfrak{T}$ is stronger than $\mathfrak{T}_1$.]

Using Theorems 4.1.3 and 4.1.8, we can derive the following characterization of the topology induced by the s-pseudometric on $\mathfrak{G}$.

Theorem 4.1.9. Let G be a linear topological space, let $\Omega = (G, \mathfrak{B}, \mu)$ be a regular finite measure space, let $\mathfrak{G}$ be a maximal quasi-invariant quasi-continuous linear subspace associated with Ω, and let $\mathfrak{G}_0$ be a maximal[10] invariant linear subspace relative to Ω, such that $\mathfrak{G}_0 \subset \mathfrak{G}$. Let ρ be an invariant pseudometric on $\mathfrak{G}$, with the following properties. (i) $\mathfrak{G}$ is a complete linear pseudometric space with respect to ρ; (ii) $\mathfrak{G}_0 = \{x \mid \rho(x, 0) = 0\}$; (iii) the topology $\mathfrak{T}$ induced by ρ is stronger than the topology on $\mathfrak{G}$ induced by G. Then, $\mathfrak{T}$ is just the topology $\mathfrak{T}_1$ induced by the s-pseudometric on $\mathfrak{G}$.

PROOF. Consider the quotient space $\mathfrak{G}_1 = \mathfrak{G}/\mathfrak{G}_0$, and for any $x \in \mathfrak{G}$, let $\hat{x}$ denote the residue class $x + \mathfrak{G}_0$. Since $\mathfrak{G}_0 = \{x \mid \rho_1(x, 0) = 0\}$, we may regard ρ and ρ_1 as metrics defined on $\mathfrak{G}_1$:

$$\rho(\hat{x}, \hat{y}) = \rho(x, y), \quad x \in \hat{x}, \quad y \in \hat{y}; \qquad \rho_1(\hat{x}, \hat{y}) = \rho_1(x, y), \quad x \in \hat{x}, \quad y \in \hat{y}.$$

Moreover, $\mathfrak{G}_1$ is also a complete linear topological space with respect to both ρ and ρ_1. Now, by a well-known theorem, a complete linear pseudometric space is of the second category; hence, $\mathfrak{T}$ satisfies the conditions of Theorem 4.1.8, and therefore $\mathfrak{T}$ is stronger than $\mathfrak{T}_1$. From this, it follows easily that the topology $\mathfrak{T}^*$ induced by ρ on $\mathfrak{G}_1$ is stronger than the topology $\mathfrak{T}_1^*$ induced by ρ_1 on $\mathfrak{G}_1$. This means that the identity mapping $\hat{x} \to \hat{x}$, regarded as a linear operator from $(\mathfrak{G}_1, \mathfrak{T}^*)$ onto $(\mathfrak{G}_1, \mathfrak{T}_1^*)$, is continuous, and therefore (by the Banach inverse operator theorem), its inverse is also continuous, that is, $\mathfrak{T}_1^*$ is stronger than $\mathfrak{T}^*$. Thus, $\mathfrak{T}^* = \mathfrak{T}_1^*$, from which we easily deduce that $\mathfrak{T} = \mathfrak{T}_1$.]

Notice that the conditions (i), (ii), and (iii) in Theorem 4.1.9 are satisfied by ρ_1 itself, so that these conditions characterize the topology $\mathfrak{T}_1$. Thus, the topology $\mathfrak{T}_1$ is completely determined by $\mathfrak{G}$, $\mathfrak{G}_0$ and the relative topology of $\mathfrak{G}$ as a subset of G. In other words, to determine $\mathfrak{T}_1$, one need not know the explicit form of μ; it suffices to know the corresponding maximal quasi-invariant quasi-continuous linear subspace and maximal invariant linear subspace.

[10] *Translator's note:* As indicated in footnote 7, both $\mathfrak{G}$ and $\mathfrak{G}_0$ are uniquely determined, being in fact the *greatest* subspaces in their respective classes, and the relation $\mathfrak{G}_0 \subset \mathfrak{G}$ always holds.

Corollary 4.1.10. Let G be a linear topological space satisfying some separation axiom, let $(G, \mathfrak{B}, \mu)$ be a regular finite measure space, and let $\mathfrak{G}$ be the corresponding maximal quasi-invariant linear subspace. Let ρ be a metric on $\mathfrak{G}$, such that, (i) $\mathfrak{G}$ is a complete linear metric space with respect to ρ; (ii) the topology $\mathfrak{T}$ induced by ρ is stronger than the topology on $\mathfrak{G}$ induced by G. Then $\mathfrak{T}$ is just the topology $\mathfrak{T}_1$ induced by the s-pseudometric ρ_1 on $\mathfrak{G}$.

PROOF. By virtue of Corollary 4.1.7, we have $\{x \mid \rho_1(x, 0) = 0\} = \{0\}$. Hence, the maximal invariant linear subspace $\mathfrak{G}_0$ reduces to $\{0\}$, and the desired conclusion follows immediately from Theorem 4.1.9.]

Next, we consider conditions for the quasi-continuity of the functional $M_\varphi(h)$.

Lemma 4.1.11. Let G be a linear topological space, $\mathfrak{G}$ a linear subspace of G, and $\Omega = (G, \mathfrak{B}, \mu)$ a regular finite measure space which is quasi-invariant and quasi-continuous with respect to $\mathfrak{G}$. Then, for each $\varphi \in L^2(\Omega)$, the functional $M_\varphi(h)$ is quasi-continuous on $\mathfrak{G}$, moreover, the topology on $\mathfrak{G}$ induced by the pseudometric $R_\varphi(h_1 - h_2)$ is weaker than the s-topology $\mathfrak{T}_1$.

PROOF. By Theorem 4.1.6, the topology on $\mathfrak{G}$ induced by G is weaker than $\mathfrak{T}_1$. Hence, by Lemma 3.1.7, $\mathfrak{T}_1$ is an admissible topology on $\mathfrak{G}$. Furthermore, by Theorem 4.1.4, the functional $M_1(h)$ is continuous on $(\mathfrak{G}, \mathfrak{T}_1)$, hence, it follows by Theorem 3.1.30 that $\mathfrak{T}_1$ is stronger than the μ-topology on $\mathfrak{G}$. Consequently, for each $\varphi \in L^2(\Omega)$, the functional $M_\varphi(h)$ is continuous on $(\mathfrak{G}, \mathfrak{T}_1)$; since $(\mathfrak{G}, \mathfrak{T}_1)$ is a linear topological space, this obviously implies that $M_\varphi(h)$ is quasi-continuous. Now, suppose that $\{h_n\} \subset \mathfrak{G}$, $\lim_{n\to\infty} \rho_1(h_n, 0) = 0$; then

$$\lim_{n\to\infty} M_\varphi(th_n) = 0, \qquad -\infty < t < \infty,$$

and therefore

$$\lim_{n\to\infty} R_\varphi(h_n) = 0.$$

This shows that $\mathfrak{T}_1$ is stronger than the topology induced by $R_\varphi(h_1 - h_2)$, $h_1, h_2 \in \mathfrak{G}$.]

Theorem 4.1.12. Let G be a linear topological space, let $(G, \mathfrak{B}_k, \mu_k)$, $k = 1, 2$, be regular finite measure spaces, with the respective maximal quasi-invariant quasi-continuous linear subspaces $\mathfrak{G}_k$, $k = 1, 2$. Let ρ_k denote the s-pseudometric on $\mathfrak{G}_k$ corresponding to the measure space $(G, \mathfrak{B}_k, \mu_k)$. If $\mathfrak{G}_2 \subset \mathfrak{G}_1$, then the topology $\mathfrak{T}_2$ on $\mathfrak{G}_2$ induced by ρ_2 is stronger than the topology $\mathfrak{T}_1$ on $\mathfrak{G}_2$ induced by ρ_1.

PROOF. By Theorem 4.1.3, $\mathfrak{G}_2$ is a complete linear pseudometric space with respect to ρ_2, hence $(\mathfrak{G}_2, \mathfrak{T}_2)$ is of the second category. Furthermore, $(G, \mathfrak{B}_1, \mu_1)$ is quasi-invariant and quasi-continuous with respect to $\mathfrak{G}_2$, and, by Theorem 4.1.6, $\mathfrak{T}_2$ is stronger than the topology on $\mathfrak{G}_2$ induced by G. Hence, the desired conclusion follows immediately from Theorem 4.1.8 (with $\mu = \mu_1$).]

4° Lebesgue-Type Measure Spaces

We proceed to investigate a type of measure space which is frequently encountered in the applications, and whose quasi-invariant linear subspaces always satisfy the quasi-continuity condition.

Definition 4.1.3. Let $\Omega = (G, \mathfrak{B}, \mu)$ be a measure space such that there exists a sequence of linearly independent real measurable functions $\{f_k\}$ which form a determining set on Ω, and such that, for each n, the finite-dimensional measure corresponding to $f_1, \ldots, f_n$ is equivalent to Lebesgue measure, that is, the measure $\tilde{\mu}_n$ on n-dimensional real space R_n, defined by

$$\tilde{\mu}_n(A) = \mu(\{g \mid (f_1(g), \ldots, f_n(g)) \in A\})$$

(where A denotes an arbitrary Borel set in R_n) is weakly equivalent to the Lebesgue measure on R_n. We then say that $(G, \mathfrak{B}, \mu)$ is a *Lebesgue-type* measure space.

For example, let G be the Cartesian product of a sequence of real lines $\{R_\alpha \mid \alpha \in \mathfrak{A}\}$, let $\mathfrak{B}$ be the σ-algebra generated by the totality of Borel cylinders in G, and let μ be a probability measure on $(G, \mathfrak{B})$. For each $\alpha \in \mathfrak{A}$, $g \in G$, let g_α denote the αth coordinate of g, and let g_α be regarded as a random variable on $(G, \mathfrak{B}, \mu)$. If, for any finite set of indices $\alpha_1, \ldots, \alpha_n$, the joint distribution of the random variables $g_{\alpha_1}, \ldots, g_{\alpha_n}$ is equivalent to Lebesgue measure (e.g., if $\{g_\alpha \mid \alpha \in \mathfrak{A}\}$ is a family of nondegenerate Gaussian variables), then $(G, \mathfrak{B}, \mu)$ is a Lebesgue-type measure space.

The following lemma gives a sufficient condition for a measure space to be of Lebesgue type.

Lemma 4.1.13. Let G be a linear space, $\mathfrak{G}$ a linear subspace of G, and $f_1, \ldots, f_n, \ldots$ a sequence of real linear functionals on G, with the following property: for each n, there exist vectors $h_1, \ldots, h_n \in \mathfrak{G}$ such that

$$f_k(h_l) = \delta_{kl}, \qquad k, l = 1, 2, \ldots, n. \tag{4.1.14}$$

Let $\mathfrak{B}$ be the smallest σ-algebra of subsets of G which makes all the functions f_k measurable, and let μ be a σ-finite measure on $(G, \mathfrak{B})$ which is quasi-invariant under $\mathfrak{G}$. Then, $(G, \mathfrak{B}, \mu)$ is a Lebesgue-type measure space.

PROOF. For each n, we define a Borel measure $\tilde{\mu}_n$ on n-dimensional Euclidean space R_n, as follows: if A is any Borel set in R_n, then

$$\tilde{\mu}_n(A) = \mu(\{g \mid (f_1(g),\ldots, f_n(g)) \in A\}).$$

Now, for any $y = (y_1, \ldots, y_n) \in R_n$, we see from (4.1.14) that there is a vector $h \in \mathfrak{G}$ such that $f_k(h) = y_k$, $k = 1, 2,\ldots, n$. Therefore,

$$\tilde{\mu}_n(A + y) = \mu(\{g \mid (f_1(g),\ldots, f_n(g)) \in A\} + h).$$

Since μ is quasi-invariant under $\mathfrak{G}$, it follows that $\tilde{\mu}_n(A + y) = 0$ if and only if $\tilde{\mu}_n(A) = 0$. Thus, $\tilde{\mu}_n$ is a Borel measure on R_n which is quasi-invariant under translation. According to Theorem 3.1.5, $\tilde{\mu}_n$ is equivalent to Lebesgue measure.[11]]

Theorem 4.1.14. Let G be a linear space, and let $\Omega = (G, \mathfrak{B}, \mu)$ be a finite Lebesgue type measure space which is quasi-invariant under a linear subspace $\mathfrak{G}$ of G. Then, for any $\varphi \in L^2(\Omega)$, the functional $M_\varphi(h) (h \in \mathfrak{G})$ is quasi-continuous. In particular, Ω is quasi-continuous with respect to $\mathfrak{G}$.

PROOF. For convenience, we take $\mu(G) = 1$. By hypothesis, there is a sequence of measurable functions $\{f_k \mid k = 1, 2,\ldots\}$ on Ω such that, for each n, the finite dimensional measure $\tilde{\mu}_n$, corresponding to $f_1, \ldots, f_n$, is equivalent to Lebesgue measure. Let $F_n(x_1, \ldots, x_n)$ denote the Radon–Nikodym derivative of $\tilde{\mu}_n$ with respect to the Lebesgue measure on R_n; we may suppose that F_n is a Baire function such that $0 < F_n < \infty$. Then, for any $y = (y_1, \ldots, y_n)$, the Radon–Nikodym derivative of the measure $\tilde{\mu}_n(A + y)$ with respect to $\tilde{\mu}_n(A)$ is

$$\frac{F_n(x_1 + y_1, \ldots, x_n + y_n)}{F_n(x_1, \ldots, x_n)}. \tag{4.1.15}$$

Let $\mathfrak{B}_n$ denote the smallest σ-algebra in G which makes the functions $f_1, \ldots, f_n$ measurable, and let μ_n be the probability measure obtained by restricting μ to $\mathfrak{B}_n$. Then, using expression (4.1.15), we have

$$\frac{d\mu_n(g + th)}{d\mu_n(g)} = \frac{F_n(f_1(g) + tf_1(h),\ldots, f_n(g) + tf_n(h))}{F_n(f_1(g),\ldots, f_n(g))}$$

[11] *Translator's note:* The hypothesis of Theorem 3.1.5 requires that both of the measures concerned be σ-finite. In the present instance it is not obvious that μ_n is σ-finite. Of course, if μ is finite, then there is no difficulty.

for any fixed $h \in \mathfrak{G}$. Let $\mathfrak{F}_1$ denote the totality of Borel subsets of the real line R_1. Then, the function

$$\frac{d\mu_n(g + th)}{d\mu_n(g)}, \qquad (t, g) \in R_1 \times G,$$

is measurable with respect to $\mathfrak{F}_1 \times \mathfrak{B}$. Now, by Theorem 1.1.20,[12] for each t, there is a μ-null set $E_t \in \mathfrak{B}$ such that $\lim_{n\to\infty} d\mu_n(g + th)/d\mu_n(g)$ exists and

$$\frac{d\mu(g + th)}{d\mu(g)} = \lim_{n\to\infty} \frac{d\mu_n(g + th)}{d\mu_n(g)} \tag{4.1.16}$$

for all $g \in G - E_t$. Let E be the set of all pairs (t, g) such that $\lim_{n\to\infty} d\mu_n(g + th)/d\mu_n(g)$ does not exist. Since $d\mu_n(g + th)/d\mu_n(g)$ is a measurable function on $(R_1 \times G, \mathfrak{F}_1 \times \mathfrak{B})$, it follows that E is a measurable set. Also, for each t_0, we have $\{g \mid (t_0, g) \in E\} \subset E_{t_0}$. Hence, by Fubini's theorem, E is a null set relative to the measure $\mu_1 \times \mu$ (where μ_1 denotes the Lebesgue measure on R_1). Therefore, $\lim_{n\to\infty} d\mu_n(g + th)/d\mu_n(g)$ exists for almost all (t, g) (relative to the measure $\mu_1 \times \mu$), moreover, this limit is a measurable function on $(R_1 \times G, \mathfrak{F}_1 \times \mathfrak{B}, \mu_1 \times \mu)$, and by (4.1.16), it may be regarded as the Radon–Nikodym derivative $d\mu(g + th)/d\mu(g)$. Thus, $d\mu(g + th)/d\mu(g)$ is a measurable function on $(R_1 \times G, \mathfrak{F}_1 \times \mathfrak{B}, \mu_1 \times \mu)$. If the function $\xi(g)$ is $\mathfrak{B}$-measurable, one can prove that $\xi(g + th)$ is a measurable function[13] on $(R_1 \times G, \mathfrak{F}_1 \times \mathfrak{B})$. Using Fubini's theorem, we deduce that, if $\xi, \eta \in L^2(\Omega)$,

$$\int \xi(g + th)\, \overline{\eta(g)} \left(\frac{d\mu(g + th)}{d\mu(g)}\right)^{1/2} d\mu(g), \qquad t \in R_1 \tag{4.1.17}$$

is a measurable function on $(R_1, \mathfrak{F}_1)$.

Consider the one-parameter group of unitary operators

$$\{U(th) \mid -\infty < t < \infty\}$$

where $U(\cdot)$ is defined as in §3.1. Since (4.1.17) is just the inner product $(\xi, U(th)\,\eta)$, the unitary operator group $\{U(th) \mid -\infty < t < \infty\}$ is

[12] *Translator's note:* Theorem 1.1.20 is directly applicable only if one assumes that $\{f_k\}$ is a determining set for the measurable space $(G, \mathfrak{B})$, and not merely for the measure space $(G, \mathfrak{B}, \mu)$ (see Definitions 1.1.6 and 1.1.7).

[13] *Translator's note:* This requires some additional assumption which will ensure the measurability of the mapping $(g, t) \to g + th$, for example, the measurability of the linear operations $(g, g') \to g + g'$ and $(g, t) \to gt$.

weakly measurable. But since Ω has a countable determining set, $L^2(\Omega)$ is separable. Hence, by Lemma II.3.2 of Appendix II, $\{U(th) \mid -\infty < t < \infty\}$ is strongly continuous with respect to the parameter t. In particular,

$$M_\varphi(th) = \| (U(th) - I) \varphi \|$$

is a continuous function of t for each $\varphi \in L^2(\Omega)$.]

5° Strongly Cyclic Quasi-Invariant Measure Spaces

Lemma 4.1.15. Let $\Omega = (G, \mathfrak{B}, \mu)$ be a localizable measure space which is quasi-invariant under a linear space of measurable transformations $\mathfrak{G}$. Suppose that Ω is strongly cyclic relative to $\mathfrak{G}$, with the $\mathfrak{G}$-cyclic element $\varphi_0 \in L^2(\Omega)$, and that the functional $M_{\varphi_0}(h)$, $h \in \mathfrak{G}$, is quasi-continuous. Then, for any $\varphi \in L^2(\Omega)$, the functional $R_\varphi(h)$, $h \in \mathfrak{G}$, is continuous with respect to the topology induced by the pseudometric $R_{\varphi_0}(h_1 - h_2)$, $h_1, h_2 \in \mathfrak{G}$.

PROOF. According to Corollary 3.4.17, if $\varphi \in L^2(\Omega)$, then the functional $M_\varphi(h)$, $h \in \mathfrak{G}$, is continuous with respect to the topology induced by $M_{\varphi_0}(h)$, $h \in \mathfrak{G}$. Therefore, since $M_{\varphi_0}(th)$, $-\infty < t < \infty$, is continuous in t, it follows that, for any $\varphi \in L^2(\Omega)$, the functional $M_\varphi(th)$, $-\infty < t < \infty$, is also continuous in t, so that $R_\varphi(h)$ is well defined, moreover, by Lemma 4.1.1, $\mathfrak{G}$ is a linear pseudometric space with respect to the pseudometric $R_\varphi(h_1 - h_2)$, $h_1, h_2 \in \mathfrak{G}$.

Consider any sequence $\{h_n\} \subset \mathfrak{G}$ such that

$$\lim_{n\to\infty} R_{\varphi_0}(h_n) = 0. \tag{4.1.18}$$

We shall prove that

$$\lim_{n\to\infty} R_\varphi(h_n) = 0. \tag{4.1.19}$$

If (4.1.19) were false, there would exist a subsequence $\{h_{n'}\}$ of $\{h_n\}$ such that

$$\lim_{n'\to\infty} R_\varphi(h_{n'}) = c > 0, \tag{4.1.20}$$

where possibly $c = +\infty$. Now, (4.1.18) implies that, on any finite interval $[0, T]$, the sequence of functions $\{M_{\varphi_0}(th_{n'})\}$ converges to zero in Lebesgue measure. Consequently, there is a null set $\Delta \subset [0, \infty)$, and a subsequence $\{n''\} \subset \{n'\}$, such that

$$\lim_{n''\to\infty} M_{\varphi_0}(th_{n''}) = 0 \tag{4.1.21}$$

for every $t \in \Delta$. Since $M_\varphi(h)$ is continuous with respect to the topology induced by $M_{\varphi_0}(h)$, it follows from (4.1.21) that

$$\lim_{n'' \to \infty} M_\varphi(th_{n''}) = 0 \tag{4.1.22}$$

for every $t \in \Delta$. Using the Lebesgue dominated convergence theorem, we deduce that

$$\lim_{n'' \to \infty} R_\varphi(h_{n''}) = \lim_{n'' \to \infty} \left(\int_0^\infty e^{-t} M_\varphi(th_{n''})^2\, dt\right)^{1/2} = 0, \tag{4.1.23}$$

which contradicts (4.1.20). Thus, we conclude that (4.1.19) holds.]

Corollary 4.1.16. Under the hypotheses of Lemma 4.1.15,[14] the pseudometric $R_{\varphi_0}(h_1 - h_2)$, $h_1, h_2 \in \mathfrak{G}$, induces the s-topology on $\mathfrak{G}$.

PROOF. By Lemma 4.1.15, the s-topology is weaker than the topology $\mathfrak{T}_0$ induced by $R_{\varphi_0}(h_1 - h_2)$, and by Lemma 4.1.11, $\mathfrak{T}_0$ is weaker than the s-topology. Hence, these two topologies coincide.]

Using the proof of Lemma 4.1.15, we may rewrite Corollary 3.4.18 in the following form.

Corollary 4.1.17. Let $\Omega_k = (G, \mathfrak{B}, \mu_k)$, $k = 1, 2$, be localizable measure spaces which are quasi-invariant under and strongly cyclic relative to the linear space of measurable transformations $\mathfrak{G}$, and let $\varphi_k \in L^2(\Omega_k)$, $k = 1, 2$, be corresponding $\mathfrak{G}$-cyclic elements. Suppose that the functions

$$M^{(k)}(h) = \left(\int_G |\varphi_k(g+h)(d\mu_k(g+h))^{1/2} - \varphi_k(g)(d\mu_k(g))|^2\right)^{1/2}, \qquad h \in \mathfrak{G}$$

are quasi-continuous on $\mathfrak{G}$, and let

$$R^{(k)}(h) = \left(\int_0^\infty e^{-t} M^{(k)}(th)^2\, dt\right)^{1/2}.$$

If μ_1 is absolutely continuous with respect to μ_2, then the pseudometric $R^{(1)}(h_1 - h_2)$, $h_1, h_2 \in \mathfrak{G}$, is continuous with respect to the topology induced by the pseudometric $R^{(2)}(h_1 - h_2)$. In particular, if μ_1 and μ_2 are equivalent, then $R^{(1)}$ and $R^{(2)}$ are topologically equivalent.

Under certain circumstances, one can obtain results similar to the

[14] *Translator's note:* Since Lemma 4.1.11 is used in the proof, it appears that the hypotheses of Lemma 4.1.11 should also be included here. In particular, the s-topology may not be defined if μ is not totally finite.

above, but not explicitly involving cyclic elements. For example, we have the following theorem.

Theorem 4.1.18. Let G be a linear topological space, let $\mathfrak{G}$ be a linear subspace of G, and let $\Omega_k = (G, \mathfrak{B}, \mu_k)$, $k = 1, 2$, be regular finite measure spaces which are quasi-invariant, quasi-continuous and strongly cyclic relative to $\mathfrak{G}$. Let

$$R_1^{(k)}(h) = \left(\int_0^\infty \int_G e^{-t}((d\mu_k(g+h))^{1/2} - (d\mu_k(g))^{1/2})^2\, dt\right)^{1/2}, \qquad h \in \mathfrak{G}.$$

If μ_1 is absolutely continuous with respect to μ_2, then $R_1^{(1)}(h)$ is continuous with respect to the topology induced by $R_1^{(2)}(h)$ on $\mathfrak{G}$. If μ_1 and μ_2 are equivalent, then $R_1^{(1)}(h)$ and $R_1^{(2)}(h)$ induce the same topology on $\mathfrak{G}$.

PROOF. It follows from Lemma 4.1.15 that the functions $M_{\varphi_k}(h)$ are quasi-continuous, where φ_k is a cyclic element corresponding to μ_k, $k = 1, 2$. Hence, Corollary 4.1.16 is applicable, so that $R_1^{(k)}(h)$ and $R^{(k)}(h)$ induce the same topology on $\mathfrak{G}$. Thus, the conclusions of Theorem 4.1.18 follow immediately from Corollary 4.1.17.]

Theorem 4.1.18 may be used as a criterion for the inequivalence of two quasi-invariant measures.

Example 4.1.2. Let R_α, $\alpha = 1, 2,..., n,...$ be a sequence of copies of the real line, let $\mathfrak{B}_\alpha$ be the totality of Borel sets in R_α, and let $\mathfrak{T}_\alpha$ denote the topology on R_α defined by the usual metric. Let $l = \mathsf{X}_{\alpha=1}^\infty R_\alpha$, $\mathfrak{B}_\omega = \mathsf{X}_{\alpha=1}^\infty \mathfrak{B}_\alpha$. We regard l as the totality of real number sequences $x = \{x_1,..., x_n,...\}$, and define linear operations in l as follows. If $x = \{x_1,..., x_n,...\}$, $y = \{y_1,..., y_n,...\} \in l$, and a, b are real numbers, then

$$ax + by = \{ax_1 + by_1,..., ax_n + by_n,...\}.$$

Obviously, l is a linear space with respect to these operations. Let $(l, \mathfrak{T})$ denote the topological product of the spaces $\{(l_\alpha, \mathfrak{T}_\alpha), \alpha = 1, 2,...\}$. The topological space $(l, \mathfrak{T})$ is metrizable, in fact, $\mathfrak{T}$ is just the topology induced by the metric

$$\rho(x, y) = \sum_{\nu=1}^\infty \frac{1}{2^\nu} \frac{|x_\nu - y_\nu|}{1 + |x_\nu - y_\nu|}, \qquad x = \{x_\nu\}, \qquad y = \{y_\nu\} \in l.$$

For each α, let μ_α be a probability measure on $(R_\alpha, \mathfrak{B}_\alpha)$, equivalent to Lebesgue measure, with the probability density f_α^2, $0 < f_\alpha < \infty$:

$$\mu_\alpha(E) = \int_E f_\alpha^2(t)\, dt, \qquad E \in \mathfrak{B}_\alpha.$$

Form the product measure $\mu = \times_{\alpha=1}^{\infty} \mu_\alpha$. Then $(l, \mathfrak{B}_\omega, \mu)$ is a regular Lebesgue-type probability measure space.

For any $y = \{y_1, ..., y_n, ...\} \in l$, define a measure μ_y by

$$\mu_y(E) = \mu(E + y), \qquad E \in \mathfrak{B}_\omega .$$

Using Theorem 1.4.4, one easily calculates that

$$\rho(\mu_y, \mu) = \prod_{k=1}^{\infty} \int_{-\infty}^{\infty} f_k(u + y_k) f_k(u)\, du. \tag{4.1.24}$$

Let $\hat{f}_k$ denote the ordinary Fourier transform of f_k:

$$\hat{f}_k(t) = \operatorname*{l.i.m.}_{T\to\infty} \int_{-T}^{T} f_k(u)\, e^{iut}\, du.$$

Since f_k is a real function, we have $\overline{\hat{f}_k(t)} = \hat{f}_k(-t)$. Using the Plancheral theorem, we get

$$\rho(\mu_y, \mu) = \prod_{k=1}^{\infty} \frac{1}{2\pi} \int_{-\infty}^{\infty} |\hat{f}_k(t)|^2 \cos ty_k\, dt. \tag{4.1.25}$$

By another well-known theorem, a necessary and sufficient condition for the convergence of the infinite product (4.1.25) is the convergence of the series

$$\sum \left(1 - \frac{1}{2\pi} \int_{-\infty}^{\infty} |\hat{f}_k(t)|^2 \cos ty_k\, dt\right). \tag{4.1.26}$$

Again, by the Plancheral theorem,

$$\frac{1}{2\pi} \int_{-\infty}^{\infty} |\hat{f}_k(t)|^2\, dt = \int_{-\infty}^{\infty} f_k(t)^2\, dt = 1,$$

hence, the series (4.1.26) may be rewritten as

$$\sum \frac{1}{\pi} \int_{-\infty}^{\infty} |\hat{f}_k(t)|^2 \sin^2 \frac{ty_k}{2}\, dt. \tag{4.1.27}$$

Let $\mathfrak{G}_\mu$ denote the totality of points in l such that $(l, \mathfrak{B}_\omega, \mu)$ is quasi-invariant under the corresponding translation. Then, $\mathfrak{G}_\mu$ is just the totality of sequences $y = \{y_1, ..., y_k, ...\}$ such that (4.1.27) converges.

For any measure μ of the type under discussion, $\mathfrak{G}_\mu$ obviously includes the subspace $\mathfrak{G}$ consisting of all points of the form $y = \{y_1, ..., y_n, 0, 0, ...\}$ (i.e., for some n, the coordinate $y_k = 0$ for all $k \geqslant n + 1$). Using Theorems 3.1.32, 4.1.5, and so on, one may verify that the measure space $\Omega = (l, \mathfrak{B}_\omega, \mu)$ is quasi-invariant, quasi-continuous, ergodic, and strongly cyclic[15] relative to $\mathfrak{G}$.

In particular, if f_k is absolutely continuous on every finite interval, and

$$a_k = \int_{-\infty}^{\infty} |f_k'(t)|^2\, dt < \infty, \qquad k = 1, 2, \ldots, \tag{4.1.28}$$

then, since

$$\hat{f}_k(t)\, t = i \operatorname*{l.i.m.}_{T\to\infty} \int_{-T}^{T} e^{iut} f_k'(u)\, du,$$

we obtain

$$\frac{1}{2\pi}\int_{-\infty}^{\infty} |\hat{f}_k(t)|^2\, t^2\, dt = \int_{-\infty}^{\infty} |f_k'(u)|^2\, du.$$

Therefore, since $\sin^2(ty_k/2) \leqslant (ty_k/2)^2$, it follows that (4.1.27) converges whenever

$$\sum_{k=1}^{\infty} a_k y_k^2 < \infty. \tag{4.1.29}$$

Thus, under condition (4.1.28), $\mathfrak{G}_\mu$ contains the Hilbert space $l(\{a_k\})$, which is defined as the totality of all sequences $y = \{y_\nu\}$ in l which satisfy (4.1.29), with the inner product $(y, z) = \sum a_k y_k z_k$, where $y = \{y_1, \ldots, y_k, \ldots\}$, $z = \{z_1, \ldots, z_k, \ldots\} \in l(a_k)$.

§4.2. Linear and Quasi-Linear Functionals on Linear Spaces

1° Linear Functionals and Quasi-Continuous Characters on Linear Spaces

Let Φ be a real linear space, and let Φ^Δ denote the totality of real linear functionals on Φ. Define linear operations in Φ^Δ as follows: if $f, g \in \Phi^\Delta$, and α, β are real numbers, then $\alpha f + \beta g$ is the linear functional

$$(\alpha f + \beta g)(\varphi) = \alpha f(\varphi) + \beta g(\varphi), \qquad \varphi \in \Phi.$$

Thus, Φ^Δ forms a linear space, which we call the *algebraic dual* of Φ. In particular, we note that Φ^Δ forms a group with respect to addition. The role of Φ^Δ in the present chapter is analogous to that of the algebraic dual of a group in Chapter III.

[15] *Translator's note:* More details regarding the verification of strong cyclicity would be welcome.

Definition 4.2.1. Let Φ be a linear space and let α be a character on Φ. If, for every fixed $\varphi \in \Phi$, $\alpha(t\varphi)$ is a continuous function of t, $-\infty < t < \infty$, then we say that α is quasi-continuous. Let Φ' denote the multiplicative group of all quasi-continuous characters on Φ.

For each $f \in \Phi^\Lambda$, define

$$f'(\varphi) = e^{if(\varphi)}, \qquad \varphi \in \Phi. \tag{4.2.1}$$

Since f is a real-valued linear functional, f' is a quasi-continuous character. Thus, we obtain a mapping from Φ^Λ to Φ':

$$\Lambda : f \to f'. \tag{4.2.2}$$

Obviously, Λ is a homomorphism. Moreover, Λ is monomorphic; in fact, if $f, g \in \Phi^\Lambda$ and $f' = g'$, then, for each $\varphi \in \Phi$, we have $e^{itf(\varphi)} = e^{itg(\varphi)}$ for all real t, hence $f(\varphi) = g(\varphi)$.

Lemma 4.2.1. Let Φ be a linear space and let α be any quasi-continuous character on Φ. Then, there is an $f \in \Phi^\Lambda$ such that

$$\alpha(\varphi) = e^{if(\varphi)}, \qquad \varphi \in \Phi.$$

Proof. From our hypothesis, it is obvious that for each fixed $\varphi \in \Phi$, $\alpha(t\varphi)$ is a continuous character on the additive group of real numbers R (with the Euclidean topology). Hence, by Example 3.2.4, there is a unique real number $f(\varphi)$ such that

$$\alpha(t\varphi) = e^{itf(\varphi)}, \qquad -\infty < t < \infty. \tag{4.2.3}$$

It only remains to prove that the correspondence $\varphi \to f(\varphi)$ is a linear functional on Φ. If $s \in R$ and $\varphi \in \Phi$, then

$$e^{itf(s\varphi)} = \alpha(ts\varphi) = e^{itsf(\varphi)} \tag{4.2.4}$$

for all $t \in R$, hence $sf(\varphi) = f(s\varphi)$. Similarly, if $\varphi, \psi \in \Phi$, then

$$e^{itf(\varphi+\psi)} = \alpha(t(\varphi + \psi)) = \alpha(t\varphi)\,\alpha(t\psi) = e^{it(f(\varphi)+f(\psi))}$$

for all $t \in R$, hence $f(\varphi + \psi) = f(\varphi) + f(\psi)$.]

Thus, the mapping Λ is an isomorphism from Φ^Λ onto Φ'.

Definition 4.2.2. Let Φ be a linear topological space, and let $\Phi^\dagger$ denote the linear subspace of Φ^Λ consisting of all continuous real linear functionals on Φ. We call $\Phi^\dagger$ the *conjugate space* of Φ.

In this case, Φ is, in particular, a topological group with respect to addition. The role of $\Phi^\dagger$ in the present chapter is analogous to that of the dual group Φ^* in Chapter III.

Theorem 4.2.2. Let Φ be a linear topological space. Then, the restriction to $\Phi^\dagger$ of the mapping Λ is an isomorphism from $\Phi^\dagger$ onto Φ^*.

PROOF. Obviously, $\Lambda\Phi^\dagger \subset \Phi^*$. Moreover, if $\alpha \in \Phi^*$, then, by Lemma 4.2.1, there is an $f \in \Phi^\Lambda$ such that $\alpha(\varphi) = e^{if(\varphi)}$, $\varphi \in \Phi$. Thus, we need only prove that f is continuous. Now, by the continuity of α, the set

$$D_t = \{\varphi \mid \alpha(t\varphi) = 1\}$$

is closed in Φ for every $t \in R$. Consequently, the set $D = \bigcap_{-\infty<t<\infty} D_t$ is also closed. Furthermore, it is easily seen that D is just the null space $\{\varphi \mid f(\varphi) = 0\}$ of the linear functional f.

In proving continuity, we may obviously assume that $f \not\equiv 0$. Pick any $\varphi_0 \bar{\in} D$. Since D is closed, there exists a neighborhood V of zero in Φ such that $\varphi_0 + V \subset \Phi - D$. Choose a neighborhood W of zero such that $\lambda W \subset V$ whenever $|\lambda| \leqslant 1$. Thus, if $\varphi \in W$, $|\lambda| \leqslant 1$, then $\varphi_0 + \lambda\varphi \subset \Phi - D$, that is,

$$f(\varphi_0 + \lambda\varphi) \neq 0,$$

or, $\lambda f(\varphi) \neq f(\varphi_0)$. It obviously follows that $|f(\varphi)| < |f(\varphi_0)|$ whenever $\varphi \in W$. Therefore, given any $\psi \in G$ and $\epsilon > 0$, we have

$$|f(\varphi) - f(\psi)| < \epsilon$$

provided φ belongs to the neighborhood $\psi + (\epsilon/|f(\varphi_0)|)\, W$. This shows that f is continuous. Hence, $\Lambda\Phi^\dagger = \Phi^*$. Having already proved that Λ is monomorphic, we conclude that Λ maps $\Phi^\dagger$ isomorphically onto Φ^*.]

Definition 4.2.3. Let Φ be a linear space, and let $\mathfrak{B}$ be a σ-ring consisting of certain subsets of Φ. Let $\mathfrak{B}_1$ denote the σ-algebra of all Borel subsets of the real line R_1. Suppose that $\mathfrak{B}$ satisfies the following conditions: (i) for each fixed $\varphi \in \Phi$, the correspondence $t \to t\varphi$ is a measurable mapping from $(R_1, \mathfrak{B}_1)$ to $(\Phi, \mathfrak{B})$; (ii) for each fixed real number t, the correspondence $\varphi \to t\varphi$ is a measurable mapping from $(\Phi, \mathfrak{B})$ to $(\Phi, \mathfrak{B})$. Then, we say that $(\Phi, \mathfrak{B})$ is a *linear measurable space*, and if μ is a measure on $(\Phi, \mathfrak{B})$, then we say that $(\Phi, \mathfrak{B}, \mu)$ is a *linear measure space*.

We now proceed to consider the relation between measurable characters and measurable linear functionals.

If $(\Phi, \mathfrak{B})$ is a linear measurable space, we let $\Phi_{\mathfrak{B}}$ denote the totality of real $\mathfrak{B}$-measurable linear functionals on Φ. Obviously, $\Phi_{\mathfrak{B}}$ forms a linear space with respect to the ordinary linear operations.

Theorem 4.2.3. Let $(\Phi, \mathfrak{B})$ be a linear measurable space. Then, the mapping

$$\Lambda: f \to e^{if}$$

is an isomorphism from the additive group $\Phi_{\mathfrak{B}}$ onto $\Phi^{\mathfrak{B}}$ (the multiplicative group of all measurable characters on Φ).

PROOF. Clearly, Λ is a monomorphism from $\Phi_{\mathfrak{B}}$ into $\Phi^{\mathfrak{B}}$; we need only prove that $\Lambda\Phi_{\mathfrak{B}} = \Phi^{\mathfrak{B}}$. First, we observe that, if $\alpha \in \Phi^{\mathfrak{B}}$, then α is quasi-continuous. In fact, by virtue of property (i) in Definition 4.2.3, it is obvious that, for each $\varphi \in \Phi$, the correspondence $t \to \alpha(t\varphi)$ is a $\mathfrak{B}_1$-measurable character on the additive group of real numbers R_1. Hence, by Example 3.2.4, $\alpha(t\varphi)(t \in R_1)$ is a continuous function of t, that is, $\alpha \in \Phi'$. Therefore, given any $\alpha \in \Phi^{\mathfrak{B}}$, it follows by Lemma 4.2.1 that there is an $f \in \Phi^{\Lambda}$ such that

$$\alpha(\varphi) = e^{if(\varphi)}, \qquad \varphi \in \Phi, \tag{4.2.5}$$

that is, $\alpha = \Lambda f$.

Now, if A is any Borel subset of the complex plane, then, by the measurability of α, we know that $\{\varphi \mid \alpha(\varphi) \in A\} \in \mathfrak{B}$, hence, by property (ii) in Definition 4.2.3, for each fixed $t \neq 0$, we have

$$\{\varphi \mid \alpha(t\varphi) \in A\} = \frac{1}{t}\{\varphi \mid \alpha(\varphi) \in A\} \in \mathfrak{B},$$

that is, $\alpha(t\varphi)$ is a measurable function on $(\Phi, \mathfrak{B})$. But, from (4.2.5), we get

$$f(\varphi) = \lim_{n\to\infty} in \left(1 - \alpha\left(\frac{1}{n}\varphi\right)\right).$$

Thus, $f(\varphi)$ is the limit of a sequence of measurable functions on $(\Phi, \mathfrak{B})$, and is therefore measurable, that is, $f \in \Phi_{\mathfrak{B}}$.]

In analogy with Definition 3.2.4, we have:

Definition 4.2.4. Let $\Omega = (\Phi, \mathfrak{B}, \mu)$ be a linear measure space, with $\mu(\Phi) < \infty$, and let J be a linear subspace of $\Phi_{\mathfrak{B}}$. The function

$$\psi(f) = \int_{\Phi} e^{if(\varphi)}\, d\mu(\varphi), \qquad f \in J$$

will be called the *characteristic function* of Ω (on J).

2° Quasi-Linear Functionals

In analogy with the quasi-characters of Chapter III, we now introduce quasi-linear functionals.

Definition 4.2.5. Let G be a real linear space, $\mathfrak{G}$ a linear subspace of G, and $(G, \mathfrak{B}, \mu)$ a measure space quasi-invariant under $\mathfrak{G}$. Let f be a measurable real-valued function on $(G, \mathfrak{B})$, satisfying the following conditions: (i) $f(tg) = tf(g)$ for every real number t and every $g \in G$; (ii) for each $h \in \mathfrak{G}$, there exists a real number $\tilde{f}(h)$ and a set $E_h \in \mathfrak{B}$, $\mu(G - E_h) = 0$, such that[16]

$$f(h + tg) = tf(g) + \tilde{f}(h) \tag{4.2.6}$$

for every $g \in E_h$, $-\infty < t < \infty$. Then, we call f a *quasi-linear functional* on $(G, \mathfrak{B}, \mu)$ (relative to $\mathfrak{G}$). The totality of such functions will be denoted by G_μ.

Observe that $\tilde{f}$ is a linear functional on $\mathfrak{G}$. In fact, the additivity of $\tilde{f}$ can be shown as in §3.2; as for homogeneity, if s is a real number, $s \neq 0$, $g \in E_h$, then, by (4.2.6), we have

$$f(sh + tg) = sf\left(h + \frac{t}{s} g\right) = tf(g) + s\tilde{f}(h),$$

hence $\tilde{f}(sh) = s\tilde{f}(h)$.

We say that $\tilde{f}$ is the linear functional *induced by* f, or that f is a quasi-linear functional *corresponding to* $\tilde{f}$. The totality of such induced functionals $\tilde{f}$ will be denoted by $\tilde{\mathfrak{G}}_\mu$.

Clearly, two quasi-linear functionals which are equal almost everywhere induce the same linear functional on $\mathfrak{G}$. In the ensuing discussion, quasi-linear functionals which are equal almost everywhere will be regarded as identical.

In analogy with property III in §3.2, 2°, we have the following result.

Lemma 4.2.4. Let G be a real linear space, $\mathfrak{G}$ a linear subspace of G, and $(G, \mathfrak{B}, \mu)$ a measure space which is quasi-invariant and ergodic with respect to $\mathfrak{G}$. If $f_1, f_2 \in G_\mu$ and $\tilde{f}_1 = \tilde{f}_2$, then $f_1 - f_2$ is almost everywhere equal to some constant.

PROOF. Arguing as in the proof of §3.2, 2°, III, we see at once that, for any two real numbers c_1, c_2, $c_1 < c_2$, the set $\{g \mid c_1 < f_2(g) - f_1(g) < c_2\}$

[16] *Translator's note:* The notation $\tilde{f}(h)$ seems redundant, since, by setting $t = 0$ in (4.2.6), one gets $f(h) = \tilde{f}(h)$. Perhaps the author's original intention was to consider arbitrary commutative groups (or linear spaces) operating on $(G, \mathfrak{B}, \mu)$, and not merely linear subspaces of G.

is quasi-invariant. By the ergodicity of μ, it then follows that there exists a real number c such that the set $\{g \mid f_2(g) - f_1(g) = c\}$ differs from G by a set of measure zero.]

In particular, suppose that $\mathfrak{B}$ *does not* contain any set E such that (i) E is quasi-invariant, (ii) $\mu(G - E) = 0$, and (iii) for each $g \in G$, $g \neq 0$, the line $\{tg \mid -\infty < t < \infty\}$ intersects E in at most one point. Then, from the proof of Lemma 4.2.4, it can be seen that the above-mentioned constant c must be zero, that is, f_1 and f_2 are equal almost everywhere.

Let $\mathfrak{V}$ denote the totality of elements of G_μ corresponding to the linear functional 0 in $\tilde{\mathfrak{G}}^\mu$. Then, we have the natural isomorphism between the quotient space $G_{\mu 0} = G_\mu/\mathfrak{V}$ and $\tilde{\mathfrak{G}}^\mu$, that is,

$$\mathfrak{L}\colon \eta \to \tilde{f}, \qquad \eta \in G_{\mu 0}, \tag{4.2.7}$$

where η denotes the residue class consisting of all the quasi-linear functionals corresponding to the same linear functional $\tilde{f}$.

Next, we derive another corollary of Theorem 3.2.3.

Corollary 4.2.5. Let G be a linear topological space, and let $\mathfrak{G}$ be a linear subspace of G which is itself a linear topological space of the second category with respect to a topology $\mathfrak{T}$ which is stronger than the topology on $\mathfrak{G}$ induced by G. Let $(G, \mathfrak{B}, \mu)$ be a regular measure space which is quasi-invariant under $\mathfrak{G}$. Then, every quasi-linear functional on $(G, \mathfrak{B}, \mu)$ (relative to $\mathfrak{G}$) induces a continuous linear functional on $\mathfrak{G}$.

In particular, every real $\mathfrak{B}$-measurable linear functional on G, when restricted to $\mathfrak{G}$, is continuous with respect to the topology $\mathfrak{T}$.

PROOF. Let $f \in G_\mu$, and consider the function

$$\alpha(g) = e^{if(g)}, \qquad g \in G.$$

Clearly, α is a quasi-character on G, and by Corollary 3.2.4, α induces a continuous character on $\mathfrak{G}$. However, by (4.2.6), if $h \in \mathfrak{G}$, then

$$e^{if(g+h)} = e^{i\tilde{f}(h)} e^{if(g)}$$

holds for almost all g, hence

$$\tilde{\alpha}(h) = e^{i\tilde{f}(h)}.$$

Since $\tilde{\alpha}$ is continuous, it follows by Theorem 4.2.2 that $\tilde{f}$ is continuous.]

One can also give a direct proof of Corollary 4.2.5 by the method used in the proof of Theorem 3.2.3.

3° Topologies on Spaces of Linear and Quasi-Linear Functionals

Paralleling §3.2, 3°, we proceed to describe a number of useful topologies for spaces of linear functionals.

I. Let G be a linear space, and let J be a linear space consisting of certain linear functionals on G. Given any $f_0 \in J$, any finite set of elements $g_1, \ldots, g_n \in G$, and any $\epsilon > 0$, form the set

$$U(f_0\,; g_1, \ldots, g_n\,; \epsilon) = \{f \mid |f(g_k) - f_0(g_k)| < \epsilon, \quad k = 1, 2, \ldots, n\}.$$

Taking the totality of sets of this form as a neighborhood basis, we obtain a topology on J, which we call the *weak topology*. It is easily verified that, with this topology, J forms a linear topological space.

If one also takes the weak topology on ΛJ, then Λ is a continuous mapping[17] from J onto ΛJ; this fact follows readily from the inequality $|1 - e^{if(g)}| \leqslant |f(g)|$. However, in general, Λ is not a homeomorphism.

II. Let G be a linear topological space, J a linear subspace of $G^\dagger$. Given any $f_0 \in J$, any bounded subset Q of G, and any $\epsilon > 0$, form the set

$$V(f_0\,; Q, \epsilon) = \{f \mid \sup_{g \in Q} |f(g) - f_0(g)| < \epsilon\}.$$

Taking the totality of sets of this form as a neighborhood basis, we obtain a topology on J, which we call the *strong topology*. It is easily verified that, under this topology, J forms a linear topological space. The strong topology is obviously stronger than the weak topology; in general, these two topologies do not coincide.

If G is a linear topological space satisfying the first axiom of countability, then, under the strong topology, $G^\dagger$ forms a complete Hausdorff linear topological space satisfying the first axiom of countability.[18]

III. Again, let G be a linear topological space, J a linear subspace of $G^\dagger$, and $\mathfrak{U}$ a neighborhood of zero in G such that every functional in J is bounded on $\mathfrak{U}$. Define

$$|f|_{\mathfrak{U}} = \sup_{g \in \mathfrak{U}} |f(g)|.$$

[17] For the definition of Λ, see (4.2.1) and (4.2.2).

[18] *Translator's note:* To obtain the stated conclusions, the hypothesis that G satisfies the first axiom of countability should be replaced by "there exists a bounded neighborhood of zero in G." Note that when $G^\dagger$ satisfies the first axiom of countability, it follows from the theory of linear topological spaces that $G^\dagger$ is pseudometrizable, which, combined with the Hausdorff property, implies that $G^\dagger$ is metrizable.

Clearly, J is a normed linear space with respect to the norm $|f|_{\mathfrak{U}}$; the corresponding topology $\mathfrak{T}_{\mathfrak{U}}$ has the neighborhood basis

$$W(f_0\,;\mathfrak{U},\epsilon)=\{f\mid |f-f_0|_{\mathfrak{U}}<\epsilon\},\qquad \epsilon>0$$

at the point f_0. We call $\mathfrak{T}_{\mathfrak{U}}$ the $\mathfrak{U}$-topology. Observe that the $\mathfrak{U}$-topology on J is equivalent to the topology induced by the homogeneous convex function[19]

$$\sup_{h\in\mathfrak{U}}\frac{|f(h)|}{(1+|f(h)|^2)^{1/2}},\qquad f\in J$$

Given any bounded subset Q of G, there exists a positive number δ such that $\delta Q\subset\mathfrak{U}$, hence

$$W(f_0\,;\mathfrak{U},\delta\epsilon)\subset V(f_0\,,Q,\epsilon).$$

Consequently, the $\mathfrak{U}$-topology is stronger than the strong topology; in general, these two topologies do not coincide.

However, if G is a normed linear space, and if we take $\mathfrak{U}=\{x\mid \|x\|<1\}$ (where $\|x\|$ denotes the given norm on G), then all the functionals in $G^{\dagger}$ are bounded on $\mathfrak{U}$. In this case, the $\mathfrak{U}$-topology on J is equivalent to the strong topology, and is, in fact, just the topology induced by the usual norm $\sup_{x\in\mathfrak{U}}|f(x)|=\|f\|$.

IV. Let G be a linear space, $\mathfrak{G}$ a linear subspace of G, and $(G,\mathfrak{B},\mu)$ a measure space quasi-invariant under $\mathfrak{G}$. Let J be a linear space consisting of certain quasi-linear functionals on $(G,\mathfrak{B},\mu)$ (relative to $\mathfrak{G}$). For each set $A\in\mathfrak{B}$ such that $\mu(A)<\infty$, define

$$R_A(f)=\Big(\int_A\frac{f(g)^2}{1+f(g)^2}\,d\mu(g)\Big)^{1/2}.$$

Then, $R_A(\cdot)$ is a homogeneous convex function on J. The family of functions $\{R_A(\cdot)\}$ induce a topology on J as indicated in §I.1 of Appendix I, namely, for each $f_0\in J$, we take, as a neighborhood basis at f_0, the totality of sets of the form

$$Y(f_0\,,A;\epsilon)=\{f\mid R_A(f-f_0)<\epsilon\},$$

where $A\in\mathfrak{B}$, $\mu(A)<\infty$ and $\epsilon>0$. We call this the *μ-topology*. Under the μ-topology, J forms a linear topological space. In particular, if μ

[19] See Definition I.1.3 of Appendix I.

is totally finite, then the μ-topology is just the topology induced by the metric

$$R(f - \varphi) = \left(\int_G \frac{(f(g) - \varphi(g))^2}{1 + (f(g) - \varphi(g))^2}\, d\mu(g)\right)^{1/2}, \qquad f, \varphi \in J,$$

and the sequence $\{f_n\}$ converges to f in the metric R if and only if $\{f_n\}$ converges to f in measure.

If $Q \subset G_{\mu 0}$, $A \in \mathfrak{B}$, and $\mu(A) < \infty$, let

$$R_A(\eta) = \inf_{f \in \eta} R_A(f), \qquad \eta \in Q.$$

The topology on Q induced by the family of convex functions $\{R_A(\cdot), A \in \mathfrak{B}, \mu(A) < \infty\}$ will also be called the μ-topology. In particular, if $\mu(G) < \infty$, then the μ-topology is just the topology induced by the single convex function $R(\eta) = R_G(\eta)$, $\eta \in Q$.

If $\mathfrak{H} \subset \tilde{\mathfrak{G}}^{\mu}$, $A \in \mathfrak{B}$, and $\mu(A) < \infty$, let

$$R_A(\varphi) = R_A(\mathfrak{L}^{-1}\varphi), \qquad \varphi \in \mathfrak{H}.$$

Again, the topology on $\mathfrak{H}$ induced by the family of convex functions $\{R_A(\varphi), A \in \mathfrak{B}, \mu(A) < \infty\}$ will be called the μ-topology. In particular, if $\mu(G) < \infty$, then the μ-topology is induced by the single convex function $R(\varphi) = R_A(\varphi)$, $\varphi \in \mathfrak{H}$.

Next, the concept of quasi-characteristic function, introduced in §3.2, will be transferred to the present context.

Definition 4.2.6. Let G be a real linear space, $\mathfrak{G}$ a linear subspace of G, $\Omega = (G, \mathfrak{B}, \mu)$ a finite measure space quasi-invariant under $\mathfrak{G}$, and J a linear subspace of G_μ. The function

$$\psi(f) = \int_G e^{if(g)}\, d\mu(g), \qquad f \in J,$$

will be called the *quasi-characteristic function* (on J) corresponding to Ω and $\mathfrak{G}$.

The homogeneous convex function $R(f)$ and the quasi-characteristic function $\psi(f)$ on J are related by the formula

$$R(f)^2 = \int_0^\infty (\mu(G) - \Re\psi(tf))\, e^{-t}\, dt. \tag{4.2.8}$$

This may be proved by using the identity

$$\frac{a^2}{1+a^2} = \int_0^\infty (1 - \Re e^{ita})\, e^{-t}\, dt, \qquad -\infty < a < \infty$$

and Fubini's Theorem.

Lemma 4.2.6. Under the hypotheses of Definition 4.2.6, let $\mathfrak{T}$ be any topology on J such that $(J, \mathfrak{T})$ is a linear topological space. Then, the quasi-characteristic function $\psi(f)$ is continuous on $(J, \mathfrak{T})$ if and only if $\mathfrak{T}$ is stronger than the μ-topology. In other words, the μ-topology is the weakest topology on J such that (i) $(J, \mathfrak{T})$ is a linear topological space and (ii) the quasi-characteristic function ψ is continuous on $(J, \mathfrak{T})$.

PROOF. The necessity of the condition follows easily from (4.2.8) and the Lebesgue dominated convergence theorem. Conversely, assume that $\mathfrak{T}$ is stronger than the μ-topology, and let $\{f_\alpha\}$ be a generalized sequence converging to f in $(J, \mathfrak{T})$. Then $\{f_\alpha\}$ also converges to f in the μ-topology, that is, $\{f_\alpha\}$ converges to f in measure. Since $| e^{if(g)} | = 1$, it follows by the Lebesgue dominated convergence theorem[20] that $\psi(f_\alpha) \rightarrow \psi(f)$.]

Lemma 4.2.6 shows that, if μ is finite, then the μ-topology is completely determined by the quasi-characteristic function.

Lemma 4.2.7. Let G be a linear space, $\mathfrak{G}$ a linear subspace of G, and $(G, \mathfrak{B}, \mu)$ a finite measure space quasi-invariant under $\mathfrak{G}$. Then, (i) G_μ is complete with respect to the metric $R(f - \varphi)$, $f, \varphi \in G_\mu$, (ii) $G_{\mu 0}$ is complete with respect to the metric $R(\xi - \eta)$, $\xi, \eta \in G_{\mu 0}$, and (iii) $\tilde{\mathfrak{G}}^\mu$ is complete with respect to the metric $R(\tilde{f} - \tilde{\varphi})$, $\tilde{f}, \tilde{\varphi} \in \tilde{\mathfrak{G}}^\mu$.

PROOF. Let $\{f_n\}$ be an R-fundamental sequence, that is,

$$\lim_{m,n\to\infty} R(f_m - f_n) = 0.$$

Since the totality of measurable functions on $(G, \mathfrak{B}, \mu)$ forms a complete metric space with respect to R, it follows that there is a subsequence $\{f_{n_k}\}$ which converges almost everywhere on G. We define a function f as follows: let E denote the set of all points $g \in G$ such that $\{f_{n_k}(g)\}$ converges, let $f(g) = \lim_{k\to\infty} f_{n_k}(g)$ for $g \in E$, and let $f(g) = 0$ for $g \in G - E$. We assert that $f \in G_\mu$. In fact, f is a real measurable function, and since every f_{n_k} is homogeneous, we obviously have $f(tg) = tf(g)$ for every $g \in G$ and every real number t. Furthermore, for every positive

[20] The Lebesgue dominated convergence theorem is valid for generalized sequences.

integer k and every $h \in \mathfrak{G}$, there exists a set $E_h^{(k)} \in \mathfrak{B}$ satisfying the following three conditions[21]: (i) $tE_h^{(k)} \subset E_h^{(k)}$, $-\infty < t < \infty$; (ii) $\mu(G - E_h^{(k)}) = 0$; (iii) if $g \in E_h^{(k)}$, $-\infty < t < \infty$, then

$$f_{n_k}(tg + h) = \tilde{f}_{n_k}(h) + tf_{n_k}(g). \tag{4.2.9}$$

Let $\bigcap_{k=1}^{\infty} E_h^{(k)} \cap E = E_h$. Obviously, $tE_h \subset E_h$ and $\mu(G - E_h) = 0$; moreover, if $g \in E_h$, $-\infty < t < \infty$, then, from (4.2.9), we get[22]

$$f(tg + h) - tf(g) = \lim_{k\to\infty}(f_{n_k}(tg + h) - tf_{n_k}(g)) = \lim_{k\to\infty} \tilde{f}_{n_k}(h).$$

Taking $\tilde{f}(h) = \lim_{k\to\infty} \tilde{f}_{n_k}(h)$, we see that $f \in G_\mu$. Since we obviously have $R(f_{n_k} - f) \to 0$, this proves that G_μ is complete with respect to R.

Finally, since $\mathfrak{B}$ is a closed subspace of G_μ, the quotient space $G_{\mu 0} = G_\mu/\mathfrak{B}$ is also complete with respect to R.]

Corollary 4.2.8. If $\{f_n\}$, $f \in G_\mu$, and $\lim_{n\to\infty} R(f_n - f) = 0$, then

$$\lim_{n\to\infty} \tilde{f}_n(h) = \tilde{f}(h)$$

for all $h \in \mathfrak{G}$.

Lemma 4.2.9. Let G be a linear topological space, $\mathfrak{G}$ a linear subspace of G, and $(G, \mathfrak{B}, \mu)$ a regular measure space quasi-invariant under $\mathfrak{G}$. Let $\mathfrak{T}$ be a topology on $\mathfrak{G}$ such that $(\mathfrak{G}, \mathfrak{T})$ is a linear topological space of the second category, satisfying the first axiom of countability,[23] and suppose that $\mathfrak{T}$ is stronger than the topology on $\mathfrak{G}$ induced by G.

Then, for every set $A \in \mathfrak{B}$ such that $0 < \mu(A) < \infty$, there is a neighborhood V_A of zero in $(\mathfrak{G}, \mathfrak{T})$, such that (i) every functional $\tilde{f}$ in $\tilde{\mathfrak{G}}^\mu$ is bounded on V_A, and (ii) the topology on $\tilde{\mathfrak{G}}^\mu$ induced by the convex function $R_A(\eta)$ is stronger than the V_A-topology on $\tilde{\mathfrak{G}}^\mu$.

PROOF. Let $A \in \mathfrak{B}$, $0 < \mu(A) < \infty$. For each $f \in G_\mu$, the two functions

$$p(g) = \frac{|f(g)|}{(1 + f(g)^2)^{1/2}}, \qquad g \in G,$$

$$\tilde{p}(h) = \frac{|\tilde{f}(h)|}{(1 + \tilde{f}(h)^2)^{1/2}}, \qquad h \in \mathfrak{G}$$

[21] *Translator's note:* If μ is a complete measure space, we may choose

$$E_h^{(k)} = \{g \mid f_{n_k}(h + tg) = tf_{n_k}(g) + \tilde{f}_{n_k}(h), \; -\infty < t < \infty\}.$$

But if μ is not complete, then the existence of the required $E_h^{(k)}$ is not obvious.

[22] *Translator's note:* It is not obvious why $tg + h \in E$, that is, why $f(tg + h) = \lim_{k\to\infty} f_{n_k}(tg + h)$.

[23] For example, $(\mathfrak{G}, \mathfrak{T})$ is a complete linear pseudometric space.

constitute a quasi-convex pair. Therefore, by Corollary 3.1.18, there exists a neighborhood V_A of zero in $(\mathfrak{G}, \mathfrak{T})$, and a positive number c_A, such that

$$\sup_{h\in V_A} \frac{|\tilde{f}(h)|}{(1+\tilde{f}(h)^2)^{1/2}} \leqslant c_A \int_A \frac{|f(g)|}{(1+f(g)^2)^{1/2}}\, d\mu(g) \leqslant c_A{}'R_A(f) \qquad (4.2.10)$$

[where $c_A{}' = c_A(\mu(A))^{1/2}$] holds for all $f \in G_\mu$.

Now, given any $f \in G_\mu$, we choose a sufficiently small real number t, such that $c_A{}'R_A(tf) < 1$. Then, from (4.2.10), we get

$$\sup_{h\in V_A} \frac{t\,|\tilde{f}(h)|}{(1+t^2\tilde{f}(h)^2)^{1/2}} < 1,$$

and it easily follows that $\sup_{h\in V_A} |\tilde{f}(h)| < \infty$. Thus, every $\tilde{f} \in \tilde{\mathfrak{G}}^\mu$ is bounded on V_A.

Next, let the element f in (4.2.10) vary over the residue class $\eta = \{\varphi \mid \varphi \in G_\mu, \tilde{\varphi} = \tilde{f}\}$. Taking the infimum of the right-hand side of (4.2.10), we obtain

$$\sup_{h\in V_A} \frac{|\tilde{f}(h)|}{(1+\tilde{f}(h)^2)^{1/2}} \leqslant c'_A R_A(\eta). \qquad (4.2.11)$$

Hence, given any positive number $\epsilon < 1$, if we choose a positive number $\delta < \epsilon/c_A{}'(1+\epsilon^2)^{1/2}$, then, from (4.2.11), we see that $\tilde{f} \in \{\tilde{f} \mid |\tilde{f}|_{V_A} < \epsilon\}$ whenever $\eta \in \{\eta \mid R_A(\eta) < \delta\}$. This proves that the topology induced by R_A is stronger than the V_A-topology.]

Incidentally, if, in (4.2.10), we replace f by tf, $0 < t < \infty$, divide both sides by t, then let $t \to 0$, we obtain

$$\sup_{h\in V_A} |\tilde{f}(h)| \leqslant c_A \int_A |f(g)|\, d\mu(g). \qquad (4.2.12)$$

Corollary 4.2.10. Under the hypotheses of Lemma 4.2.9, the μ-topology on $\tilde{\mathfrak{G}}^\mu$ is stronger than the strong topology.

PROOF. Since any V-topology is stronger than the strong topology, the conclusion follows at once from Lemma 4.2.9.]

Theorem 4.2.11. Let G be a linear topological space, and let $\mathfrak{G}$ be a linear subspace of G which is itself a complete linear metric space, such that the corresponding metric topology of $\mathfrak{G}$ is stronger than the relative topology induced by G on $\mathfrak{G}$; let $\mathfrak{T}$ denote the strong topology on $\mathfrak{G}^\dagger$. Let $\Omega = (G, \mathfrak{B}, \mu)$ be a regular finite measure space which is quasi-invariant under $\mathfrak{G}$. Then, $\tilde{\mathfrak{G}}^\mu = \mathfrak{G}^\dagger$ implies that $\mathfrak{T}$ coincides with the μ-topology. Moreover, if $\mathfrak{H}$ is a linear subspace of $\tilde{\mathfrak{G}}^\mu$ which is dense in $(\mathfrak{G}^\dagger, \mathfrak{T})$, then the following two statements are equivalent.

(i) $\tilde{\mathfrak{G}}^\mu = \mathfrak{G}^\dagger$.

(ii) The relative topology $\mathfrak{T}_0$ on $\mathfrak{H}$ induced by the μ-topology of $\tilde{\mathfrak{G}}^\mu$ is weaker than the relative topology $\mathfrak{T}_1$ on $\mathfrak{H}$ induced by $\mathfrak{T}$.

PROOF. According to Lemma 4.2.7, $G_{\mu 0}$ is a complete linear metric space under the μ-topology, hence $\tilde{\mathfrak{G}}^\mu$ is also a complete linear metric space under the μ-topology $\mathfrak{T}_\mu$. Furthermore, $\mathfrak{G}^\dagger$, under the strong topology $\mathfrak{T}$, is a complete Hausdorff linear topological space satisfying the first axiom of countability, and is therefore metrizable.[24] If $\tilde{\mathfrak{G}}^\mu = \mathfrak{G}^\dagger$, then, by Corollary 4.2.10, we have $\mathfrak{T}_\mu \supset \mathfrak{T}$, hence, by the Banach inverse operator theorem, $\mathfrak{T}_\mu = \mathfrak{T}$. Statement (ii) then follows trivially from (i).

Conversely, assume that (ii) holds. Given any $f \in \mathfrak{G}^\dagger$, there exists a sequence $\{f_n\} \subset \mathfrak{H}$ which converges to f in the topology $\mathfrak{T}$. Obviously, $\{f_n\}$ is fundamental relative to $\mathfrak{T}_1$, hence also fundamental relative to $\mathfrak{T}_0$. Since $\tilde{\mathfrak{G}}^\mu$ is complete with respect to the μ-topology (by Lemma 4.2.7), there exists an element $\varphi \in \tilde{\mathfrak{G}}^\mu$ such that $\{f_n\}$ converges to φ in the μ-topology. Hence, by Corollary 4.2.10, $\{f_n\}$ also converges to φ in the topology $\mathfrak{T}$. Since $(\mathfrak{G}^\dagger, \mathfrak{T})$ is Hausdorff, we conclude that $f = \varphi \in \tilde{\mathfrak{G}}^\mu$. This shows that $\tilde{\mathfrak{G}}^\mu \supset \mathfrak{G}^\dagger$, hence, by Corollary 4.2.5, we obtain (i).]

When $\tilde{\mathfrak{G}}^\mu = \mathfrak{G}^\dagger$, we say that Ω is a *standard quasi-invariant measure space* associated with $\mathfrak{G}$.

Definition 4.2.7. Let $\Omega = (G, \mathfrak{B}, \mu)$ be a linear measure space which is quasi-invariant under the linear subspace $\mathfrak{G}$ of G. If $f \in L^1(\Omega) \cap G_\mu = G_\mu{}^1$, then the integral

$$E(f) = \int_G f(g)\, d\mu(g)$$

will be called the *mean value* of f on Ω; if $\varphi, f \in L^2(\Omega) \cap G_\mu = G_\mu{}^2$, then

$$c(f, \varphi) = \int_G f(g)\, \varphi(g)\, d\mu(g)$$

will be called the *correlation number* of f and φ, or, when regarded as a function of the two variables $f, \varphi \in G_\mu{}^2$, the *correlation functional.* Clearly, the correlation functional is an inner product on $G_\mu{}^2$.

Lemma 4.2.12. Let $\Omega = (G, \mathfrak{B}, \mu)$ be a linear measure space

[24] *Translator's note:* As the translator remarked earlier, this conclusion requires an additional hypothesis regarding $\mathfrak{G}$, such as local boundedness (see footnote 18 relating to the definition of the strong topology).

which is quasi-invariant under the linear subspace $\mathfrak{G}$ of G. If $\mu(G) < \infty$, then $G_\mu{}^2$ is complete with respect to the inner product $c(\varphi, f)$, $\varphi, f \in G_\mu{}^2$.

PROOF. Suppose that the sequence $\{f_n\} \subset G_\mu{}^2$ is fundamental relative to $c(\varphi, f)$, that is, fundamental in $L^2(\Omega)$. Since $L^2(\Omega)$ is complete, there exists an element $\varphi_0 \in L^2(\Omega)$ such that $\{f_n\}$ converges to φ_0 in $L^2(\Omega)$. Moreover, since

$$R(f) \leqslant (c(f, f))^{1/2},$$

$\{f_n\}$ is also fundamental relative to the metric $R(f - \varphi)$, f, $\varphi \in G_\mu$. Therefore, by Lemma 4.2.7, there exists an $f_0 \in G_\mu$ such that $R(f_n - f_0) \to 0$. But it is easily proved that φ_0 and f_0 are equal almost everywhere, hence $f_0 \in L^2(\Omega) \cap G_\mu$.]

Let us write $\tilde{\mathfrak{G}}_2{}^\mu = \{\tilde{f} \mid f \in G_\mu{}^2\}$, and, when $\psi \in \tilde{\mathfrak{G}}_2{}^\mu$, let

$$c(\psi, \psi) = \inf_{\tilde{f}=\psi} c(f, f).$$

It is easily seen that the norm $(c(\psi, \psi))^{1/2}$ is also derivable from an inner product, which we also denote by $c(\varphi, \psi)$. Again, we call it the *correlation functional* on $\tilde{\mathfrak{G}}_2{}^\mu$.

Theorem 4.2.13. Under the hypotheses of Theorem 4.2.11, if $\tilde{\mathfrak{G}}_2{}^\mu = \mathfrak{G}^\dagger$, then $\mathfrak{T}$ is just the topology induced by the norm $(c(\psi, \psi))^{1/2}$, $\psi \in \tilde{\mathfrak{G}}_2{}^\mu$. If $\mathfrak{H} \subset \tilde{\mathfrak{G}}_2{}^\mu$, and $\mathfrak{H}$ is dense in $(\mathfrak{G}^\dagger, \mathfrak{T})$, then the following two statements are equivalent.

(i) $\tilde{\mathfrak{G}}_2{}^\mu = \mathfrak{G}^\dagger$.

(ii) The topology $\mathfrak{T}_2$ induced on $\mathfrak{H}$ by the norm $(c(\psi, \psi))^{1/2}$, $\psi \in \mathfrak{H}$, is weaker than the relative topology induced by $\mathfrak{T}$ on $\mathfrak{H}$.

PROOF. Assume that $\tilde{\mathfrak{G}}_2{}^\mu = \mathfrak{G}^\dagger$. Let $\mathfrak{B}_2 = L^2(\Omega) \cap \mathfrak{B}$ (where, again, $\mathfrak{B}$ denotes the null space of the mapping $f \to \tilde{f}$). By Lemma 4.2.12, $G_\mu{}^2$ is complete with respect to the norm $(c(\varphi, \varphi))^{1/2}$, hence the quotient space $\tilde{\mathfrak{G}}_2{}^\mu = G_\mu{}^2/\mathfrak{B}_2$ is also complete with respect to the norm$(c(\psi, \psi))^{1/2}$. Now, it follows easily from (4.2.12) that there exists a neighborhood V_G of zero in $\mathfrak{G}$ such that

$$\sup_{h \in V_G} |\xi(h)| \leqslant c_G(\mu(G))^{-1/2}(c(\xi, \xi))^{1/2}, \qquad \xi \in \tilde{\mathfrak{G}}_2^\mu.$$

Consequently, the topology induced by the norm $(c(\psi, \psi))^{1/2}$ on $\mathfrak{G}^\dagger = \tilde{\mathfrak{G}}_2{}^\mu$ is stronger than $\mathfrak{T}$, and it follows by the Banach inverse operator theorem that these two topologies coincide. Statement (ii) now follows trivially from (i).

Conversely, following the argument used in the corresponding part of the proof of Theorem 4.2.11, one can show that (ii) implies (i).]

Similar considerations apply to the subspaces $G_\mu \cap L^p(\Omega)$, $p \geqslant 1$; we omit the details.

The following theorem is entirely analogous to Theorem 3.2.10.

Theorem 4.2.14. Let G be a linear topological space, $\mathfrak{G}$ a linear subspace of G, and $\Omega = (G, \mathfrak{B}, \mu)$ a regular measure space which is quasi-invariant and ergodic with respect to $\mathfrak{G}$. Suppose that $\mathfrak{G}$ itself is a linear topological space of the second category under a certain topology which is stronger than the topology induced by G. Choose any nonzero vector $\psi \in L^1(\Omega)$, $\psi(g) \geqslant 0$, and form the homogeneous convex function

$$R_\psi(f) = \left(\int_G \frac{f(g)^2}{1 + f(g)^2} \psi(g)\, d\mu(g)\right)^{1/2}, \qquad f \in G_\mu . \tag{4.2.13}$$

Then, the topology on G_μ induced by $R_\psi(f)$ is independent of the choice of ψ, and coincides with the μ-topology.

In fact, $\mathfrak{G}$, being a linear topological space, is connected. Hence, the argument used in proving Theorem 3.2.10 is also applicable here; we omit the details.

We observe that the functional defined by (4.2.13) and the functional defined in Theorem 3.2.10 are related by the formula

$$R_\psi(f)^2 = \frac{1}{2} \int_0^\infty N_\psi(e^{itf(\cdot)})^2\, e^{-t}\, dt, \tag{4.2.14}$$

which may be proved in the same manner as (4.2.8).

Corollary 4.2.15. Under the hypotheses of Theorem 4.2.14, let μ' be another measure on $(G, \mathfrak{B})$, choose any nonzero vector $\psi' \in L^1(G, \mathfrak{B}, \mu')$, $\psi'(g) \geqslant 0$, and form the homogeneous convex function

$$R'_{\psi'}(f) = \left(\int_G \frac{f(g)^2}{1 + f(g)^2} \psi'(g)\, d\mu'(g)\right)^{1/2}, \qquad f \in G_\mu . \tag{4.2.15}$$

If μ' and μ are strongly equivalent,[25] then R_ψ and $R'_{\psi'}$ induce the same topology on G_μ .

PROOF. Let $E = \{g \mid \psi'(g) > 0\}$; then, E is σ-finite relative to μ', hence also σ-finite relative to μ. By virtue of the (weak) equivalence of μ' and μ, there exists a measurable function

$$0 < \frac{d\mu'(g)}{d\mu(g)} < \infty, \qquad g \in E,$$

[25] When μ' and μ are strongly equivalent, $G_{\mu'}$ and G_μ coincide.

such that $\varphi = \psi'(d\mu'/d\mu) \in L^1(\Omega)$, $\varphi \geqslant 0$, φ does not vanish almost everywhere, and

$$R_\varphi = R'_{\psi'} .$$

But according to Theorem 4.2.14, R_ψ and R_φ induce the same topology. Therefore, R_ψ and $R'_{\psi'}$ induce the same topology.]

Corollary 4.2.15 may also be used to establish the inequivalence of two given measures.

Example 4.2.1. We again consider the measure space described in Example 4.1.2. Let $l^\dagger$ denote the totality of linear functionals on l having the form

$$f(x) = \sum_1^n f_m x_m , \qquad x = \{x_m\} \in l$$

for some natural number n and some set of real numbers $f_1 ,..., f_n$ (if we take the product topology $\mathfrak{T}$ on l, then $l^\dagger$ is just the totality of continuous linear functionals on l). Let l_0 denote the linear subspace of l consisting of all sequences of the form $\{f_1 ,..., f_n , 0, 0,...\}$. Then, the correspondence $f \rightarrow \{f_1 ,..., f_m ,...\}$ is a one-to-one mapping of $l^\dagger$ onto l_0. Also, since every f in $l^\dagger$ is a measurable linear functional on $(l, \mathfrak{B}_\omega)$, we have $l^\dagger \subset G_\mu$.

We now consider the following special case. If $1 \leqslant a \leqslant 2$, the function

$$\exp(-| t |^a), \qquad -\infty < t < \infty,$$

is the characteristic function of an infinitely divisible distribution (see, e.g., Doob [1]), furthermore, since this function is Lebesgue square integrable, there exists a function $g \in L^2(-\infty, \infty)$ such that

$$\int_{-\infty}^{\infty} e^{itx} g(x)^2 \, dx = \exp(-| t |^a), \qquad -\infty < t < \infty,$$

moreover, this function g can be chosen positive almost everywhere.[26] Now, in Example 4.1.2, we choose $f_\alpha = g$, $\alpha = 1, 2,...$, thus obtaining

[26] See, for example, Zhao Zhong-zhe, Shuxue Xuebao (Acta Mathematica Sinica) **3** (1953), 177.

a quasi-invariant measure μ_a. Let us calculate the corresponding functional $R_1(f)$, $f \in l^\dagger$. We first calculate $N_1(e^{itf(\cdot)})$.

$$N_1(e^{itf(\cdot)})^2 = 2 \int_l (1 - \Re e^{itf(x)})\, d\mu(x)$$
$$= 2 \left(1 - \Re \prod_{m=1}^{\infty} \int_{-\infty}^{\infty} \exp(itf_m(x_m))\, g(x_m)^2\, dx_m\right)$$
$$= 2 \left(1 - \exp\left\{-\sum_{m=1}^{\infty} |f_m|^a |t|^a\right\}\right), \qquad f \in l^\dagger.$$

Hence,

$$R_1(f)^2 = \int_0^{\infty} \left(1 - \exp\left\{-\sum_{m=1}^{\infty} |f_m|^a |t|^a\right\}\right) e^{-t}\, dt.$$

4° Some σ-Algebras on Linear Spaces

In this subsection, we describe certain σ-algebras which will often be used in the ensuing discussion, and examine the relations between them.

Definition 4.2.8. Let Φ be a linear space and $\mathfrak{H}$ a linear subspace of Φ^A. If E is a Borel set in real n-dimensional space, and $\varphi_1, \ldots, \varphi_n \in \Phi$, the set

$$\{f \mid (f(\varphi_1), \ldots, f(\varphi_n)) \in E, f \in \mathfrak{H}\}$$

will be called the *Borel cylinder* with *base E* corresponding to $\{\varphi_1, \ldots, \varphi_n\}$. If the elements $\varphi_1, \ldots, \varphi_n$ generate the linear subspace M of Φ, then we also call the above set a Borel cylinder corresponding to M, or a Borel M-cylinder. The totality of Borel cylinders corresponding to a fixed M forms a σ-algebra, which we denote by $S(M)$ [or $S(\varphi_1, \ldots, \varphi_n)$], and the totality of all Borel cylinders forms an algebra S. Let $\mathfrak{B}$ denote the smallest σ-algebra containing S; we call the elements of $\mathfrak{B}$ *weak Borel sets corresponding to* Φ, or Φ-weak Borel sets. In particular, if Φ is a linear topological space and $\mathfrak{H} \subset \Phi^\dagger$, then we simply refer to the elements of $\mathfrak{B}$ as weak Borel sets. In any case, it is clear that $(\mathfrak{H}, \mathfrak{B})$ is a linear measurable space.

Let Φ be a linear space and $\mathfrak{H}$ a linear subspace of Φ^A. Let $\mathfrak{F}$ be the smallest σ-algebra of subsets of Φ which contains all sets of the form

$$\{\varphi \mid F(\varphi) < a\}, \qquad -\infty < a < \infty, \quad F \in \mathfrak{H}.$$

The elements of $\mathfrak{F}$ will be called *weak Borel sets corresponding to* $\mathfrak{H}$, or

$\mathfrak{H}$-weak Borel sets. In particular, if Φ is a linear topological space, and $\mathfrak{H} \subset \Phi^\dagger$, then the elements of $\mathfrak{F}$ will simply be called weak Borel sets.

If $\mathfrak{H}$ distinguishes[27] the points of Φ, then we may imbed Φ in $\mathfrak{H}^A$ by identifying each point $\varphi \in \Phi$ with the element $\varphi_1 \in \mathfrak{H}^A$ defined by

$$\varphi_1(f) = f(\varphi), \qquad f \in \mathfrak{H}.$$

Let $\Phi_1 = \{\varphi_1 \mid \varphi \in \Phi\}$. Then the weak Borel sets in $\mathfrak{H}$ corresponding to Φ_1 are just the weak Borel sets corresponding to Φ. Thus, in this case, the two concepts of "weak Borel set" described above essentially coincide.

The following lemma shows that the weak Borel sets constitute a sufficiently wide class of sets.

Lemma 4.2.16.[28] If Φ is a separable countably normed space, then every open (or closed) subset of Φ is a weak Borel set.

PROOF. Let

$$\| \varphi \|_1 \leqslant \| \varphi \|_2 \leqslant \cdots \leqslant \| \varphi \|_n \leqslant \cdots$$

be the given sequence of norms on Φ, and let $S_n(r) = \{\varphi \mid \| \varphi \|_n \leqslant r\}$. We shall first prove that $S_n = S_n(1)$ is a weak Borel set. Since Φ is separable, there is a sequence $\{\varphi_m\} \subset \Phi - S_n$, such that $\{\varphi_m\}$ is dense in $\Phi - S_n$ relative to the norm $\| \varphi \|_n$. By the Hahn–Banach theorem, there exists a linear functional $f_m \in \Phi^\dagger$, such that

$$\| f_m \|_{-n} = \sup_{\|\varphi\|_n \leqslant 1} | f_m(\varphi)| = 1, \qquad f_m(\varphi_m) = \| \varphi_m \|_n . \tag{4.2.16}$$

Form the set

$$E = \bigcap_{m=1}^{\infty} \{\varphi \mid | f_m(\varphi)| \leqslant 1\}.$$

Clearly, E is a weak Borel set. We assert that $E = S_n$. In fact, from (4.2.16), we obviously have $E \supset S_n$. On the other hand, if $\varphi \,\overline{\in}\, S_n$, choose a positive number $\delta < \frac{1}{2}(\| \varphi \|_n - 1)$. Then, there exists an index m such that $\| \varphi_m - \varphi \|_n < \delta$, hence $| f_m(\varphi_m - \varphi)| < \delta$, and therefore

$$| f_m(\varphi)| \geqslant | f_m(\varphi_m)| - \delta = \| \varphi_m \|_n - \delta \geqslant \| \varphi \|_n - 2\delta > 1.$$

Thus, $\varphi \,\overline{\in}\, E$. This proves that S_n is a weak Borel set. In a similar fashion, one may prove that, for every $\varphi \in \Phi$, $S_n(r) + \varphi$ is a weak Borel set.

[27] That is, for each nonzero $\varphi \in \Phi$, there exists $f \in \mathfrak{H}$ such that $f(\varphi) \neq 0$.

[28] Regarding the terminology used in the statement and proof of Lemmas 4.2.16 and 4.2.17, see Appendix I, §I.3.

Now, let W be any open set in Φ. Then, since Φ is separable, there exists a sequence $\{\varphi_n\} \subset \Phi$, a sequence of positive numbers $\{r_n\}$, and a sequence of natural numbers $\{m_n\}$ such that

$$W = \bigcup_{n=1}^{\infty} \{\varphi_n + S_{m_n}(r_n)\}.$$

Hence, W is a weak Borel set. Since, in particular, Φ itself is a weak Borel set, it follows that every closed set $\Phi - W$ is also a weak Borel set.]

Thus, under the hypotheses of Theorem 4.2.16, the totality of weak Borel sets in Φ is just the σ-algebra generated by the totality of open sets in Φ.

Lemma 4.2.17. Let Φ be a separable countably normed space, with the norm sequence $\{\| \varphi \|_n\}$. Then, $S_{-n}(r) = \{f \mid \| f \|_{-n} \leqslant r\}$ is a weak Borel set in $\Phi^\dagger$.

PROOF. Choose a sequence $\{\varphi_m\}$ which is dense in S_n, and consider the set

$$B = \bigcap_{m=1}^{\infty} \{f \mid | f(\varphi_m)| \leqslant r\}.$$

Obviously, B is a weak Borel set in $\Phi^\dagger$. We shall prove that $B = S_{-n}(r)$. First, since $\{\varphi_m\} \subset S_n$, we clearly have $B \supset S_{-n}(r)$. Conversely, let $f \in B$; if $\varphi \in S_n$, there exists a subsequence $\{\varphi_{m_k}\}$ such that $\lim_{k\to\infty} \varphi_{m_k} = \varphi$ (in the topology of Φ). Consequently,

$$| f(\varphi)| = \lim_{k\to\infty} |f(\varphi_{m_k})| \leqslant r,$$

that is, $f \in S_{-n}(r)$. Thus, $S_{-n}(r) = B$ is a weak Borel set.]

In particular, we see that $\Phi_n^\dagger = \bigcup_{m=1}^{\infty} S_{-n}(m)$ is a weak Borel set.

5° The Existence of Quasi-Invariant Measures

Now, we shall show how inequality (4.2.12) may be used to derive necessary conditions for the existence of quasi-invariant measures. Here, we only consider Hilbert spaces and number sequence spaces, however, the methods used can, in fact, be applied in an even more general context. Moreover, all the necessary conditions given here are actually sufficient conditions, as will be shown in §5.3.

Theorem 4.2.18. Let G be a separable Hilbert space, with the inner product (x, y), and let $\mathfrak{B}$ be the totality of weak Borel sets in G. Let $\mathfrak{G}$ be a linear subspace of G, and suppose that $\mathfrak{G}$ itself is a complete countably inner product space[29] with respect to the sequence of inner products $(\varphi, \psi)_n$, $n = 1, 2, \ldots$, where $(\varphi, \varphi)_1 \leqslant (\varphi, \varphi)_2 \leqslant \cdots$. Also, suppose that the inclusion mapping T from $\mathfrak{G}$ into G is continuous.[30] For each n, let $\mathfrak{G}_n$ denote the completion of $\mathfrak{G}$ with respect to the inner product $(\varphi, \psi)_n$. Then, the existence of a $\mathfrak{G}$-quasi-invariant finite measure μ on $(G, \mathfrak{B})$ implies that, for some natural number n, the mapping T can be extended to a Hilbert–Schmidt type operator from $\mathfrak{G}_n$ to G.

PROOF. We may assume that μ satisfies the condition

$$M_2 = \int_G (x, x)\, d\mu(x) < \infty,$$

for otherwise, we could replace μ by the equivalent measure

$$\mu'(E) = \int_E e^{-(x,x)}\, d\mu(x), \qquad E \in \mathfrak{B},$$

which certainly satisfies this condition. [Notice here that (x, x) is a measurable function on $(G, \mathfrak{B})$].

Now, by (4.2.12), there exists a positive number c and a neighborhood V of zero in $\mathfrak{G}$, such that, for each continuous linear functional f on G,

$$\sup_{h \in V} |f(h)| \leqslant c \int_G |f(g)|\, d\mu(g). \tag{4.2.12'}$$

Of course, V must contain some neighborhood of the form $\{h \mid (h, h)_n \leqslant \delta^2\}$. Choose any complete orthonormal set $\{e_m\}$ in G. Then, by (4.2.12′), we obtain

$$\sum_m \sup_{(h,h)_n \leqslant \delta^2} |(h, e_m)|^2 \leqslant c^2 \mu(G) \sum_m \int_G |(g, e_m)|^2\, d\mu(g) = c^2 \mu(G) \int_G (g, g)\, d\mu(g).$$

Clearly, we may assume that n is sufficiently large, so that T is continuous with respect to the inner product $(\varphi, \psi)_n$, and may therefore be extended to a bounded linear operator T_n from $\mathfrak{G}_n$ to G. We then have[31]

$$\begin{aligned} \sup_{(h,h)_n \leqslant \delta^2} |(h, e_m)|^2 &= \sup_{(h,h)_n \leqslant \delta^2} |(h, T_n{}^* e_m)_n|^2 \\ &\geqslant (\delta / \| T_n \|)(T_n{}^* e_m, T_n{}^* e_m)_n . \end{aligned} \tag{4.2.17}$$

[29] Regarding the concept of a countably inner product space, consult Appendix I.

[30] In other words, the topology of $\mathfrak{G}$ is stronger than the relative topology induced by G on $\mathfrak{G}$.

[31] *Translator's note:* The author has incorrectly squared the absolute values appearing in this inequality. It is not clear how this gap can be fixed up.

It follows from (4.2.12′), (4.2.17), and Lemma II.1.2 of Appendix II that $T_n{}^*$ is a Hilbert–Schmidt-type operator. Hence, by Lemma II.1.3 of Appendix II, T_n is Hilbert–Schmidt.]

The following two cases are of greatest interest to us.

Corollary 4.2.19. Let $\Phi \subset H \subset \Phi^\dagger$ be a separable rigged Hilbert space, and let $\mathfrak{F}^\dagger$ denote the totality of weak Borel sets in $\Phi^\dagger$. If there exists a Φ-quasi-invariant regular finite measure μ on $(\Phi^\dagger, \mathfrak{F}^\dagger)$, then there must be a pair of natural numbers n, $m (n \geqslant m)$ such that the imbedding of Φ_n (i.e., the completion of Φ relative to the nth inner product) in $\Phi_m{}^\dagger$ is a Hilbert–Schmidt-type operator.

PROOF. Since $\Phi_m{}^\dagger \in \mathfrak{F}^\dagger$ (see Lemma 4.2.17), the sequence of sets $\{\Phi_m{}^\dagger\}$ is monotonic increasing, and $\Phi^\dagger = \bigcup_{m=1}^{\infty} \Phi_m{}^\dagger$, it follows that $\mu(\Phi_m{}^\dagger) \to \mu(\Phi^\dagger)$, hence, there must be an m such that $\mu(\Phi_m{}^\dagger) > 0$.

Restricting $\mathfrak{F}^\dagger$ and μ to $\Phi_m{}^\dagger$, we obtain a Φ-quasi-invariant measure space $(\Phi_m{}^\dagger, \mathfrak{F}_m{}^\dagger, \mu_m)$. Now, in Theorem 4.2.18, we take Φ for $\mathfrak{G}$ and $\Phi_m{}^\dagger$ for G, and the desired conclusion follows at once.[32]]

Corollary 4.2.20. Let G and $\mathfrak{G}$ be separable Hilbert spaces, where $\mathfrak{G}$ is a linear subspace of G such that the inclusion map T of $\mathfrak{G}$ into G is continuous. Let $\mathfrak{B}$ be the totality of weak Borel sets in G. If there exists a $\mathfrak{G}$-quasi-invariant finite measure μ on $(G, \mathfrak{B})$, then T is a Hilbert–Schmidt operator.

PROOF. This corollary is just a special case of Theorem 4.2.18.]

Example 4.2.2. Let $1 \leqslant p < \infty$, let $\{a_n\}$ be a sequence of positive numbers, and let $l^p(\{a_n\})$ denote the totality of real number sequences $\xi = \{\xi_n\}$ which satisfy the condition

$$\| \xi \| = \left(\sum a_n \mid \xi_n \mid^p\right)^{1/p} < \infty. \tag{4.2.18}$$

$l^p(\{a_n\})$ forms a Banach space with respect to the usual coordinatewise linear operations and the norm (4.2.18). In particular, we write l^p in place of $l^p(\{1\})$. Let $\mathfrak{B}$ be the σ-algebra in $l^p(\{a_n\})$ generated by the totality of Borel cylinders

$$\{\xi \mid (\xi_1, \xi_2, \ldots, \xi_n) \in B\}$$

(where B represents an arbitrary finite-dimensional Borel set). If

[32] *Translator's note:* The definition of rigged Hilbert space given in Appendix I does not include the completeness of Φ, which is required here.

$1 \leqslant q < \infty$, then the existence of an l^q-quasi-invariant[33] finite measure space $(l^p(\{a_n\}), \mathfrak{B}, \mu)$ implies that

$$\sum a_n < \infty. \tag{4.2.19}$$

In fact, since $\| \xi \| = \lim_{n\to\infty} (\sum_{\nu=1}^{n} a_\nu \mid \xi_\nu \mid^p)^{1/p}$ is a measurable function, we may suppose that

$$M = \int_{l^p(\{a_n\})} \| \xi \|^p \, d\mu(\xi) < \infty, \tag{4.2.20}$$

for otherwise, we could replace μ by the equivalent measure

$$\mu'(E) = \int_E \exp(-\| \xi \|^p) \, d\mu(E), \qquad E \in \mathfrak{B}.$$

For each n, consider the $\mathfrak{B}$-measurable linear functional

$$f_n(\xi) = \xi_n\,, \qquad \xi = \{\xi_n\} \in l^p(\{a_n\}).$$

By (4.2.12), there exists a neighborhood V of zero in l^q, and a positive number c, such that

$$\sup_{\xi\in V} | f_n(\xi)| \leqslant c \int_{l^p(\{a_n\})} | f_n(\xi)| \, d\mu(\xi), \qquad n = 1, 2,\ldots. \tag{4.2.12$''$}$$

We may suppose that $V = \{\xi \mid | \xi | < \delta\}$, $\delta > 0$, $| \xi | = (\sum_{n=1}^{\infty} | \xi_n |^q)^{1/q}$, so that (4.2.12″) becomes

$$1 \leqslant \frac{c}{\delta} \int_{l^p(\{a_n\})} | \xi_n | \, d\mu(\xi).$$

Using the Hölder inequality, we deduce that

$$1 \leqslant \left(\frac{c}{\delta}\right)^p \mu(l^p(\{a_n\}))^{p-1} \int_{l^p(\{a_n\})} | \xi_n |^p \, d\mu(\xi).$$

Multiplying both sides of the above inequality by a_n, and summing with respect to n, we get

$$\sum a_n \leqslant \left(\frac{c}{\delta}\right)^p \mu(l^p(\{a_n\}))^{p-1} M < \infty.$$

[33] *Translator's note:* More properly stated, quasi-invariant under the intersection of l^q and $l^p(\{a_n\})$. However, if (4.2.19) holds, then $l^q \subset l^p(\{a_n\})$.

6° The L^2-Fourier Transform

Naturally, the theory of L^2-Fourier transforms on groups, already discussed in §3.4, can also be transferred to the context of linear spaces; the relevant definitions and results require only slight modification. We do not propose to restate the entire theory in detail. Instead, we merely give the definition of the L^2-Fourier transform for linear measure spaces, leaving the rest to the reader. The results will subsequently be applied without further preliminary explanations.

Definition 4.2.9. Let $\Omega = (G, \mathfrak{B}, \mu)$ be a linear measure space which is localizable and quasi-invariant under a linear subspace $\mathfrak{G}$ of G. Let $\hat{\Omega} = (\hat{G}, \hat{\mathfrak{B}}, \hat{\mu})$ be a linear measure space which is quasi-invariant under a linear subspace $\hat{\mathfrak{G}}$ of $\hat{G}$, and suppose that the following two conditions are satisfied. (i) There exists an isomorphic mapping

$$\tilde{f} \to \hat{f}$$

from $\tilde{\mathfrak{G}}^{\mu}$ onto $\hat{\mathfrak{G}}$, and an isomorphic mapping

$$h \to h(\hat{g}), \qquad \hat{g} \in \hat{G},$$

from $\mathfrak{G}$ into $\hat{G}_{\hat{\mu}}$, such that $\hat{\epsilon} = \{h(\cdot) \mid h \in \mathfrak{G}\}$ is a determining set of functions on $\hat{\Omega}$, moreover, for each $h \in \mathfrak{G}$, the linear functional induced by $h(\cdot)$ on $\hat{\mathfrak{G}}$ is

$$\tilde{h}(\hat{f}) = \tilde{f}(h), \qquad \hat{f} \in \hat{\mathfrak{G}}.$$

(ii) There exists a unitary operator F from $L^2(\Omega)$ onto $\mathfrak{L}_k{}^2(\hat{\Omega})$ such that, for each $h \in \mathfrak{G}$, the corresponding operator $\hat{U}(h) = F(U(h))F^{-1}$ takes the form

$$(\hat{U}(h)\,\varphi)(\hat{g}) = \exp(ih(\hat{g}))\,\varphi(\hat{g}), \qquad \hat{g} \in \hat{G}, \quad \varphi \in \mathfrak{L}_k{}^2(\hat{\Omega}).$$

Then, we say that $(\hat{\Omega}, \hat{\mathfrak{G}})$ is a *dual* of $(\Omega, \mathfrak{G})$ (or that $\hat{\Omega}$ is a *dual measure space* of Ω), and that F is an associated *L^2-Fourier transform* [from $L^2(\Omega)$ onto $\mathfrak{L}_k{}^2(\hat{\Omega})$].

§4.3. Continuous Positive Definite Functions on Linear Topological Spaces

In this section, we shall, as before, consider only real linear spaces, and shall regard these spaces, in particular, as additive groups. Since any finite-dimensional T_0 linear topological space is topologically isomorphic to a Euclidean space of the same dimension, the Bochner theorem provides the general form for continuous positive definite functions on finite-dimensional linear topological spaces. However, for infinite-dimensional linear topological spaces, there still exists no comprehensive theorem concerning the representation of continuous positive definite functions. Nevertheless, there are a number of important results along this line. First, we shall consider the situation for linear spaces in general.

1° Continuous Positive Definite Functions on Linear Spaces. Cylinder Measures

The usual notion of measure is somewhat too strong for the formulation of certain preliminary results required in the following discussion. Accordingly, we begin by introducing a weaker concept, that is, cylinder measure.

Definition 4.3.1. Let $\mathfrak{G}$ be a linear space, let $\mathfrak{H}$ be a linear space consisting of certain linear functionals[34] on $\mathfrak{G}$, and let S be the algebra of all Borel cylinders in $\mathfrak{H}$. Suppose that P is a set function on S having the following property: if Φ is any finite-dimensional linear subspace of $\mathfrak{G}$, and $S(\Phi)$ is the σ-algebra of Borel cylinders corresponding to Φ, then the restriction of P to $S(\Phi)$ is a probability measure. Then, we call P a *cylinder measure*[35] on $\mathfrak{H}$.

Clearly, any cylinder measure P also has the following properties: (i) $P(z) \geqslant 0$ for all $z \in S$; (ii) $P(\mathfrak{H}) = 1$; (iii) P is finitely additive. In fact, (i) and (ii) are obvious from the definition, while (iii) may be seen as follows. If $z_1, \ldots, z_n$ are disjoint sets in S, then it is clear from the definition of a Borel cylinder that there exists a finite-dimensional subspace Φ of $\mathfrak{G}$ such that $z_1, \ldots, z_n \in S(\Phi)$, hence, by the additivity of P on $S(\Phi)$, we have

$$\mu(z_1 + \cdots + z_n) = \mu(z_1) + \cdots + \mu(z_n),$$

that is, property (iii). However, P is not, in general, countably additive.

But if it happens that P is countably additive, then, using a well-known technique (see, e.g., Halmos [1]), we can extend P to a probability measure (which we also denote by P) on the σ-algebra $\mathfrak{F}$ generated by S.

Now, let P be any cylinder measure on $\mathfrak{H}$. Let $\mathfrak{U}$ be a complex-valued function on $\mathfrak{H}$, and suppose there exists a finite-dimensional linear subspace Φ of $\mathfrak{G}$ such that $\mathfrak{U}$ is integrable on the probability measure

[34] With the usual linear operations.

[35] Notice that this is, in fact, a special case of the cylinder measure concept introduced in §1.3.

space $(\mathfrak{H}, S(\Phi), P)$.[36] We then say that $\mathfrak{U}$ is *integrable with respect to the cylinder measure P*. In this case, the integral of $\mathfrak{U}$ over $(\mathfrak{H}, S(\Phi), P)$ is clearly independent of the choice of Φ, and will be called the *integral of* $\mathfrak{U}$ *with respect to the cylinder measure P*; we denote it by

$$\int_{\mathfrak{H}} \mathfrak{U}(\xi)\, dP(\xi).$$

In particular, when P is countably additive, this integral coincides with the ordinary integral of $\mathfrak{U}$ over the measure space $(\mathfrak{H}, \mathfrak{F}, P)$.

Definition 4.3.2. Let $\mathfrak{G}$ be a linear space and let f be a complex-valued function on $\mathfrak{G}$ such that, for every finite set $\varphi_1, ..., \varphi_n \in \mathfrak{G}$, the value $f(t_1\varphi_1 + \cdots + t_n\varphi_n)$ is a continuous function of the n real variables $t_1, ..., t_n$, that is, f is continuous on every finite-dimensional linear subspace[37] of $\mathfrak{G}$. We then say that f is *pseudocontinuous* on $\mathfrak{G}$.

Lemma 4.3.1. Let $\mathfrak{G}$ be a linear space, let $\mathfrak{H}$ be a linear space consisting of certain real linear functionals on $\mathfrak{G}$, and let P be a cylinder measure on $\mathfrak{H}$. Then, the integral

$$f(g) = \int_{\mathfrak{H}} e^{i\xi(g)}\, dP(\xi), \qquad g \in \mathfrak{G}, \tag{4.3.1}$$

is a pseudocontinuous positive definite function on $\mathfrak{G}$, and $f(0) = 1$.

PROOF. $P(\mathfrak{H}) = 1$, hence $f(0) = 1$. Furthermore, since

$$\sum_{j,k} f(g_j - g_k)\, \eta_j \bar{\eta}_k = \int_{\mathfrak{H}} \left| \sum_j \exp(i\xi(g_j))\, \eta_j \right|^2 dP(\xi) \geqslant 0,$$

f is positive definite. Finally, given any $\varphi_1, ..., \varphi_n \in \mathfrak{G}$, we observe that

$$f\left(\sum_{j=1}^n t_j\varphi_j\right) = \int_{\mathfrak{H}} \exp\left(i \sum_{j=1}^n t_j\xi(\varphi_j)\right) dP(\xi), \tag{4.3.2}$$

hence, applying the Lebesgue dominated convergence theorem to the probability measure space $(\mathfrak{H}, S(\varphi_1, ..., \varphi_n), P)$, we see that (4.3.2) is a continuous function of the real variables $t_1, ..., t_n$.]

If $\mathfrak{H}$ contains "sufficiently many" functionals, then the converse of Lemma 4.3.1 also holds, that is, we have the following result.

[36] Here, P means the restriction of P to $S(\Phi)$.

[37] Normally, it is always assumed that finite-dimensional linear spaces are given the Euclidean topology.

Lemma 4.3.2. Let $\mathfrak{G}$ be a linear space and let $\mathfrak{H}$ be a linear space consisting of certain real linear functionals on $\mathfrak{G}$, moreover, suppose that $\mathfrak{H}$ distinguishes the points of $\mathfrak{G}$. Then, given any pseudocontinuous positive definite function f on $\mathfrak{G}$ [with $f(0) = 1$], there exists a unique cylinder measure P on $\mathfrak{H}$ such that

$$f(g) = \int_{\mathfrak{H}} e^{i\xi(g)}\, dP(\xi), \qquad g \in \mathfrak{G}. \tag{4.3.3}$$

PROOF. Take any linearly independent finite set of vectors $\varphi_1, \ldots, \varphi_n$ in $\mathfrak{G}$, and consider the value $f(t_1\varphi_1 + \cdots + t_n\varphi_n)$, regarded as a function of the real variables $t_1, \ldots, t_n$. Since f is positive definite and pseudocontinuous, and $f(0) = 1$, it is easily verified that $f(t_1\varphi_1 + \cdots + t_n\varphi_n)$ satisfies the conditions of the Bochner–Khinchin theorem, hence, there exists a unique probability measure $Q_{\{\varphi_j\}}$ on real n-dimensional space R_n, such that

$$f(t_1\varphi_1 + \cdots + t_n\varphi_n) = \int \cdots \int \exp i(t_1\xi_1 + \cdots + t_n\xi_n)\, dQ_{\{\varphi_j\}}(\xi_1, \ldots, \xi_n).$$

Consider the correspondence $T_{\{\varphi_j\}} : \xi \to (\xi(\varphi_1), \ldots, \xi(\varphi_n))$, $\xi \in \mathfrak{H}$. Since $\mathfrak{H}$ distinguishes the points of Φ, the image of $T_{\{\varphi_j\}}$ is the entire space R_n. Define a set function $P_{\{\varphi_j\}}$ on $S(\{\varphi_j\})$ as follows: if A is any Borel set in R_n, let

$$P_{\{\varphi_j\}}(T^{-1}_{\{\varphi_j\}}A) = Q_{\{\varphi_j\}}(A). \tag{4.3.4}$$

Clearly, $(\mathfrak{H}, S(\{\varphi_j\}), P_{\{\varphi_j\}})$ is a probability measure space. Moreover, a straightforward calculation shows that $P_{\{\varphi_j\}}$ is independent of the particular choice of the basis $\varphi_1, \ldots, \varphi_n$ for the finite-dimensional linear subspace under consideration.

We now proceed to prove that the family of probability measure spaces constructed in this manner is consistent, that is, if $B \in S(\{\psi_j\}) \cap S(\{\omega_j\})$, $\psi_1, \ldots, \psi_m, \omega_1, \ldots, \omega_k \in \mathfrak{G}$, then

$$P_{\{\psi_j\}}(B) = P_{\{\omega_j\}}(B). \tag{4.3.5}$$

Choose a linearly independent set $\varphi_1, \ldots, \varphi_n \in \mathfrak{G}$, such that

$$S(\{\psi_j\}) \cup S(\{\omega_j\}) \subset S(\{\varphi_j\}).$$

We need only prove that

$$P_{\{\varphi_j\}}(B) = P_{\{\psi_j\}}(B), \tag{4.3.6}$$

for we may then deduce by the same reasoning that $P_{\{\varphi_j\}}(B) = P_{\{\omega_j\}}(B)$, thereby obtaining (4.3.5). Obviously, we may assume that $\varphi_j = \psi_j$, $j = 1, 2, \ldots, m$, $n \geqslant m$. Since

$$\int \cdots \int \exp i(t_1\xi_1 + \cdots + t_m\xi_m)\, dQ_{\{\psi_j\}}(\xi_1, \ldots, \xi_m)$$

$$= f(t_1\psi_1 + \cdots + t_m\psi_m)$$

$$= f(t_1\varphi_1 + \cdots + t_m\varphi_m + 0\varphi_{m+1} + \cdots + 0\varphi_n)$$

$$= \int \cdots \int \exp i(t_1\xi_1 + \cdots + t_m\xi_m)\, dQ_{\{\varphi_j\}}(\xi_1, \ldots, \xi_n),$$

it follows that

$$Q_{\{\varphi_j\}}(\{(\xi_1, \ldots, \xi_n) \mid (\xi_1, \ldots, \xi_m) \in D\}) = Q_{\{\psi_j\}}(D)$$

for every Borel set D in R_m, that is,

$$P_{\{\varphi_j\}}(\{\xi \mid (\xi(\varphi_1), \ldots, \xi(\varphi_m)) \in D\}) = P_{\{\psi_j\}}(\{\xi \mid (\xi(\psi_1), \ldots, \xi(\psi_m)) \in D\}). \quad (4.3.7)$$

But every set B in $S_{(\psi_j)}$ is of the form $\{\xi \mid (\xi(\psi_1), \ldots, \xi(\psi_m)) \in D\}$, hence (4.3.7) is equivalent to (4.3.6). Thus, we have shown that the family of measures $\{P_{(\varphi_j)}\}$ is consistent.

Define a set function P on S, as follows: if $A \in S(\{\varphi_j\})$, let

$$P(A) = P_{\{\varphi_j\}}(A).$$

Using the consistency of the family of measure spaces $(\mathfrak{H}, S(\{\varphi_j\}), P_{\{\varphi_j\}})$, $\{\varphi_j\} \subset \mathfrak{G}$, and following the reasoning of §1.3, we see that P is unambiguously defined. It is then obvious that P is a cylinder measure.

Next, we verify that P satisfies (4.3.3). If $g = 0$, then (4.3.3) is obvious. If $g \neq 0$, then

$$f(g) = \int e^{iu}\, dQ_{\{g\}}(u) = \int e^{i\xi(g)}\, dP_{\{g\}}(\xi) = \int e^{i\xi(g)}\, dP(\xi).$$

Finally, the uniqueness of the cylinder measure in (4.3.3) may be seen as follows. The values of P on each $S(\{\varphi_j\})$, $\varphi_1, \ldots, \varphi_n \in \mathfrak{G}$, are determined by the measure $Q_{(\varphi_j)}$ in (4.3.4), and by the Bochner–Khinchin theorem, $Q_{(\varphi_j)}$ is uniquely determined by f.]

2° Continuity of Cylinder Measures

In analogy with Definition 3.2.6, we shall now introduce a notion of continuity for cylinder measures.[38]

[38] In particular, applicable to probability measures on σ-algebras of the type described in Definition 4.2.8.

Definition 4.3.3. Let $\mathfrak{G}$ be a linear topological space with topology $\mathfrak{T}$. Let $\mathfrak{H}$ be a linear space consisting of certain linear functionals on $\mathfrak{G}$, and let P be a cylinder measure on $\mathfrak{H}$. Suppose that, given any positive number ϵ, there exists a neighborhood V of zero in $\mathfrak{G}$ such that

$$P(\{\xi \mid |\xi(x)| > 1, \xi \in \mathfrak{H}\}) < \epsilon$$

whenever $x \in V$. Then, we say that P is *continuous with respect to the topology* $\mathfrak{T}$ (or simply *continuous*).

Next, we establish the connection between continuity of positive definite functions and continuity of cylinder measures (cf. §3.3).

Lemma 4.3.3. Let $\mathfrak{G}$ be a linear topological space, $\mathfrak{H}$ a linear space consisting of certain real linear functionals on $\mathfrak{G}$, and P a cylinder measure on $\mathfrak{H}$. Then, the function

$$f(g) = \int_{\mathfrak{H}} e^{i\xi(g)}\, dP(\xi), \qquad g \in \mathfrak{G}, \tag{4.3.8}$$

is continuous if and only if P is continuous.

PROOF. Assume that f is continuous. Then, given any positive number ϵ, there exists a neighborhood U of zero in $\mathfrak{G}$, such that

$$1 - \Re f(g) < \frac{\epsilon}{4} \tag{4.3.9}$$

for all $g \in U$ [since $f(0) = 1$]. Choose a positive number a such that $e^{-a} < \epsilon/8$. Since $\mathfrak{G}$ is a linear topological space, there is a neighborhood V of zero such that $tV \subset U$ whenever $|t| \leqslant a$. Thus, if $g \in V$, then, from (4.3.9), we get

$$\int_0^\infty (1 - \Re f(gt))e^{-t}\, dt < \frac{\epsilon}{4} + \int_a^\infty 2e^{-t}\, dt < \frac{\epsilon}{2}. \tag{4.3.10}$$

However, using Fubini's theorem, we have

$$\begin{aligned} \int_0^\infty (1 - \Re f(gt))e^{-t}\, dt &= \int_{\mathfrak{H}} \int_0^\infty (1 - \cos t\xi(g))\, e^{-t}\, dt\, dP(\xi) \\ &= \int_{\mathfrak{H}} \frac{\xi(g)^2}{1 + \xi(g)^2}\, dP(\xi). \end{aligned} \tag{4.3.11}$$

Combining (4.3.10) and (4.3.11), we deduce that

$$P(\{\xi \mid |\xi(g)| > 1\}) \leqslant 2 \int_{\mathfrak{H}} \frac{\xi(g)^2}{1 + \xi(g)^2}\, dP(\xi) < \epsilon$$

whenever $g \in V$. This shows that P is continuous.

Conversely, assume that P is continuous. Then, given any positive number ϵ, there exists a neighborhood V of zero in $\mathfrak{G}$ such that

$$P(\{\xi \mid |(x)| > 1\}) < \frac{\epsilon}{4}$$

whenever $x \in V$. Choose a positive number $\delta < \pi/2$ such that $1 - \cos\delta < \epsilon/2$, and let $U = \delta V$. Obviously, U is a neighborhood of zero, and if $g \in U$, then $x = (1/\delta)\, g \in V$, hence

$$P(\{\xi \mid |\xi(g)| > \delta\}) = P(\{\xi \mid |\xi(x)| > 1\}) < \frac{\epsilon}{4}.$$

Therefore, if $g \in U$, then

$$\begin{aligned} 1 - \Re f(g) &= \int_{\mathfrak{H}} (1 - \cos \xi(g))\, dP(\xi) \\ &\leqslant (1 - \cos\delta) + 2P(\{\xi \mid |\xi(g)| > \delta\}) < \epsilon. \end{aligned}$$

This proves that $\Re f(g)$ is continuous at $g = 0$. But by Lemma 4.3.1, f is positive definite, hence, it follows by inequality (3.3.3) that f is continuous.]

The following theorem is entirely analogous to Lemma 3.2.19.

Theorem 4.3.4. Let $\mathfrak{G}$ be a linear topological space satisfying the first axiom of countability, $\mathfrak{H}$ a linear space consisting of certain continuous linear functionals on $\mathfrak{G}$, and $\mathfrak{F}$ the smallest σ-algebra containing all the Borel cylinders in $\mathfrak{H}$. Then, every probability measure P on $(\mathfrak{H}, \mathfrak{F})$ is continuous.

The proof is left to the reader. The conclusion of Theorem 4.3.4 may not be valid if $\mathfrak{G}$ does not satisfy the first axiom of countability (see Xia Dao-xing [3]).

3° Representation of Pseudocontinuous Positive Definite Functions as Fourier Transforms of Measures

Let $\mathfrak{G}$ be a linear space, $\mathfrak{H}$ a linear space consisting of certain linear functionals on $\mathfrak{G}$, and $\mathfrak{F}$ the smallest σ-algebra containing all the Borel cylinders in $\mathfrak{H}$. In Lemma 4.3.2, we have already obtained a result concerning the representation of pseudocontinuous positive definite functions as Fourier transforms of cylinder measures. Now, we proceed to investigate, under what circumstances a positive definite function f can be expressed as an integral

$$f(g) = \int_{\mathfrak{H}} e^{i\xi(g)}\, dP(\xi), \qquad g \in \mathfrak{G}, \tag{4.3.1}$$

with respect to a measure P on $(\mathfrak{H}, \mathfrak{F})$. By virtue of Lemma 4.3.1, we know that the pseudocontinuity of f is a necessary condition for the existence of a representation of the form (4.3.1). It can also be proved that[39] if (4.3.1) holds, then $\mathfrak{H}$ distinguishes the points of $\mathfrak{G}$. However, these two necessary conditions, taken together, still do not constitute a sufficient condition (see §5.3). But we do have the following result.

Theorem 4.3.5. Let $\mathfrak{G}$ be a linear space, $\mathfrak{G}^{\Delta}$ the totality of real linear functionals on $\mathfrak{G}$, and $\mathfrak{F}^{\Delta}$ the smallest σ-algebra containing all the Borel cylinders in $\mathfrak{G}^{\Delta}$. Then, for any pseudocontinuous positive definite function f on $\mathfrak{G}$ [with $f(0) = 1$], there exists a unique probability measure P^{Δ} on $(\mathfrak{G}^{\Delta}, \mathfrak{F}^{\Delta})$ such that

$$f(g) = \int_{\mathfrak{G}^{\Delta}} e^{i\xi(g)}\, dP^{\Delta}(\xi) \tag{4.3.12}$$

for all $g \in \mathfrak{G}$.

Proof. Choose a linear basis $\{\omega_\alpha \mid \alpha \in \mathfrak{A}\}$ for $\mathfrak{G}$. For each finite set of indices $\alpha_1, \ldots, \alpha_n \in \mathfrak{A}$, we form a function of n real variables $t_{\alpha_1}, \ldots, t_{\alpha_n}$, as follows:

$$f_{\{\alpha_j\}}(t_{\alpha_1}, \ldots, t_{\alpha_n}) = f(t_{\alpha_1}\omega_{\alpha_1} + \cdots + t_{\alpha_n}\omega_{\alpha_n}). \tag{4.3.13}$$

Since f is pseudocontinuous, positive definite, and satisfies $f(0) = 1$, it follows at once that $f_{(\alpha_j)}$ is continuous, positive definite, and satisfies $f_{\{\alpha_j\}}(0, \ldots, 0) = 1$. Moreover, it is easily verified that the family $\{f_{\{\alpha_j\}}\}$ of all functions constructed in this manner is consistent. For each $\alpha \in \mathfrak{A}$, let R_α be a copy of the real line, and let $\mathfrak{F}_\alpha$ denote the totality of Borel sets in R_α. By the Kolmogorov theorem (Corollary 1.3.5′), there exists a unique probability measure P^{Δ} on $(\times_{\alpha\in\mathfrak{A}} R_\alpha, \times_{\alpha\in\mathfrak{A}} \mathfrak{F}_\alpha)$ such that, for any finite set of indices $\alpha_1, \ldots, \alpha_n \in \mathfrak{A}$,

$$\begin{aligned} f(t_{\alpha_1}\omega_{\alpha_1} + \cdots + t_{\alpha_n}\omega_{\alpha_n}) &= f_{\{\alpha_j\}}(t_{\alpha_1}, \ldots, t_{\alpha_n}) \\ &= \int_{\times R_\alpha} \exp\left(i \sum_{j=1}^{n} t_{\alpha_j} x_{\alpha_j}\right) dP^{\Delta}(x), \end{aligned} \tag{4.3.14}$$

where x_α denotes the αth coordinate of the point x in $\times_{\alpha\in\mathfrak{A}} R_\alpha$.

Now, for each $x \in \times_{\alpha\in\mathfrak{A}} R_\alpha$, define a functional $\hat{x}$ on $\mathfrak{G}$, as follows:

$$\hat{x}(t_{\alpha_1}\omega_{\alpha_1} + \cdots + t_{\alpha_n}\omega_{\alpha_n}) = t_{\alpha_1}x_{\alpha_1} + \cdots + t_{\alpha_n}x_{\alpha_n}.$$

[39] *Translator's note:* This remark as it stands is obviously false. Perhaps the author actually meant that if representation (4.3.1) is possible for *all* members of a certain class of positive definite functions on $\mathfrak{G}$, then $\mathfrak{H}$ must distinguish the points of $\mathfrak{G}$.

Clearly, $\hat{x}$ is a linear functional on $\mathfrak{G}$, and the correspondence $x \to \hat{x}$ is a one-to-one mapping of $\times_{\alpha \in \mathfrak{A}} R_\alpha$ onto $\mathfrak{G}^A$. Thus, if we identify x and $\hat{x}$, (4.3.14) becomes (4.3.12). Moreover, the uniqueness of P^A in (4.3.12) follows immediately from the uniqueness of P^A in (4.3.14).]

If $\mathfrak{H}$ is a linear subspace of $\mathfrak{G}^A$ which distinguishes the points of $\mathfrak{G}$, then the conclusion of Lemma 4.3.2 can be deduced from Theorem 4.3.5. In fact, if B is a Borel cylinder in $\mathfrak{G}^A$, then $B \cap \mathfrak{H}$ is a Borel cylinder in $\mathfrak{H}$, and one may verify that the set function

$$P(B \cap \mathfrak{H}) = P^A(B)$$

is a cylinder measure which fulfills the requirements of Lemma 4.3.2.

The following theorem may be regarded as an alternative form of the Kolmogorov theorem.

Theorem 4.3.5′. Let $\mathfrak{G}$ be a linear space and $\mathfrak{G}^A$ the totality of real linear functionals on $\mathfrak{G}$. Then, every cylinder measure on $\mathfrak{G}^A$ is countably additive.

PROOF. Let P be a cylinder measure on $\mathfrak{G}^A$. Taking $\mathfrak{H} = \mathfrak{G}^A$ in formula (4.3.8), we define a corresponding pseudocontinuous positive definite function f. By Theorem 4.3.5, there exists a probability measure P^A on $(\mathfrak{G}^A, \mathfrak{F}^A)$ such that (4.3.12) holds for this f. Of course, P^A is also a cylinder measure on $\mathfrak{G}^A$. Applying Lemma 4.3.2 (with $\mathfrak{H} = \mathfrak{G}^A$), it follows from the uniqueness property that $P(B) = P^A(B)$ for every Borel cylinder B. Therefore, since P^A is countably additive, the cylinder measure P is also countably additive.]

Incidentally, notice that the measure P^A in the above proof is the unique extension of P to $\mathfrak{F}^A$.

If $\mathfrak{H} \neq \mathfrak{G}^A$, then cylinder measures on $\mathfrak{H}$ are not, in general, countably additive. However, we are primarily interested in countably additive cylinder measures, since they can be extended to a larger family of sets (see §4.2, 4°), and provide a more effective analytical tool. We shall conclude this subsection by deriving some conditions for the countable additivity of continuous cylinder measures.

The countable additivity of continuous cylinder measures is intimately connected with the notion of a spectral linear space, which we now proceed to define.

Definition 4.3.4. Let $(\mathfrak{G}, \mathfrak{T})$ be a linear topological space, $\mathfrak{H}$ a linear subspace of $\mathfrak{G}^A$, and $\mathfrak{F}$ the totality of weak Borel sets in $\mathfrak{H}$. Then, $\mathfrak{H}$ is called a *spectral* (linear) *space* for $(\mathfrak{G}, \mathfrak{T})$, provided that, for each continuous positive definite function f on $(\mathfrak{G}, \mathfrak{T})$ [with $f(0) = 1$], there exists a unique probability measure P on $(\mathfrak{H}, \mathfrak{F})$, such that

$$f(h) = \int_{\mathfrak{H}} e^{i\xi(h)}\, dP(\xi)$$

for all $h \in \mathfrak{G}$.

Lemma 4.3.6. Let $(\mathfrak{G}, \mathfrak{T})$ be a linear topological space, and let $\mathfrak{H}$ be a linear subspace of $\mathfrak{G}^{\Delta}$ which distinguishes the points of $\mathfrak{G}$. Then $\mathfrak{H}$ is a spectral space for $(\mathfrak{G}, \mathfrak{T})$ if and only if every continuous cylinder measure on $\mathfrak{H}$ is countably additive.

PROOF. Assume that every continuous cylinder measure on $\mathfrak{H}$ is countably additive. Now, according to Lemmas 4.3.2 and 4.3.3, for each continuous positive definite function f on $(\mathfrak{G}, \mathfrak{T})$ [with $f(0) = 1$], there exists a cylinder measure P on $\mathfrak{H}$ which is continuous with respect to $\mathfrak{T}$ and such that (4.3.3) holds. By our assumption, P is countably additive and therefore defines a probability measure which also satisfies relation (4.3.3) for the given function f. Thus, $\mathfrak{H}$ is a spectral space for $(\mathfrak{G}, \mathfrak{T})$.

Conversely, assume that $\mathfrak{H}$ is a spectral space for $(\mathfrak{G}, \mathfrak{T})$. Given any continuous cylinder measure P on $\mathfrak{H}$, we form the corresponding pseudocontinuous positive definite function f in accordance with formula (4.3.1). By Lemma 4.3.3, f is continuous, hence, by our assumption, there exists a probability measure P' on $(\mathfrak{H}, \mathfrak{F})$ such that

$$f(g) = \int_{\mathfrak{H}} e^{i\xi(g)}\, dP'(\xi), \qquad g \in \mathfrak{G}.$$

By the uniqueness property of the cylinder measure in Lemma 4.3.2, it follows that $P'(B) = P(B)$ for every Borel cylinder B in $\mathfrak{H}$. Since P' is countably additive, P is also countably additive.]

Lemma 4.3.6 clarifies our earlier remark concerning the intimate connection between spectral spaces and countable additivity of continuous cylinder measures.

If $\mathfrak{G}$ is a linear space and $\mathfrak{H} \subset \mathfrak{G}^{\Delta}$, then, by a *weakly open cylinder*, we mean a set of the form

$$\{f \mid (f(\varphi_1),\dots, f(\varphi_n)) \in V, f \in \mathfrak{H}\},$$

where $\varphi_1, \dots, \varphi_n \in \mathfrak{G}$ and V is an open subset of R_n. Using this definition, we introduce a sort of regularity property for cylinder measures.

Lemma 4.3.7. Let $\mathfrak{G}$ be a linear space, $\mathfrak{H}$ a linear subspace of $\mathfrak{G}^{\Delta}$, and P a cylinder measure on $\mathfrak{H}$. Given any Borel cylinder A in $\mathfrak{H}$, and any positive number ϵ, there exists a weakly open cylinder U such that $U \supset A$ and

$$P(A) > P(U) - \epsilon.$$

PROOF. Let $A \in S(\Phi)$, where Φ is an m-dimensional linear subspace of $\mathfrak{G}$. Choose a basis $\varphi_1, \ldots, \varphi_m$ for Φ, and let $\mathfrak{F}$ denote the totality of Borel subsets of R_m. For any $B \in \mathfrak{F}$, let

$$\tilde{B} = \{f \mid (f(\varphi_1), \ldots, f(\varphi_m)) \in B, \quad f \in \mathfrak{H}\},$$

and define a set function Q on $\mathfrak{F}$ as follows:

$$Q(B) = P(\tilde{B}), \qquad B \in \mathfrak{F}. \tag{4.3.15}$$

Since $(\mathfrak{H}, S(\Phi), P)$ is a probability measure space, it is easily verified that $(R_m, \mathfrak{F}, Q)$ is also a probability measure space. Therefore, if $D \in \mathfrak{F}$ and $\tilde{D} = A$, then there exists an open set V in R_m such that $V \supset D$ and

$$Q(D) > Q(V) - \epsilon. \tag{4.3.16}$$

Thus, $\tilde{V}$ is a weakly open cylinder, $\tilde{V} \supset \tilde{D} = A$, and, by (4.3.15) and (4.3.16), we have

$$P(A) > P(\tilde{V}) - \epsilon. \quad]$$

Lemma 4.3.8. Let $\mathfrak{G}$ be a linear space, $\mathfrak{H}$ a linear subspace of $\mathfrak{G}^A$, and P a cylinder measure on $\mathfrak{H}$. Suppose that, given any positive number ϵ, there exists a weakly compact set C_ϵ in $\mathfrak{H}$, such that

$$P(\mathfrak{L}) < \epsilon \tag{4.3.17}$$

for every Borel cylinder $\mathfrak{L}$ which does not intersect C_ϵ. Then, P is countably additive.

PROOF. We shall first prove that, if $Z_1, \ldots, Z_k, \ldots$ is any sequence of Borel cylinders in $\mathfrak{H}$ such that $\bigcup_{k=1}^{\infty} Z_k = \mathfrak{H}$, then

$$\sum_{k=1}^{\infty} P(Z_k) \geqslant 1. \tag{4.3.18}$$

Given any positive number ϵ, we know by Lemma 4.3.7 that, for each k, there is a weakly open cylinder U_k such that $U_k \supset Z_k$ and

$$P(Z_k) > P(U_k) - \frac{\epsilon}{2^k}.$$

Since $\bigcup_{k=1}^{\infty} U_k \supset \bigcup_{k=1}^{\infty} Z_k = \mathfrak{H} \supset C_\epsilon$, it follows from the weak compactness of C_ϵ that $\bigcup_{k=1}^m U_k \supset C_\epsilon$ for some m. Thus, the Borel cylinder $\mathfrak{H} - \bigcup_{k=1}^m U_k$ does not intersect C_ϵ. Therefore, by (4.3.17), we have

$$P\left(\mathfrak{H} - \bigcup_{k=1}^{m} U_k\right) < \epsilon,$$

and so, by the finite additivity of P,

$$\sum_{k=1}^{m} P(U_k) \geqslant P\left(\bigcup_{k=1}^{m} U_k\right) > 1 - \epsilon.$$

Hence,

$$\sum_{k=1}^{\infty} P(Z_k) \geqslant \sum_{k=1}^{m} P(Z_k) > \sum_{k=1}^{m} \left(P(U_k) - \frac{\epsilon}{2^k}\right) > 1 - 2\epsilon.$$

Letting $\epsilon \to 0$, we obtain (4.3.18).

Now, let $\{E_n\}$ be any disjoint sequence of Borel cylinders in $\mathfrak{H}$, such that $\bigcup_{n=1}^\infty E_n = E$ is also a Borel cylinder. Let $E_0 = \mathfrak{H} - E$; then E_0 is also a Borel cylinder, and $\bigcup_{n=0}^\infty E_n = \mathfrak{H}$. By (4.3.18), we have

$$\sum_{n=0}^{\infty} P(E_n) \geqslant 1.$$

Hence, by the finite additivity of P, we get

$$P(E) = P(\mathfrak{H}) - P(E_0) \leqslant \sum_{n=1}^{\infty} P(E_n).$$

Again, by the finite additivity of P,

$$\sum_{n=1}^{N} P(E_n) + P\left(E - \bigcup_{n=1}^{N} E_n\right) = P(E),$$

hence,

$$P(E) \geqslant \sum_{n=1}^{N} P(E_n).$$

Letting $N \to \infty$, we obtain

$$P(E) \geqslant \sum_{n=1}^{\infty} P(E_n).$$

Therefore,

$$P(E) = \sum_{n=1}^{\infty} P(E_n),$$

that is, P is countably additive.]

The following special case of Lemma 4.3.8 will be used in the ensuing discussion.

Corollary 4.3.9. Let $\mathfrak{G}$ be a separable countably normed space, $\mathfrak{G}^\dagger$ the conjugate space of $\mathfrak{G}$, and P a cylinder measure on $\mathfrak{G}^\dagger$. Let $\| \xi \|_{-n}$ denote the minus nth norm in $\mathfrak{G}^\dagger$, and let $S_n(R) = \{\xi \mid \| \xi \|_{-n} \leqslant R\}$ (see Appendix I). Suppose that, given any positive number ϵ, there exist n and R such that

$$P(\mathfrak{L}) < \epsilon$$

for every Borel cylinder $\mathfrak{L}$ which does not intersect $S_n(R)$. Then P is countably additive.

PROOF. According to a proposition in the theory of linear topological spaces, $S_n(R)(R < \infty)$ is weakly compact. Hence, the conclusion follows immediately from Lemma 4.3.8.]

4° Representation of Continuous Positive Definite Functions on Linear Topological Spaces

Throughout the present subsection, $\mathfrak{G}$ will denote a linear topological space, $\mathfrak{G}^\dagger$ the conjugate space of $\mathfrak{G}$, and $\mathfrak{F}^\dagger$ the smallest σ-algebra containing all Borel cylinders in $\mathfrak{G}^\dagger$.

First, we remark that an arbitrary continuous positive definite function f on $\mathfrak{G}$ [with $f(0) = 1$] cannot, in general, be expressed as a Fourier transform

$$f(g) = \int_{\mathfrak{G}^\dagger} e^{i\xi(g)}\, dP^\dagger(\xi), \qquad g \in \mathfrak{G}, \tag{4.3.19}$$

where $P^\dagger$ is a probability measure on $(\mathfrak{G}^\dagger, \mathfrak{F}^\dagger)$ (see §5.2). Therefore, as in §3.3, we shall impose additional conditions on the positive definite function f. We begin by bringing Theorem 3.3.7 into the present context.

Theorem 4.3.10. Let G be a linear topological space, and let $\mathfrak{B}$ be a σ-algebra consisting of certain subsets of G, including all the closed subsets of G. Suppose there exists a locally finite regular measure μ on $(G, \mathfrak{B})$. Then, given any continuous positive definite function f on G

[with $f(0) = 1$], there is a probability measure $P_{\mathfrak{B}}$ on $(G_{\mathfrak{B}}, \mathfrak{F}_{\mathfrak{B}})$, such that

$$f(g) = \int_{G_{\mathfrak{B}}} e^{ix(g)}\, dP_{\mathfrak{B}}(x)$$

for all $g \in G$.

The proof of Theorem 4.3.10 is similar to that of Theorem 4.3.11, and is therefore omitted.

The analog of Theorem 3.3.8 is as follows.

Theorem 4.3.11. Let G be a linear topological space, and let $\mathfrak{G}$ be a linear subspace of G which is itself a linear topological space of the second category with respect to a topology $\mathfrak{T}$ which is stronger than the topology induced by G on $\mathfrak{G}$. Let $\mathfrak{B}$ be the σ-algebra generated by the totality of closed subsets of G, and suppose there exists a strongly $\mathfrak{G}$-quasi-invariant locally finite regular measure μ on $(G, \mathfrak{B})$. Then, for each continuous positive definite function f on G [with $f(0) = 1$], there is a unique probability measure $P^\dagger$ on $(\mathfrak{G}^\dagger, \mathfrak{F}^\dagger)$, such that

$$f(g) = \int_{\mathfrak{G}^\dagger} e^{i\xi(g)}\, dP^\dagger(\xi), \qquad g \in \mathfrak{G}. \tag{4.3.19'}$$

Moreover, $P^\dagger$ is continuous with respect to the topology induced by G on $\mathfrak{G}$.

PROOF. According to Theorem 4.2.2, the mapping

$$\Lambda\colon \xi \to \hat{\xi} = e^{i\xi}$$

is an isomorphism of $\mathfrak{G}^\dagger$ onto $\mathfrak{G}^*$. Moreover, Λ clearly maps each set in $\mathfrak{F}^\dagger$ into a set in $\mathfrak{F}^*$ (the totality of weak Borel sets in $\mathfrak{G}^*$), and Λ^{-1} maps each set in $\mathfrak{F}^*$ into a set in $\mathfrak{F}^\dagger$. By Theorem 3.3.8, there is a probability measure P^* on $(\mathfrak{G}^*, \mathfrak{F}^*)$ such that the given function f is expressed in the form (3.3.41). Define

$$P^\dagger(E) = P^*(\Lambda E), \qquad E \in \mathfrak{F}^\dagger.$$

Obviously, $P^\dagger$ is a probability measure on $(\mathfrak{G}^\dagger, \mathfrak{F}^\dagger)$, and (3.3.41) is equivalent to (4.3.19′) via the measurable transformation Λ. The uniqueness of $P^\dagger$ is easily deduced from the uniqueness of P^*. Finally, since f is continuous on G, the required continuity property of $P^\dagger$ follows by Lemma 4.3.3.]

We shall be especially concerned with the case where $\mathfrak{G}$ is a complete linear metric space, and $(G, \mathfrak{B}, \mu)$ is a $\mathfrak{G}$-quasi-invariant regular finite measure space.

Let $\mathfrak{G}$ be a (real) countably Hilbert space, with the sequence of inner products $(\varphi, \psi)_n$, $n = 1, 2, \ldots$, and let $\mathfrak{G}_n$ denote the completion of $\mathfrak{G}$ with respect to the norm $\| \varphi \|_n = ((\varphi, \varphi)_n)^{1/2}$. We introduce a new topology on $\mathfrak{G}$, as follows. Let n be any natural number, let T be any positive nuclear operator on $\mathfrak{G}_n$, and let

$$U(T, n) = \{\varphi \mid (T\varphi, \varphi)_n < 1, \varphi \in \mathfrak{G}\}.$$

We take the totality of sets of the form $U(T, n)$ as a neighborhood basis at zero, and so obtain a topology $\mathfrak{S}$ on $\mathfrak{G}$. The topology $\mathfrak{S}$ is clearly weaker than the original topology of $\mathfrak{G}$, moreover, $(\mathfrak{G}, \mathfrak{S})$ is a linear topological space.[40]

Theorem 4.3.12. Let $\mathfrak{G}$ be a real separable countably Hilbert space, and let f be a positive definite function on $\mathfrak{G}$, with $f(0) = 1$. If f is continuous with respect to the topology $\mathfrak{S}$, then there exists a unique probability measure $P^\dagger$ on $(\mathfrak{G}^\dagger, \mathfrak{F}^\dagger)$, such that f is expressible in the form (4.3.19), moreover, this measure $P^\dagger$ is continuous with respect to $\mathfrak{S}$.

The proof of this theorem depends upon Lemma 4.3.13 below. We shall employ the following notation. Let R_m be an m-dimensional real inner product space, let $e_1, \ldots, e_m$ be an orthonormal basis in R_m, and if $x \in R_m$, let $x_l = (x, e_l)$. Let $\mathfrak{B}$ be the totality of Borel sets in R_m, let $(R_m, \mathfrak{B}, Q)$ be a probability measure space, and denote the characteristic function of Q by

$$\chi(y) = \int e^{i(y,x)}\, dQ(x).$$

Lemma 4.3.13. Let $\sum_{k,l=1}^{m} a_{kl} y_k y_l$ be a positive definite quadratic form, and let $A^2 = \sum_{k=1}^{m} a_{kk}$. If

$$\eta \geqslant \sup_{\sum a_{kl} y_k y_l \leqslant 1} (1 - \Re \chi(y)),$$

then

$$Q(\{x \mid (x, x) \geqslant R^2\}) \leqslant \frac{\sqrt{e}}{\sqrt{e} - 1}\left(\eta + 2\frac{A^2}{R^2}\right). \tag{4.3.20}$$

PROOF. It suffices to prove (4.3.20) for $R = 1$. First, we note the following property of Gaussian measures (see §5.1):

$$\Re\left(\frac{1}{(2\pi)^{m/2}} \int_{R_m} \exp\left(i(y, x) - \frac{1}{2}\| y \|^2\right) dy\right) = \exp\left(-\frac{1}{2}\| x \|^2\right).$$

[40] *Translator's note:* A detailed proof of this assertion would not be amiss.

Using this identity, we obtain

$$Q(\{x \mid (x, x) \geqslant 1\})$$

$$\leqslant \frac{1}{1-e^{-1/2}} \int_{\|x\| \geqslant 1} \left\{1 - \exp\left(-\frac{1}{2}\| x \|^2\right)\right\} dQ(x)$$

$$\leqslant \frac{1}{1-e^{-1/2}} \int_{R_m} \left\{1 - \exp\left(-\frac{1}{2}\| x \|^2\right)\right\} dQ(x)$$

$$= \frac{1}{1-e^{-1/2}} \int_{R_m} \left(\frac{1}{(2\pi)^{m/2}} \int_{R_m} (e^{-\frac{1}{2}\|y\|^2} - \Re e^{i(y,x)-\frac{1}{2}\|y\|^2})\, dy\right) dQ(x)$$

$$= \frac{1}{1-e^{-1/2}} \frac{1}{(2\pi)^{m/2}} \int_{R_m} (1 - \Re\chi(y)) \exp\left(-\frac{1}{2}\| y \|^2\right) dy. \tag{4.3.21}$$

Now, if $\sum a_{kl} y_k y_l \leqslant 1$, then $1 - \Re\chi(y) \leqslant \eta$, and if $\sum a_{kl} y_k y_l \geqslant 1$, then $1 - \Re\chi(y) \leqslant 2 \leqslant 2 \sum a_{kl} y_k y_l$. Hence, since $\sum a_{kl} y_k y_l$ is positive definite, the inequality

$$1 - \Re\chi(y) \leqslant \eta + 2 \sum a_{kl} y_k y_l \tag{4.3.22}$$

holds for all $y \in R_m$. Using (4.3.22) in (4.3.21), and evaluating a simple Gaussian integral, we obtain (4.3.20) (for $R = 1$). The calculation for arbitrary R is entirely similar.]

PROOF OF THEOREM 4.3.12. Obviously, $\mathfrak{G}^\dagger$ distinguishes the points of $\mathfrak{G}$. Hence, applying Lemma 4.3.2 to the function f, we obtain a cylinder measure P on $\mathfrak{G}^\dagger$ such that f and P satisfy (4.3.3). We need only prove that P is countably additive, for then P can be extended to a probability measure $P^\dagger$ on $(\mathfrak{G}^\dagger, \mathfrak{F}^\dagger)$, and (4.3.3) becomes (4.3.19). Let

$$\| F \|_{-k} = \sup_{\|x\|_k = 1} |F(x)|.$$

By virtue of Corollary 4.3.9, it suffices to prove that, for any positive number ϵ, there exists a ball

$$S_n(R) = \{F \mid \| F \|_{-n} \leqslant R\}$$

such that the cylinder measure of any Borel cylinder lying outside of $S_n(R)$ is less than ϵ. First, choose a number η satisfying

$$0 < \eta < \frac{\sqrt{e} - 1}{\sqrt{e}} \frac{\epsilon}{2}. \tag{4.3.23}$$

Since f is continuous with respect to the topology $\mathfrak{S}$, there is a natural number n and a positive nuclear operator T on $\mathfrak{G}_n$ such that, if $x \in U(T, n)$, that is, if $(Tx, x)_n < 1$, then

$$1 - \Re f(x) < \eta. \tag{4.3.24}$$

Let $\| T \|_1$ denote the trace norm of the nuclear operator T (see §II.1), and choose a positive number R such that

$$\frac{2\sqrt{e}}{\sqrt{e} - 1} \frac{\| T \|_1}{R^2} < \frac{\epsilon}{2}. \tag{4.3.25}$$

We shall prove that the ball $S_n(R)$ fulfills the above-mentioned requirement.

Let L be any Borel cylinder in $\mathfrak{G}^\dagger$, say,

$$L = \{\xi \mid (\xi(e_1),..., \xi(e_m)) \in B\},$$

where $e_1 ,..., e_m \in \mathfrak{G}$ and B is a Borel subset of R_m. We may assume that the vectors $e_1 ,..., e_m$ form an orthonormal set in $\mathfrak{G}_n$. If L and $S_n(R)$ are disjoint, then the base B of L and the ball $x_1{}^2 + \cdots + x_m{}^2 \leqslant R^2$ are disjoint subsets of R_m. In fact, if there were a point $(x_1 ,..., x_m) \in B$ such that $\sum x_j{}^2 \leqslant R^2$, then the functional

$$\xi_0(\varphi) = \left(\varphi, \sum_{j=1}^{m} x_j e_j\right)_n$$

on $\mathfrak{G}_n$ would satisfy

$$\| \xi_0 \|_{-n} = \left\| \sum x_j e_j \right\|_n = \left(\sum x_j{}^2\right)^{1/2} \leqslant R,$$

hence $\xi_0 \in S_n(R)$. But, on the other hand, we get

$$(\xi_0(e_1),..., \xi_0(e_m)) = (x_1 ,..., x_m) \in B,$$

that is, $\xi_0 \in L$, which is impossible.

Define a probability measure Q on R_m as follows: for any Borel set D in R_m, let

$$Q(D) = P(\{\xi \mid (\xi(e_1),..., \xi(e_m)) \in D\}).$$

We shall apply Lemma 4.3.13 to the measure Q, the function $\chi(y) = f(\sum y_k e_k)$, and the quadratic form with coefficients

$$a_{kl} = (Te_k , e_l)_n .$$

By the properties of the trace norm, we have $A^2 = \sum_{k=1}^{m} a_{kk} \leqslant \| T \|_1$. Moreover,

$$\left(T\left(\sum y_k e_k\right), \sum y_k e_k\right)_n = \sum a_{kl} y_k y_l ,$$

which shows that the quadratic form $\sum a_{kl} y_k y_l$ is positive definite, and that $\sum y_k e_k \in U(T, n)$ whenever $\sum a_{kl} y_k y_l \leqslant 1$. Hence, if $\sum a_{kl} y_k y_l \leqslant 1$, then, by (4.3.24),

$$1 - \Re f\left(\sum y_k e_k\right) < \eta.$$

Now, applying Lemma 4.3.13, and using (4.3.23), (4.3.25), we get

$$Q\left(\left\{(x_1, \ldots, x_m) \,\Big|\, \sum x_j^2 \geqslant R^2\right\}\right) \leqslant \frac{\sqrt{e}}{\sqrt{e} - 1}\left(\eta + 2\frac{A^2}{R^2}\right) < \epsilon.$$

But B lies in the region $\sum x_j^2 > R$, therefore

$$P(L) = Q(B) \leqslant Q\left(\left\{(x_1, \ldots, x_m) \,\Big|\, \sum x_j^2 \geqslant R^2\right\}\right) < \epsilon.$$

Thus, $S_n(R)$ satisfies the required condition, and we conclude that P is countably additive. The uniqueness and continuity of $P^\dagger$ may be deduced by applying Lemmas 4.3.2 and 4.3.3, respectively.]

Corollary 4.3.14. Let $\mathfrak{G}$ be a real nuclear space, and let f be a continuous positive definite function on $\mathfrak{G}$, with $f(0) = 1$. Then, there exists a unique probability measure $P^\dagger$ on $(\mathfrak{G}^\dagger, \mathfrak{F}^\dagger)$, such that f and $P^\dagger$ are related by (4.3.19).

PROOF. Let $\mathfrak{T}$ denote the original topology of $\mathfrak{G}$. It suffices to prove that $\mathfrak{T} \subset \mathfrak{S}$, for this implies that f is also continuous with respect to $\mathfrak{S}$, and the desired conclusion then follows from Theorem 4.3.12. Consider any basic neighborhood of zero in the topology $\mathfrak{T}$, say,

$$V(k, \epsilon) = \{\varphi \mid (\varphi, \varphi)_k < \epsilon\}.$$

Since $(\mathfrak{G}, \mathfrak{T})$ is a nuclear space, there is an $n > k$ such that the imbedding map $T_k^{\,n}$ is a nuclear operator. Now,

$$(\varphi, \varphi)_k = ((T_k^{\,n})^* \, T_k^{\,n}\varphi, \varphi)_n ,$$

and, by Lemma II.1.1 of Appendix II, $T = (T_k^{\,n})^* \, T_k^{\,n}$ is a nuclear operator on $\mathfrak{G}_n$. Thus, $V(k, \epsilon) = U(T/\epsilon, n)$, which shows that $T \subset \mathfrak{S}$.]

Corollary 4.3.15. Let G be a real Hilbert space, and let $\mathfrak{G}$ be a linear subspace of G which is itself a real separable Hilbert space with respect to an inner product (h, x), $h \in \mathfrak{G}$, $x \in \mathfrak{G}$. Suppose that the imbedding A of $\mathfrak{G}$ into G is a Hilbert–Schmidt type operator. Then, given any continuous positive definite function f on G, there exists a unique finite measure $P^\dagger$ on $(\mathfrak{G}^\dagger, \mathfrak{F}^\dagger)$, such that

$$f(h) = \int_{\mathfrak{G}} e^{i(h,x)}\, dP^\dagger(x), \qquad h \in \mathfrak{G},$$

moreover, $P^\dagger$ is continuous with respect to the topology $\mathfrak{T}$ induced by G on $\mathfrak{G}$.

Proof. Let $[h, g]$ denote the inner product on G. Then, for any $h, g \in \mathfrak{G}$,

$$[h, g] = (A^*Ah, g),$$

and by Lemma II.1.1, A^*A is a nuclear operator. This shows that $\mathfrak{T}$ is weaker than the topology $\mathfrak{S}$ on $\mathfrak{G}$ (here, we regard a Hilbert space as a special case of a countably Hilbert space).]

Now, we give the converse of Theorem 4.3.12.

Theorem 4.3.16. Let $\mathfrak{G}$ be a real separable countably Hilbert space. If $P^\dagger$ is a probability measure on $(\mathfrak{G}^\dagger, \mathfrak{F}^\dagger)$, then the function

$$f(g) = \int_{\mathfrak{G}^\dagger} e^{i\xi(g)}\, dP^\dagger(\xi), \qquad g \in \mathfrak{G}, \tag{4.3.26}$$

is continuous with respect to the topology $\mathfrak{S}$.

Proof. Consider the sequence of sets $S_n = \{F \mid \|F\|_{-n} \leqslant n\}$ in $\mathfrak{G}^\dagger$. By Lemma 4.2.17, each S_n is measurable, and obviously

$$\mathfrak{G}^\dagger = \bigcup_{n=1}^{\infty} S_n .$$

Therefore, given any positive number ϵ, there is an n such that $P^\dagger(\mathfrak{G}^\dagger - S_n) < \epsilon/4$. Since

$$1 - \cos \xi(g) \leqslant \tfrac{1}{2}\, \xi(g)^2,$$

we obtain

$$(1 - \Re f(g)) = \int_{\mathfrak{G}^\dagger} (1 - \cos \xi(g))\, dP^\dagger(\xi) \leqslant \frac{1}{2} \int_{S_n} \xi(g)^2\, dP^\dagger(\xi) + \frac{\epsilon}{2}. \tag{4.3.27}$$

Define a positive definite Hermitian bilinear functional T on $\mathfrak{G}_n$, as follows:

$$T(g, h) = \frac{1}{\epsilon} \int_{S_n} \xi(g)\, \xi(h)\, dP^\dagger(\xi).$$

Since $|\xi(g)| \leqslant n \| g \|_n$ whenever $\xi \in S_n$, we easily obtain the estimate

$$| T(g, h)| \leqslant \frac{1}{\epsilon} n^2 \| g \|_n \| h \|_n .$$

Thus, $T(g, h)$ is continuous with respect to the topology of $\mathfrak{G}_n$. Choosing any complete orthonormal system $\{g_k\}$ in $\mathfrak{G}_n$, we get

$$\sum T(g_k , g_k) = \frac{1}{\epsilon} \int_{S_n} \sum \xi(g_k)^2\, dP^\dagger(\xi) = \frac{1}{\epsilon} \int_{S_n} \| \xi \|_{-n}^2\, dP^\dagger(\xi) \leqslant \frac{n^2}{\epsilon} .$$

Hence, using Theorem II.1.6, we see that there exists a positive nuclear operator T on $\mathfrak{G}_n$ such that $T(g, h) = (Tg, h)_n$. If $g \in U(T, n)$, that is, $(Tg, g)_n < 1$, then, from (4.3.27), we get

$$1 - \Re f(g) < \epsilon.$$

This shows that $\Re f$ is continuous with respect to the topology $\mathfrak{S}$. Since f is positive definite, it follows that f itself is continuous with respect to $\mathfrak{S}$.]

5° Representation of Continuous Positive Definite Functions on Exact Unions of Nuclear Spaces

We now proceed to establish a further generalization of Corollary 4.3.14.

Definition 4.3.5. Let Φ be a linear topological space, and let $\Phi^{(1)}, \Phi^{(2)},..., \Phi^{(m)},...$ be a sequence of closed linear subspaces of Φ; for convenience, we assume that

$$\Phi^{(1)} \subset \Phi^{(2)} \subset \cdots \subset \Phi^{(m)} \subset \cdots.$$

Suppose that the following conditions are satisfied.

(i) $\Phi = \bigcup_{m=1}^{\infty} \Phi^{(m)}$.

(ii) Every continuous linear functional on $\Phi^{(m)}$ can be extended to a continuous linear functional on Φ.

(iii) If, for each m, there is given a continuous linear functional $F^{(m)}$ on $\Phi^{(m)}$, and if $F^{(m)}(\varphi) = F^{(n)}(\varphi)$ whenever $m \geqslant n$ and $\varphi \in \Phi^{(n)}$, then there exists a continuous linear functional F on Φ such that $F(\varphi) = F^{(n)}(\varphi)$ when $\varphi \in \Phi^{(n)}$.

We then say that Φ is the *exact union*[41] of the sequence of subspaces $\{\Phi^{(m)}, m = 1, 2,\ldots\}$.

Let $\Phi^\dagger$ and $\Phi^{(m)\dagger}$ denote the conjugate spaces of Φ and $\Phi^{(m)}$, respectively. Take the weak topology on $\Phi^\dagger$ and $\Phi^{(m)\dagger}$, $m = 1, 2,\ldots$. If $n \geqslant m$, then every element $\xi^{(n)}$ of $\Phi^{(n)\dagger}$, when restricted to $\Phi^{(m)}$, defines a continuous linear functional $\xi^{(m)}$ on $\Phi^{(m)}$. Thus, we have a mapping

$$P_m{}^n \xi^{(n)} = \xi^{(m)}$$

from $\Phi^{(n)\dagger}$ to $\Phi^{(m)\dagger}$. Condition (ii) in Definition 4.3.4 implies that $P_m{}^n \Phi^{(n)\dagger} = \Phi^{(m)\dagger}$, moreover, it is easily seen that $P_m{}^n$ is continuous. Let Λ denote the totality of positive integers directed by the natural ordering; then $\{P_m{}^n, n \geqslant m, n, m \in \Lambda\}$ is a consistent family of projection operators in the sense of Definition 1.3.5.

If $\xi \in \Phi^\dagger$, the restriction of ξ to $\Phi^{(m)}$ is a continuous linear functional $\xi^{(m)}$ on $\Phi^{(m)}$. We thus obtain a projection operator

$$P_m \colon \xi \to \xi^{(m)}$$

from $\Phi^\dagger$ onto $\Phi^{(m)\dagger}$. It is easily verified that the projections $\{P_m, m \in \Lambda\}$ and $\{P_m{}^n, n \geqslant m, n, m \in \Lambda\}$ satisfy the mutual consistency condition (1.3.2), and that the linear topological space $\Phi^\dagger$ is thereby represented as a projective limit (with respect to the family of projections $\{P_m{}^n, n \geqslant m, n, m \in \Lambda\}$) of the sequence of linear topological spaces $\{\Phi^{(m)\dagger}, m = 1, 2,\ldots\}$, in accordance with Definition 1.3.5.

By virtue of condition (iii), the projective limit $\Phi^\dagger$ satisfies the conditions of Lemma 1.3.2, and is therefore projectively complete.

Let $\mathfrak{F}^{(m)\dagger}$ and $\mathfrak{F}^\dagger$ denote the σ-algebras consisting of all weak Borel sets in $\Phi^{(m)\dagger}$ and $\Phi^\dagger$, respectively. It is easily proved that $\{P_n{}^m, m \geqslant n\}$ is a consistent family of projections in the family of measurable spaces $\{(\Phi^{(m)\dagger}, \mathfrak{F}^{(m)\dagger}), m = 1, 2,\ldots\}$. Let $\mathfrak{F}$ be the algebra of all sets of the form $P_m^{-1}(A)$, $A \in \mathfrak{F}^{(m)\dagger}$, $m = 1, 2,\ldots$. Again, using the projections $\{P_m, m \in \Lambda\}$, one may represent $(\Phi^\dagger, \mathfrak{F})$ as the projective limit (with respect to the family of projections $\{P_n{}^m, m \geqslant n\}$) of the sequence of measurable spaces $\{(\Phi^{(m)\dagger}, \mathfrak{F}^{(m)\dagger}), m = 1, 2,\ldots\}$, in accordance with Definition 1.3.1. Moreover, $\mathfrak{F}^\dagger$ is clearly the smallest σ-algebra containing $\mathfrak{F}$.

Theorem 4.3.17. If the linear topological space Φ is the exact union of a sequence of nuclear spaces, then, for each continuous positive

[41] *Translator's note:* The corresponding term used in Gel'fand–Vilenkin [1] has been translated as *union space*.

definite function f on Φ [with $f(0) = 1$], there exists a unique probability measure $P^\dagger$ on $(\Phi^\dagger, \mathfrak{F}^\dagger)$ such that

$$f(\varphi) = \int_{\Phi^\dagger} e^{i\xi(\varphi)}\, dP^\dagger(\xi) \tag{4.3.28}$$

for all $\varphi \in \Phi$.

Proof. Let Φ be the exact union of the sequence of nuclear spaces $\Phi^{(1)} \subset \Phi^{(2)} \subset \cdots \subset \Phi^{(n)} \subset \cdots$. By Corollary 4.3.14, for each m, there is a unique probability measure $P^{(m)\dagger}$ on $(\Phi^{(m)\dagger}, \mathfrak{F}^{(m)\dagger})$ such that

$$f(\varphi) = \int_{\Phi^{(m)\dagger}} e^{i\xi(\varphi)}\, dP^{(m)\dagger}(\xi) \tag{4.3.29}$$

for all $\varphi \in \Phi^{(m)}$. By Lemma 1.1.18, $(\Phi^{(m)}, \mathfrak{F}^{(m)\dagger}, P^{(m)\dagger})$ is a regular measure space. Moreover, since the family of measure spaces $\{(\Phi^{(m)\dagger}, \mathfrak{F}^{(m)\dagger}, P^{(m)\dagger}), m = 1, 2, \ldots\}$ is consistent with the projections $\{P_n{}^m, m \geqslant n\}$, it follows by Lemma 1.3.1 that there exists a cylinder measure μ on $(\Phi^\dagger, \mathfrak{F})$ such that $(\Phi^\dagger, \mathfrak{F}, \mu)$ is a projective limit of the family $\{(\Phi^{(m)\dagger}, \mathfrak{F}^{(m)\dagger}, P^{(m)\dagger}), m = 1, 2, \ldots\}$. Furthermore, by Theorem 1.3.4, μ is countably additive, and may therefore be extended to a probability measure $P^\dagger$ on $(\Phi^\dagger, \mathfrak{F}^\dagger)$. If $\varphi \in \Phi^{(m)}$, then, from (4.3.29), we get

$$\int_{\Phi^\dagger} e^{i\xi(\varphi)}\, dP^\dagger(\xi) = \int_{\Phi^\dagger} e^{i\xi(\varphi)}\, d\mu(\xi) = \int_{\Phi^{(m)\dagger}} e^{i\xi(\varphi)}\, dP^{(m)\dagger}(\xi) = f(\varphi).$$

Therefore, since $\Phi = \bigcup_{m=1}^\infty \Phi^{(m)}$, relation (4.3.28) holds for all $\varphi \in \Phi$. Finally, the uniqueness of $P^\dagger$ follows from the uniqueness of the $P^{(m)\dagger}$.]

Corollary 4.3.18. Let K be the space of test functions defined in Example I.3.3 of Appendix I. Then, the space of generalized functions $K^\dagger$ is a spectral space for K.

Proof. Since K is the exact union of the sequence of subspaces $\{K([-n, n]), n = 1, 2, \ldots\}$, and since each $K([-n, n])$ is a nuclear space, our conclusion follows immediately from Theorem 4.3.17.]

6° Generalized Stochastic Processes

As we know, an ordinary stochastic process is just a family of random variables $\{x_t(\cdot), t \in T\}$ defined on a probability measure space $\Omega = (G, \mathfrak{B}, P)$; here, t represents the "time parameter," and the indexing set T is a set of real numbers representing the time range under consideration. For example, the coordinate of a particle performing one-dimensional Brownian motion may be described by a stochastic process of just this kind, namely, the Wiener process. However, phenomena governed by statistical laws cannot all be described by this sort of model, for example, the velocity of a particle performing one-dimensional Brownian motion cannot be described by an ordinary stochastic process. It frequently happens that one cannot directly observe the values of $x_t(\omega)$, rather, one observes a kind of average value

$$X(\omega; \varphi) = \int_T x_t(\omega)\, \varphi(t)\, dt,$$

where the function φ is determined by the instrument used for the observation. This clearly necessitates the use of a function space. In this case, the random process is not manifested in the form $\{x_t(\omega), t \in T, \omega \in G\}$, instead, we have a family of random variables $\{X(\cdot; \varphi), \varphi \in \Phi\}$, and one expects $X(\cdot; \varphi)$ to satisfy certain linearity conditions. Formalizing these ideas, we are led to the concept of a generalized stochastic process (see Definition 4.3.6).

Let $\Omega = (G, \mathfrak{B}, P)$ be a probability measure space, and let $M(\Omega)$ denote the linear space consisting of all random variables on Ω, with the ordinary linear operations; here, any two random variables which are equal almost everywhere are regarded as identical. We introduce the metric

$$\rho(\varphi, \psi) = \int_G \frac{|\varphi(\omega) - \psi(\omega)|}{1 + |\varphi(\omega) - \psi(\omega)|}\, dP(\omega), \qquad \varphi, \psi \in M(\Omega), \quad (4.3.30)$$

which makes $M(\Omega)$ a complete linear topological space.

Definition 4.3.6. Let Φ be a linear (or linear topological) space, let Ω be a probability measure space, and let $\{X(\cdot; \varphi), \varphi \in \Phi\}$ be a family of random variables on Ω, such that the correspondence

$$U: \varphi \to X(\cdot; \varphi), \qquad \varphi \in \Phi$$

is a linear (or continuous linear) mapping from Φ into $M(\Omega)$. Then, we say that $\{X(\cdot; \varphi), \varphi \in \Phi\}$ is a *linear* (or *generalized*) *stochastic process* on (Ω, Φ).

In the applications, Φ is usually some space of the type $K_n(\Delta)$, $K(\Delta)$, K, and so on (see Examples I.3.1 and I.3.3 in Appendix I).

Definition 4.3.7. Let $\{X(\cdot; \varphi), \varphi \in \Phi\}$ be a linear stochastic process on (Ω, Φ), let Ψ be a linear subspace of Φ^{Δ} which distinguishes the points of Φ, let $\mathfrak{F}$ be the totality of Φ-weak Borel sets in Ψ, and let μ be a probability measure on $(\Psi, \mathfrak{F})$. For each $\varphi \in \Phi$, we define a random variable on $\Omega' = (\Psi, \mathfrak{F}, \mu)$, as follows:

$$X'(\psi; \varphi) = \psi(\varphi), \qquad \psi \in \Psi.$$

Clearly, $\{X'(\cdot; \varphi), \varphi \in \Phi\}$ is a linear stochastic process on (Ω', Φ). Suppose that, for every finite set $\varphi_1, \ldots, \varphi_n \in \Phi$, the two sets of random variables

$$\begin{gathered} X(\cdot; \varphi_1), \ldots, X(\cdot; \varphi_n) \\ X'(\cdot; \varphi_1), \ldots, X'(\cdot; \varphi_n) \end{gathered} \tag{4.3.31}$$

have the same probability distribution, that is, the equality

$$\begin{aligned} &P(\{\omega \mid (X(\omega; \varphi_1), \ldots, X(\omega; \varphi_n)) \in A, \omega \in G\}) \\ &\quad = \mu(\{\psi \mid (X'(\psi; \varphi_1), \ldots, X'(\psi; \varphi_n)) \in A, \psi \in \Psi\}) \end{aligned}$$

holds for every Borel set A in n-dimensional space. Then, we say that Ψ, Ω' and $\{X'(\cdot; \varphi), \varphi \in \Phi\}$ are, respectively, a *sample space*, a *sample probability space*, and a *sample process* for the linear stochastic process $\{X(\cdot; \varphi), \varphi \in \Phi\}$.

To facilitate our study of sample spaces of linear stochastic processes, we introduce the notion of a characteristic functional, as follows.

Definition 4.3.8. If $\{X(\cdot; \varphi), \varphi \in \Phi\}$ is a linear stochastic process on (Ω, Φ), the integral

$$L(\varphi) = \int_G e^{iX(\omega;\varphi)}\, dP(\omega), \qquad \varphi \in \Phi \tag{4.3.32}$$

will be called the *characteristic functional* of the process.

Using the linearity of $X(\cdot; \varphi)$ with respect to φ, one easily verifies that the characteristic functional (4.3.32) is a pseudocontinuous positive definite function on Φ.

Theorem 4.3.19. Any linear stochastic process $\{X(\cdot; \varphi), \varphi \in \Phi\}$ has Φ^Λ as a sample space.

PROOF. Since the characteristic functional $L(\varphi)$ of $\{X(\cdot; \varphi), \varphi \in \Phi\}$ is a pseudocontinuous positive definite function on Φ, it follows by Theorem 4.3.5 that there exists a probability measure P^Λ on $(\Phi^\Lambda, \mathfrak{F}^\Lambda)$ such that

$$L(\varphi) = \int_{\Phi_\Lambda} e^{i\xi(\varphi)}\, dP^\Lambda(\xi), \qquad \varphi \in \Phi. \tag{4.3.33}$$

Define a linear stochastic process $\{X'(\cdot;\varphi), \varphi \in \Phi\}$ on $(\Phi^A, \mathfrak{F}^A, P^A, \Phi)$, as follows. For each $\varphi \in \Phi$, let

$$X'(\xi;\varphi) = \xi(\varphi), \qquad \xi \in \Phi^A. \tag{4.3.34}$$

Then $L(\varphi)$ is also the characteristic functional of $\{X'(\cdot;\varphi), \varphi \in \Phi\}$. Consequently, by substituting $\varphi = t_1\varphi_1 + \cdots + t_n\varphi_n$ $(-\infty < t_\nu < \infty)$ into (4.3.32), (4.3.33), and (4.3.34), one may deduce that the two sets of random variables (4.3.31) have the same probability distribution. Thus, Φ^λ is a sample space for $\{X(\cdot;\varphi), \varphi \in \Phi\}$.]

Using an argument similar to the proof of Lemma 4.3.3, we obtain the following result.

Lemma 4.3.20. If Φ is a linear topological space and Ω is a probability measure space, then a linear stochastic process on (Ω, Φ) is a generalized stochastic process if and only if its characteristic functional is continuous.

Using Lemma 4.3.20, Definition 4.3.4, and the proof of Theorem 4.3.19, we obtain the following result.

Theorem 4.3.21. Let Φ be a linear topological space, and let Ψ be a spectral space for Φ. Then, every generalized stochastic process $\{X(\cdot;\varphi), \varphi \in \Phi\}$ has Ψ as a sample space.

Using Corollaries 4.3.14 and 4.3.18, we immediately obtain the following two corollaries.

Corollary 4.3.22. Every generalized stochastic process

$$\{X(\cdot;\varphi), \varphi \in K(\Delta)\}$$

has $K(\Delta)^\dagger$ as a sample space (see Example II.1.1 of Appendix II).

Corollary 4.3.23. Every generalized stochastic process

$$\{X(\cdot;\varphi), \varphi \in K\}$$

has $K^\dagger$ as a sample space (see Example I.3.3 of Appendix I).

We end this chapter by pointing out the relation between linear stochastic processes and cylinder measures. The following theorem is a consequence of Lemmas 4.3.2 and 4.3.3.

Theorem 4.3.24. Let $\{X(\cdot;\varphi), \varphi \in \Phi\}$ be a linear stochastic process on (Ω, Φ), and let Ψ be a linear subspace of Φ^A which distinguishes the points of Φ. Then, there exists a cylinder measure μ on Ψ, such that, for any finite set $\varphi_1, \dots, \varphi_n \in \Phi$, and any Borel set A in n-dimensional Euclidean space,

$$P(\{\omega \mid (X(\omega; \varphi_1), \dots, X(\omega; \varphi_n)) \in A\})$$
$$= \mu(\{f \mid (f(\varphi_1), \dots, f(\varphi_n)) \in A, f \in \Psi\}.$$

Moreover, if Φ is a linear topological space, then $\{X(\cdot; \varphi), \varphi \in \Phi\}$ is a generalized stochastic process if and only if μ is continuous with respect to the topology of Φ.

We say that the cylinder measure μ in Theorem 4.3.24 is *generated* by the linear stochastic process $\{X(\cdot; \varphi), \varphi \in \Phi\}$. Conversely, cylinder measures give rise to linear stochastic processes, as shown by the following theorem.

Theorem 4.3.25. Let Φ be a linear (or linear topological) space, let Ψ be a linear subspace of Φ^A which distinguishes the points of Φ, and let μ be a cylinder measure (or continuous cylinder measure) on Ψ. Then, there exists a probability measure space $\Omega = (G, \mathfrak{B}, P)$ and a linear (or generalized) stochastic process $\{X(\cdot; \varphi), \varphi \in \Phi\}$ on (Ω, Φ), such that μ is generated by $\{X(\cdot; \varphi), \varphi \in \Phi\}$.

PROOF. Consider the positive definite function

$$L(\varphi) = \int_\Psi e^{i\xi(\varphi)} \, d\mu(\xi),$$

and apply the reasoning used in proving Theorems 4.3.19 and 4.3.21.]

CHAPTER

V

GAUSSIAN MEASURES

Gaussian measures are the most classical category of quasi-invariant measures. In §5.1, we introduce the basic properties of Gaussian measures, and give an example of an ergodic quasi-invariant measure which is not equivalent to any Gaussian measure; we also present an introductory discussion of the most commonly used type of Gaussian measure, that is, Wiener measure. In §5.2 we derive a criterion for the equivalence or mutual singularity of Gaussian measures. A very significant problem is that of determining those types of measures on a given linear measurable space which are quasi-invariant relative to a prescribed linear subspace; the investigation of equivalence and singularity of Gaussian measures constitutes a preliminary step in this direction. In §5.3, we describe a class of Gaussian measure spaces which are useful in connection with harmonic analysis on linear topological spaces; we also explain some important problems in the theory of quasi-invariant measures, illustrating the situation by reference to Gaussian measures. In §5.4, we discuss in detail the L^2-Fourier transform for Gaussian measure spaces, and, in particular, the Fourier transform on Wiener measure spaces.

§5.1. Some Properties of Gaussian Measures

1° The Joint Distribution of Finitely Many Gaussian Variables

First, we describe the nondegenerate case. Let R_n be the ordinary real n-dimensional Euclidean space, denote the points of R_n by $x = (x_1, \ldots, x_n)$, and write $(x, y) = \sum x_l y_l$. Given any positive-definite real quadratic form $\sum_{l,m} b_{lm} x_l x_m$ $(b_{lm} = b_{ml})$, we define another inner product in R_n by

$$(x, y)_b = \sum_{l,m=1}^{n} b_{lm} x_l y_m, \qquad x = (x_1, \ldots, x_n), \quad y = (y_1, \ldots, y_n).$$

Let (c_{lm}) denote the inverse of the matrix (b_{lm}). For any Borel set E in R_n, define

$$N(E) = \frac{1}{(2\pi)^{n/2}(\det(b_{lm}))^{1/2}} \int_E e^{-(x-a, x-a)_c/2}\, dx, \tag{5.1.1}$$

where $(x, x)_c = \sum c_{lm} x_l x_m$, $a = (a_1, \ldots, a_n) \in R_n$, and dx denotes the Lebesgue measure $dx_1 \cdots dx_n$ induced by the original inner product (x, y). Clearly, (5.1.1) defines a probability[1] measure on R_n. Any measure N defined in this way is said to be *Gaussian.*

Likewise, a set of random variables $x_1, \ldots, x_n$, whose joint distribution is of the form (5.1.1), is said to be Gaussian. Note that the mathematical expectations, correlations and characteristic function of a set of Gaussian variables are given, respectively, by

$$E(x_l) = \int_{R_n} x_l\, dN(x) = a_l, \tag{5.1.2}$$

$$E(x_l x_m) = \int_{R_n} x_l x_m\, dN(x) = b_{lm} + a_l a_m, \tag{5.1.3}$$

$$E(\exp i(t_1 x_1 + \cdots + t_n x_n)) = \int_{R_n} e^{i\sum t_\nu x_\nu}\, dN(x) = e^{-(t,t)_b + i(t,a)/2}, \quad t = (t_1, \ldots, t_n). \tag{5.1.4}$$

In the degenerate case, the quadratic form $\sum b_{lm} x_l x_m$ is only non-negative definite. We may then use (5.1.4) to define the corresponding Gaussian measure, since any characteristic function uniquely determines a measure. The explicit expression for the measure is, in this case,

[1] To verify that $N(R_n) = 1$, apply an orthogonal transformation which diagonalizes the matrix (c_{lm}), thus separating the variables in the integrand. The integral may then be evaluated by using formula (5.1.5).

slightly more complicated than (5.1.1). If $b_{lm} = 0$, $l, m = 1, 2, \ldots, n$, then the measure is concentrated at the point a, that is, $N(E) = 1$ if $a \in E$ and $N(E) = 0$ if $a \bar{\in} E$. If the rank of (b_{lm}) is r, $0 < r \leqslant n$, then there is an orthogonal transformation u of R_n [relative to the inner product (x, y)] such that $(x, x)_b = (x', x')_{b'}$, where $x' = ux = (x_1', \ldots, x_n')$ and

$$(x', y')_{b'} = \sum_{\nu=1}^{r} \lambda_\nu x_\nu' y_\nu', \qquad \lambda_\nu > 0.$$

Let $M_r = R_r + a'$, where R_r denotes the r-dimensional linear subspace of R_n consisting of all vectors of the form $(x_1, x_2, \ldots, x_r, 0, 0, \ldots, 0)$, and $a' = (a_1', \ldots, a_n') = ua$. Then a calculation shows that

$$N(E) = \frac{1}{(2\pi)^{n/2}(\prod_{\nu=1}^{r} \lambda_\nu)^{1/2}} \int \cdots \int_{uE \cap M_r} \exp\left(-\frac{1}{2} \sum_{\nu=1}^{r} \frac{(x_\nu - a_\nu')^2}{\lambda_\nu}\right) dx_1 \cdots dx_r,$$

and that (5.1.2) and (5.1.3) are still valid.

Relations (5.1.2)–(5.1.4) show that a Gaussian measure is completely determined by its means and correlations (or covariances).

The following formula is often useful when calculating integrals with respect to Gaussian measures. If $\Re p > 0$ and m is a nonnegative integer,

$$\int_{-\infty}^{\infty} \lambda^m \exp(-p\lambda^2 + q\lambda)\, d\lambda = (\pi/p)^{1/2}(d/dq)^m \exp(q^2/4p). \qquad (5.1.5)$$

Notice that two Gaussian variables X, Y are stochastically independent if and only if

$$E((X - E(X))(Y - E(Y))) = 0,$$

and that X is a degenerate Gaussian variable if and only if

$$E((X - E(X))^2) = 0.$$

Furthermore, using (5.1.5), one can calculate that a nondegenerate Gaussian variable X satisfies[2]

$$\begin{aligned} &E((X^2 - E(X^2))^2) \\ &\quad = \frac{1}{\sqrt{2\pi\sigma(X)}} \int_{-\infty}^{\infty} (X^2 - E(X^2))^2 \exp\left(-\frac{1}{2}\frac{(X - E(X))^2}{\sigma(X)}\right) dX \\ &\quad = 2(E(X^2)^2 - E(X)^4). \end{aligned} \qquad (5.1.6)$$

Next, we consider limits of sequences of Gaussian random variables.

[2] Here, $\sigma(X) = E((X - E(X))^2)$ is the variance of X.

Lemma 5.1.1. Let $S = (\Omega, \mathfrak{B}, P)$ be a probability measure space, and let $\{X_n(\cdot)\}$ be a sequence of Gaussian variables on Ω which converges in $L^2(S)$ to a random variable X. Then X is also a Gaussian variable.

PROOF. Since $X_n(\cdot) \in L^2(S)$, naturally, $X(\cdot) \in L^2(S)$, therefore

$$\overline{\lim_{n\to\infty}} \mid E(X_n) - E(X)| \leqslant \lim_{n\to\infty} E((X_n - X)^2)^{1/2} = 0,$$

$$\overline{\lim_{n\to\infty}} \mid E(X_n{}^2) - E(X^2)| = \overline{\lim_{n\to\infty}} \mid E((X_n - X)(X_n + X))|$$

$$\leqslant \lim_{n\to\infty} \sqrt{E(| X_n - X |^2)}\, (\sqrt{E(X_n{}^2)} + \sqrt{E(X^2)}) = 0,$$

$$\overline{\lim_{n\to\infty}} \mid E(\exp(iX_n t)) - E(\exp(iXt))| \leqslant \lim_{n\to\infty} E(| X_n - X |^2)^{1/2} \cdot | t | = 0.$$

However, by (5.1.4), the characteristic function of X_n is

$$E(\exp(iX_n t)) = \exp(-\tfrac{1}{2}\sigma(X_n)\, t^2 + iE(X_n)\, t).$$

Letting $n \to \infty$ on both sides, we get

$$E(\exp(iXt)) = \exp(-\tfrac{1}{2}\sigma(X)\, t^2 + iE(X)\, t).$$

Hence, X is a Gaussian variable.]

2° Gaussian Stochastic Processes

Definition 5.1.1. Let $S = (\Omega, \mathfrak{B}, P)$ be a probability space, $\{X_\alpha, \alpha \in \mathfrak{A}\}$ a family of random variables on S. If the joint distribution of any finite subfamily $X_{\alpha_1}, \ldots, X_{\alpha_n}$ is Gaussian, then the family $\{X_\alpha, \alpha \in \mathfrak{A}\}$ is said to be *Gaussian*, and is also called a *Gaussian process* on $\mathfrak{A}$.

As we have seen, the means and second moments, that is,

$$a_\alpha = E(X_\alpha), \qquad E(X_\alpha X_\beta) = b_{\alpha\beta} + a_\alpha a_\beta \tag{5.1.7}$$

completely determine the finite-dimensional distributions of such a process. Notice that the family of real numbers $\{b_{\alpha\beta}, \alpha, \beta \in \mathfrak{A}\}$ is non-negative definite, that is, for any finite set of complex numbers $\xi_{\alpha_1}, \ldots, \xi_{\alpha_n}$,

$$\sum_{\nu,\nu'=1}^{n} b_{\alpha_\nu \alpha_{\nu'}} \xi_{\alpha_\nu} \bar{\xi}_{\alpha_{\nu'}} = E\left(\left| \sum \xi_{\alpha_\nu}(X_{\alpha_\nu} - a_{\alpha_\nu}) \right|^2\right) \geqslant 0.$$

Conversely, given any family of real numbers $\{a_\alpha, \alpha \in \mathfrak{A}\}$ and any nonnegative definite family of real numbers $\{b_{\alpha\beta}, \alpha, \beta \in \mathfrak{A}\}$, there exists a Gaussian process $\{X_\alpha, \alpha \in \mathfrak{A}\}$ satisfying (5.1.7). In fact, for any finite set of indices $\alpha_1, \ldots, \alpha_n \in \mathfrak{A}$, form the function

$$\varphi_{\alpha_1,\ldots,\alpha_n}(t_{\alpha_1}, \ldots, t_{\alpha_n}) = \exp\left\{-\frac{1}{2}\sum_{\nu,\nu'} b_{\alpha_\nu\alpha_{\nu'}} t_{\alpha_\nu} t_{\alpha_{\nu'}} + i\sum_{\nu} t_{\alpha_\nu} a_{\alpha_\nu}\right\}. \tag{5.1.8}$$

By (5.1.4), this is the characteristic function of a Gaussian measure. Obviously, the family of all such functions $\{\varphi_{\alpha_1,\ldots,\alpha_n}\}$ is consistent. For every $\alpha \in \mathfrak{A}$, let R_α be a copy of the real line, and form the Cartesian product $\Gamma = \times_{\alpha\in\mathfrak{A}} R_\alpha$. Let $\mathfrak{B}$ be the smallest σ-algebra in Γ which contains all Borel cylinders. By the Kolmogorov theorem (Corollary 1.3.5′), there exists a probability measure P on $(\Gamma, \mathfrak{B})$ such that the characteristic functions of the finite-dimensional distributions of the stochastic process $\{X_\alpha(\omega), \alpha \in \mathfrak{A}\}$ [where $X_\alpha(\omega)$ denotes the αth coordinate of the point $\omega \in \Gamma$] are given by (5.1.8). This process is clearly Gaussian and satisfies (5.1.7). We call a measure space $(\Gamma, \mathfrak{B}, P)$ constructed in this manner a *canonical Gaussian measure space.*

Using the Kakutani inner product, we first proceed to consider the equivalence or perpendicularity of a special type of canonical Gaussian measure space.

Lemma 5.1.2. Let $\times_{\alpha\in\mathfrak{A}} R_\alpha = \Gamma$ be the Cartesian product of a family of real lines, $\mathfrak{B}$ the smallest σ-algebra in Γ containing all the Borel cylinders. Let N_k, $k = 1, 2$, be probability measures on $(\Gamma, \mathfrak{B})$ such that the coordinate functions $\{x_\alpha(\omega), \alpha \in \mathfrak{A}\}$ [here, $x_\alpha(\omega)$ represents the αth coordinate of the point $\omega \in \Gamma$] are independent Gaussian random variables on $(\Gamma, \mathfrak{B}, N_k)$, with mathematical expectations $m_\alpha^{(k)}$ and variances $\sigma_\alpha^{(k)} > 0$ relative to N_k, $k = 1, 2$. Then N_1 and N_2 are either equivalent or mutually singular. A necessary and sufficient condition for the equivalence of N_1 and N_2 is

$$\sum_\alpha \left(\frac{\sigma_\alpha^{(1)}}{\sigma_\alpha^{(2)}} - \frac{\sigma_\alpha^{(2)}}{\sigma_\alpha^{(1)}}\right)^2 < \infty \quad \text{and} \quad \sum_\alpha \frac{(m_\alpha^{(1)} - m_\alpha^{(2)})^2}{\sigma_\alpha^{(1)} + \sigma_\alpha^{(2)}} < \infty. \tag{5.1.9}$$

Moreover, when N_1 and N_2 are equivalent,[3]

$$\frac{dN_1(\omega)}{dN_2(\omega)} = \prod_\alpha \left\{\sqrt{\frac{\sigma_\alpha^{(2)}}{\sigma_\alpha^{(1)}}} \exp\left[-\frac{1}{2}\left(\frac{m_\alpha^{(1)^2}}{\sigma_\alpha^{(1)}} - \frac{m_\alpha^{(2)^2}}{\sigma_\alpha^{(2)}}\right) + \left(\frac{m_\alpha^{(1)}}{\sigma_\alpha^{(1)}} - \frac{m_\alpha^{(2)}}{\sigma_\alpha^{(2)}}\right) x_\alpha(\omega) - \frac{1}{2}\left(\frac{1}{\sigma_\alpha^{(1)}} - \frac{1}{\sigma_\alpha^{(2)}}\right) x_\alpha(\omega)^2\right]\right\}. \tag{5.1.10}$$

[3] By virtue of condition (5.1.9), there are but countably many indices α such that $\sigma_\alpha^{(1)} \neq \sigma_\alpha^{(2)}$, $m_\alpha^{(1)} \neq m_\alpha^{(2)}$, hence this product is essentially just an ordinary infinite product of countably many factors.

PROOF. Suppose that (5.1.9) holds. Then, the set $\mathfrak{A}'$ of indices α such that $m_\alpha^{(1)} \neq m_\alpha^{(2)}$ or $\sigma_\alpha^{(1)} \neq \sigma_\alpha^{(2)}$ is at most countable; for convenience, we shall denote these indices α by the numerals $1, 2, \ldots$. Let $\mathfrak{A}'' = \mathfrak{A} - \mathfrak{A}'$. First, we point out that one may assume $\mathfrak{A}' = \mathfrak{A}$. In fact, $\mathsf{X}_{\alpha \in \mathfrak{A}} R_\alpha$ is the direct product of $\mathsf{X}_{\alpha \in \mathfrak{A}'} R_\alpha$ and $\mathsf{X}_{\alpha \in \mathfrak{A}''} R_\alpha$. Since $\{x_\alpha , \alpha \in \mathfrak{A}\}$ are independent Gaussian variables, each measure N_k is the product of the Gaussian measures N_k' and N_k'' on $\mathsf{X}_{\alpha \in \mathfrak{A}'} R_\alpha$ and $\mathsf{X}_{\alpha \in \mathfrak{A}''} R_\alpha$, respectively. Since $m_\alpha^{(1)} = m_\alpha^{(2)}$ and $\sigma_\alpha^{(1)} = \sigma_\alpha^{(2)}$ for all $\alpha \in \mathfrak{A}''$, the variables $\{x_\alpha , \alpha \in \mathfrak{A}''\}$ have the same finite-dimensional distributions relative to N_1'' and N_2'' , hence $N_1'' = N_2''$. Consequently, in view of the facts mentioned in §1.1, 4°, it suffices to prove that N_1' and N_2' are equivalent. If $\mathfrak{A}'$ should contain only a finite number of indices, then, since $\sigma_\alpha \neq 0$, N_1' and N_2' would be nondegenerate Gaussian measures on a finite-dimensional space, and so they would obviously be equivalent. Hence, we shall assume that $\mathfrak{A} = \mathfrak{A}'$ is just the totality of natural numbers. Let $N_\alpha^{(k)}$ be the Gaussian measure on R_α with mean $m_\alpha^{(k)}$ and variance $\sigma_\alpha^{(k)} > 0$. Then $N_\alpha^{(1)}$ and $N_\alpha^{(2)}$ are equivalent, and their Kakutani inner product is

$$\begin{aligned}
&\rho(N_\alpha^{(1)}, N_\alpha^{(2)}) \\
&= \frac{1}{[2\pi(\sigma_\alpha^{(1)}\sigma_\alpha^{(2)})^{1/2}]^{1/2}} \int_{-\infty}^{\infty} \exp\left\{-\frac{1}{4}\left[\frac{(x-m_\alpha^{(1)})^2}{\sigma_\alpha^{(1)}} + \frac{(x-m_\alpha^{(2)})^2}{\sigma_\alpha^{(2)}}\right]\right\} dx \\
&= \left[\frac{2(\sigma_\alpha^{(1)}\sigma_\alpha^{(2)})^{1/2}}{\sigma_\alpha^{(1)} + \sigma_\alpha^{(2)}}\right]^{1/2} \exp\left[-\frac{1}{4}\frac{(m_\alpha^{(1)} - m_\alpha^{(2)})^2}{\sigma_\alpha^{(1)} + \sigma_\alpha^{(2)}}\right].
\end{aligned}$$

Now, N_k' is the product of the measures $\{N_\alpha^{(k)}, \alpha = 1, 2, \ldots\}$. Hence, by Theorem 1.4.4,

$$\begin{aligned}
\rho(N_1', N_2') &= \prod_{\alpha=1}^{\infty} \rho\,(N_\alpha^{(1)}, N_\alpha^{(2)}) \\
&= \lim_{n\to\infty} \left\{\prod_{\alpha=1}^{n}\left[1 + \frac{1}{4}\left(\left(\frac{\sigma_\alpha^{(1)}}{\sigma_\alpha^{(2)}}\right)^{1/2} - \left(\frac{\sigma_\alpha^{(2)}}{\sigma_\alpha^{(1)}}\right)^{1/2}\right)^2\right]\right\}^{-1/4} \\
&\quad \cdot \exp\left[-\frac{1}{4}\sum_{\alpha=1}^{n}\frac{(m_\alpha^{(1)} - m_\alpha^{(2)})^2}{\sigma_\alpha^{(1)} + \sigma_\alpha^{(2)}}\right]. \qquad (5.1.11)
\end{aligned}$$

Since $\sum (a_n - (1/a_n))^2 < \infty$ is equivalent to $\sum (a_n^{1/2} - (1/a_n^{1/2}))^2 < \infty$, it follows from (5.1.11) and condition (5.1.9) that $\prod_{\alpha=1}^{\infty} \rho(N_\alpha^{(1)}, N_\alpha^{(2)}) > 0$. Therefore, by Theorem 1.4.4, N_1' and N_2' are equivalent. Moreover,

$$\frac{dN_\alpha^{(1)}(x)}{dN_\alpha^{(2)}(x)} = \left(\frac{\sigma_\alpha^{(2)}}{\sigma_\alpha^{(1)}}\right)^{1/2} \exp\left[-\frac{1}{2}\left(\frac{m_\alpha^{(1)^2}}{\sigma_\alpha^{(1)}} - \frac{m_\alpha^{(2)^2}}{\sigma_\alpha^{(2)}}\right) + \left(\frac{m_\alpha^{(1)}}{\sigma_\alpha^{(1)}} - \frac{m_\alpha^{(2)}}{\sigma_\alpha^{(2)}}\right) x - \frac{1}{2}\left(\frac{1}{\sigma_\alpha^{(1)}} - \frac{1}{\sigma_\alpha^{(2)}}\right) x^2\right],$$

hence, using (1.4.15) and Theorem 1.1.20, we obtain (5.1.10).

Conversely, suppose that condition (5.1.9) does not hold, for example, suppose that

$$\sum_\alpha \left(\frac{\sigma_\alpha^{(1)}}{\sigma_\alpha^{(2)}} - \frac{\sigma_\alpha^{(2)}}{\sigma_\alpha^{(1)}}\right)^2 = \infty;$$

then there exists a sequence of indices $\alpha \in \mathfrak{A}$, which we denote by $\alpha = 1, 2, \ldots$, such that

$$\sum_{\alpha=1}^{\infty} \left(\frac{\sigma_\alpha^{(1)}}{\sigma_\alpha^{(2)}} - \frac{\sigma_\alpha^{(2)}}{\sigma_\alpha^{(1)}}\right)^2 = \infty. \tag{5.1.12}$$

Now, in view of (5.1.11), we have

$$\prod_{\alpha=1}^{\infty} \rho(N_\alpha^{(1)}, N_\alpha^{(2)}) \leqslant \left\{\prod_{\alpha=1}^{\infty} \left(1 + \frac{1}{4}\left(\left(\frac{\sigma_\alpha^{(1)}}{\sigma_\alpha^{(2)}}\right)^{1/2} - \left(\frac{\sigma_\alpha^{(2)}}{\sigma_\alpha^{(1)}}\right)^{1/2}\right)^2\right)\right\}^{-1/4}. \tag{5.1.13}$$

However, by (5.1.12),

$$\prod_{\alpha=1}^{\infty} \left(1 + \frac{1}{4}\left(\left(\frac{\sigma_\alpha^{(1)}}{\sigma_\alpha^{(2)}}\right)^{1/2} - \left(\frac{\sigma_\alpha^{(2)}}{\sigma_\alpha^{(1)}}\right)^{1/2}\right)^2\right) \geqslant \frac{1}{4}\sum\left(\left(\frac{\sigma_\alpha^{(1)}}{\sigma_\alpha^{(2)}}\right)^{1/2} - \left(\frac{\sigma_\alpha^{(2)}}{\sigma_\alpha^{(1)}}\right)^{1/2}\right)^2 = \infty.$$

Hence, by (5.1.13), we conclude that $\prod_{\alpha=1}^{\infty} \rho(N_\alpha^{(1)}, N_\alpha^{(2)}) = 0$, that is, that N_1' and N_2' are mutually singular. The case

$$\sum_\alpha \frac{(m_\alpha^{(1)} - m_\alpha^{(2)})^2}{\sigma_\alpha^{(1)} + \sigma_\alpha^{(2)}} = \infty$$

may be handled in a similar fashion.]

Example 5.1.1. Let $(l, \mathfrak{B}_\omega, \mu)$ be the linear measure space discussed in Example 4.1.2, with probability densities as follows:

$$f_\alpha(t) = \frac{1}{(2\pi\sigma_\alpha)^{1/4}} \exp\left(-\frac{1}{4}\frac{t^2}{\sigma_\alpha}\right), \qquad \sigma_\alpha > 0, \quad \alpha = 1, 2, \ldots.$$

Then, the totality of quasi-invariant points of $(l, \mathfrak{B}_\omega, \mu)$ is just $l^2(\{1/\sigma_\alpha\})$.

In fact, let $y \in l$, $y = \{y_1, ..., y_n, ...\}$, and form the measure

$$\mu_y(E) = \mu(E - y), \qquad E \in \mathfrak{B}_\omega .$$

Then $\{x_\alpha(\omega), \omega \in l\}(\omega = \{x_1(\omega),..., x_n(\omega),...\})$ are mutually independent Gaussian variables on $(l, \mathfrak{B}_\omega, \mu_y)$, with mean values y_α and variances σ_α. Using (5.1.9), we see that a necessary and sufficient condition for the equivalence of μ_y and μ is

$$\sum_{\alpha=1}^{\infty} \frac{y_\alpha^2}{\sigma_\alpha} < \infty,$$

that is, $y \in l^2(\{1/\sigma_\alpha\})$. Moreover, if $y \in l^2(\{1/\sigma_\alpha\})$, then, by (5.1.10),

$$\frac{d\mu_y(\omega)}{d\mu(\omega)} = \prod_{\alpha=1}^{\infty} \exp\left[-\frac{1}{2}\frac{y_\alpha^2}{\sigma_\alpha} + \frac{y_\alpha}{\sigma_\alpha} x_\alpha(\omega)\right]. \tag{5.1.14}$$

Now, let $(l, \mathfrak{B}_\omega, P)$ be a canonical Gaussian measure space, with Gaussian variables $x_\alpha(\omega)$, $\omega \in l$ [$x_\alpha(\omega)$ denotes the αth coordinate of ω]. The characteristic function $\varphi(f)$ of $(l, \mathfrak{B}_\omega, P)$ may be constructed as follows. For each $f \in l^\dagger$ [$f(x) = \sum f_\alpha x_\alpha$, $x = \{x_\alpha\} \in l$], applying (5.1.4), we obtain

$$\varphi(f) = E(e^{if(\omega)}) = \exp\left\{-\frac{1}{2}\sum_{\alpha,\beta} E((x_\alpha - E(x_\alpha))(x_\beta - E(x_\beta)) f_\alpha f_\beta + i\sum_\alpha E(x_\alpha) f_\alpha\right\}.$$

Write $c(f, g) = E((f(\omega) - E(f(\omega)))(g(\omega) - E(g(\omega))))$; this is a positive definite bilinear functional on $l^\dagger$. Also, write

$$m(f) = E(f(\omega)) = \sum f_\alpha E(x_\alpha),$$

which defines a linear functional on $l^\dagger$. Then, the characteristic function may be rewritten as

$$\varphi(f) = e^{-c(f,f)+im(f)/2}, \qquad f \in l^\dagger. \tag{5.1.15}$$

Lemma 5.1.3. Let l_0 be the totality of real number sequences $x = \{x_n\}$ such that only a finite number of the x_n are not zero; l_0 forms a linear space with respect to the usual linear operations (see Example 4.2.1). Let $1 \leqslant \alpha < 2$, and define a norm on l_0 by

$$|x|_\alpha = \left(\sum_{n=1}^{\infty} |x_n|^\alpha\right)^{1/\alpha}.$$

Let $[x, y]$ be any positive definite bilinear functional on l_0. Then, the norms $| x | = [x, x]^{1/2}$ and $| x |_\alpha$ cannot induce the same topology on l_0.

PROOF. Suppose that $| x |$ and $| x |_\alpha$ did induce the same topology. Let l_α denote the completion of l_0 relative to the norm $| x |_\alpha$; clearly, l_α may be identified with the Banach space consisting of all real number sequences $\{x_n\}$ such that $| x |_\alpha = (\sum_{n=1}^{\infty} | x_n |^\alpha)^{1/\alpha} < \infty$. Since $[x, x]$ and $| x |_\alpha$ induce the same topology on l_0, it follows that $[x, y]$ can be extended to a continuous inner product on l_α; we denote this extension also by $[x, y]$. Moreover, $| x | = [x, x]^{1/2}$ and $| x |_\alpha$ induce the same topology on l_α, hence l_α is a Hilbert space relative to $[x, y]$. Therefore, there exists a positive number c such that, for all $x \in l_\alpha$,

$$\frac{1}{c} | x | \leqslant | x |_\alpha \leqslant c | x |. \tag{5.1.16}$$

Now, for any n vectors $\xi_1, \ldots, \xi_n \in l_\alpha$, since $[x, y]$ is an inner product, we have

$$\sum_{\epsilon_j = \pm 1} \left| \sum_{j=1}^{n} \xi_j \epsilon_j \right|^2 = 2^n \sum_{j=1}^{n} | \xi_j |^2,$$

whence, using (5.1.16), we get

$$\frac{2^n}{c^4} \sum_{j=1}^{n} | \xi_j |_\alpha^2 \leqslant \sum_{\epsilon_j = \pm 1} \left| \sum_{j=1}^{n} \xi_j \epsilon_j \right|_\alpha^2 \leqslant c^4 2^n \sum_{j=1}^{n} | \xi_j |_\alpha^2 . \tag{5.1.17}$$

In particular, taking ξ_j to be the vector $\{0, 0, \ldots, 0, x_j, 0, \ldots, 0, \ldots\}$, and substituting into (5.1.17), we find that

$$\frac{1}{c^2} \left(\sum_{j=1}^{n} | x_j |^2 \right)^{1/2} \leqslant \left(\sum_{j=1}^{n} | x_j |^\alpha \right)^{1/\alpha} \leqslant c^2 \left(\sum_{j=1}^{n} | x_j |^2 \right)^{1/2} \tag{5.1.18}$$

for any $n, x_1, \ldots, x_n$. But since $\alpha \neq 2$, this is obviously impossible. Hence, we conclude that $| x |$ and $| x |_\alpha$ induce different topologies on l_0.]

Theorem 5.1.4. If $1 \leqslant \alpha < 2$, then the ergodic measure μ_α in Example 4.2.1 is not equivalent to any Gaussian measure.[4]

PROOF. Suppose there existed a measure N on $(l, \mathfrak{B}_\omega)$, equivalent to μ_α and such that the variables $\{x_\nu(\omega), \nu = 1, 2, \ldots\}$ are Gaussian. Form

[4] *Translator's note:* Meaning, in this case, a measure relative to which the coordinates $\{x_\nu(\omega)\}$ are Gaussian variables.

the characteristic function $\varphi(f)$ of N as in (5.1.15); then, according to (4.2.14), we have

$$R_1(f) = \left(\int_l \frac{f(\omega)^2}{1+f(\omega)^2}\, dN(\omega)\right)^{1/2}$$
$$= \left(\int_0^\infty \Re\left(1 - \exp\left(-\frac{1}{2}c(f,f)\,t^2 + im(f)\,t\right)\right) e^{-t}\,dt\right)^{1/2}, \quad f \in l^\dagger. \tag{5.1.19}$$

Since we have assumed that N and μ_α are equivalent, it follows from Corollary 4.2.15[5] and Example 4.2.1 that the functionals

$$R^\alpha(f) = \left(\int_0^\infty (1 - \exp(-(|f|_\alpha t)^\alpha))e^{-t}\,dt\right)^{1/2}$$

and $R_1(f)$ [see (5.1.19)] induce the same topology on $l^\dagger$.

Write

$$[f,g] = c(f,g) + E(f)\,E(g), \qquad f, g \in l^\dagger.$$

We shall now prove that $R_1(f)$, $f \in l^\dagger$, and $[f,f]^{1/2}$, $f \in l^\dagger$, induce the same topology. To this end, consider the function

$$u(a,b) = \Re\int_0^\infty (1 - \exp(-at^2 + ibt))\,e^{-t}\,dt, \quad 0 \leqslant a < \infty, \ -\infty < b < \infty.$$

Obviously, $\lim_{a+|b|\to 0} u(a,b) = 0$. Conversely, if $u(a_n, b_n) \to 0$, then $a_n + |b_n| \to 0$. In fact, suppose that, for some subsequence, we had $a_n \to a$, where $0 < a \leqslant \infty$; then

$$\underline{\lim}_{n\to\infty} u(a_n, b_n) \geqslant 1 - \int_0^\infty \exp(-at^2 - t)\,dt > 0.$$

On the other hand, if $a_n \to 0$, but $b_n \to b$, where $b \neq 0$, $b \neq \pm\infty$, then

$$\underline{\lim}_{n\to\infty} u(a_n, b_n) \geqslant 1 - \int_0^\infty \cos bt e^{-t}\,dt > 0.$$

The case $a_n \to 0$, $b_n \to \pm\infty$ gives a similar result. To summarize, whenever $\lim u(a,b) = 0$, we have $\lim(a + |b|) = 0$. Consequently, in view of (5.1.19), $R_1(f_n) \to 0$ is equivalent to $c(f_n, f_n) + |E(f_n)| \to 0$,

[5] *Translator's note:* There seems to be a snag here, that is, the transformation group $\mathfrak{G} = l_0$ is *not* of the second category in the topology induced by l; it is not obvious that there exists any stronger topology on l_0 relative to which l_0 is a linear topological space of the second category.

that is, equivalent to $[f_n, f_n] \to 0$. In a similar fashion, we can show that $R^\alpha(f)$ and $|f|_\alpha$ induce the same topology on $l^\dagger$.

Thus, it follows that $|f|_\alpha$ and $[f, f]^{1/2}$ induce the same topology on $l^\dagger$; but this contradicts Lemma 5.1.3. Therefore, we conclude that μ_α is not equivalent to any Gaussian measure.]

Thus, we have given an example of an ergodic measure which is not equivalent to any Gaussian measure.[6]

3° A Special Class of Gaussian Processes: Wiener Processes

There is an important class of stochastic processes, called *Wiener processes*, which are Gaussian, Markov, and have stationary increments. For convenience, we shall only consider Brownian motion on the real line.

Example 5.1.2. Let $C_0[0, T]$ be the totality of continuous functions $x(t)$ on the interval $[0, T]$ satisfying the condition $x(0) = 0$. Clearly, $C_0[0, T]$ forms a Banach space with respect to the usual linear operations and the norm

$$|x| = \max_{0 \leqslant t \leqslant T} |x(t)|.$$

Let $\mathfrak{B}$ denote the totality of weak Borel sets in $C_0[0, T]$. Actually, $\mathfrak{B}$ can be constructed in the following manner: take any partition $0 < t_1 < t_2 < \cdots < t_n \leqslant T$, any Borel set E in real n-dimensional space, and call the set

$$\hat{E} = \{x \mid (x(t_1), \dots, x(t_n)) \in E\} \tag{5.1.20}$$

the *Borel cylinder* with base E associated with the points $t_1, \dots, t_n$. Then $\mathfrak{B}$ is just the σ-algebra generated by the totality of Borel cylinders. Let $0 < c < \infty$, and let W_c be a probability measure on $(C_0[0, T], \mathfrak{B})$ having the following property: for any set $\hat{E}$ of the form (5.1.20),

$$W_c(\hat{E}) = \frac{1}{(c\pi)^{n/2}(\prod_{\nu=1}^{n} (t_\nu - t_{\nu-1}))^{1/2}} \int \cdots \int_E \exp\left(-\frac{1}{c}\sum_{\nu=1}^{n} \frac{(x_\nu - x_{\nu-1})^2}{t_\nu - t_{\nu-1}}\right) dx_1 \cdots dx_n, \tag{5.1.21}$$

where $t_0 = x_0 = 0$. Then, $\mathscr{W}_c = (C_0[0, T], \mathfrak{B}, W_c)$ is called a *Wiener measure space*.

For each $t \in [0, T]$, the function $x(t)$, $x \in C_0[0, T]$, is a measurable

[6] *Translator's note:* This assertion must be interpreted in the context of the translator's note appended to the statement of Theorem 5.1.4.

function on the measurable space $(C_0[0, T], \mathfrak{B})$, and is therefore a random variable on $\mathscr{W}_c$. Moreover, we see from (5.1.21) that the variables $x(t)$ are Gaussian. The Gaussian stochastic process $\{x(t), t \in [0, T]\}$ on $\mathscr{W}_c$ is called a *Wiener process*. Using (5.1.21), one easily calculates that the mean and correlation functions of the Wiener process are

$$E(x(t)) = 0, \tag{5.1.22}$$

$$E(x(t)\, x(s)) = \frac{c}{2} \min(t, s). \tag{5.1.23}$$

Now, let $0 \leqslant \alpha \leqslant 1$ and let $C_0^{(\alpha)}[0, T]$ denote the totality of functions in $C_0[0, T]$ which satisfy the Hölder continuity condition

$$\sup_{0 \leqslant t_1 < t_2 \leqslant T} \frac{|\, x(t_1) - x(t_2)|}{|\, t_1 - t_2\,|^\alpha} < \infty.$$

Theorem 5.1.5. The Wiener measure space $\mathscr{W}_c = (C_0[0, T], \mathfrak{B}, W_c)$ does exist. Moreover, if $0 \leqslant \alpha < \frac{1}{2}$, then the Wiener measure W_c is concentrated on $C_0^{(\alpha)}[0, T]$.

Proof. Since $C_0^{(0)}[0, T] = C_0[0, T]$, it suffices to prove the theorem for $\alpha > 0$. Clearly, we may assume that $T = 1$, $c = 2$. Let I denote the totality of numbers of the form

$$\frac{m}{2^n}, \qquad m = 1, 2, \ldots, 2^n; \quad n = 1, 2, \ldots.$$

Obviously, I is a denumerable set. For each $t \in I$, let R_t be a copy of the real line, and form the Cartesian product $R(I) = \times_{t \in I} R_t$, that is, the set of all real-valued functions on I. Let $\mathfrak{B}_1$ be the σ-algebra generated by the totality of Borel cylinders in $R(I)$. By virtue of the remarks at the beginning of 2°, there exists a canonical Gaussian measure space $(R(I), \mathfrak{B}_1, P)$ such that the coordinates have mathematical expectation (5.1.22) and correlations (5.1.23), where $t, s \in I$. Observe that, if $0 < t_1 < t_2 < \cdots < t_n \leqslant 1$, where $t_j \in I$ for $j = 1, \ldots, n$, and $\hat{E}$ is of the form (5.1.20), then the value of $P(\hat{E})$ is indeed given by (5.1.21).

Write $I_1 = I + \{0\}$, and extend the functions in $R(I)$ to I_1 by defining $x(0) = 0$. For $0 < \alpha \leqslant 1$, $K \geqslant 0$, let $C_{\alpha K}(I_1)$ denote the totality of functions $x \in R(I)$ which satisfy the condition

$$|\, x(t_1) - x(t_2)| \leqslant K\, |\, t_1 - t_2\,|^\alpha, \qquad t_1, t_2 \in I_1.$$

Let $C_\alpha(I_1) = \bigcup_{k=1}^{\infty} C_{\alpha k}(I_1)$.

Lemma 5.1.6. If $0 < \alpha < \frac{1}{2}$, then

$$P(R(I) - C_\alpha(I_1)) = 0. \tag{5.1.24}$$

PROOF. For any positive number K, positive integer n, and non-negative integer m, $0 \leqslant m < 2^n$, let

$$S(K, n, m) = \left\{x \mid x \in R(I), \left| x\left(\frac{m+1}{2^n}\right) - x\left(\frac{m}{2^n}\right)\right| > K\left(1 - \frac{1}{2^\alpha}\right) \cdot \frac{1}{2^{\alpha n+1}}\right\}.$$

Clearly, $S(K, m, n) \in \mathfrak{B}_1$. We shall now prove that

$$R(I) - C_{\alpha K}(I_1) \subset \bigcup_{n=1}^{\infty} \bigcup_{m=0}^{2^n-1} S(K, m, n). \tag{5.1.25}$$

Take any $x \in R(I) - C_{\alpha K}(I_1)$; then, for some $t_1 < t_2$, $t_1, t_2 \in I_1$, we have

$$| x(t_2) - x(t_1)| > K \mid t_2 - t_1 \mid^\alpha.$$

Let

$$j = \min\left\{n \,\middle|\, t_2 - t_1 > \frac{1}{2^{n+1}}\right\}. \tag{5.1.26}$$

Since

$$\frac{1}{2^{(j+1)\alpha}} = \sum_{\nu=j+1}^{\infty} \frac{1}{2^{\alpha\nu}}\left(1 - \frac{1}{2^\alpha}\right),$$

we get

$$| x(t_2) - x(t_1)| > K\left(1 - \frac{1}{2^\alpha}\right) \sum_{\nu=j+1}^{\infty} \frac{1}{2^{\alpha\nu}}. \tag{5.1.27}$$

We shall show that $x \in \bigcup_{n=j+1}^{\infty} \bigcup_{m=0}^{2n-1} S(K, n, m)$. Suppose this were not true; then, $x \bar{\in} S(K, m, n)$ for all $n \geqslant j + 1$, $0 \leqslant m < 2^n$, that is,

$$\left| x\left(\frac{m+1}{2^n}\right) - x\left(\frac{m}{2^n}\right)\right| \leqslant K\left(1 - \frac{1}{2^\alpha}\right) \cdot \frac{1}{2^{\alpha n+1}},$$

$$0 \leqslant m < 2^n, \quad n \geqslant j + 1. \tag{5.1.28}$$

Let us refer to an interval of the form

$$\left(\frac{m}{2^n}, \frac{m+1}{2^n}\right]$$

as an interval of order m. By (5.1.26), we have $t_2 - t_1 \leqslant (1/2^j)$. Therefore, $(t_1, t_2]$ contains at most two intervals of order $j + 1$, which we denote by $I_{j+1}^{(1)}$, $I_{j+1}^{(2)}$, with the understanding that if $(t_1, t_2]$ should contain only one interval of order $j+1$, or none at all, then one or both of $I_{j+1}^{(1)}$, $I_{j+1}^{(2)}$ shall denote the empty set. Now, since the set $(t_1, t_2] - (I_{j+1}^{(1)} \cup I_{j+1}^{(2)})$ intersects at most two intervals of order $j+1$, it contains at most two intervals of order $j+2$. Continuing in this fashion, we obtain mutually disjoint sets $I_{j+l}^{(\nu)}$ ($\nu = 1, 2; l = 1, 2, \ldots$), where $I_{j+l}^{(\nu)}$ is either an interval of order $j+l$ or the empty set, such that

$$(t_1, t_2] = \bigcup_{l=1}^{m} (I_{j+l}^{(1)} + I_{j+l}^{(2)}). \tag{5.1.29}$$

However, when $n \geqslant j+1$, we have the estimate (5.1.28) for the difference of the values of $x(t)$ at the two endpoints of $I_n^{(\nu)}$ [if $I_n^{(\nu)}$ is nonempty]. Therefore, using (5.1.29), we get

$$\begin{aligned} |x(t_1) - x(t_2)| &\leqslant K\left(1 - \frac{1}{2^\alpha}\right) \cdot 2 \sum_{n=j+1}^{\infty} \frac{1}{2^{\alpha n+1}} \\ &= K\left(1 - \frac{1}{2^\alpha}\right) \sum_{n=j+1}^{\infty} \frac{1}{2^{\alpha n}}, \end{aligned} \tag{5.1.30}$$

which contradicts (5.1.27). We conclude that (5.1.25) is true.

Write $b = \frac{1}{2}(1 - 2^{-\alpha})$. Using (5.1.21), we calculate that, for $m > 0$,

$$\begin{aligned} &P(S(K, m, n)) \\ &\quad = \frac{1}{\dfrac{\pi}{2^{n-1}} m^{1/2}} \iint_{|x_1 - x_2| > bK/2^{\alpha n}} \exp\left[-\frac{1}{2}\left(\frac{2^n x_1^2}{m} + 2^n (x_2 - x_1)^2\right)\right] dx_1\, dx_2 \\ &\quad = \beta(bK 2^{n(\frac{1}{2} - \alpha)}), \end{aligned} \tag{5.1.31}$$

where $\beta(t)$ denotes the function $(2/\pi)^{1/2} \int_t^\infty \exp(-x^2/2)\, dx$. Similarly, we get

$$P(S(K, 0, n)) = \beta(bK 2^{n(\frac{1}{2} - \alpha)}). \tag{5.1.31$'$}$$

However, when $t \geqslant 2$, it is easily seen that $\beta(t) \leqslant (2/\pi)^{1/2} e^{-t}$. Hence, by (5.1.25), (5.1.31), and (5.1.31′), it follows that, if $K > 2^{\frac{1}{2}+\alpha}/b$, then

$$\begin{aligned} P(R(I) - C_{\alpha K}(I_1)) &\leqslant \sum_{n=1}^{\infty} \sum_{m=0}^{2^n - 1} P(S(K, m, n)) \\ &\leqslant \sum_{n=1}^{\infty} \sum_{m=0}^{2^n - 1} \beta(bK 2^{n(\frac{1}{2} - \alpha)}) \\ &\leqslant (2/\pi)^{1/2} \sum_{n=1}^{\infty} 2^n \exp(-bK 2^{n(\frac{1}{2} - \alpha)}). \end{aligned} \tag{5.1.32}$$

Clearly, the right-hand side of (5.1.32) tends to zero as $K \to \infty$. But $R(I) - C_\alpha(I_1) \subset R(I) - C_{\alpha K}(I_1)$, hence, letting $K \to \infty$ in (5.1.32), we obtain (5.1.24).

If $x \in C_\alpha(I_1)$, $\alpha > 0$, then x is uniformly continuous on I_1, hence x can be uniquely extended to a continuous function on [0, 1], which we also denote by x, moreover, $x \in C_0^{(\alpha)}[0, 1]$. Identifying $C_\alpha(I_1)$ with $C_0{}^\alpha[0, 1]$ in this manner, it is easily seen that the restriction $\mathfrak{F}^{(\alpha)}$ of $\mathfrak{B}_1$ to $C_\alpha(I_1)$ is just the restriction of $\mathfrak{B}$ to $C_0^{(\alpha)}[0, 1]$. Let $P^{(\alpha)}$ denote the restriction of P to $C_\alpha(I_1)$ (i.e., to $C_0^{(\alpha)}[0, 1]$). We thus obtain a Gaussian measure space $S^{(\alpha)} = (C_0^{(\alpha)}[0, 1], \mathfrak{F}^{(\alpha)}, P^{(\alpha)})$, $0 < \alpha < \frac{1}{2}$. Moreover, $\{x(t), t \in [0, 1]\}$ are clearly random variables on $S^{(\alpha)}$; when $t \in I_1$, they are Gaussian variables with mathematical expectation and covariance function given by (5.1.22) and (5.1.23), where $t, s \in I_1$.

We shall now prove that, for any $t \in [0, 1]$, $x(t)$ is also a Gaussian variable on $S^{(\alpha)}$, and that (5.1.22) and (5.1.23) are valid for any $t, s \in [0, 1]$. Let $t_0 \in [0, 1] - I_1$, and choose a sequence $\{t_n\} \subset I_1$ such that $t_n \to t_0$. By (5.1.23), we have

$$E((x(t_n) - x(t_m))^2) = | t_n - t_m |,$$

hence $\{x(t_n)\}$, $n = 1, 2,\ldots$, is a fundamental sequence in $L^2(S^{(\alpha)})$. Moreover, since the functions x in $C_0^{(\alpha)}[0, 1]$ are continuous, $\lim_{n\to\infty} x(t_n) = x(t_0)$, consequently, $\{x(t_n)\}$ converges to $x(t_0)$ in $L^2(S^{(\alpha)})$. Hence, it follows by Lemma 5.1.1 that $x(t_0)$ is a Gaussian variable, furthermore, $E((x(t_n) - x(t_0))^2) \to 0$, whence it easily follows that (5.1.22) holds for $t = t_0$. Now, let $s_0 \in [0, 1]$, and choose $\{s_n\} \subset I_1$ such that $s_n \to s_0$; then we also have $E((x(s_n) - x(s_0))^2) \to 0$, therefore, since

$$| E(x(s_0)\, x(t_0)) - E(x(s_n)\, x(t_n))|$$
$$\leqslant [E((x(s_0) - x(s_n))^2)\, E(x(t_n)^2)]^{1/2} + [E(s_0{}^2)\, E((x(t_0) - x(t_n))^2)]^{1/2},$$

it follows that (5.1.23) holds for $s = s_0$, $t = t_0$.

If $\tilde{E} \in \mathfrak{B}$, then obviously $\tilde{E} \cap C_0^{(\alpha)}[0, 1] \in \mathfrak{F}^{(\alpha)}$. Define

$$W_c(\tilde{E}) = P^{(\alpha)}(\tilde{E} \cap C_0^{(\alpha)}[0, 1]).$$

Clearly, W_c is a probability measure on $(C_0[0, 1], \mathfrak{B})$ which is concentrated on $C_0^{(\alpha)}[0, 1]$, and is such that (5.1.22) and (5.1.23) hold. Hence, W_c satisfies (5.1.21), and is therefore the required Wiener measure.]

§5.2. Equivalence and Perpendicularity of Gaussian Measures

1° The Basic Theorem

In §5.1, we considered the question of equivalence for a rather special class of Gaussian measures. In the present section, we shall consider a more general problem. Let $(\Omega, \mathfrak{B})$ be a measurable space, and $\{\xi_\alpha, \alpha \in \mathfrak{A}\}$ a family of measurable functions on $(\Omega, \mathfrak{B})$. Let P_1, P_2 be two probability measures on $(\Omega, \mathfrak{B})$ such that $\{\xi_\alpha, \alpha \in \mathfrak{A}\}$ is a joint determining set for the two measure spaces $(\Omega, \mathfrak{B}, P_\nu) = S_\nu$, $\nu = 1, 2$. Moreover, suppose that $\{\xi_\alpha(\cdot), \alpha \in \mathfrak{A}\}$ is a Gaussian stochastic process with respect to both of the measure spaces S_1 and S_2. We shall investigate the equivalence and perpendicularity of P_1 and P_2.

Henceforth, we shall write

$$E_k(x) = \int_\Omega x(\omega)\, dP_k(\omega), \qquad D_k(x) = E_k((x - E_k(x))^2), \quad k = 1, 2.$$

Let $\mathscr{L}$ denote the totality of linear combinations of functions $\{\xi_\alpha, \alpha \in \mathfrak{A}\}$ and real constant functions on Ω; obviously, $\mathscr{L}$ forms a linear space. Define two nonnegative definite bilinear functionals on $\mathscr{L}$, as follows:

$$(x, y)_k = E_k(xy), \qquad k = 1, 2.$$

Lemma 5.2.1. If the Gaussian measures P_1 and P_2 are not mutually singular, then there exist positive numbers c_1 and c_2 such that, for all $x \in \mathscr{L}$,

$$(x, x)_2 \leqslant c_1(x, x)_1, \tag{5.2.1}$$

and

$$(x, x)_1 \leqslant c_2(x, x)_2. \tag{5.2.2}$$

PROOF. Suppose there were no $c_1 > 0$ such that (5.2.1) holds; then there is a sequence $\{x_n\} \subset \mathscr{L}$ such that

$$(x_n, x_n)_2 = 1, \qquad (x_n, x_n)_1 \to 0\ (n \to \infty).$$

Write $E_2(x_n) = a_n$. Consider the sets $A_n = \{\omega \mid |x_n| \leqslant (x_n, x_n)_1^{1/4}\}$ and $\Omega - A_n = \{\omega \mid |x_n| > (x_n, x_n)_1^{1/4}\}$. If $(x_n, x_n)_1 > 0$, then, by the Chebyshev inequality[7]

$$P_1(\Omega - A_n) \leqslant \frac{1}{(x_n, x_n)_1^{1/2}} \int_\Omega x_n^2\, dP_1(\omega) \leqslant (x_n, x_n)_1^{1/2}.$$

[7] Let $P(\omega)$ be a probability measure, $X(\omega)$ a random variable and $E(X) = \int X(\omega) dP(\omega)$; then

$$P(\{\omega \mid |X(\omega) - E(X)| \geqslant a\}) \leqslant \frac{1}{a^2} \int_{|X(\omega) - E(X)| \geqslant a} (X(\omega) - E(X))^2\, dP(\omega).$$

If $(x_n, x_n)_1 = 0$, then $x_n = 0$ almost everywhere (P_1), hence $P_1(\Omega - A_n) = 0$. Thus, we have

$$\lim_{n\to\infty} P_1(\Omega - A_n) = 0. \tag{5.2.3}$$

Now, let $\{n_k\}$ be any increasing sequence of positive integers. If $\lim_{k\to\infty} a_{n_k}^2 = 1$, then, writing n in place of n_k, we have $E_2((x_n - a_n)^2) \to 0$, whence it follows that $x_n - a_n$ converges to zero in measure (P_2). Let N be an integer such that $(x_n, x_n)_1^{1/4} < \frac{1}{4}$ and $|1 - a_n{}^2| < \frac{1}{4}$ for all $n \geqslant N$. Then, for $\omega \in A_n$, that is, $|x|_n \leqslant (x_n, x_n)_1^{1/4}$, we have $|x_n - a_n| \geqslant \frac{1}{2}$ for all $n \geqslant N$. But $P_2(\{\omega \mid |x_n - a_n| \geqslant \frac{1}{2}\}) \to 0$ as $n \to \infty$, hence $P_2(A_n) \to 0$. On the other hand, if $a_{n_k}^2$ does not converge to 1 as $k \to \infty$, then there is a subsequence $\{n_k'\}$ of $\{n_k\}$ such that $\lim_{k\to\infty} a_{n_k'}^2 < 1$. Since any linear combination of Gaussian variables is also Gaussian, it follows by (5.1.1) that (for brevity, we write n in place of n_k')

$$P_2(A_n) = \frac{1}{[2\pi(1 - a_n{}^2)]^{1/2}} \int_{|\lambda| \leqslant (x_n, x_n)_1^{1/4}} \exp\left(-\frac{(\lambda - a_n)^2}{2(1 - a_n{}^2)}\right) d\lambda$$

$$= \frac{1}{(2\pi)^{1/2}} \int_{\frac{-a_n - (x_n, x_n)_1^{1/4}}{(1 - a_n{}^2)^{1/2}}}^{\frac{-a_n + (x_n, x_n)_1^{1/4}}{(1 - a_n{}^2)^{1/2}}} \exp(-\lambda^2/2)\, d\lambda \to 0.$$

Thus, $\{n_k\}$ always contains a subsequence $\{n_k'\}$ such that $P_2(A_{n_k'}) \to 0$. Therefore, we must have

$$\lim_{n\to\infty} P_2(A_n) = 0. \tag{5.2.4}$$

But then it follows from (5.2.3), (5.2.4), and Lemma 1.1.21 that P_1 and P_2 are mutually singular, contrary to hypothesis. We conclude that (5.2.1) holds for some $c_1 > 0$. Similarly, (5.2.2) holds for some $c_2 > 0$.]

We continue to assume that P_1 and P_2 are not mutually singular. By virtue of (5.2.1) and (5.2.2), equality almost everywhere (P_1) and equality almost everywhere (P_2) are equivalent for functions in $\mathscr{L}$. If we identify functions in $\mathscr{L}$ which coincide almost everywhere (P_1), or, what is the same, coincide almost everywhere (P_2), then $\mathscr{L}$ becomes an inner product space relative to each of the inner products $(x, y)_k$, $k = 1, 2$. Thus, $\mathscr{L}$ may be regarded as a linear subspace of each of the Hilbert spaces $L^2(S_1)$ and $L^2(S_2)$. Let H_1 and H_2 denote the closures of $\mathscr{L}$ in $L^2(S_1)$ and $L^2(S_2)$, respectively; by Lemma 5.1.1, the elements of H_k are also Gaussian variables on S_k, $k = 1, 2$. Suppose that $\xi \in H_2$, and let $\{x_n\} \subset \mathscr{L}$ be a sequence of functions converging to ξ in the norm $(x, x)_2$. Since $\{x_n\}$ is a Cauchy sequence relative to $(x, x)_2$, it follows by (5.2.2) that it is also a Cauchy sequence relative to $(x, x)_1$, and hence converges in the norm $(x, x)_1$ to an element of H_1, which we call $T\xi$. It is easily seen that the correspondence $\xi \to T\xi$ defines a one-to-one linear mapping T from H_2 onto H_1 such that the restriction $T \mid \mathscr{L}$ is the identity mapping on $\mathscr{L}$. Using the mapping T, we shall identify the point-sets of H_1 and H_2, i.e., we shall regard H_1 and H_2 as the Hilbert spaces associated with the respective inner products $(x, y)_1$ and $(x, y)_2$ defined on the same underlying linear space[8], which we call H. With this understanding, inequalities (5.2.1) and (5.2.2) are valid for all $x \in H$.

Lemma 5.2.2. If P_1 and P_2 are not mutually singular, then there exists a positive number K such that, for any finite set of vectors $\eta_1, \ldots, \eta_n$ in H satisfying the conditions

$$E_2(\eta_k) = 0, \qquad (\eta_k, \eta_l)_2 = \delta_{kl},$$

$$(\eta_k - E_1(\eta_k), \eta_l - E_1(\eta_l))_1 = D_1(\eta_k)\, \delta_{kl},$$

(where δ_{kl} is the Kronecker delta) the following inequality holds,

$$\sum_{k=1}^{n} (D_1(\eta_k) - 1)^2 + \sum_{k=1}^{n} E_1(\eta_k)^2 \leqslant K. \tag{5.2.5}$$

Proof. For convenience, we write $E_1(\eta_k) = m_k$, $D_1(\eta_k) = \sigma_k$. Now, given any set $\{\eta_1, \ldots, \eta_n\}$ satisfying the conditions of the lemma, we form the measurable function[9]

$$f(\omega) = \sum_{k=1}^{n} \left[\frac{(\eta_k - m_k)^2}{\sigma_k} - \eta_k^2\right]$$

[8] Actually, for each $\xi \in H_2$, one can choose a measurable function x on $(\Omega, \mathfrak{B})$ such that x represents ξ in H_2 and represents $T\xi$ in H_1. In fact, let $\{x_n\}$ be a sequence of functions in $\mathscr{L}$ such that x_n converges in the norm $(\cdot, \cdot)_2$ to a function x' representing ξ, and also converges in the norm $(\cdot, \cdot)_1$ to a function x'' representing $T\xi$. Choosing subsequences, if necessary, we may assume that x_n converges almost everywhere (P_2) to x' and also converges almost everywhere (P_1) to x''. Let $x(\omega) = \lim x_n(\omega)$ if this limit exists, and $x_n(\omega) = 0$ otherwise. Then the function x has the desired properties.

[9] Here, we choose functions representing $\{\eta_1, \ldots, \eta_n\}$ in the manner indicated in footnote 8.

on $(\Omega, \mathfrak{B})$. It is easy to calculate that[10]

$$
\begin{aligned}
E_1(f) &= \sum_k (1 - \sigma_k - m_k^2), \\
E_2(f) &= \sum_k \left[\frac{1}{\sigma_k} - 1 + \frac{m_k^2}{\sigma_k}\right], \\
D_1(f) &= \sum_k [2(\sigma_k - 1)^2 + 4\sigma_k m_k^2], \\
D_2(f) &= \sum_k \left[2\,\frac{(\sigma_k - 1)^2}{\sigma_k^2} + 4\,\frac{m_k^2}{\sigma_k^2}\right].
\end{aligned}
\tag{5.2.6}
$$

Write

$$
\begin{aligned}
a(f) &= \frac{1}{2}(E_2(f) - E_1(f)) \\
&= \frac{1}{2}\sum \left[\frac{(1 - \sigma_k)^2}{\sigma_k} + \left(1 + \frac{1}{\sigma_k^2}\right) m_k^2\right],
\end{aligned}
\tag{5.2.7}
$$

and let

$$
\begin{aligned}
A_1 &= \{\omega \mid f(\omega) > \tfrac{1}{2}(E_1(f) + E_2(f))\}, \\
A_2 &= \{\omega \mid f(\omega) \leqslant \tfrac{1}{2}(E_1(f) + E_2(f))\}.
\end{aligned}
$$

Then A_1 and A_2 are disjoint, moreover, $A_1 \cup A_2 = \Omega$. By the Chebyshev inequality, we get

$$
P_1(A_1) \leqslant P_1(\{\omega \mid |f(\omega) - E_1(f)| > a(f)\}) \leqslant \frac{D_1(f)}{a(f)^2}, \tag{5.2.8}
$$

$$
P_2(A_2) \leqslant P_2(\{\omega \mid |f(\omega) - E_2(f)| \geqslant a(f)\}) \leqslant \frac{D_2(f)}{a(f)^2}. \quad 5.2.9)
$$

[10] For example, $D_1(f)$ may be calculated as follows: since η_k and $\eta_{k'}$ are mutually independent for $k \neq k'$, we have

$$
\begin{aligned}
E_1((f - E_1(f))^2) &= \sum E_1\left(\frac{(\eta_k - m_k)^2}{\sigma_k} - (\eta_k - m_k)^2 + 2m_k(\eta_k - m_k) - (1 - \sigma_k)\right)^2 \\
&= \sum \Bigg[\left(\frac{1}{\sigma_k} - 1\right)^2 E_1((\eta_k - m_k)^4) \\
&\quad + \left(4m_k^2 - 2\,\frac{(1 - \sigma_k)^2}{\sigma_k}\right) E_1((\eta_k - m_k)^2) + (1 - \sigma_k)^2\Bigg] \\
&= \sum_k [2(\sigma_k - 1)^2 + 4m_k^2\sigma_k].
\end{aligned}
$$

On the other hand, by (5.2.1) and (5.2.2), we have

$$\begin{gathered}\sigma_k \leqslant (\eta_k , \eta_k)_1 \leqslant c_2(\eta_k , \eta_k)_2 \leqslant c_2 , \\ \frac{1}{c_1} \leqslant \frac{1}{c_1}(\eta_k - m_k , \eta_k - m_k)_2 \leqslant (\eta_k - m_k , \eta_k - m_k)_1 = \sigma_k .\end{gathered} \tag{5.2.10}$$

If we write

$$m = \sum_{k=1}^{n} [(1 - \sigma_k)^2 + m_k^{\,2}],$$

then it follows from (5.2.6), (5.2.7), and (5.2.10) that there exist positive numbers b_1 , b_2 and b (independent of n and $\eta_1 ,..., \eta_n$) such that

$$D_k(f) \leqslant b_k m, \qquad k = 1, 2, \quad a(f) \geqslant bm.$$

Consequently, by (5.2.8) and (5.2.9),

$$P_\nu(A_\nu) \leqslant \frac{b_\nu}{b^2}\frac{1}{m} . \tag{5.2.11}$$

Now, as n and $\{\eta_1 ,..., \eta_n\}$ vary, if the supremum of m were $+\infty$, then, for any given positive number ϵ, there would exist some $\{\eta_1 ,..., \eta_n\}$ such that the corresponding number m satisfies $m \geqslant (1/b^2\epsilon) \max(b_1 , b_2)$, and so, by (5.2.11), the corresponding set A_1 would satisfy

$$P_1(A_1) < \epsilon, \qquad P_2(\Omega - A_1) < \epsilon.$$

It would then follow from Lemma 1.1.21 that P_1 and P_2 are mutually singular, contrary to hypothesis. Therefore, we conclude that m has a finite upper bound K.]

Now, let H_0 denote the totality of vectors of H which are orthogonal to the vector 1 relative to the inner product $(\cdot, \cdot)_2$, that is, H_0 comprises the totality of functions in H having mean value zero relative to the measure P_2 . Then H_0 is a closed linear subspace of H_2 , and hence is itself a Hilbert space with respect to the norm $(\cdot, \cdot)_2$. We form a non-negative definite bilinear functional on H_0 , as follows:

$$B(\xi, \eta) = (\xi - E_1(\xi), \eta - E_1(\eta))_1 .$$

By (5.2.2),

$$B(\eta, \eta) = (\eta, \eta)_1 - E_1(\eta)^2 \leqslant c_2(\eta, \eta)_2 ,$$

which shows that $B(\xi, \eta)$ is bounded. By a well-known theorem, there exists a bounded self-adjoint operator B on H_0 such that

$$B(\xi, \eta) = (B\xi, \eta)_2$$

for any $\xi, \eta \in H_0$.

Lemma 5.2.3. If P_1 and P_2 are not mutually singular, then $B - I$ (where I is the identity operator in H_0) is a Hilbert–Schmidt operator on H_0.

PROOF. According to Lemma 5.2.2, there exists a positive number K such that

$$\sum (1 - \sigma_k)^2 = \sum_k (1 - B(\eta_k, \eta_k))^2 = \sum_k ((B - I)\, \eta_k, \eta_k)_2{}^2 \leqslant K$$

for any finite set of vectors $\eta_1, \ldots, \eta_n \in H_0$ which is orthonormal relative to the inner product $(\cdot, \cdot)_2$, and satisfies

$$B(\eta_k, \eta_{k'}) = 0$$

whenever $k \neq k'$, that is, $((B - I)\, \eta_k, n_{k'})_2 = 0$. But then it follows immediately by Lemma II.1.5 of Appendix II that $B - I$ is a Hilbert–Schmidt type operator from H_0 to H_0.]

By virtue of the Hilbert–Schmidt property of $B - I$, there exists (see §II.1) a complete orthonormal system $\{\eta_\alpha, \alpha \in \mathfrak{A}\}$ in H_0 [relative to the inner product $(\cdot, \cdot)_2$] consisting of eigenvectors of $B - I$, or, what is the same, eigenvectors of B, moreover, if we denote the corresponding eigenvalues by σ_α, that is, $B\eta_\alpha = \sigma_\alpha \eta_\alpha$, then

$$\sum_\alpha (\sigma_\alpha - 1)^2 \leqslant K < \infty. \tag{5.2.12}$$

In particular, there are at most countably many indices α, say $\alpha = 1, 2, \ldots$, such that $\sigma_\alpha \neq 1$. Furthermore, by Lemma 5.2.2,

$$\sum_\alpha E_1(\eta_\alpha)^2 \leqslant K < \infty. \tag{5.2.13}$$

Lemma 5.2.4. If P_1 and P_2 are not mutually singular, then they are equivalent.

PROOF. Since $\mathscr{L}$ is a joint determining set for S_1 and S_2, and since each function in $\mathscr{L}$ can be expressed as a sum of a function in H_0 and a constant function, it follows that the totality of functions in H_0 also forms a joint determining set for S_1 and S_2. Moreover, since $\{\eta_\alpha, \alpha \in \mathfrak{A}\}$ is a complete orthonormal set in H_0, the totality of linear combinations of the η_α is dense in H_0 relative to the inner product $(\cdot, \cdot)_2$, and hence [by (5.2.2)] also dense in H_0 relative to the inner product $(\cdot, \cdot)_1$. Using this fact, it is not difficult to deduce that $\{\eta_\alpha, \alpha \in \mathfrak{A}\}$[11] is a joint determining set for S_1 and S_2. Let $\Omega_0 = \mathsf{X}_{\alpha\in\mathfrak{A}} R_\alpha$, where R_α is the real line for every $\alpha \in \mathfrak{A}$, and let $\mathfrak{B}_0$ be the σ-algebra generated by the totality of Borel cylinders in Ω_0. Now, from the discussion at the beginning of §5.1, 2°, we see that there exists a probability measure N_k on $(\Omega_0, \mathfrak{B}_0)$ such that $(\Omega_0, \mathfrak{B}, N_k)$ is a sample measure space for the Gaussian process $\{\eta_\alpha(\cdot), \alpha \in \mathfrak{A}\}$ on $(\Omega_0, \mathfrak{B}_0, N_k)$, $k = 1, 2$.

Let p_α denote the αth coordinate of the point p of Ω_0; then the stochastic process $\{p_\alpha(p), \alpha \in \mathfrak{A}\}$ on $(\Omega_0, \mathfrak{B}_0, N_k)$ has the same finite-dimensional distributions as the Gaussian process $\{\eta_\alpha(\omega), \alpha \in \mathfrak{A}\}$ on S_k. Consequently, the variables $\{p_\alpha, \alpha \in \mathfrak{A}\}$ are also mutually independent and Gaussian, moreover, their respective mathematical expectations and variances with respect to N_1, N_2 are $E_1(\eta_\alpha)$, 0; σ_α, 1. Using conditions (5.2.12) and (5.2.13), it follows by Lemma 5.1.2 that N_1 and N_2 are equivalent. Hence, by Lemma 1.3.7, we deduce that P_1 and P_2 are equivalent.]

We are now ready to state and prove the basic theorem on the equivalence of Gaussian measures.

Theorem 5.2.5. Let $S_\nu = (\Omega, \mathfrak{B}, P_\nu)$, $\nu = 1, 2$, be probability measure spaces, and let $\{\xi_\alpha, \alpha \in \mathfrak{A}\}$ be a joint determining set for S_1 and S_2, which is also a Gaussian process with respect to both S_1 and S_2. Then P_1 and P_2 are either equivalent or mutually singular.

Furthermore, a necessary and sufficient condition for the equivalence of P_1 and P_2 is as follows. Let $\mathscr{L}$ be the linear space generated by the functions $\{\xi_\alpha, \alpha \in \mathfrak{A}\}$ and the real-valued constant functions. Then two functions in $\mathscr{L}$ are equal almost everywhere (P_1) if and only if they are equal almost everywhere (P_2), and the identity correspondence between the two types of equivalence classes in $\mathscr{L}$ is induced by an equivalence operator[12] T between the closure H_1 of the P_1-equivalence classes in $L^2(S_1)$ and the closure H_2 of the P_2-equivalence classes in $L^2(S_2)$.[13]

PROOF. We have already proved that P_1 and P_2 are either equivalent or mutually singular (Lemma 5.2.4). Thus, it remains to prove that the above condition is necessary and sufficient for equivalence.

If P_1 and P_2 are equivalent, then it follows from the remarks preceding

[11] Or, more precisely, a set of functions representing the elements η_α, chosen as indicated in footnote 8.

[12] See Appendix II for the definition and properties of equivalence operators.

[13] *Translator's note:* In the original, the author states that the spaces H_1 and H_2 coincide, and calls T the identity mapping from H_1 to H_2; this is not accurate.

Lemma 5.2.2 that T is a bounded linear operator from H_1 onto H_2 such that T^{-1} exists and is bounded. Let Q denote the projection operator from H_2 onto H_0. Then $I - Q$ is the projection operator from H_2 onto the one-dimensional subspace $\{\lambda \mid -\infty < \lambda < \infty\}$. Also, let Q_1 denote the projection operator from H_1 onto the one-dimensional subspace $\{\lambda \mid -\infty < \lambda < \infty\}$; obviously, $Q_1(\xi)$ is just the constant function $E_1(\xi)$. Now, we assert that

$$QT^*TQ = QBQ + QT^*Q_1TQ. \tag{5.2.14}$$

In fact, take any $\xi, \eta \in H_2$, and write $\xi_1 = Q\xi$, $\eta_1 = Q\eta \in H_0$; then

$$\begin{aligned}((QBQ + QT^*Q_1TQ)\,\xi, \eta)_2 &= ((B + T^*Q_1T)\,\xi_1, \eta_1)_2 \\ &= B(\xi_1, \eta_1) + (Q_1\xi_1, Q_1\eta_1)_1 \\ &= (\xi_1, \eta_1)_1 = (T^*T\xi_1, \eta_1)_2 \\ &= (QT^*TQ\xi, \eta)_2,\end{aligned}$$

hence (5.2.14) holds. Using (5.2.14), it is easy to calculate that

$$T^*T - I = (B - I)Q + A, \tag{5.2.15}$$

where

$$A = (T^*T - I)(I - Q) + (I - Q)(T^*T - I)Q + QT^*Q_1TQ.$$

Since $I - Q$ and Q_1 are operators of rank one, A is also of rank one, moreover, by Lemma 5.2.3, $(B - I)Q$ is Hilbert–Schmidt. Hence, the left-hand side of (5.2.15) is a Hilbert–Schmidt operator, that is, T is an equivalence operator.

Conversely, if the condition stated in the Theorem is satisfied, we form the space H_0 and the operator B just as before. By (5.2.14), we have $(B - I)Q = Q(T^*T - I)Q - QT^*Q_1TQ$, whence it follows that $B - I$ is a Hilbert–Schmidt type operator on H_0. Again, we choose the random variables $\{\eta_\alpha, \alpha \in \mathfrak{A}\}$ in the manner indicated following Lemma 5.2.3, and (5.2.12) holds as before. Since T is bounded and Q_1 is of rank one, it is clear that

$$\sum_\alpha E_1(\eta_\alpha)^2 = \sum_\alpha \| Q_1T\eta_\alpha \|_1^2 \leqslant K < \infty,$$

that is, (5.2.13) still holds. Now, proceeding as in the proof of Lemma 5.2.4, we deduce that P_1 and P_2 are equivalent.]

Remark. Examining the proof of Theorem 5.2.5, it is easily seen that, if $E_k(\xi_\alpha) = 0$, $k = 1, 2$, $\alpha \in \mathfrak{A}$, then one may define $\mathscr{L}$ as the totality of linear combinations of the functions $\{\xi_\alpha, \alpha \in \mathfrak{A}\}$, without introducing the constant functions, and with respect to this $\mathscr{L}$, the condition of Theorem 5.2.5 is still necessary and sufficient for the equivalence of P_1 and P_2.

2° Some Concrete Examples

We shall now give an example illustrating the application of Theorem 5.2.5 to certain Gaussian processes having the same mathematical expectations but different correlation functions.

Example 5.2.1. Consider the class of Gaussian processes whose mathematical expectation is identically zero, and whose correlation function $r(s, t)$, $a \leqslant s, t \leqslant b$, has the properties (i)–(vi) listed below.

(i) $r(s, t)$ is a continuous function of the variables $s, t \in [a, b]$.

(ii) For every fixed $t \in [a, b]$, $r(s, t)$ is an absolutely continuous function of s, $a \leqslant s \leqslant b$, moreover,

$$\int_a^b \left| \frac{\partial r(s, a)}{\partial s} \right|^2 ds < \infty.$$

(iii) For almost every (in Lebesgue measure) fixed $s \in [a, b]$, the function $\partial r(s, t)/\partial s$ is absolutely continuous on the intervals $[a, s)$ and $(s, b]$ [whence it follows that $\partial^2 r(s, t)/\partial s\, \partial t$ exists almost everywhere and is measurable on the square $a \leqslant s$, $t \leqslant b$].

(iv) $$\int_a^b \int_a^b \left| \frac{\partial^2 r(s, t)}{\partial s\, \partial t} \right|^2 ds\, dt < \infty.$$

By virtue of condition (iii), the limits

$$q_-(s) = \lim_{t \to s-0} \frac{\partial r(s, t)}{\partial s}, \qquad q_+(s) = \lim_{t \to s+0} \frac{\partial r(s, t)}{\partial s} \tag{5.2.16}$$

exist and are finite for almost all $s \in [a, b]$, moreover, it is easily seen that $q_-(s)$ and $q_+(s)$ are measurable functions. We further assume the following property.

(v) There exist positive numbers c_1, c_2 such that the function $D_r(s) = q_+(s) - q_-(s)$ satisfies the relation

$$c_1 \leqslant D_r(s) \leqslant c_2 \tag{5.2.17}$$

almost everywhere.

We call $D_r(s)$ the *discriminator*[14] corresponding to the correlation

[14] *Translator's note:* The Chinese term used by the author is the same as that expressing "characteristic function," which seems rather inappropriate here.

function r. And finally, we assume that $\partial^2 r(s, t)/\partial s\, \partial t$ is a positive definite integral kernel, that is, the following property holds.

(vi) For every $f \in L^2[a, b]$,

$$\int_a^b \int_a^b f(t) f(s) \frac{\partial^2 r(s, t)}{\partial s\, dt}\, ds\, dt \geqslant 0.$$

We say that a Gaussian process satisfying the conditions listed above is of *type* $S[a, b]$, and its correlation function $r(s, t)$ is said to be of *type* $s[a, b]$.

Suppose that $u(s)$, $v(s)$ are real-valued absolutely continuous functions on $[a, b]$ such that $v(s) > 0$ $(s \in [a, b])$,

$$\int_a^b u'(s)^2\, ds < \infty, \qquad \int_a^b v'(s)^2\, ds < \infty,$$

and that there exist positive numbers c_1, c_2 such that

$$c_1 \leqslant v(s)\, u'(s) - u(s)\, v'(s) \leqslant c_2$$

for almost all s. Then, it is easily verified that the function

$$r(s, t) = \begin{cases} u(s)\, v(t), & s \leqslant t, \\ u(t)\, v(s), & s \geqslant t \end{cases}$$

is of type[15] $s[a, b]$, and that

$$D_r(s) = v(s)\, u'(s) - u(s)\, v'(s).$$

In particular, for $u(s) = cs/2$, $v(s) \equiv 1$, we have

$$r(s, t) = \frac{c}{2} \min(s, t),$$

which is just the correlation function for the Wiener process (see §5.1).

Theorem 5.2.6. Let $\{\xi(t, \omega), t \in [a, b]\}$ be a joint determining set of functions on the probability measure spaces $S_k = (\Omega, \mathfrak{B}, P_k)$, $k = 1, 2$, forming a Gaussian process of type $S[a, b]$ relative to both S_1 and S_2,

[15] *Translator's note:* The assumptions made here are, in fact, not sufficient to ensure that condition (vi) above is satisfied. It would suffice, however, to replace the requirement that $v(s) > 0$ by, for example, the following condition:

$$\int_a^b \int_a^t f(t) f(s)\, v'(t)\, v'(s)\, ds\, dt \geqslant 0, \qquad \text{for all } f \in L^2[a, b].$$

with the respective correlation functions $r_k(s, t)$, $k = 1, 2$. Then P_1 and P_2 are equivalent if and only if

(i) the discriminators $D_{r_1}(t)$, $D_{r_2}(t)$ are equal almost everywhere on $[a, b]$, and

(ii) $r_1(a, a)$ and $r_2(a, a)$ are either both zero or both different from zero.

PROOF. In order to use Theorem 5.2.5, we shall explicitly construct the closure H_k of $\mathscr{L}$ in $L^2(S_k)$ $(k = 1, 2)$.

1. Let $V[a, b]$ denote the totality of real-valued functions f of bounded variation on $[a, b]$ which are also continuous on the left at every point of (a, b) and satisfy $f(b) = 0$.

For every $f \in V[a, b]$ and every sequence of partitions

$$a = t_0^{(n)} < t_1^{(n)} < \cdots < t_n^{(n)} = b$$

of the interval $[a, b]$, we form a sequence of functions $\{\xi_n\}$ in $\mathscr{L}$, as follows:

$$\xi_n(\omega) = \sum_{k=1}^{n} \xi(t_k^{(n)}, \omega)(f(t_k^{(n)}) - f(t_{k-1}^{(n)})).$$

Then, one easily calculates that

$$(\xi_n, \xi_m)_\nu = \sum_{k=1}^{n} \sum_{l=1}^{m} r_\nu(t_k^{(n)}, t_l^{(m)})(f(t_k^{(n)}) - f(t_{k-1}^{(n)}))(f(t_l^{(m)}) - f(t_{l-1}^{(m)})), \qquad \nu = 1, 2,$$

whence it follows immediately that

$$(\xi_n, \xi_m)_\nu \to \int_a^b \int_a^b r_\nu(s, t)\, df(s)\, df(t), \qquad \nu = 1, 2, \tag{5.2.18}$$

as $n, m \to \infty$. Consequently, $\{\xi_n\}$ is a fundamental sequence relative to the inner product $(\cdot, \cdot)_\nu$, and hence converges to a vector in H_ν, which we denote by $\xi_f^{(\nu)}$. Using (5.2.18), we easily calculate that, if $f, g \in V[a, b]$, then

$$(\xi_f^{(\nu)}, \xi_g^{(\nu)})_\nu = \int_a^b \int_a^b r_\nu(s, t)\, df(s)\, dg(t). \tag{5.2.19}$$

Let $K_\nu = \{\xi_f^{(\nu)} \mid f \in V[a, b]\}$; then K_ν is dense in H_ν. In fact, for any $t_0 \in (a, b)$, we form the function

$$f(t) = \begin{cases} 0, & t > t_0, \\ -1, & t \leqslant t_0. \end{cases}$$

Clearly, $f \in V[a, b]$, and $\xi_f(\cdot) = \xi(t_0\,; \cdot)$, hence $\xi(t_0\,; \cdot) \in K_\nu$. Moreover, by the continuity of $r(s, t)$, it is obvious that $\xi(a, \cdot) = \lim_{t_0 \to a} \xi(t_0 ; \cdot)$, and therefore $\xi(a; \cdot)$ belongs to the closure $K_\nu{}^0$ of K_ν in H_ν. Similarly, $\xi(b; \cdot) \in K_\nu{}^0$. Since K_ν is a linear space, it follows that $\mathscr{L} \subset K_\nu{}^0$, and so

$$K_\nu{}^0 = H_\nu . \tag{5.2.20}$$

By condition (ii), $r_\nu(s, t)$ is absolutely continuous with respect to s, therefore, using integration by parts, we find that

$$\int_a^b r_\nu(s, t)\, df(s) = -f(a)\, r(a, t) - \int_a^b f(s) \frac{\partial r_\nu(s, t)}{\partial s}\, ds \tag{5.2.21}$$

holds for all $f \in V[a, b]$.

According to property (iii) in Example 5.2.1, for almost all fixed $s \in [a, b]$, the partial derivative $\partial r_\nu(s, t)/\partial s$ is an absolutely continuous function of t on $[a, s)$, $(s, b]$. Consequently, when $t < s$,

$$\frac{\partial r_\nu(s, t)}{\partial s} = \int_a^t \frac{\partial^2 r_\nu(s, \eta)}{\partial \eta\, \partial s}\, d\eta + \frac{\partial r_\nu(s, a)}{\partial s} . \tag{5.2.22}$$

Let $q_+^{(\nu)}$, $q_-^{(\nu)}$ denote the functions (5.2.16) corresponding to r_ν. Then, letting $t \to s - 0$ in (5.2.22), we get

$$q_-^{(\nu)}(s) = \int_a^s \frac{\partial^2 r_\nu(s, \eta)}{\partial \eta\, \partial s}\, d\eta + \frac{\partial r_\nu(s, a)}{\partial s} . \tag{5.2.23}$$

Similarly, when $t > s$, using (5.2.23), we obtain

$$\begin{aligned} \frac{\partial r_\nu(s, t)}{\partial s} &= \int_s^t \frac{\partial^2 r_\nu(s, \eta)}{\partial \eta\, \partial s}\, d\eta + q_+^{(\nu)}(s) \\ &= \int_a^t \frac{\partial^2 r_\nu(s, \eta)}{\partial \eta\, \partial s}\, d\eta + D_{r_\nu}(s) + \frac{\partial r_\nu(s, a)}{\partial s} . \end{aligned} \tag{5.2.24}$$

Substituting (5.2.22) and (5.2.24) into (5.2.21), and applying Fubini's theorem to interchange the order of integration, we have

$$\begin{aligned} \int_a^b r_\nu(s, t)\, df(s) = -f(a)\, r_\nu(a, t) &- \int_a^t \left(\int_a^b \frac{\partial^2 r_\nu(s, \eta)}{\partial \eta\, \partial s} f(s)\, ds \right) d\eta \\ &- \int_a^b \frac{\partial r_\nu(s, a)}{\partial s} f(s)\, ds - \int_a^t D_{r_\nu}(s) f(s)\, ds \end{aligned}$$

Finally, using integration by parts, we arrive at the formula

$$\int_a^b \int_a^b r_\nu(s, t)\, df(s)\, dg(t)$$

$$= f(a)\, g(a)\, r_\nu(a, a) + f(a) \int_a^b \frac{\partial r_\nu(a, t)}{\partial t} g(t)\, dt$$

$$+ \int_a^b \frac{\partial r_\nu(s, a)}{\partial s} f(s)\, ds\, g(a) + \int_a^b D_{r_\nu}(s) f(s)\, g(s)\, ds$$

$$+ \int_a^b \int_a^b \frac{\partial^2 r_\nu(s, t)}{\partial t\, \partial s} f(s)\, g(t)\, ds\, dt. \tag{5.2.25}$$

2. We now construct function spaces $L_2^{(\nu)}[a, b]$, $\nu = 1, 2$, in the following manner. The functions in $L_2^{(\nu)}[a, b]$ are the same as the functions comprising $L_2[a, b]$, however, two such functions f, g are to be identified if and only if $f(t) = g(t)$ almost everywhere on $(a, b]$, and also

$$f(a)\, r_\nu(a, a) = g(a)\, r_\nu(a, a).$$

The linear operations in $L_2^{(\nu)}[a, b]$ are defined in the usual way, while the inner product is defined by

$$[f, g]_\nu = \begin{cases} \displaystyle\int_a^b f(t)\, g(t)\, D_{r_\nu}(t)\, dt + f(a)\, g(a), & r_\nu(a, a) > 0, \\ \displaystyle\int_a^b f(t)\, g(t)\, D_{r_\nu}(t)\, dt, & r_\nu(a, a) = 0. \end{cases}$$

Using condition (v) of Example 5.2.1, one may verify that $L_2^{(\nu)}[a, b]$ is indeed a Hilbert space relative to the inner product $[f, g]_\nu$.

Next, we form a linear operator A_ν from $L_2^{(\nu)}[a, b]$ into itself, as follows. Given $f \in L_2^{(\nu)}[a, b]$, let $A_\nu f$ be the function defined by

$$(A_\nu f)(t) = \frac{1}{D_{r_\nu}(t)} \int_a^b \frac{\partial^2 r_\nu(s, t)}{\partial t\, \partial s} f(s)\, ds + \frac{f(a)}{D_{r_\nu}(t)} \frac{\partial r_\nu(a, t)}{\partial t}$$

for $t \in (a, b]$, and, in the case $r_\nu(a, a) \neq 0$, let

$$(A_\nu f)(a) = \int_a^b \frac{\partial r_\nu(s, a)}{\partial s} f(s)\, ds + f(a)(r_\nu(a, a) - 1).$$

We proceed to prove that A_ν is a Hilbert–Schmidt operator. (1) If

$r_\nu(a, a) = 0$, then $\xi(a, \omega) \equiv 0$, hence $r(a, t) \equiv 0$, $a \leqslant t \leqslant b$. In this case, $L_2^{(\nu)}[a, b]$ is just $L_2[a, b]$, and

$$(A_\nu f)(t) = \frac{1}{D_{r_\nu}(t)} \int_a^b \frac{\partial^2 r_\nu(s, t)}{\partial t\, \partial s} f(s)\, ds. \tag{5.2.26}$$

By virtue of conditions (iv) and (v) of Example 5.2.1, the correspondence $f \to \int_a^b [\partial^2 r_\nu(s, t)/\partial s\, \partial t] f(s)\, ds$ is a Hilbert–Schmidt operator, and $f \to 1/D_{r_\nu}$ is a bounded linear operator, hence A_ν, being the product of these two operators, is Hilbert–Schmidt. (2) If $r_\nu(a, a) \neq 0$, then A_ν is the sum of a Hilbert–Schmidt operator of the form (5.2.26) and a bounded linear operator of finite rank, hence A_ν is also Hilbert–Schmidt.

We observe that, if $f, g \in L_2^{(\nu)}[a, b]$, the right-hand side of (5.2.25) is just $[(I + A_\nu) f, g]_\nu$. In particular, when $f, g \in V[a, b]$, it follows by (5.2.19) that

$$(\xi_f^{(\nu)}, \xi_g^{(\nu)})_\nu = [(I + A_\nu) f, g]_\nu\,. \tag{5.2.27}$$

Now, for any $f \in L_2^{(\nu)}[a, b]$, since $V[a, b]$ is dense in $L_2^{(\nu)}[a, b]$, one can choose a sequence $\{f_n\} \subset V[a, b]$ such that $\{f_n\}$ converges to f in $L_2^{(\nu)}[a, b]$. Using (5.2.27), we see at once that $\{\xi_{f_n}^{(\nu)}\}$ is a fundamental sequence in K_ν. Therefore, $\{\xi_{f_n}^{(\nu)}\}$ converges in H_ν to a vector which we call $\xi_f^{(\nu)}$, moreover, (5.2.27) then holds for all $f, g \in L_2^{(\nu)}[a, b]$. Since $(\xi_f^{(\nu)}, \xi_f^{(\nu)}) \geqslant 0$, it follows from (5.2.27) that $B_\nu = I + A_\nu$ is a nonnegative self-adjoint operator on $L_2^{(\nu)}[a, b]$.

Let us examine the null space of B_ν. Again, we distinguish two cases.

(1) Suppose that $r_\nu(a, a) = 0$. If $B_\nu f = 0$, then, by (5.2.26), we have

$$\begin{aligned} 0 &= (B_\nu f, f) = (f, f) + (A_\nu f, f) \\ &= \int_a^b D_{r_\nu}(t) f(t)^2\, dt + \int_a^b \int_a^b \frac{\partial^2 r_\nu(s, t)}{\partial t\, \partial s} f(s) f(t)\, ds\, dt. \end{aligned} \tag{5.2.28}$$

However, by condition (vi), the double integral in (5.2.28) is nonnegative, whence it follows by condition (v) that $f = 0$. This shows that B_ν has an inverse operator. Thus, -1 is not an eigenvalue of A_ν, therefore, since A_ν is completely continuous, -1 is a regular value of A_ν. Hence, B_ν^{-1} is defined as a bounded operator on the entire space $L_2^{(\nu)}[a, b]$. In other words, B_ν is an equivalence operator from $L_2^{(\nu)}[a, b]$ onto itself.

(2) Suppose that $r_\nu(a, a) > 0$. In this case, we assert that, if φ, ψ belong to the null space of B_ν and $\varphi(a) = \psi(a)$, then $\varphi = \psi$. In fact, under these assumptions, the vector $f = \varphi - \psi$ also belongs to the null space of B_ν and $f(a) = 0$. Consequently, it can be seen from the expression for A_ν that f also satisfies (5.2.28), and hence $f = 0$, that is, $\varphi = \psi$.

It follows that the null space E_ν of A_ν is one-dimensional.[16] Using the fact that A_ν is a completely continuous self-adjoint operator, it is easy to prove that the restriction of B_ν to the orthogonal complement F_ν of E_ν is a homeomorphism of F_ν onto itself. Hence, $B_\nu \mid F_\nu$ is an equivalence operator from F_ν onto F_ν. In order to combine cases (1) and (2) throughout the remainder of this proof, we agree that F_ν is to denote $L_2^{(\nu)}[a, b]$ if $r_\nu(a, a) = 0$.

Now, by virtue of (5.2.27), we know that the correspondence

$$U_\nu : f \to \xi_f^{(\nu)}$$

maps F_ν homeomorphically onto $U_\nu F_\nu$. However, if $f \in E_\nu$, then $B_\nu f = 0$, that is, $\xi_f^{(\nu)} = 0$; consequently, $U_\nu F_\nu = U_\nu L_2^{(\nu)}[a, b] \supset K_\nu$. Therefore, since F_ν is complete, it follows from (5.2.20) that $U_\nu F_\nu = H_\nu$. Moreover, by (5.2.27), we have

$$U_\nu^* U_\nu = B_\nu ,$$

hence U_ν is an equivalence operator from F_ν onto H_ν.

(3) We can now demonstrate the necessity of conditions (i) and (ii). Assume that P_1 and P_2 are equivalent. Then $T' = U_1^{-1} T U_2$ is the identity operator[17]: $T'f = f, f \in F_2$. Consequently, either (1) $F_\nu = L_2^{(\nu)}[a, b]$ for both $\nu = 1, 2$, whence[18] $r_1(a, a) = r_2(a, a) = 0$, or (2) $F_\nu \neq L_2^{(\nu)}[a, b]$ for both $\nu = 1, 2$, whence $r_1(a, a) \neq 0$, $r_2(a, a) \neq 0$. This proves the necessity of condition (ii). Furthermore, since U_1, T and U_2 are all equivalence operators, it follows by Theorem II.1.9 of Appendix II that T' is also an equivalence operator. In case (1) [i.e., $r_1(a, a) = r_2(a, a) = 0$] we have

$$(T')^* f = \frac{D_{r_1}}{D_{r_2}} f,$$

hence,

$$((T')^* T' - I) f = \frac{D_{r_1} - D_{r_2}}{D_{r_2}} f.$$

[16] *Translator's note:* In actual fact, it has only been proved that E_ν is *at most* one dimensional. If $E_\nu = (0)$, then it follows just as in case (1) that B_ν is an equivalence operator. However, in this case, $F_\nu = L_2^{(\nu)}[a, b]$, and this seems to invalidate the already questionable reasoning used in the proof of sufficiency (see below).

[17] *Translator's note:* The reasoning behind this assertion is, perhaps, based upon the inaccurate statement of Theorem 5.2.5 in the original text (see footnote 13).

[18] *Translator's note:* See, however, footnote 16.

However, it is easily shown that the above operator cannot be completely continuous unless $D_{r_1} = D_{r_2}$ almost everywhere. Thus, condition (i) is also necessary. A similar argument applies to case (2).

(4) Conversely, assume that conditions (i) and (ii) are satisfied. Then the spaces $L_2^{(1)}[a, b]$ and $L_2^{(2)}[a, b]$, as well as their respective inner products, are completely identical. Hence, F_1 and F_2 coincide.[19] Consider the operator

$$T = U_1 U_2^{-1}$$

from H_2 to H_1. By Theorem II.1.9, T is an equivalence operator; moreover, if $\xi \in \mathscr{L}$, then $T\xi = \xi$. Thus, T satisfies the conditions of Theorem 5.2.5. Hence, P_1 and P_2 are equivalent.]

Corollary 5.2.7. Let $\{\xi(t, \omega), a \leqslant t \leqslant b\}$ be a Gaussian process of type $S[a, b]$ on the probability space $(\Omega, \mathfrak{F}, P)$; let $r(s, t)$ be the correlation function of this process, and $D_r(t)$ $(a \leqslant t \leqslant b)$ its discriminator. Assume also that $\{\xi(t, \omega), a \leqslant t \leqslant b\}$ is a determining set of functions on $(\Omega, \mathfrak{F}, P)$. Then, there exists a probability measure P' on $(\Omega, \mathfrak{F})$, equivalent to P, such that $\{\xi(t, \omega), t \in [a, b]\}$ is a Gaussian process of type $S[a, b]$ on $(\Omega, \mathfrak{F}, P')$, with correlation function

$$\rho(s, t) = \int_a^{\min(s,t)} D_r(\eta)\, d\eta + c_r\,, \tag{5.2.29}$$

where $c_r = 0$ if $r(a, a) = 0$, and $c_r = 1$ if $r(a, a) > 0$.

Proof. Let $(\Gamma, \mathfrak{B}, \mu)$ be the sample probability space associated with the given Gaussian process (see §1.3, 3°) and denote the corresponding mapping of Ω into Γ by $\varphi : \omega \to \xi = \{\xi(t, \omega), t \in [a, b]\}$. It is easily verified that ρ is a positive definite function, that $\rho \in s[a, b]$, and that $D_\rho(\eta) = D_r(\eta)$. In view of the remarks following Definition 5.1.1, there exists a probability measure ν on $(\Gamma, \mathfrak{B})$ relative to which $\{\xi_t, t \in [a, b]\}$ is a Gaussian process with identically zero mean and correlation function $\rho(s, t)$. Since the process $\{\xi_t, t \in [a, b]\}$ has zero means and correlation function $r(s, t)$ relative to the measure μ, and since $D_\rho(\eta) = D_r(\eta)$, $\rho(a, a) = c_r$, it follows by Theorem 5.2.6 that μ and ν are equivalent. Now, for any set $F \in \mathfrak{F}$, there exists a set B in $\mathfrak{B}$ such that $P(F \Delta \varphi^{-1}(B)) = 0$. Define

$$P'(F) = \nu(B).$$

[19] *Translator's note:* The justification for this statement is not clear. Certainly $L_2^{(1)}[a, b]$ and $L_2^{(2)}[a, b]$ are identical, but B_1 and B_2 need not be identical.

It is easily verified that $P'(F)$ is independent of the choice of B, and that P' is a probability measure having the required properties.]

Corollary 5.2.7 provides a canonical form[20] (5.2.29) for Gaussian processes of type $S[a, b]$. As a special case, if $(\Omega, \mathfrak{B})$ is the measurable space $(C_0[0, 1], \mathfrak{B})$ of Example 5.1.2, then any of these canonical Gaussian processes is obtainable from the Wiener process $\{x(s, \omega), s \in [0, 1]\}$ (taking, say, $c = 2$) by a transformation, that is,

$$\xi(t, \omega) = mx(q(t), \omega) + \xi,$$

where

$$q(t) = \frac{\int_a^t D_r(\eta)\, d\eta}{\int_a^b D_r(\eta)\, d\eta}, \qquad m = \left(\int_a^b D_r(\eta)\, d\eta\right)^{1/2},$$

and ξ is a Gaussian variable, independent of all the $x(s, \omega)$, such that $E(\xi) = 0$, $E(\xi^2) = c_r$.

§5.3. Gaussian Measures on Linear Spaces

1° Standard Gaussian Measure Spaces

When studying problems of harmonic analysis related to Gaussian measures, frequent use is made of the following type of measure space.

Definition 5.3.1. Let Ω be a real linear space, H a linear subspace of Ω, and $(\cdot, \cdot)$ an inner product defined on H. Let $\mathfrak{B}$ be a σ-algebra formed by certain subsets of Ω, and let N be a probability measure on $(\Omega, \mathfrak{B})$ such that $(\Omega, \mathfrak{B}, N)$ is quasi-invariant with respect to H. Moreover, suppose that, for every $x \in H$, there corresponds an H-quasi-linear functional $x(\omega)$ on $(\Omega, \mathfrak{B}, N)$ such that the linear functional on H induced by $x(\omega)$ is just the inner product (x, y), $y \in H$. In other words, the identity $x(t\omega) = tx(\omega)$ holds for all $\omega \in \Omega$, $-\infty < t < \infty$, and, for any given $y \in H$, the relation $x(t\omega + y) = tx(\omega) + (y, x)$ holds for all t, $-\infty < t < \infty$, provided that ω lies in a certain set $E_y \in \mathfrak{B}$, $N(E_y) = 1$. We also suppose that the correspondence between x and $x(\omega)$ is linear, that is, for any $x, y \in H$, $\alpha, \beta \in R$ and $\omega \in \Omega$, we have $(\alpha x + \beta y)(\omega) = \alpha x(\omega) + \beta y(\omega)$.

Furthermore, suppose that the family of functions $\{x(\omega), x \in H\}$

[20] *Translator's note:* That is to say, canonical only with reference to a fixed indexed set of variables $\{\xi(t, \omega), t \in [a, b]\}$.

constitutes both a determining set and a Gaussian process on $(\Omega, \mathfrak{B}, N)$, the mathematical expectation of this process being identically zero and its correlation function being

$$\int x(\omega)\, y(\omega)\, dN(\omega) = \frac{c}{2}(x, y), \qquad c > 0.$$

We then say that $(\Omega, \mathfrak{B}, N)$ is a *standard Gaussian measure space* (with parameter c) relative to H, and that $\{x(\cdot), x \in H\}$ is a *standard Gaussian process* on $(\Omega, \mathfrak{B}, N)$. To indicate the parameter, we shall sometimes write N_c instead of N.

Theorem 5.3.1. Let H, Ω be real separable Hilbert spaces such that H forms a dense linear subspace of Ω and the imbedding of H in Ω is an operator of Hilbert–Schmidt type. Let $\mathfrak{B}$ be the σ-algebra of weak Borel sets in Ω. Then, there exists a standard Gaussian measure space $(\Omega, \mathfrak{B}, N)$ relative to H.

PROOF. Denote the inner products in H and Ω by $(\cdot, \cdot)$ and $(\cdot, \cdot)_1$, respectively. Since the imbedding operator $T: H \to \Omega$ is of Hilbert–Schmidt type, the operator $B = T^*T$ is nuclear. Obviously, B has the following property: for any $x, y \in H$,

$$(x, y)_1 = (Bx, y).$$

Let $\{e_n, n = 1, 2,\ldots\}$ be a complete orthonormal system in H, consisting of eigenvectors of B, and let λ_n denote the eigenvalue corresponding to e_n. Since B is a nuclear operator, $\sum_{m=1}^{\infty} \lambda_m < \infty$; also,

$$(e_m, e_n)_1 = \lambda_m \delta_{mn},$$

and $(e_m, e_m)_1 > 0$, hence $\lambda_m > 0$, $m = 1, 2,\ldots$. Let $l = \times_{n=1}^{\infty} R_n$, where R_n is the real line for every n, and let l^2 denote the totality of elements $x = \{x_n\}$ in l which satisfy the condition $\sum x_n^2 < \infty$. Then, l^2 is a Hilbert space with respect to the inner product

$$(x, y) = \sum_{\nu=1}^{\infty} x_\nu y_\nu$$

(where $y = \{y_\nu\}$). Also, let $l^2(\{\lambda_m\})$ denote the totality of elements $x = \{x_n\}$ in l which satisfy the condition $\sum \lambda_n x_n^2 < \infty$ (see Example 4.2.2). Then, $l^2(\{\lambda_m\})$ is a Hilbert space with respect to the inner product

$$(x, y)_1 = \sum_{m=1}^{\infty} \lambda_m x_m y_m .$$

Define a mapping U from Ω to $l^2(\{\lambda_m\})$, as follows:

$$U : x \to \{(x, e_1)_1/\lambda_1 , (x, e_2)_1/\lambda_2 ,..., (x, e_n)_1/\lambda_n ,...\}.$$

Since the points of Ω are characterized by the condition

$$(x, x)_1 = \sum_{m=1}^{\infty} \lambda_m^{-1} |(x, e_m)_1|^2 < \infty,$$

it follows that U is a unitary operator from Ω to $l^2(\{\lambda_m\})$, moreover, U maps H unitarily onto l^2. Consequently, we may as well assume that $\Omega = l^2(\{\lambda_m\})$ and $H = l^2$.

Let $\mathfrak{B}_\omega$ be the smallest σ-algebra containing all Borel cylinders in l. According to §5.1, 2°, there exists a Gaussian measure N on $(l, \mathfrak{B}_\omega)$ such that the coordinates $\{x_\nu\}$ (regarded as functions of x) are mutually independent Gaussian variables with mathematical expectation zero and variance $\frac{1}{2}c$ $(c > 0)$.

Given any $y \in l$, consider the measure N_y on $(l, \mathfrak{B}_\omega)$ defined by

$$N_y(E) = N(E - y), \qquad E \in \mathfrak{B}_\omega . \tag{5.3.1}$$

Then, the functions $\{x_\nu , \nu = 1, 2,...\}$ are also mutually independent Gaussian variables on $(l, \mathfrak{B}_\omega , N_y)$, with variance $\frac{1}{2}c$. However, the mathematical expectation of x_ν with respect to the measure N_y is given by

$$\int x_\nu \, dN_y(x) = \int x_\nu \, dN(x - y) = \int (x_\nu + y_\nu) \, dN(x) = y_\nu .$$

By Lemma 5.1.2, the measures N_y and N are equivalent if and only if $\sum y_\nu{}^2 < \infty$, that is, $y \in l^2$. Thus, the translations under which $(l, \mathfrak{B}_\omega , N)$ is quasi-invariant are precisely those corresponding to the elements of l^2.

Next, we assert that the measure N is concentrated on $l^2(\{\lambda_m\})$. In fact, using the Levi Lemma, we have

$$\int (x, x)_1 \, dN(x) = \lim_{n\to\infty} \int \sum_{\nu=1}^{n} \lambda_\nu x_\nu{}^2 \, dN(x) = \frac{c}{2} \sum_{\nu=1}^{\infty} \lambda_\nu < \infty,$$

hence $N(\{x \mid (x, x)_1 = \infty\}) = 0$, which means that the measure N is concentrated on $l^2(\{\lambda_m\})$. The restrictions of $\mathfrak{B}_\omega$ and N to $l^2(\{\lambda_m\})$ will also be denoted by $\mathfrak{B}_\omega$ and N, respectively. Thus, we obtain a Gaussian measure space $(l^2(\{\lambda_m\}), \mathfrak{B}_\omega , N)$ which is quasi-invariant relative to l^2.

For each $y \in l^2$, we form a sequence of measurable functions $\{y^{(n)}(x)\}$ on $(l^2(\{\lambda_m\}), \mathfrak{B}_\omega, N)$, as follows:

$$y^{(n)}(x) = \sum_{\nu=1}^{n} y_\nu x_\nu, \qquad x \in l^2(\{\lambda_m\}).$$

Then, when $n \geqslant m$, $n, m \to \infty$,

$$\int [y^{(n)}(x) - y^{(m)}(x)]^2 \, dN(x) = \frac{c}{2} \sum_{\nu=m+1}^{n} y_\nu^2 \to 0.$$

Hence, $\{y^{(n)}(x)\}$ converges in $L^2(N)$, and therefore there exists a subsequence $\{y^{(n_k)}(x)\}$ which converges almost everywhere.[21] Let A_y denote the set of all points $x \in l^2(\{\lambda_m\})$ such that $\lim_{k\to\infty} y^{(n_k)}(x)$ exists; then $l^2(\{\lambda_m\}) - A_y$ is an N-null set. Now, define a measurable function $y(x)$ on $(l^2(\{\lambda_m\}), \mathfrak{B}_\omega, N)$ by letting

$$y(x) = \lim_{k\to\infty} y^{(n_k)}(x)$$

when $x \in A_y$, and $y(x) = 0$ when $x \bar{\in} A_y$. Then, for any $z \in l^2$, we have $y^{(n_k)}(x + z) = y^{(n_k)}(x) + \sum_{\nu=1}^{n_k} y_\nu z_\nu$, consequently, if $x \in A_y$, then $x + z \in A_y$, and

$$y(x + z) = y(x) + (y, z).$$

Moreover, if $x \in A_y$ and t is any real number, then $tx \in A_y$, and so it follows easily that $y(\cdot)$ is a quasi-linear functional.

By Lemma 5.1.1, the function $y(x)$, $x \in l^2(\{\lambda_m\})$, being the limit of the sequence of Gaussian variables $\{y^{(n_k)}(x)\}$, is also a Gaussian variable, moreover, if $y, z \in l^2$, then,

$$\int y(x)\, z(x)\, dN(x) = \lim_{k\to\infty} \int y^{(n_k)}(x)\, z^{(n_k)}(x)\, dN(x)$$

$$= \frac{c}{2} \sum_{\nu=1}^{\infty} y_\nu z_\nu = \frac{c}{2} (y, z).$$

Also,

$$\int y(x)\, dN(x) = \lim \int y^{(n_k)}(x)\, dN(x) = 0.$$

[21] *Translator's note:* According to Kakutani [1], Lemma 6, the right-hand side of (5.1.14) converges μ-almost everywhere to a nonzero limit. In the present situation, this implies that $\{y^{(n)}(x)\}$ actually converges N-almost everywhere, so that the selection of a subsequence is not strictly necessary.

If we consider the elements $y \in l^2$ defined by $y_\nu = \delta_{\nu n}$ for $n = 1, 2, \ldots$, we see that the family of functions $\{y(\cdot), y \in l^2\}$ includes all the coordinate functions $\{x_n\}$, and is therefore a determining set on $(l^2(\{\lambda_m\}), \mathfrak{B}_\omega, N)$. Thus, we have shown that $\{y(\cdot), y \in l^2\}$ is a standard Gaussian process on $(l^2(\{\lambda_m\}), \mathfrak{B}_\omega, N)$. Finally, we transfer the above construction back to Ω and H *via* the unitary operator U^{-1}, and so obtain the desired result.]

Notice that, in the case just considered, (5.1.14) yields

$$\frac{dN_y(x)}{dN(x)} = \lim_{k\to\infty} \exp\left(\frac{1}{c}\sum_{\nu=1}^{n_k} [x_\nu^2 - (x_\nu - y_\nu)^2]\right) = \exp\left(\frac{2}{c}\, y(x) - \frac{1}{c}\,(y, y)\right),$$

whence, returning to Ω, we get

$$\frac{dN_y(\omega)}{dN(\omega)} = \exp\left(\frac{2}{c}\, y(\omega) - \frac{1}{c}\,(y, y)\right), \qquad y \in H, \quad \omega \in \Omega. \tag{5.3.2}$$

Furthermore, from the proof of Theorem 5.3.1, we easily obtain the following result.

Corollary 5.3.2. In Theorem 5.3.1, the totality of points x such that the Gaussian measure N is quasi-invariant with respect to the translation $\omega \to \omega + x$ is precisely H.[22]

Corollary 5.3.3. Let H, Ω be real separable Hilbert spaces such that H is a linear subspace of Ω and the imbedding $H \to \Omega$ is an operator of Hilbert–Schmidt type. Let $\mathfrak{B}$ be the σ-algebra of weak Borel sets in Ω. Then, there exists a (Gaussian[23]) probability measure N on $(\Omega, \mathfrak{B})$ such that the translations under which $(\Omega, \mathfrak{B}, N)$ is quasi-invariant are precisely those corresponding to the elements of H.

PROOF. Let Ω_1 be the closure of H in Ω; then H is dense in Ω_1. Let $\mathfrak{B}_1$ be the restriction of $\mathfrak{B}$ to Ω_1. Applying Theorem 5.3.1 and Corollary 5.3.2 to H and $(\Omega_1, \mathfrak{B}_1)$, we obtain a probability measure N_1 on $(\Omega_1, \mathfrak{B}_1)$ such that N_1 is quasi-invariant under translation by elements of H (and no others). Also, let N_2 be the probability measure on $\Omega \ominus \Omega_1$ which is concentrated at the zero vector. Then, the product measure $N_1 \times N_2$ has the desired properties.]

[22] *Translator's note:* This is certainly true with regard to the specific measure N constructed in the proof of Theorem 5.3.1. However, it is not immediately clear that it need be true for *any* measure N which satisfies the requirements of the theorem.

[23] *Translator's note:* Presumably, the term "Gaussian measure" as used here means that there exists some Gaussian process which forms a determining set for the measure space.

The following corollary is an immediate consequence of Corollary 5.3.3.

Corollary 5.3.4. Let G be a real separable Hilbert space, and $\mathfrak{B}$ the totality of weak Borel sets in G. Let $\mathfrak{G}$ be a countably Hilbert space which is imbedded as a linear subspace of G, and let $\mathfrak{G}_n$ denote the completion of $\mathfrak{G}$ relative to the nth inner product, $n = 1, 2, \ldots$. Suppose that, for some n, the imbedding $\mathfrak{G} \to G$ extends to a Hilbert–Schmidt operator[24] from $\mathfrak{G}_n$ to G. Then, there exists a finite measure μ on $(G, \mathfrak{B})$ which is quasi-invariant with respect to $\mathfrak{G}$.

Corollary 5.3.4 provides a converse to Theorem 4.2.18. Similarly, one can state a converse of Corollary 4.2.19; we omit the details.

Under the hypotheses of Corollary 5.3.4, does there necessarily exist a finite measure space $(G, \mathfrak{B}, \mu)$ such that the totality of quasi-invariant translations is precisely $\mathfrak{G}$? This question still remains unanswered.

2° On the Equivalence of Three Conditions

Theorem 5.3.5. Let $\mathfrak{G}$ and G be real separable Hilbert spaces such that $\mathfrak{G}$ is a linear subspace of G and the imbedding operator $T : \mathfrak{G} \to G$ is continuous. Let $\mathfrak{B}$ denote the totality of weak Borel sets in G, and $\mathfrak{F}^\dagger$ the totality of weak Borel sets in the conjugate space $\mathfrak{G}^\dagger$ of $\mathfrak{G}$. Then, the following three statements are equivalent.

(i) T is of Hilbert–Schmidt type.

(ii) There exists a finite measure on $(G, \mathfrak{B})$ which is quasi-invariant with respect to $\mathfrak{G}$.

(iii) For any positive definite continuous function f on G, there exists a unique measure $P^\dagger$ on $(\mathfrak{G}^\dagger, \mathfrak{F}^\dagger)$ such that, for any $h \in \mathfrak{G}$,

$$f(h) = \int_{\mathfrak{G}^\dagger} e^{i\xi(h)}\, dP^\dagger(\xi). \tag{5.3.3}$$

Proof. We present several ways of deducing these conditions from one another, as follows.[25]

(1) By Corollary 5.3.3, (i) implies (ii).

(2) Statement (iii) may be deduced from (i) by using Theorem 4.3.12. In fact, let $(\cdot, \cdot)_1$ and $(\cdot, \cdot)$ denote the inner products in G and $\mathfrak{G}$, respectively; then, when $\varphi, \psi \in \mathfrak{G}$,

$$(\varphi, \psi)_1 = (T^*T\varphi\, , \psi). \tag{5.3.4}$$

[24] *Translator's note:* Apparently, one must also assume that this extension is injective.

[25] *Translator's note:* Obviously, parts (2) and (3) below could be omitted from the proof; presumably, their inclusion was merely intended to shed additional light on the situation.

By (i) and Lemma II.1.1, the positive operator T^*T is nuclear. Now $\mathfrak{G}$, being a Hilbert space, is, in particular, a countably Hilbert space. Suppose that the positive definite function f is continuous relative to the topology of G, that is, continuous relative to the inner product (5.3.4). Then, given any $\epsilon > 0$, there exists a $\delta > 0$ such that $|f(\varphi) - f(0)| < \epsilon$ whenever $\varphi \in \{\varphi \mid ((T^*T/\delta)\,\varphi, \varphi) < 1\}$. Since T^*T/δ is also a positive nuclear operator, this shows that $f \mid \mathfrak{G}$ is continuous relative to the $\mathfrak{S}$-topology (see §4.3). Hence, by Theorem 4.3.12, there exists a unique finite measure $P^\dagger$ on $(\mathfrak{G}^\dagger, \mathfrak{F}^\dagger)$ such that (5.3.3) holds.

(3) By Corollary 4.2.20, (ii) implies (i).

(4) By Theorem 4.3.11, (ii) implies (iii).

(5) We now proceed to deduce (i) from (iii). Assume that (iii) holds, and consider the continuous function

$$f(h) = \exp(-\tfrac{1}{2}(h, h)_1), \qquad h \in G.$$

It is easily verified that this function is positive definite (in fact, it is the characteristic function of a certain Gaussian measure space). Hence, by condition (iii), there exists a unique measure $P^\dagger$ on $(\mathfrak{G}^\dagger, \mathfrak{F}^\dagger)$ such that f satisfies (5.3.3).

Next, replace h in (5.3.3) by $\sum_{\nu=1}^{n} t_\nu h_\nu$, $h_1, \ldots, h_n \in \mathfrak{G}$, and define an n-dimensional Borel measure P by the equation

$$P(E) = P^\dagger(\{\xi \mid (\xi(h_1), \ldots, \xi(h_n)) \in E\}).$$

We then have

$$\exp\left(-\frac{1}{2}\sum (h_l, h_m)_1\, t_l t_m\right) = \int \exp i(t_1 x_1 + \cdots + t_n x_n)\, dP(x_1, \ldots, x_n),$$

which shows that P is the probability distribution of a Gaussian variable. By (5.1.3),

$$\int_{\mathfrak{G}^\dagger} \xi(h_k)^2\, dP^\dagger(\xi) = \int x_k^2\, dP(x) = (h_k, h_k)_1, \qquad k = 1, 2, \ldots, n,$$

and if the vectors $\{h_k\}$ are orthogonal relative to the inner product $(\cdot, \cdot)_1$, then, by (5.1.6), we get

$$\int_{\mathfrak{G}^\dagger} (\xi(h_k)^2 - (h_k, h_k)_1)(\xi(h_l)^2 - (h_l, h_l)_1)\, dP^\dagger(\xi) = 2(h_k, h_k)_1^2\, \delta_{k,l}, \qquad k, l = 1, 2, \ldots, n, \tag{5.3.5}$$

where $\delta_{k,l}$ is the Kronecker delta.

If the operator T is not of Hilbert–Schmidt type, then, by

Lemma II.1.1, T^*T is not nuclear. Hence, by Lemma II.1.6, given any positive integer K, there exist vectors $h_1, \dots, h_n \in \mathfrak{G}$, satisfying $(h_k, h_l) = \delta_{k,l}$ and $(h_k, h_l)_1 = \delta_{k,l}(h_k, h_k)_1$, such that

$$\lambda = \sum_{k=1}^{n} (h_k, h_k)_1 > K. \tag{5.3.6}$$

Let $Q(\xi) = \sum_{k=1}^{n} \xi(h_k)^2$. Then, using (5.3.5), we calculate that

$$\int_{\mathfrak{G}^\dagger} (Q(\xi) - \lambda)^2 \, dP^\dagger(\xi) = 2 \sum_{k=1}^{m} (h_k, h_k)_1^2,$$

moreover, $(h_k, h_k)_1 < \| T \|^2$, hence, by the Chebyshev inequality,

$$P^\dagger(\{\xi \mid | Q(\xi) - \lambda | > 2\sqrt{\lambda} \| T \|\}) \leqslant \frac{1}{4\lambda \| T \|^2} \int_{\mathfrak{G}^\dagger} (Q(\xi) - \lambda)^2 \, dP^\dagger(\xi)$$

$$= \frac{\sum (h_k, h_k)_1^2}{2\lambda \| T \|^2} \leqslant \frac{1}{2}. \tag{5.3.7}$$

However, since the $\{h_k\}$ are orthonormal relative to $(\cdot, \cdot)$, the Bessel inequality implies

$$(\xi, \xi) \geqslant Q(\xi). \tag{5.3.8}$$

From (5.3.7) and (5.3.8), we get

$$\begin{aligned} &P^\dagger(\{\xi \mid (\xi, \xi) \geqslant \lambda - 2 \| T \| \sqrt{\lambda}\}) \\ &\quad \geqslant P^\dagger(\{\xi \mid Q(\xi) \geqslant \lambda - 2 \| T \| \sqrt{\lambda}\}) \\ &\quad \geqslant P^\dagger(\{\xi \mid | Q(\xi) - \lambda | \leqslant 2 \| T \| \sqrt{\lambda}\}) \geqslant \tfrac{1}{2}. \end{aligned} \tag{5.3.9}$$

On the other hand,

$$\bigcap_{K=1}^{\infty} \{\xi \mid (\xi, \xi) \geqslant K - 2 \| T \| \sqrt{K}\} = \text{empty set},$$

hence

$$\lim_{K \to \infty} P^\dagger(\{\xi \mid (\xi, \xi) \geqslant K - 2 \| T \| \sqrt{K}\}) = 0. \tag{5.3.10}$$

But by (5.3.6), $\lambda > K$, therefore (5.3.9) and (5.3.10) are contradictory. Hence (i) holds.]

We shall now establish the converse of Corollary 4.3.14.

Corollary 5.3.6. Let $\mathfrak{G}$ be a real separable countably Hilbert space,

$\mathfrak{F}^\dagger$ the totality of weak Borel sets in $\mathfrak{G}^\dagger$. Suppose that, for any positive definite continuous function f on $\mathfrak{G}$ [with $f(0) = 1$] there exists a probability measure $P^\dagger$ on $(\mathfrak{G}^\dagger, \mathfrak{F}^\dagger)$ such that (5.3.3) holds for all $h \in \mathfrak{G}$. Then $\mathfrak{G}$ is a nuclear space.

PROOF. Let $\{\| \cdot \|_m\}$ be the given sequence of norms on $\mathfrak{G}$, and let $\mathfrak{G}_m$ be the completion of $\mathfrak{G}$ relative to $\| \cdot \|_m$. Let m be an arbitrary natural number, and consider the function

$$f(h) = \exp(-\tfrac{1}{2}(h, h)_m).$$

Now, suppose that, for every $n > m$, the imbedding $T_m{}^n : \mathfrak{G}_n \to \mathfrak{G}_m$ is not of Hilbert–Schmidt type. Then, using part (5) of the proof of Theorem 5.3.5 (replacing $\mathfrak{G}$ by $\mathfrak{G}_n$ and G by $\mathfrak{G}_m$),[26] we obtain

$$P^\dagger(\{\xi \mid \| \xi \|_{-n} = \infty\}) = \lim_{k\to\infty} P^\dagger(\{\xi \mid \| \xi \|_{-n}^2 \geqslant K - 2 \| T_m{}^n \| \sqrt{K}\}) \geqslant \tfrac{1}{2},$$

that is, for all $n > m$,

$$P^\dagger(\mathfrak{G}^\dagger - \mathfrak{G}_n{}^\dagger) \geqslant \tfrac{1}{2}.$$

But this contradicts the fact that $\bigcap_{n=m+1}^{\infty} (\mathfrak{G}^\dagger - \mathfrak{G}_n{}^\dagger)$ is the empty set. We conclude that, for some $n > m$, the operator $T_m{}^n$ is of Hilbert–Schmidt type. Therefore, $\mathfrak{G}$ is a nuclear space.]

Next, for number sequence spaces, we establish a partial result analogous to Theorem 5.3.5.

Example 5.3.1. Under the hypotheses of Example 4.2.2, if we also assume that $1 \leqslant q \leqslant 2$, then the following two statements are equivalent.

(i) $\sum a_n < \infty$.

(ii) There exists a finite measure on $(l^p(\{a_n\}), \mathfrak{B})$ which is quasi-invariant with respect to l^q.

Furthermore, let $l^{q^\dagger}$ denote the conjugate space of l^q, and $\mathfrak{F}^\dagger$ the totality of weak Borel sets in $l^{q^\dagger}$. Then, if either (i) or (ii) holds, we have the following conclusion.

(iii) For each positive definite continuous function f on $l^p(\{a_n\})$, there is a unique finite measure $P^\dagger$ on $(l^{q^\dagger}, \mathfrak{F}^\dagger)$ such that, when $h \in l^q$,

$$f(h) = \int_{l^{q^\dagger}} e^{i\xi(h)}\, dP^\dagger(\xi).$$

[26] Here, we bear in mind the density of $\mathfrak{G}$ in $\mathfrak{G}_n$, and the remark following Lemma II.1.6.

PROOF. (1) It was shown in Example 4.2.2 that (ii) implies (i).

(2) Assume that (i) holds. In the construction of Example 4.1.2, choose

$$f_a(x) = \frac{1}{(2\pi)^{1/4}} \exp\left(-\frac{t^2}{4}\right),$$

and form the corresponding Gaussian measure μ. Clearly, $\Omega = (l, \mathfrak{B}_\omega, \mu)$ is quasi-invariant with respect to l^2. However, since $q \leqslant 2$, we have $l^q \subset l^2$, hence Ω is quasi-invariant with respect to l^q.

Write $\| \xi \| = (\sum a_n \mid \xi_n \mid^p)^{1/p}$, $\xi = \{\xi_1, \ldots, \xi_n, \ldots\}$. Using the Levi Lemma, one easily verifies that

$$\int_l \| \xi \|^p \, d\mu(\xi) = \lim_{n\to\infty} \int_l \sum_{\nu=1}^{n} a_\nu \mid \xi_\nu \mid^p d\mu(\xi)$$

$$= \frac{1}{(2\pi)^{1/2}} \sum_{\nu=1}^{\infty} a_\nu \int_{-\infty}^{\infty} t^p \exp\left(-\frac{1}{2} t^2\right) dt < \infty.$$

Consequently, $\mu(\{\xi \mid \| \xi \| = \infty\}) = 0$, that is,

$$\mu(l^p(\{a_n\})) = \mu(\{\xi \mid \| \xi \| < \infty\}) = 1.$$

Thus, restricting μ to $(l^p(\{a_n\}), \mathfrak{B})$, we obtain a finite measure which is quasi-invariant with respect to l^q, that is, (ii) holds.

(3) Conclusion (iii) may be deduced from (ii) by applying Theorem 4.3.11.]

It is still not known under what circumstances conditions (ii) and (iii) of Example 5.3.1 are equivalent.

Problem. Can one establish a theorem analogous to Theorem 5.3.5 when $\mathfrak{G}$ and G belong to some suitable class of separable Banach spaces?

3° More on Standard Gaussian Measure Spaces

We shall now give a procedure for constructing standard Gaussian processes from Gaussian measures on linear topological spaces.

Theorem 5.3.7. Let Ω be a real linear topological space such that the conjugate space $\Omega^\dagger$ distinguishes points of Ω. Let $\mathfrak{B}$ be the σ-algebra of all weak Borel sets in Ω, and let $S = (\Omega, \mathfrak{B}, N)$ be a regular probability measure space such that $\{f(\cdot), f \in \Omega^\dagger\}$ is a Gaussian process on S with mathematical expectation zero. Write

$$\int f(\omega)\, g(\omega)\, dN(\omega) = \frac{c}{2}(f, g),$$

and assume that $(f, f) > 0$ whenever $f \neq 0$. Let H denote the set of all vectors x in Ω such that

$$\| x \| = \sup_{f \in \Omega^\dagger, (f,f) \leqslant 1} |f(x)| < \infty.$$

Then H is a linear subspace of Ω, and constitutes a Hilbert space relative to the above norm[27] $\| \cdot \|$. Moreover, S is quasi-invariant and quasi-continuous with respect to the translations H.

Furthermore, if H is dense in Ω, then there exists a system of vectors $\{f_n\} \subset \Omega^\dagger$ which is orthonormal relative to the inner product $(\cdot, \cdot)$ and has the following property: when $\lambda \in H$, there is a function $\lambda(\omega)$ such that

$$\lambda(\omega) = \sum_n f_n(\lambda) f_n(\omega) \tag{5.3.11}$$

in $L^2(S)$, and S forms a standard Gaussian measure space relative to the process $\{\lambda(\cdot), \lambda \in H\}$.

Proof. Without loss of generality, we may suppose that $c = 2$. Let $\lambda \in \Omega$, and define a measure N_λ on $(\Omega, \mathfrak{B})$ as follows:

$$N_\lambda(A) = N(A - \lambda), \qquad A \in \mathfrak{B}.$$

We shall apply Theorem 5.2.5 to the Gaussian measures $P_1 = N$, $P_2 = N_\lambda$. Accordingly, we consider the totality $\mathscr{L}$ of linear combinations of elements of $\Omega^\dagger$ and constant functions. For any constants a, b and any $f, g \in \Omega^\dagger$,

$$(a + f, b + g)_1 = ab + (f, g),$$

$$\begin{aligned}(a + f, b + g)_2 &= \int (a + f(\omega))(b + g(\omega))\, dN_\lambda(\omega) \\ &= \int [(a + f(\lambda)) + f(\omega)][(b + g(\lambda)) + g(\omega)]\, dN(\omega) \\ &= (a + f(\lambda))(b + g(\lambda)) + (f, g).\end{aligned}$$

If N_λ and N are equivalent, then the two inner products $(\cdot, \cdot)_1$ and $(\cdot, \cdot)_2$ are at least topologically equivalent. Hence, there is a constant K such that

$$f(\lambda)^2 \leqslant (f, f)_2 \leqslant K(f, f)_1 = K(f, f) \tag{5.3.12}$$

for all $f \in \Omega^\dagger$, that is, $\| \lambda \| \leqslant K$. Thus, $\lambda \in H$.

[27] It is not difficult to see that the norm $\| \cdot \|$ is induced by an inner product.

Choose any complete orthonormal system[28] $\{f_n\}$ in $\Omega^\dagger$. Then the functions $\{f_n(\omega)\}$ are mutually independent Gaussian variables on S, with variance 1 and mathematical expectation zero, hence the $\{f_n(\omega)\}$ are also mutually independent Gaussian variables on $(\Omega, \mathfrak{B}, N_\lambda)$, with variance 1 and respective mathematical expectations $\{f_n(\lambda)\}$. It follows by (5.1.11) that the Kakutani inner product of N and N_λ is

$$\rho(N, N_\lambda) = \exp\left(-\frac{1}{8}\sum_n f_n(\lambda)^2\right). \tag{5.3.13}$$

However, when $\lambda \in H$, it is easily seen from the Parseval equality that

$$\| \lambda \|^2 = \sum_n | f_n(\lambda)|^2, \tag{5.3.14}$$

and so $\rho(N, N_\lambda) > 0$, that is, N_λ and N are equivalent. This, combined with the result of the preceding paragraph, shows that the totality of quasi-invariant points of S is precisely H. It is obvious that H is a linear subspace of Ω. Furthermore, by (1.4.4), (4.1.1), and (5.3.13), we have

$$M_1(\lambda)^2 = 2(1 - \exp(-\tfrac{1}{8} \| \lambda \|^2))$$

for any $\lambda \in H$, which shows that S is quasi-continuous with respect to the translations H. By Theorem 4.1.3, H is a complete linear pseudometric space relative to the pseudometric defined by the convex functional

$$R_1(\lambda) = 2\left(\int_0^\infty e^{-t}\left(1 - \exp\left(-\frac{1}{8}\| \lambda \|^2 t^2\right)\right) dt\right)^{1/2}.$$

But it is easily seen that $R_1(\cdot)$ and $\| \cdot \|$ induce the same topology on H, therefore, H is complete relative to $\| \cdot \|$.

According to Theorem 4.1.6, the topology on H induced by $\| \cdot \|$ [i.e., induced by $R_1(\cdot)$] is stronger than the relative topology induced by the topology of Ω. Hence, for every $f \in \Omega^\dagger$, the restriction f' of f to H is a continuous linear functional on H, that is, the correspondence

$$f \to f'$$

is a mapping of $\Omega^\dagger$ into $H^\dagger$. If H is dense in Ω, this mapping is injective.

[28] *Translator's note:* Let $\bar{\Omega}^\dagger$ denote the completion of $\Omega^\dagger$ with respect to the inner product $(\cdot, \cdot)$. If $\Omega^\dagger$ is separable, then one can use the Gram–Schmidt process to construct an orthonormal system in $\Omega^\dagger$ which is complete relative to $\bar{\Omega}^\dagger$. However, if $\Omega^\dagger$ is not separable, it is not clear whether such a system can be found.

Thus, identifying f with f', we obtain an imbedding[29]:

$$\Omega^\dagger \subset H^\dagger.$$

Now, $\Omega^\dagger \subset \Omega_N{}^2$, and clearly[30] $\Omega_N{}^2$ is complete relative to the inner product $(\cdot, \cdot)$. Hence, the closure of $\Omega^\dagger$ in $\Omega_N{}^2$ is also complete relative to $(\cdot, \cdot)$. But $\{f_n\}$ is a complete orthonormal system in $\Omega^\dagger$ relative to the inner product $(\cdot, \cdot)$, hence, for any $\lambda \in H$, since $\sum |f_n(\lambda)|^2 < \infty$ by (5.3.14), it follows by the Riesz–Fischer theorem that there is an element of $\Omega_N{}^2$, which we call $\lambda(\cdot)$, such that (5.3.11) holds in $L^2(S)$. Since $\lambda(\cdot) \in \Omega_N$, therefore, for any $y \in H$, the relation[31]

$$\lambda(\omega + y) = \lambda(\omega) + \sum f_n(\lambda) f_n(y) = \lambda(\omega) + (y, \lambda) \qquad (5.3.15)$$

holds for almost all ω in Ω. Also, $\{\lambda(\cdot) \mid \lambda \in H\}$ is a determining set on S, since it contains[32] $\Omega^\dagger$. It is now clear that $\{\lambda(\cdot) \mid \lambda \in H\}$ is a standard Gaussian process on S.]

Example 5.3.2. Consider the case of the Wiener process. We use the notation of Example 5.1.2.

With respect to the ordinary linear operations and the norm

$$|x| = \max_{0 \leqslant t \leqslant 1} |x(t)|,$$

the set of functions $C_0[0, 1]$ forms a normed linear space, which we denote by Ω. Let $V_0[0, 1]$ be the totality of real-valued functions f on $[0, 1]$ which are of bounded variation, continuous on the right, and such that $f(1) = 0$. The set $V_0[0, 1]$ forms a normed linear space relative to the ordinary linear operations and the norm

$$\|f\| = \bigvee_0^1 (f)$$

[29] *Translator's note:* Under the given assumptions, this imbedding is continuous relative to the norms $[(\cdot, \cdot)]^{1/2}$ and $\|\cdot\|$ on $\Omega^\dagger$ and $H^\dagger (= H)$, respectively; however, I see no proof that it is isometric. It is not clear how this imbedding is to be utilized in the proof, and the density of H in Ω is used only in obtaining this imbedding. Likewise, it is not clear how the assumption that $\Omega^\dagger$ distinguishes points of Ω enters into the proof.

[30] *Translator's note:* Presumably, the author has in mind an application of Lemma 4.2.12.

[31] *Translator's note:* To establish (5.3.15), it seems that one must produce an argument, perhaps similar to that used in the proof of Lemma 4.2.7, showing that (5.3.11) can be assumed to hold for all $\omega \in H$ [using the fact that $\Sigma_n f_n(\lambda) f_n(\omega)$ converges to (λ, ω) for all $\omega \in H$].

[32] *Translator's note:* This last assertion can be proved if one makes the assumption that every linear functional ξ on $\Omega^\dagger$, which is continuous relative to the norm $[(\cdot, \cdot)]^{1/2}$, is of the form $\xi(f) = f(x)$ for some $x \in H$ (this condition is satisfied in Example 5.3.2). It is certainly not obvious that this condition can be deduced from the original hypotheses.

(i.e., the total variation of f on $[0, 1]$). For each $\alpha \in V_0[0, 1]$, we define a linear functional F_α on $C_0[0, 1]$ as follows:

$$F_\alpha(x) = \int_0^1 x(t)\, d\alpha(t). \tag{5.3.16}$$

It is easily proved that the correspondence $\alpha \to F_\alpha$ is an isomorphism of $V_0[0, 1]$ onto $\Omega^\dagger$. Let $\alpha, \beta \in V_0[0, 1]$; then, applying Fubini's theorem to interchange the order of integration, and using (5.1.23), we have

$$\begin{aligned}(F_\alpha, F_\beta) &= \int_{C_0[0,1]} F_\alpha(x)\, F_\beta(x)\, dW_c(x) \\ &= \frac{c}{2} \int_0^1 \int_0^1 \min(t, s)\, d\alpha(t)\, d\beta(s) \\ &= \frac{c}{2} \int_0^1 \alpha(t)\, \beta(t)\, dt.\end{aligned}$$

If we write

$$(\alpha, \beta) = \int_0^1 \alpha(t)\, \beta(t)\, dt,$$

then the space H of Theorem 5.3.7 is, in the present case, the totality of functions x in $C_0[0, 1]$ which satisfy the condition

$$\| x \| = \sup_{\substack{\alpha \in V_0[0,1] \\ (\alpha,\alpha) \leq 1}} \left| \int_0^1 x(t)\, d\alpha(t) \right| < \infty. \tag{5.3.17}$$

It is easily verified that condition (5.3.17) holds if and only if x is absolutely continuous and satisfies

$$\| x \| = \left(\int_0^1 x'(t)^2\, dt \right)^{1/2} < \infty.$$

Consequently, the inner product in H is

$$(x, y) = \int_0^1 x'(t)\, y'(t)\, dt, \qquad x, y \in H.$$

Now, if $\alpha \in V_0[0, 1]$ and $x \in H$, then, integrating by parts in (5.3.16), we get

$$F_\alpha(x) = -\int_0^1 \alpha(t)\, x'(t)\, dt = -\left(\int_0^t \alpha(s)\, ds, x \right),$$

hence, when $\lambda \in H$ and $\lambda' \in V_0[0, 1]$,

$$\lambda(x) = -\int_0^1 x(t)\, d\lambda'(t) = \int_0^1 \lambda'(t)\, dx(t). \tag{5.3.18}$$

Choose any complete orthonormal system of vectors $\{\lambda_n\}$ in H such that $\lambda_n' \in V_0[0, 1]$ for every n. Then, by (5.3.11),

$$\lambda(x) = \sum \int_0^1 \lambda_n'(t)\, dx(t) \int_0^1 \lambda_n'(t)\, d\lambda(t). \tag{5.3.19}$$

Henceforth, we shall formally use the notation[33] (5.3.18) for any $\lambda \in H$. Summarizing, we have the following result.

Corollary 5.3.8. The Wiener measure space $\mathscr{W}_c$ is quasi-invariant under the translations corresponding to the linear subspace A of $C_0[0, 1]$ consisting of all absolutely continuous functions $\lambda \in C_0[0, 1]$ such that $\lambda'(t) \in L^2[0, 1]$. Moreover, if $\lambda(x)$ is defined by (5.3.19), then $\{\lambda(\cdot), \lambda \in A\}$ is a standard Gaussian process. When $\lambda \in A$ and $\lambda' \in V_0[0, 1]$, then $\lambda(x)$ may be expressed in the form (5.3.18).

We conclude this section by establishing the ergodicity of standard Gaussian measure spaces.

Theorem 5.3.9. Let $S = (\Omega, \mathfrak{B}, N)$ be a standard Gaussian measure space relative to a linear subspace H of Ω. Then S is ergodic with respect to H.

PROOF. For convenience, we assume that H is separable. Choose a complete orthonormal system $\{x_n\}$ in H, and define a mapping U from Ω into the totality l of real number sequences, as follows:

$$U : \omega \rightarrow \tilde{\omega} = \{x_1(\omega), \ldots, x_n(\omega), \ldots\}.$$

It is easily seen that $UH = l^2$. Construct a measure $\tilde{N}$ on $(l, \mathfrak{B}_\omega)$ such that the coordinates λ_n, $n = 1, 2, \ldots$ $(\lambda = \{\lambda_n\} \in l)$ are mutually independent Gaussian variables on $(l, \mathfrak{B}_\omega, \tilde{N})$, with mathematical expectation zero and variance $c/2$.

Since $\{x(\omega), x \in H\}$ is a determining set on S, it is easily proved that, for every $E \in \mathfrak{B}$, there exists an $\tilde{E} \in \mathfrak{B}_\omega$ such that E and $U^{-1}\tilde{E}$ differ only by an N-null set, and that

$$N(E) = \tilde{N}(\tilde{E}). \tag{5.3.20}$$

[33] *Translator's note:* Perhaps the author has in mind here his projected second volume, since no further use is made of this notation in the remainder of the present volume.

If $E \in \mathfrak{B}$ is quasi-invariant under H, that is, for every $x \in H$, $N(E - (E + x)) = 0$, then it follows easily from (5.3.20) that $\tilde{N}(\tilde{E} - (\tilde{E} + U_x)) = 0$. Hence, $\tilde{E}$ is quasi-invariant under l^2. However, by Theorem 3.1.32, $(l, \mathfrak{B}_\omega, \tilde{N})$ is ergodic with respect to $\mathfrak{G}$, the totality of elements of l having only finitely many nonzero coordinates. Obviously, $\mathfrak{G} \subset l^2$, hence we conclude that $\tilde{N}(\tilde{E})$ is either zero or one, and therefore $N(E)$ is either zero or one. Thus, S is ergodic with respect to H.]

§5.4. Fourier–Gauss Transforms

In this section, we shall consider the theory of L^2-Fourier transforms based upon Gaussian measures. For this purpose, we need only specialize the theory developed in §3.4 and subsection 5° of §4.3.

In all of the following lemmas and theorems, it is assumed that H is a separable real Hilbert space, that $S_c = (\Omega, \mathfrak{B}, N_c)$ is a standard Gaussian measure space (with parameter c) relative to H, and that $\{x(\omega), x \in H\}$ is the corresponding standard Gaussian process; we shall not bother to restate these assumptions.

Lemma 5.4.1. Let N_{ch} be the measure on $(\Omega, \mathfrak{B})$ defined by $N_{ch}(E) = N_c(E - h)$, $E \in \mathfrak{B}$, $h \in H$. Then

$$\frac{dN_{ch}(\omega)}{dN_c(\omega)} = \exp\left(\frac{2}{c}\, h(\omega) - \frac{1}{c}\,(h, h)\right), \qquad \omega \in \Omega. \tag{5.4.1}$$

Proof. Apply the mapping U defined in the proof of Theorem 5.3.9, and use formula (5.1.14).]

Lemma 5.4.2. The standard Gaussian measure space S_c is strongly cyclic (of degree one) with respect to H, with 1 as cyclic element. The corresponding adjoint function is

$$\psi_c(h) = \exp\left(-\frac{c}{4}\,(h, h)\right). \tag{5.4.2}$$

Proof. Let H_1 denote the smallest closed linear subspace of $L^2(S_c)$ containing 1 and invariant under all the operators $\{U(h) \mid h \in H\}$ (see §3.1). By virtue of Lemma 5.4.1, H_1 contains all the functions

$$U(h)1 = \left(\frac{dN_{ch}(\omega)}{dN_c(\omega)}\right)^{1/2} = \exp\left(\frac{1}{c}\, h(\omega) - \frac{1}{2c}\,(h,h)\right). \tag{5.4.3}$$

We must prove that $H_1 = L^2(S_c)$. Clearly, we may assume that $c = 1$.

Choose any $h_1, \ldots, h_n \in H$ such that $(h_l, h_k) = \delta_{lk}$, $l, k = 1, 2, \ldots, n$. We shall first show that, for any bounded continuous function $\varphi(t_1, \ldots, t_n)$ on R_n,

$$\varphi(h_1(\omega), \ldots, h_n(\omega)) \in H_1 . \tag{5.4.4}$$

Now, $h_1(\cdot), \ldots, h_n(\cdot)$ are Gaussian variables, with mathematical expectation zero and covariance $\delta_{lk}/2$. Hence, for any Baire function $\psi(t_1, \ldots, t_n)$, we have $\psi(h_1(\omega), \ldots, h_n(\omega)) \in L^2(S_1)$ if and only if

$$\begin{aligned} \| \psi \|^2 &= \frac{1}{(\pi)^{n/2}} \int_{-\infty}^{\infty} \cdots \int_{-\infty}^{\infty} \psi(t_1, \ldots, t_n)^2 \exp(-t_1^2 - \cdots - t_n^2)\, dt_1 \cdots dt_n \\ &= \int \psi(h_1(\omega), \ldots, h_n(\omega))^2 \, dN_1(\omega) < \infty. \end{aligned} \tag{5.4.5}$$

Let A denote the totality of such functions ψ; A forms a Hilbert space with respect to the obvious inner product corresponding to the norm $\| \cdot \|$ defined by (5.4.5). Let $B = \{\psi \mid \psi \in A, \psi(h_1(\omega), \ldots, h_n(\omega)) \in H_1\}$; obviously, B is a linear subspace of A. Moreover, since

$$(\psi, \varphi) = (\psi(h_1(\cdot), \ldots, h_n(\cdot)), \varphi(h_1(\cdot), \ldots, h_n(\cdot))), \tag{5.4.5'}$$

it follows that B is closed in A. We assert that $B = A$. For otherwise, there would exist $\psi_0 \in A$, $\psi_0 \neq 0$, $\psi_0 \perp B$. However, the function $U(\lambda_1 h_1 + \cdots + \lambda_n h_n)\, 1 \in H_1$, $-\infty < \lambda_1, \ldots, \lambda_n < \infty$, hence the function

$$\psi(t_1, \ldots, t_n) = \exp[-(\lambda_1 t_1 + \cdots + \lambda_n t_n)] \in B,$$

and therefore

$$\begin{aligned} &\frac{1}{(\pi)^{n/2}} \int_{-\infty}^{\infty} \exp[-(t_1^2 + \cdots + t_n^2) - (\lambda_1 t_1 + \cdots + \lambda_n t_n)]\, \psi_0(t_1, \ldots, t_n)\, dt_1 \cdots dt_n \\ &\quad = (\psi, \psi_0) = 0. \end{aligned} \tag{5.4.6}$$

For any positive numbers $\eta_1, \ldots, \eta_n$, the function

$$\exp(\eta_1 \mid t_1 \mid + \cdots + \eta_n \mid t_n \mid) \in A,$$

hence, if $\lambda_1, \ldots, \lambda_n$ are complex variables, then the function

$$\begin{aligned} &F(\lambda_1, \ldots, \lambda_n) \\ &\quad = \int_{-\infty}^{\infty} \exp[-(t_1^2 + \cdots + t_n^2) - (\lambda_1 t_1 + \cdots + \lambda_n t_n)]\, \psi_0(t_1, \ldots, t_n)\, dt_1 \cdots dt_n \end{aligned}$$

is well-defined, and is, in fact, an entire function of $\lambda_1, \ldots, \lambda_n$. But then it follows from (5.4.6) that F is identically zero. In particular, choosing $\lambda_l = ix_l$, $-\infty < x_l < \infty$, we see that the Fourier transform of the function

$$\exp[-(t_1^2 + \cdots + t_n^2)]\, \psi_0(t_1, \ldots, t_n)$$

is identically zero, and hence $\psi_0 = 0$, contrary to our assumption. We conclude that $A = B$, and therefore (5.4.4) holds.

Now, choose a complete orthonormal set $\{h_1, h_2, \ldots, h_n, \ldots\}$ in H. Let $\mathfrak{D}$ be the totality of functions of the form $\varphi(h_1(\omega), \ldots, h_n(\omega))$, where n varies over all positive integers and φ is a bounded continuous function on R_n. Obviously, $\mathfrak{D}$ forms an algebra, moreover, since $\{h(\omega) \mid h \in H\}$ is a determining set on S_1, it is not difficult to see that $\mathfrak{D}$ is also a determining set.[34] Hence, by virtue of Lemma 1.1.6, $\mathfrak{D}$ is dense in $L^2(S_1)$. But by (5.4.4), $\mathfrak{D} \subset H_1$, therefore H_1 is dense in $L^2(S_1)$. Since H_1 is closed, we have $H_1 = L^2(S_1)$.

Thus, we have proved that S_c is strongly cyclic with respect to H, with 1 as cyclic element. Furthermore, from (5.4.3) we get

$$\begin{aligned}
\psi_c(h) = (U(h)\,1, 1) &= \int \exp\left[\frac{1}{c}\, h(\omega) - \frac{1}{2c}(h, h)\right] dN_c(\omega) \\
&= \exp\left[-\frac{1}{2c}(h, h)\right] \frac{1}{[\pi c(h, h)]^{1/2}} \int \exp\left[\frac{1}{c}\, x - \frac{x^2}{c(h, h)}\right] dx \\
&= \exp\left[-\frac{1}{4c}(h, h)\right]. \quad \rbrack
\end{aligned}$$

We note that the quasi-characteristic function of S_c on the space of quasi-linear functionals $\hat{H} = \{h(\omega) \mid h \in H\}$ is

$$\begin{aligned}
\varphi_c(\eta) &= \int_\Omega e^{i\eta(\omega)}\, dN_c(\omega) \\
&= \frac{1}{[\pi c(\eta, \eta)]^{1/2}} \int_{-\infty}^{\infty} \exp\left[ix - \frac{x^2}{c(\eta, \eta)}\right] dx = \exp\left[-\frac{c}{4}(\eta, \eta)\right], \qquad \eta \in \hat{H}.
\end{aligned} \tag{5.4.7}$$

[34] *Translator's note:* The following sequence of steps may assist the reader in convincing himself of this fact. First, any trigonometric polynomial is bounded and continuous, and the totality of real trigonometric polynomials in the $h_i(\omega)$ determines the same measurable sets as $\{h_i(\omega), i = 1, 2, \ldots\}$. Second, $\{h_i(\omega), i = 1, 2, \ldots\}$ determines the same measurable sets as the totality $\mathscr{L}$ of finite linear combinations of the $h_i(\omega)$. Finally, since $\mathscr{L}$ is dense in $L^2(S_c)$, the characteristic function χ_E of any measurable set E may be approximated in $L^2(S_c)$ by a sequence of elements in $\mathscr{L}$, and one may choose a subsequence which converges to χ_E almost everywhere; from this, it can be seen E belongs to the σ-ring determined by $\mathscr{L}$, modulo a null set.

Comparing (5.4.2) and (5.4.7), we see that

$$\psi_c(h) = \varphi_{1/c}(Th), \qquad h \in H,$$

where T denotes the isomorphism

$$T : h \to h(\cdot)$$

from H onto $\hat{H}$. In other words, the adjoint function of S_c coincides with the quasi-characteristic function of $S_{1/c}$. Following the proof of Theorem 3.4.14, we obtain the following result.

Theorem 5.4.3. The standard Gaussian measure space $S_{1/c}$ is the dual[35] of S_c.

Definition 5.4.1. The Fourier transform F_c of $L^2(S_c)$ onto $L^2(S_{1/c})$, corresponding to the cyclic element 1, is called a *Fourier–Gauss transform.*

We now proceed to investigate the effect of the Fourier–Gauss transform on the elements of $L^2(S_c)$. By Lemma 5.4.2, we know that the set of all linear combinations of functions of the form

$$\exp\left(\frac{1}{c}\, h(\omega)\right), \qquad h \in H$$

is dense in $L^2(S_c)$. Consequently, in order to describe the operator F_c, it suffices to determine the functions $F_c(\exp((1/c)\, h(\cdot)))$.

Lemma 5.4.4. If $h \in H$, then

$$\begin{aligned} F_c\left(\exp\left(\frac{1}{c}\, h(\cdot)\right)\right)(\omega) &= \exp\left[ih(\omega) + \frac{1}{2c}(h, h)\right] \\ &= \int \exp\left[\frac{1}{c}\left(\sqrt{2}\, h(\omega_1) + ich(\omega)\right)\right] dN_c(\omega_1). \end{aligned} \tag{5.4.8}$$

PROOF. By the definition of F_c [see (5.4.3) and (3.4.48)], we have

$$F_c\left(\exp\left(-\frac{1}{2c}(h, h) + \frac{1}{c}\, h(\cdot)\right)\right)(\omega) = e^{ih(\omega)}.$$

[35] *Translator's note:* In order to satisfy the requirements of Definition 4.2.9, $\mathfrak{G}^{\mu}$ must be identified with H. This, in turn, requires the application of, Corollary 4.2.5. Hence, it seems necessary to add certain hypotheses, for example, that (i) Ω is a linear topological space, (ii) the imbedding $H \to \Omega$ is continuous, and (iii) the measure space S_c is regular.

Furthermore, $h(\omega)$ is a Gaussian variable, with mathematical expectation zero and variance $\frac{1}{2}c(h, h)$, therefore

$$\int \exp\left[\frac{\sqrt{2}}{c} h(\omega_1)\right] dN_c(\omega_1)$$
$$= \frac{1}{[\pi c(h, h)]^{1/2}} \int_{-\infty}^{\infty} \exp\left[\frac{\sqrt{2}}{c}\lambda - \frac{\lambda^2}{c(h, h)}\right] d\lambda = \exp\left[\frac{1}{2c}(h, h)\right],$$

whence we obtain (5.4.8). **]**

Next, we establish a useful formula for the Fourier transform of a polynomial functional.

Theorem 5.4.5. For any polynomial $P(t_1, \ldots, t_n)$ and any $g_1, \ldots, g_n \in H$, the polynomial functional $P(g_1(\cdot), \ldots, g_n(\cdot))$ belongs to $L^2(S_c)$, and

$$F_c(P(g_1(\cdot), \ldots, g_n(\cdot)))(\omega)$$
$$= \int P(\sqrt{2}\, g_1(\omega_1) + icg_1(\omega), \ldots, \sqrt{2}\, g_n(\omega_1) + icg_n(\omega))\, dN_c(\omega_1). \quad (5.4.9)$$

Moreover, the totality of such polynomial functionals is dense in $L^2(S_c)$.

PROOF. Consider an arbitrary element $g \in H$; for convenience, we may assume that $(g, g) = 1$. For any nonnegative integer m, we have $t^m \in A$ (see the proof of Lemma 5.4.2), hence $g^m(\cdot) \in L^2(S_c)$. Furthermore, for any real number λ, the sequence

$$f_n(\omega) = \sum_{\nu=0}^{n} \left(\frac{\lambda}{c}\right)^{\nu} \frac{g^{\nu}(\omega)}{\nu!}, \qquad n = 1, 2, \ldots$$

converges in the norm of $L^2(S_c)$ to the function $\exp((\lambda/c)\, g(\omega))$; in fact, this follows immediately from

$$\lim_{n\to\infty} \frac{1}{(\pi c)^{1/2}} \int_{-\infty}^{\infty} \left| \sum_{\nu=0}^{n} \left(\frac{\lambda}{c}\right)^{\nu} \frac{x^{\nu}}{\nu!} - \exp\left(\frac{\lambda}{c} x\right)\right|^2 \exp\left(-\frac{x^2}{c}\right) dx = 0$$

and (5.4.5). But F_c is a bounded linear operator, therefore, using (5.4.8), we obtain

$$\sum_{\nu=0}^{\infty} \frac{1}{\nu!} \left(\frac{\lambda}{c}\right)^{\nu} F_c(g^{\nu}(\cdot))(\omega) = F_c\left(\exp\left(\frac{\lambda}{c} g(\cdot)\right)\right)(\omega)$$
$$= \int \exp\left(\frac{\lambda}{c}(\sqrt{2}\, g(\omega_1) + icg(\omega))\right) dN_c(\omega_1)$$
$$= \sum_{\nu=0}^{\infty} \frac{1}{\nu!} \left(\frac{\lambda}{c}\right)^{\nu} \int (\sqrt{2}\, g(\omega_1) + icg(\omega))^{\nu}\, dN_c(\omega_1).$$

Comparing coefficients of like powers of λ, we see that (5.4.9) holds for all functionals of the form $g(\cdot)^\nu$. Now, for any $g_1, \ldots, g_n \in H$, let $g = \lambda_1 g_1 + \cdots + \lambda_n g_n$; then

$$g(\omega)^\nu = \sum_{\nu_1+\cdots+\nu_n=\nu} c_{\nu_1\cdots\nu_n} \lambda_1^{\nu_1} \cdots \lambda_n^{\nu_n} g_1^{\nu_1}(\omega) \cdots g_n^{\nu_n}(\omega), \tag{5.4.10}$$

where $c_{\nu_1\cdots\nu_n}$ denotes the appropriate multinomial coefficient. Setting $P(g(\omega)) = g(\omega)^\nu$ in (5.4.9), and comparing coefficients of like terms $\lambda_1^{\nu_1} \cdots \lambda_n^{\nu_n}$, we conclude that (5.4.9) holds whenever P is a monomial. Hence, by linearity, it follows that (5.4.9) is valid for all polynomials P.

As we have shown above, every functional of the form $e^{g(\omega)/c}$, $g \in H$ can be approximated in $L^2(S_c)$ by polynomial functionals, and, by Lemma 5.4.2, the set of all linear combinations of such functionals $e^{g(\omega)/c}$ is dense in $L^2(S_c)$. Consequently, the totality of polynomial functionals is dense in $L^2(S_c)$.]

Remark. By examining the proof of Lemma 5.4.2, it can be seen that the conclusion of Theorem 5.4.5 remains valid even if the choice of the elements $g_1, \ldots, g_n$ is restricted to some arbitrary but fixed orthonormal basis $\{g_1, \ldots, g_n, \ldots\}$ in H.

We shall establish another expression for the operator F_c. To this end, we consider the Hermite polynomials

$$h_n(t) = (-1)^n \exp(t^2) \frac{d^n}{dt^n} [\exp(-t^2)], \qquad n = 1, 2, \ldots; \qquad h_0(t) \equiv 1. \tag{5.4.11}$$

As is well known,

$$\int_{-\infty}^{\infty} h_n(t)\, h_m(t) \exp(-t^2)\, dt = 2^n n!\, \sqrt{\pi}\, \delta_{m,n}, \qquad m, n = 0, 1, 2, \ldots, \tag{5.4.12}$$

moreover, the functions $\{\exp(-t^2/2)\, h_n(t)\}$, $n = 0, 1, 2, \ldots$, constitute a complete orthonormal system in $L^2(-\infty, \infty)$. Furthermore, for any complex number λ and any real number t,

$$\exp\left(-\frac{t^2}{2}\right) \sum_{k=0}^{\infty} \frac{h_k(t)}{k!} \lambda^k = \exp\left(-(t-\lambda)^2 + \frac{1}{2} t^2\right). \tag{5.4.13}$$

In fact, (5.4.13) follows immediately from the Taylor expansion and (5.4.11). On the other hand, by (5.4.12),

$$\int_{-\infty}^{\infty} \left| \sum_{k=n}^{m} \frac{h_k(t)}{k!} \lambda^k \right|^2 \exp(-t^2)\, dt = \sum_{k=n}^{m} \frac{(2\lambda^2)^k}{k!} \sqrt{\pi} \to 0 \tag{5.4.14}$$

as $m, n \to \infty$. Consequently, the left-hand side of (5.4.13) converges in $L^2(-\infty, \infty)$ to the right-hand side.

For any positive integer n, and any two vectors $u = (u_1, ..., u_n)$, $v = (v_1, ..., v_n)$ in real n-dimensional space R_n, let (u, v) denote the inner product $\sum_{k=1}^{n} u_k v_k$. Let c be a positive parameter, let $k = (k_1, ..., k_n)$ be a sequence of nonnegative integers, and write $k! = k_1! \cdots k_n!$, $|k| = k_1 + \cdots + k_n$. Define

$$h_k(u; c) = \frac{1}{2^{|k|/2}\sqrt{k!}} h_{k_1}\left(\frac{u_1}{\sqrt{c}}\right) h_{k_2}\left(\frac{u_2}{\sqrt{c}}\right) \cdots h_{k_n}\left(\frac{u_n}{\sqrt{c}}\right). \tag{5.4.15}$$

Writing $du = du_1 \cdots du_n$, we have, by (5.4.12),

$$\frac{1}{(c\pi)^{n/2}} \int h_k(u; c)\, h_{k'}(u; c) \exp\left[-\frac{1}{c}(u, u)\right] du = \begin{cases} 0, & \text{when } k \neq k', \\ 1, & \text{when } k = k'. \end{cases} \tag{5.4.16}$$

Choose any orthonormal system $\{g_\nu, \nu = 1, 2, ...\}$ in H. Then, by virtue of (5.4.5′) and (5.4.16), the set of functions

$$\mathscr{H}(\{g_\nu\}; c) = \{h_k(g_1(\omega), ..., g_n(\omega); c) \mid k = (k_1, ..., k_n) \text{ a sequence of nonnegative integers, } n = 1, 2, ...\} \tag{5.4.17}$$

forms an orthonormal system in $L^2(S_c)$.

Definition 5.4.2. The set of functions $\mathscr{H}(\{g_\nu\}; c)$ is called the system of *Hermite polynomial functionals* corresponding to $\{g_\nu\}$. If $\{g_\nu\}$ is a complete orthonormal system in H, and $f \in L^2(S_c)$, then the number

$$a_{k_1 \cdots k_n}(f; c) = \int_\Omega f(\omega)\, h_k(g_1(\omega), ..., g_n(\omega); c)\, dN_c(\omega)$$

is known as the kth *Fourier–Hermite coefficient* of f relative to $\{g_\nu\}$.

Lemma 5.4.6. If $\{g_\nu\}$ is a complete orthonormal system in H, then the corresponding system of Hermite polynomial functionals (5.4.17) constitutes a complete orthonormal system in $L^2(S_c)$. Moreover, for any $k = (k_1, ..., k_n)$,

$$F_c(h_k(g_1(\cdot), ..., g_n(\cdot); c))(\omega) = (i)^{|k|}\, h_k\left(g_1(\omega), ..., g_n(\omega); \frac{1}{c}\right). \tag{5.4.18}$$

Proof. Obviously, every polynomial is a linear combination of Hermite polynomials, hence, by virtue of the remark following Theorem 5.4.5, we see that the totality of linear combinations of elements of $\mathscr{H}(\{g_\nu\}; c)$ is dense in $L^2(S_c)$, that is, $\mathscr{H}(\{g_\nu\}; c)$ is complete.

Let $\lambda_1, ..., \lambda_n$ be any set of real numbers, and denote the monomial $\lambda_1^{k_1} \cdots \lambda_n^{k_n}$ by λ^k. By (5.4.5′), (5.4.8), and (5.4.13),

$$\sum_k \frac{2^{|k|/2}\lambda^k}{(k!)^{1/2}} F_c(h_k(g_1(\cdot), ..., g_n(\cdot); c))(\omega)$$

$$= F_c\left(\prod_{\nu=1}^{n}\left(\sum_{k_\nu=0}^{\infty} \frac{h_{k_\nu}(g_\nu(\cdot)/\sqrt{c})}{k_\nu!} \lambda_\nu^{k_\nu}\right)\right)(\omega)$$

$$= F_c\left(\exp \sum_{\nu=1}^{n}\left\{\frac{2}{\sqrt{c}} \lambda_\nu g_\nu(\cdot) - \lambda_\nu^2\right\}\right)(\omega) = \exp\left(\sum_{\nu=1}^{n} (\lambda_\nu^2 + i2\sqrt{c}\,\lambda_\nu g_\nu(\omega))\right)$$

$$= \sum_k \frac{h_k\left(g_1(\omega), ..., g_n(\omega); \frac{1}{c}\right) i^{|k|} 2^{|k|/2} \lambda^k}{(k!)^{1/2}}.$$

Comparing coefficients of like terms λ^k, we obtain (5.4.18).]

It follows from Lemma 5.4.6 that, for every $f \in L^2(S_c)$, we have an orthogonal expansion

$$f(\omega) = \sum_k a_k(f; c)\, h_k(g_1(\omega), ..., g_n(\omega); c). \tag{5.4.19}$$

Since F_c is a unitary operator from $L^2(S_c)$ to $L^2(S_{1/c})$, relations (5.4.18) and (5.4.19) yield the following result.

Theorem 5.4.7. If $f \in L^2(S_c)$, then

$$F_c(f)(\omega) = \sum a_k(f; c)\, h_k\left(g_1(\omega), ..., g_n(\omega); \frac{1}{c}\right) (i)^{|k|}, \tag{5.4.20}$$

where $\{a_k\}$ are the Fourier–Hermite coefficients of f.

Corollary 5.4.8. The inverse F_c^{-1} of the Fourier–Gauss transform F_c is given by

$$F_c^{-1}(q) = \overline{F_{1/c}(\bar{q})}, \qquad q \in L^2(S_{1/c}). \tag{5.4.21}$$

PROOF. Let $f \in L^2(S_c)$; by (5.4.20), we have

$$a_k\left(F_c(f); \frac{1}{c}\right) = a_k(f; c)(i)^{|k|}. \tag{5.4.22}$$

Setting $F_c(f) = q$ in (5.4.22), we get

$$a_k\left(q; \frac{1}{c}\right) = a_k(F_c^{-1}(q); c)(i)^{|k|}. \tag{5.4.23}$$

On the other hand, replacing c by $1/c$ and f by $\bar{q}$ in (5.4.22), we get

$$a_k(F_{1/c}(\bar{q}); c) = a_k\left(\bar{q}; \frac{1}{c}\right)(i)^{|k|}. \tag{5.4.24}$$

Taking complex conjugates of both sides of (5.4.22), using (5.4.23) and noting that

$$\overline{a_k(\bar{q}; 1/c)} = a_k(q; 1/c),$$

we get

$$a_k(F_c^{-1}(q); c) = a_k(\overline{F_{1/c}(\bar{q})}; c),$$

whence (5.4.21) follows by applying (5.4.19).]

Finally, let us consider a special case, that is, Wiener measure.

Example 5.4.1. We use the notation of Examples 5.1.2 and 5.3.2; for convenience, we take $c = 1$. In this case, $\mathscr{W}_1$ is its own dual. Choose any system of functions $\{\alpha_n\}$ which is complete and orthonormal relative to the inner product

$$(\varphi, \psi) = \frac{1}{2}\int_0^1 \varphi(t)\, \psi(t)\, dt$$

in $L^2[0, 1]$, and such that every α_n is of bounded variation and satisfies $\alpha_n(1) = 0$. Form the functionals

$$g_n(x) = \int_0^1 \alpha_n(t)\, dx(t)$$

on $C_0[0, 1]$. These are the quasi-linear functionals associated with a complete orthonormal system in H. The corresponding system of Hermite polynomial functionals is

$$\left\{h_k\left(\int_0^1 \alpha_1(t)\, dx(t), \ldots, \int_0^1 \alpha_n(t)\, dx(t); 1\right) \,\middle|\, k = (k_1, \ldots, k_n),\ k_\nu \geqslant 0,\ n = 1, 2, \ldots\right\}.$$

Naturally, one may also consider Wiener measure and the corresponding Fourier–Gauss transform in the multidimensional case. However, the results are entirely similar; the details are left to the reader.

CHAPTER VI

REPRESENTATION OF COMMUTATION RELATIONS IN BOSE–EINSTEIN FIELDS

As we mentioned in Chapter III, some of the problems which led to the study of quasi-invariant measures arose in the quantum theory of fields, in particular, boson fields or Bose–Einstein fields. In the present chapter, we shall point out the connection between Bose–Einstein fields and the theory of quasi-invariant measures. In order to assist the reader in understanding the case of infinitely many degrees of freedom, we shall first (in §6.1) discuss in detail representations of the commutation relations in quantum mechanics (i.e., the case of finitely many degrees of freedom). Comparing the situation discussed in §6.1 with the corresponding situation in §6.2 and §6.3, the reader will appreciate the difficulties of the infinite-dimensional case; these are, in fact, the crucial mathematical difficulties in quantum-field theory. The discussion in §6.2 is rather general, whereas in §6.3 we study more specific representations and point out their connection with Gaussian measures, and so on.

Finally, although the present chapter is concerned only with representations of the commutation relations, we again emphasize that the applications of harmonic analysis on quasi-invariant measure spaces are by no means confined to representations of the commutation relations. Applications to interacting field equations will be discussed in Volume II.

§6.1. Representations of the Commutation Relations in Quantum Mechanics

1° Basic Properties

When dealing with quantum-mechanical systems having one degree of freedom, one must consider operators p, q satisfying the relation

$$pq - qp = I, \tag{6.1.1}$$

where I is the identity operator. First, we point out that such operators cannot be bounded. In fact, we have the following result.

Theorem 6.1.1. Let E be a Banach space. Then, (6.1.1) is not satisfied by any two bounded linear operators p, q on E.

PROOF. Suppose that p, q are bounded linear operators on E, satisfying condition (6.1.1). For any bounded linear operator B on E, let $B' = Bq - qB$, $B'' = (B')',\ldots, B^{(n)} = (B^{(n-1)})',\ldots$. Then $p' = I$ and $p'' = 0$. We shall now show that

$$(p^n)^{(n)} = n!I. \tag{6.1.2}$$

Since $(AB)' = A'B + AB'$, we have the following relation, in analogy with the Newton–Leibnitz formula for differentiation:

$$(AB)^{(n)} = \sum_{k=0}^{n} c_k{}^n A^{(k)} B^{(n-k)}. \tag{6.1.3}$$

Formula (6.1.2) obviously holds for $n = 1$. Assume that (6.1.2) holds for $n - 1$. Then, setting $A = p^{n-1}$, $B = p$ in (6.1.3), and using the fact that $p'' = 0$, we obtain (6.1.2).

Obviously, $\| B' \| \leqslant 2 \| q \| \| B \|$. Hence, by (6.1.2), we deduce that

$$n! \leqslant (2 \| q \|)^n \| p^n \| \leqslant (2 \| p \| \| q \|)^n \tag{6.1.4}$$

for all n, which contradicts the fact that $\lim_{n\to\infty}((2 \| p \| \| q \|)^n/n!) = 0.$]

In what follows, we shall confine ourselves to the case where p, q are unbounded operators on a Hilbert space; this is, in fact, the situation actually encountered in quantum mechanics, where p, q are self-adjoint operators satisfying the relation

$$qp - pq = iI. \tag{6.1.5}$$

This relation is not essentially different from (6.1.1), indeed, it may be obtained from (6.1.1) simply by replacing q by $-iq$.

By Theorem 6.1.1, the operators p, q in (6.1.5) cannot be bounded, hence (6.1.5) should more properly be written as

$$qp - pq \subset iI. \tag{6.1.6}$$

Here, the notation $A \subset B$ means that the domain $\mathfrak{D}_A$ of the operator A is contained in the domain $\mathfrak{D}_B$ of the operator B, that $\mathfrak{D}_A$ is dense in $\mathfrak{D}_B$, and that $Ax = Bx$ for all $x \in \mathfrak{D}_A$.

Now, let p be a self-adjoint operator on a Hilbert space H, and let $\{E_\lambda, -\infty < \lambda < \infty\}$ be the resolution of the identity corresponding to p, that is,

$$p = \int_{-\infty}^{\infty} \lambda \, dE_\lambda .$$

Form a one-parameter group of unitary operators $\{U(t); -\infty < t < \infty\}$, as follows:

$$U(t) = \int_{-\infty}^{\infty} e^{it\lambda} \, dE_\lambda .$$

This construction will be indicated by the concise notation $U(t) = e^{ipt}$.

Theorem 6.1.2. Let p and q be self-adjoint operators on a Hilbert space. Then, relation (6.1.6) is equivalent to

$$e^{ipt}e^{iqs} = e^{its}e^{iqs}e^{ipt}, \qquad -\infty < s, t < \infty.$$

The proof of this theorem will not be given here, since it is rather lengthy and employs methods not closely related with the subject of the present book. Henceforth, when dealing with the commutation relations, we shall always express them in the form given by Theorem 6.1.2, thus avoiding the complications connected with the manipulation of unbounded operators.

2° Quantum-Mechanical Systems Having Finitely Many Degrees of Freedom

Let H be a Hilbert space, and let $p_1, \ldots, p_n$, $q_1, \ldots, q_n$ be self-adjoint operators on H, satisfying the following commutation relations:

(i) $p_1, \ldots, p_n$ all commute with each other; (ii) $q_1, \ldots, q_n$ all commute with each other; (iii) p_ν and q_μ commute when $\mu \neq \nu$, and

$$q_\mu p_\mu - p_\mu q_\mu \subset iI.$$

We then say that $\{p_\nu\}$, $\{q_\nu\}$ satisfy the *Heisenberg commutation relations*. In view of Theorem 6.1.2, $\{p_\nu\}$, $\{q_\nu\}$ satisfy the Heisenberg commutation relations if and only if the unitary operators $\exp(ip_\nu t_\nu)$, $\exp(iq_\mu s_\mu)(-\infty < t_\nu, s_\mu < \infty)$ satisfy the so-called *Weyl commutation relations*, namely, (i) all the $\{\exp(ip_\nu t_\nu)\}$ commute with each other, (ii) all the $\{\exp(iq_\nu s_\nu)\}$ commute with each other, and (iii) $\exp(ip_\mu t_\mu)$ commutes with $\exp(iq_\nu s_\nu)$ when $\mu \neq \nu$, and

$$\exp(ip_\nu t_\nu) \exp(iq_\nu s_\nu) = \exp(it_\nu s_\nu) \exp(iq_\nu s_\nu) \exp(ip_\nu t_\nu).$$

We shall presently express these commutation relations in a more concise form.

Let R_n denote the ordinary n-dimensional real vector space of real n-tuples $t = (t_1, \ldots, t_n)$, with the inner product (t, s) defined by

$$(t, s) = \sum_{\nu=1}^{n} t_\nu s_\nu ,$$

where $s = (s_1, \ldots, s_n)$. Form two families of unitary operators, $U(t)$ and $V(s)$, parameterized by the vectors of R_n, as follows:

$$\text{for} \quad t = (t_1, \ldots, t_n), \qquad \text{let} \quad U(t) = \exp(ip_1 t_1) \cdots \exp(ip_n t_n); \tag{6.1.7}$$

$$\text{for} \quad s = (s_1, \ldots, s_n), \qquad \text{let} \quad V(s) = \exp(iq_1 s_1) \cdots \exp(iq_n s_n). \tag{6.1.8}$$

It is easily proved that, for $\{p_\mu\}$, $\{q_\nu\}$ to satisfy the Weyl commutation relations, it is necessary and sufficient that both $\{U(t), t \in R_n\}$ and $\{V(s), s \in R_n\}$ are weakly continuous unitary representations of the additive topological group R_n in H, and that

$$U(t)\, V(s) = e^{i(t,s)} V(s)\, U(t). \tag{6.1.9}$$

We shall refer to such a system $\{U(s), V(t); s, t \in R_n\}$ as a *weakly continuous unitary representation of the commutation relations*.

Actually, for any n, the Weyl commutation relations can be expressed in terms of unitary representations of a single group Γ_n. In fact, let Γ_n be the totality of triples (x, y, α), $x, y \in R_n$, $\alpha \in C$, where C is the set of all complex numbers of unit modulus, and define multiplication in Γ_n by

$$(x, y, \alpha) \cdot (x', y', \alpha') = (x + x', y + y', \alpha\alpha' e^{-i(x',y)}).$$

Clearly, Γ_n forms a group with respect to this operation. Take the Euclidean topology on R_n and C, and let $\Gamma_n = R_n \times R_n \times C$ have the corresponding product topology. Then Γ_n becomes a topological group; indeed, Γ_n is a $(2n + 1)$-dimensional Lie group.

Consider a weakly continuous unitary representation T of the topological group Γ_n in a Hilbert space H. If T satisfies the condition

$$T(0, 0, \alpha) = \alpha I,$$

where I is the identity operator, let

$$V(y) = T(0, y, 1), \qquad U(x) = T(x, 0, 1).$$

Straightforward calculation then shows that $\{U(x), x \in R_n\}$, $\{V(y), y \in R_n\}$ are both weakly continuous unitary representations of R_n in H, and satisfy relation (6.1.9). Conversely, let $\{U(x), x \in R_n\}$, $\{V(y), y \in R_n\}$ be any two weakly continuous unitary representations of R_n in H, satisfying relation (6.1.9). Then,

$$T(x, y, \alpha) = \alpha U(x)\, V(y)$$

is a weakly continuous unitary representation of Γ_n in H, and

$$T(0, 0, \alpha) = \alpha I.$$

We shall now give an important example of a pair of unitary operator groups $\{U(t)\}$, $\{V(s)\}$ satisfying the Weyl commutation relations.

Let $L^2(R_n)$ be the Hilbert space formed by the totality of quadratically integrable Lebesgue measurable complex-valued functions on R_n, with the inner product

$$(f, g) = \int_{R_n} f\bar{g}\, dx, \qquad f, g \in L^2(R_n).$$

For any $t, s \in R_n$, form unitary operators $U_0(t)$ and $V_0(s)$ on $L^2(R_n)$, as follows:

$$(U_0(t)f)(x) = e^{i(t,x)} f(x),$$
$$(V_0(s)f)(x) = f(x - s).$$

It is easily verified that $\{U_0(t), t \in R_n\}$ and $\{V_0(s), s \in R_n\}$ are weakly continuous unitary representations of R_n in $L^2(R_n)$, and that they satisfy (6.1.9). The system $\{U_0(t), V_0(s), t, s \in R\}$ is called the *Schrödinger representation* of the commutation relations, or a system of *Schrödinger operators.*

We may also express the Weyl commutation relations in another convenient form, as follows. Let C_n be the ordinary n-dimensional complex vector space, and define an inner product in C_n by

$$(z, z') = z_1\bar{z}_1' + \cdots + z_n\bar{z}_n',$$

where $z = (z_1, \ldots, z_n)$, $z' = (z_1', \ldots, z_n')$. Now, let $\{U(t), V(s), t, s \in R_n\}$ be a system of unitary operators, and let

$$W(t + is) = U(t)\, V(s) \exp\left[-\frac{i}{2}(t, s)\right]. \tag{6.1.10}$$

It is easily proved that the system $\{U(t), V(s), t, s \in R_n\}$ satisfies the Weyl commutation relations if and only if the operators $\{W(z), z \in C_n\}$ depend continuously (in the weak topology) on the parameter z, and satisfy

$$W(z)\, W(z') = \exp\left[-\frac{i}{2}\Im(z, z')\right] W(z + z'). \tag{6.1.11}$$

Conversely, given a family of unitary operators $\{W(z), z \in C_n\}$, satisfying (6.1.11), we may define $U(t) = W(t)$, $V(s) = W(is)$. Then, $\{U(t), t \in R_n\}$ and $\{V(s), s \in R_n\}$ are obviously one-parameter groups of unitary operators, satisfying (6.1.9), and are related to $W(z)$ by the equation (6.1.10). Moreover, if $W(z)$ is weakly continuous in z, then $\{U(t), t \in R_n\}$ and $\{V_s, s \in R_n\}$ are weakly continuous. Formula (6.1.11) is known as the *von Neumann form* of the commutation relations.

Theorem 6.1.3. Let H be a Hilbert space, $W = \{W(z), z \in C_n\}$ a family of unitary operators in H, satisfying (6.1.11), and such that $W(z)$ is weakly continuous with respect to z. Then, there exists a family $\{H_\alpha, \alpha \in \Lambda\}$ of mutually orthogonal subspaces of H, invariant with respect to W, such that $H = \sum_\alpha \oplus H_\alpha$ and the restriction of W to any H_α is unitarily equivalent to the Schrödinger representation.

PROOF. Define a bilinear functional $L(\xi, \eta)$ on H, as follows:

$$L(\xi, \eta) = \frac{1}{(2\pi)^n} \int \exp\left(-\frac{1}{4}\| z \|^2\right) (W(z)\xi, \eta)\, dz, \tag{6.1.12}$$

where

$$dz = \prod_{\nu=1}^{n} dx_\nu dy_\nu, \qquad z = (x_1 + iy_1, \ldots, x_n + iy_n), \qquad \| z \| = (z, z)^{1/2}.$$

Since $(W(z)\xi, \eta)$ is a bounded continuous function of $z \in C_n$, the integral (6.1.12) exists, moreover

$$| L(\xi, \eta)| \leqslant c \| \xi \| \| \eta \|,$$

where

$$c = \frac{1}{(2\pi)^n} \int \exp\left(-\frac{1}{4} \| z \|^2\right) dz = 2^n.$$

Furthermore, by (6.1.11), we have $W(z)^* = W(-z)$, whence it follows easily that $L(\xi, \eta) = \overline{L(\eta, \xi)}$. Therefore, there exists a bounded self-adjoint operator P such that

$$L(\xi, \eta) = (P\xi, \eta). \tag{6.1.13}$$

We assert that, for any $z \in C_n$,

$$PW(z)P = \exp(-\tfrac{1}{4} \| z \|^2)P. \tag{6.1.14}$$

In fact, for any $\xi, \eta \in H$,

$$\begin{aligned}
&(PW(z)\, P\xi, \eta) \\
&\quad = L(W(z)\, P\xi, \eta) \\
&\quad = \frac{1}{(2\pi)^n} \int \exp\left(-\frac{1}{4} \| z' \|^2\right) (W(z')\, W(z)\, P\xi, \eta)\, dz' \\
&\quad = \frac{1}{(2\pi)^n} \int \exp\left(-\frac{1}{4} \| z' \|^2\right) L(\xi, W(z)^*\, W(z')^* \eta)\, dz' \\
&\quad = \frac{1}{(2\pi)^{2n}} \int \exp\left(-\frac{1}{4} (\| z' \|^2 + \| z'' \|^2)\right) (W(z')\, W(z)\, W(z'')\xi, \eta)\, dz' dz''.
\end{aligned} \tag{6.1.15}$$

Again, using (6.1.11), we have

$$W(z')\, W(z)\, W(z'') = \exp\left(-\frac{i}{2} \Im[(z', z) + (z', z'') + (z, z'')]\right) W(z + z' + z'').$$

Taking $z_1 = z + z''$, $z_2 = z + z' + z''$ as the variables of integration in (6.1.15), and using the equality

$$\begin{aligned}
&\frac{1}{(2\pi)^n} \int \exp\left(-\frac{1}{4} (\| z_1 - z_2 \|^2 + \| z_1 - z \|^2) - \frac{i}{2} \Im[(z, z_1) + (z_2, z_1)]\right) dz_1 \\
&\quad = \exp\left[-\frac{1}{4} (\| z \|^2 + \| z_2 \|^2)\right] \\
&\qquad \times \prod_{\nu=1}^{n} \frac{1}{2\pi} \iint_{-\infty}^{\infty} \exp\left[-\frac{1}{2}\left(u_\nu - \frac{1}{2} p_\nu\right)^2 - \frac{1}{2}\left(v_\nu - \frac{i}{2} p_\nu\right)^2\right] du_\nu dv_\nu \\
&\quad = \exp\left[-\frac{1}{4} (\| z \|^2 + \| z_2 \|^2)\right],
\end{aligned}$$

where $(u_1 + iv_1, \ldots, u_n + iv_n) = z_1$, $(\bar{p}_1, \bar{p}_2, \ldots, \bar{p}_n) = z_2 + z$, we calculate that (6.1.15) is equal to $\exp(-\frac{1}{4} \| z \|^2)(P\xi, \eta)$. Hence, (6.1.14) is established.

In particular, setting $z = 0$ in (6.1.14), we see that P is a projection operator. Let $\mathfrak{M} = PH$, and choose a complete orthonormal system $\{\varphi_\alpha, \alpha \in \mathfrak{A}\}$ in $\mathfrak{M}$. Let H_α be the smallest closed linear subspace of H which contains φ_α and is invariant under all the operators $W(z)$, $z \in C_n$. We shall first prove that $H_\alpha \perp H_{\alpha'}$, $\alpha \neq \alpha'$. In fact, since $P\varphi_\alpha = \varphi_\alpha$, $P\varphi_{\alpha'} = \varphi_{\alpha'}$, it follows by (6.1.14) that

$$
\begin{aligned}
(W(z)\,\varphi_\alpha, W(z')\,\varphi_{\alpha'}) &= (W(z)\,P\varphi_\alpha, W(z')\,P\varphi_{\alpha'}) \\
&= (PW(-z')\,W(z)\,P\varphi_\alpha, \varphi_{\alpha'}) \\
&= \exp\left[-\frac{i}{2}\,\mathfrak{I}(z, z')\right](PW(z - z')\,P\varphi_\alpha, \varphi_{\alpha'}) \\
&= \exp\left[-\frac{1}{4}\| z - z' \|^2 - \frac{i}{2}\,\mathfrak{I}(z, z')\right](\varphi_\alpha, \varphi_{\alpha'})
\end{aligned}
\tag{6.1.16}
$$

for any $z, z' \in C_n$. Thus, $W(z)\,\varphi_\alpha \perp W(z')\,\varphi_{\alpha'}$ when $\alpha \neq \alpha'$. But H_α is the closure of the set of all linear combinations of the vectors $W(z)\,\varphi_\alpha$, $z \in C_n$, consequently $H_\alpha \perp H_{\alpha'}$ $(\alpha \neq \alpha')$. Next, we proceed to prove that the closed linear hull $\mathfrak{S}$ of $\bigcup_\alpha H_\alpha$ is the entire space H. Clearly, $\mathfrak{S}$ is invariant under all the operators $W(z)$, $z \in C_n$, and $\mathfrak{M} \subset \mathfrak{S}$. Therefore, $H \ominus \mathfrak{S} = \mathfrak{S}^\perp$ is also invariant under all the $W(z)$, $z \in C_n$, moreover, $P\varphi = 0$ for all $\varphi \in \mathfrak{S}^\perp$. Suppose that $\mathfrak{S}^\perp \neq (0)$, and choose any $\varphi \in \mathfrak{S}^\perp$, $\varphi \neq 0$. Then $W(z)\varphi$ also belongs to $\mathfrak{S}^\perp$, and so $PW(z)\varphi = 0$. Hence, by virtue of (6.1.12), (6.1.13), and (6.1.11), we obtain

$$
\begin{aligned}
&(PW(z)\varphi, W(z)\varphi) \\
&\quad = \frac{1}{(2\pi)^n}\int_{C_n} \exp\left[-\frac{1}{4}\| z' \|^2 + i\mathfrak{I}(z, z')\right](W(z')\varphi, \varphi)\,dz' = 0
\end{aligned}
$$

for all $z \in C_n$. Since $\mathfrak{I}(z', z) = \sum (y_\nu' x_\nu - y_\nu x_\nu')$, where $z = (x_1 + iy_1, \ldots, x_n + iy_n)$, $z' = (x_1' + iy_1', \ldots, x_n' + iy_n')$, the above relation shows that the Fourier transform of $\exp(-\frac{1}{4}\| z' \|^2)(W(z')\varphi, \varphi)$ (regarded as a function of the $2n$ real variables x_ν', y_ν', $\nu = 1, 2, \ldots, n$) is identically zero. Consequently, $(W(z')\varphi, \varphi) = 0$, $z' \in C_n$. In particular, taking $z' = 0$, we get $\varphi = 0$, which contradicts our assumption. Hence, we conclude that $H = \sum_\alpha \oplus H_\alpha$.

We now consider the manner in which $W(z)$ operates on H_α. Writing $W(z)\,\varphi_\alpha = \varphi_{\alpha,z}$, we have, by (6.1.11) and (6.1.16),

$$W(z')\,\varphi_{\alpha,z} = \exp\left[-\frac{i}{2}\,\Im(z', z)\right]\varphi_{\alpha,z+z'} \tag{6.1.17}$$

and

$$(\varphi_{\alpha,z}, \varphi_{\alpha,z'}) = \exp\left[-\frac{1}{4}\| z - z' \|^2 - \frac{i}{2}\,\Im(z, z')\right] \tag{6.1.18}$$

for all $z, z' \in C_n$.

Define a function $\varphi_z \in L^2(R_n)$ (with $z \in C_n$ as parameter), as follows:

$$\varphi_z(u) = \frac{1}{\pi^{n/4}} \exp\left[-\frac{1}{2}\| u \|^2 + (ix + y, u) - \frac{1}{2}\| y \|^2 - \frac{i}{2}(x, y)\right], \tag{6.1.19}$$

where $x, y \in R_n$, $z = x + iy$. It is easy to check that the system of operators $W_0(x + iy) = U_0(x)\,V_0(y)\exp[-(i/2)(x, y)]$, corresponding to the Schrödinger operators $U_0(x)$ and $V_0(y)$, satisfies

$$W_0(z')\,\varphi_z = \exp\left[-\frac{i}{2}\,\Im(z', z)\right]\varphi_{z+z'} \tag{6.1.20}$$

and

$$(\varphi_z, \varphi_{z'}) = \exp\left[-\frac{1}{4}\| z - z' \|^2 - \frac{i}{2}\,\Im(z, z')\right]. \tag{6.1.21}$$

Construct a linear operator U_α from H_α to $L^2(R_n)$ as follows: let $U_\alpha \varphi_{\alpha,z} = \varphi_z$, and extend U_α by linearity to all linear combinations of the $\varphi_{\alpha,z}$, $z \in C_n$. By virtue of (6.1.18) and (6.1.21), U_α is isometric, and since the totality of linear combinations of the $\varphi_{\alpha,z}$, $z \in C_n$, is dense in H_α, the mapping U_α can be uniquely extended to an isometry of H_α into $L^2(R_n)$. Moreover, it is clear that totality of linear combinations of the functions $\{\varphi_z, z \in C_n\}$ is dense in $L^2(R_n)$. Hence, U_α is a unitary operator, furthermore, it follows easily from (6.1.17) and (6.1.20) that

$$U_\alpha W(z)\,U_\alpha^{-1} = W_0(z),$$

that is, the restriction $W \mid H_\alpha$ is unitarily equivalent to the Schrödinger representation.]

Let H be a Hilbert space, and let $\{W(z), z \in R_n\}$ be a family of unitary operators on H, satisfying the commutation relations (6.1.11). If no closed linear subspace of H, other than (0) and H itself, is invariant under all the operators $W(z)$, then we say that $\{W(z), z \in R_n\}$ is *irreducible* in H. If $U(t) = W(t)$, $V(s) = W(is)$, then this irreducibility condition simply means that the unitary representation I of the group Γ_n [i.e., $T(x, y, \alpha) = \alpha U(x)\, V(y)$] is irreducible in the usual sense.

Theorem 6.1.4. The Schrödinger representation in $L^2(R_n)$ is irreducible.

PROOF. Let $\mathfrak{M} \neq (0)$ be a closed linear subspace of $L^2(R_n)$ which is invariant under all the operators $W_0(z)$, $z \in C_n$. By virtue of Theorem 6.1.3, we may assume that the restriction of $\{W_0(z), z \in C_n\}$ to $\mathfrak{M}$ is unitarily equivalent to the Schrödinger representation, for otherwise, we could select a nontrivial subspace of $\mathfrak{M}$ for which this is the case. Now, for the Schrödinger representation, there exists a vector φ [e.g., $\varphi(u) = \exp(-\frac{1}{2} \| u \|^2)$] such that the totality of linear combinations of the set $\{U_0(x)\varphi \mid x \in R_n\}$ is dense in the entire space. Obviously, any family of operators which is unitarily equivalent to the Schrödinger representation must also have this property. Thus, there exists a vector ψ in $\mathfrak{M}$ such that the totality of linear combinations of the set $\{U_0(x)\psi \mid x \in R_n\}$ is dense in $\mathfrak{M}$. Let $E = \{u \mid \psi(u) \neq 0\}$, and let $L^2(E)$ denote the totality of functions in $L^2(R_n)$ which vanish almost everywhere outside of E. We assert that $\mathfrak{M} = L^2(E)$. In fact, it clearly follows from the foregoing remarks that $\mathfrak{M} \subset L^2(E)$. Suppose that $\mathfrak{M} \neq L^2(E)$, that is, that there is an $f \in L^2(E) \ominus \mathfrak{M}$, $f \neq 0$. Then,

$$(U_0(x)\psi, f) = \int e^{i(x,u)}\psi(u)\,\overline{f(u)}\,du = 0$$

for all $x \in R_n$, in other words, the Fourier transform of the function $\psi(u)\,\overline{f(u)} \in L^1(R_n)$ is identically zero. Hence, $\psi(u)\,\overline{f(u)}$ vanishes almost everywhere, and so $f(u)$ vanishes almost everywhere in E. But f also vanishes almost everywhere outside of E, therefore $f = 0$ almost everywhere, which contradicts the choice of f. Thus, we must have $\mathfrak{M} = L^2(E)$.

On the other hand, $\mathfrak{M} = L^2(E)$ is also invariant under all the operators $V_0(y)$, $y \in R_n$. Consequently, for every $y \in R_n$, $\psi(u - y)$ vanishes almost everywhere outside of E. This means that $E - (E + y)$ is a null set for every $y \in R_n$, that is, E is quasi-invariant with respect to translations. But Lebesgue measure is ergodic (see Lemma 3.1.31), hence, either E or $R_n - E$ is a null set. Since $\mathfrak{M} \neq (0)$, E cannot be a null set. Therefore, $R_n - E$ is a null set, and $\mathfrak{M} = L^2(R_n)$.]

From Theorems 6.1.3 and 6.1.4, we immediately obtain the following result.

Corollary 6.1.5. Let $\{W(z), z \in C_n\}$ be a family of unitary operators on H, satisfying the commutation relations (6.1.11). Then H can be decomposed into an orthogonal sum of closed linear subspaces H_α, $\alpha \in \mathfrak{A}$, which are invariant under $\{W(z), z \in C_n\}$, and such that the restriction of $\{W(z), z \in C_n\}$ to any H_α is irreducible.

Corollary 6.1.6. If a family $\{W(z), z \in C_n\}$ of unitary operators on H satisfies the commutation relations (6.1.11) and is irreducible, then it is unitarily equivalent to the Schrödinger representation.

Thus, an irreducible representation of the commutation relations is uniquely determined up to unitary equivalence.

Let $\mathfrak{A}$ be the weakly closed operator algebra in $\mathfrak{B}(H)$ generated by the operators $\{U(x), V(x') \mid x, x' \in R_n\}$. We call $\mathfrak{A}$ a *concrete Weyl algebra* (with n degrees of freedom) on H. In general, the structure of this algebra depends upon the representation $\{U(x), V(x') \mid x, x' \in R_n\}$. However, we have the following result.

Lemma 6.1.7. If the representation $\{U(x), V(x') \mid x, x' \in R_n\}$ is irreducible in H, then the corresponding concrete Weyl algebra $\mathfrak{A}$ is just $\mathfrak{B}(H)$ (see §2.3).

Proof. Let $P \in (\mathfrak{A}')^P$. Then the subspace PH is invariant under the operators $\{U(x), V(x') \mid x, x' \in R_n\}$, whence it follows by irreducibility that either $P = 0$ or $P = I$. Therefore, by Corollary 2.3.5, $\mathfrak{A}' = \{\lambda I \mid \lambda \in F\}$ and $\mathfrak{A} = \mathfrak{B}(H)$ (see Example 2.3.2).]

Lemma 6.1.8. Let H be a Hilbert space and φ a symmetric automorphism of $\mathfrak{B}(H)$. Then, there exists a unitary operator U on H such that

$$\varphi(A) = UAU^{-1} \tag{6.1.22}$$

for all $A \in \mathfrak{B}(H)$.

Proof. For any unit vector ξ in H, let P_ξ be the projection operator defined by $P_\xi x = (x, \xi)\xi$, $x \in H$. Note that, if P is any projection operator in H, then $\varphi(P)^2 = \varphi(P^2) = \varphi(P)$, $\varphi(P)^* = \varphi(P^*) = \varphi(P)$, hence $\varphi(P)$ is also a projection operator; conversely, if $\varphi(P)$ is a projection operator, then so is P. Since P_ξ cannot be expressed as the sum of two nonzero projection operators, neither can $\varphi(P_\xi)$. Consequently, $\varphi(P_\xi)H$ is also a one-dimensional space, and so there exists a unit vector $\eta \in H$ such that $\varphi(P_\xi) = P_\eta$. Now, arbitrarily choose a fixed unit vector ξ_0, and let η_0 be a fixed unit vector such that $\varphi(P_{\xi_0}) = P_{\eta_0}$. For any vector $\xi \in H$, form an operator $P_{\xi\xi_0}$ as follows:

$$P_{\xi\xi_0}(\lambda\xi_0) = \lambda\xi,$$

$$P_{\xi\xi_0}(\xi') = 0 \qquad \text{when} \qquad \xi' \perp \xi_0 .$$

We define

$$U\xi = \varphi(P_{\xi\xi_0})\,\eta_0\,.$$

It is easily verified that U is a unitary operator, and that relation (6.1.22) is satisfied for any operator of the form P_ξ. It follows immediately that (6.1.22) is satisfied for any projection operator A which is of finite rank. Furthermore, since any projection operator is the supremum of a family of projection operators of finite rank, and since $A \leqslant B$ implies that $\varphi(B) - \varphi(A) = \varphi((B - A)^{1/2})^2 \geqslant 0$, relation (6.1.22) holds for any projection operator A. Next, observe that, for any $A \in \mathfrak{B}(H)$, the spectra of A and $\varphi(A)$ coincide, hence

$$\| A \| = (\| A^*A \|)^{1/2} = (\| \varphi(A^*A)\|)^{1/2} = \| \varphi(A)\|. \tag{6.1.23}$$

Applying (6.1.22) to linear combinations of projection operators, and then taking limits, it follows by (6.1.23) that (6.1.22) holds for any self-adjoint operator A, hence also for any $A \in \mathfrak{B}(H)$.]

Lemma 6.1.9. Let $\{U_0(x), V_0(x') \mid x, x' \in R_n\}$ be the Schrödinger representation in $L^2(R_n)$, and let φ be a symmetric automorphism of $\mathfrak{B}(L^2(R_n))$ such that

$$\varphi(U_0(x)) = U_0(x), \qquad \varphi(V_0(x')) = V_0(x'), \qquad x, x' \in R_n\,. \tag{6.1.24}$$

Then φ is just the trivial automorphism $A \rightarrow A$.

PROOF. By Lemma 6.1.8, there exists a unitary operator U on $L^2(R_n)$ such that (6.1.22) holds for all $A \in \mathfrak{B}(L^2(R_n))$, and by (6.1.24), we know that $U \in \{U_0(x), V_0(x') \mid x, x' \in R_n\}'$. Using Theorem 6.1.4 and Lemma 6.1.7, it follows by §2.3, 2⁰, (iv) and Example 2.3.2 that $U \in \mathfrak{B}(L^2(R_n))' = \{\lambda I \mid \lambda \in F\}$. Thus, $U = \lambda I$, therefore, by (6.1.22), $\varphi(A) = A$.]

Theorem 6.1.10. Let $H^{(k)}$, $k = 1, 2$, be Hilbert spaces,

$$\{U^{(k)}(x), V^{(k)}(x') \mid x, x' \in R_n\}$$

unitary representations of the commutation relations in the respective spaces $H^{(k)}$, and $\mathfrak{A}^{(k)}$ the corresponding concrete Weyl algebras. Then, there exists a unique symmetric isomorphism ψ from $\mathfrak{A}^{(1)}$ onto $\mathfrak{A}^{(2)}$ such that, for any bounded Baire function[1] f,

$$\psi(f(U^{(1)}(x))) = f(U^{(2)}(x)), \qquad x \in R_n\,, \tag{6.1.25}$$

$$\psi(f(V^{(1)}(x'))) = f(V^{(2)}(x')), \qquad x' \in R_n\,. \tag{6.1.26}$$

[1] The notation used here refers to the usual operational calculus (see, e.g., Riesz and Sz.-Nagy [1]).

PROOF. By Theorem 6.1.3, there exists a family of closed linear subspaces $\{H_\alpha^{(k)}, \alpha \in A_k\}$ in $H^{(k)}$ such that $H^{(k)} = \sum_\alpha \oplus H_\alpha^{(k)}$, each $H_\alpha^{(k)}$ is invariant under $\{U^{(k)}, V^{(k)}\}$, and the restriction of $\{U^{(k)}, V^{(k)}\}$ to $H_\alpha^{(k)}$ is unitarily equivalent to the Schrödinger representation ($k = 1, 2$). Let $Q_\alpha^{(k)}$ be the appropriate unitary operator from $L^2(R_n)$ to $H_\alpha^{(k)}$, and let $P_\alpha^{(k)}$ denote the projection operator from $H^{(k)}$ onto $H_\alpha^{(k)}$. Then

$$U^{(k)}(x) = \sum_\alpha Q_\alpha^{(k)} U_0(x) Q_\alpha^{(k)-1} P_\alpha^{(k)}, \tag{6.1.27}$$

$$V^{(k)}(x) = \sum_\alpha Q_\alpha^{(k)} V_0(x) Q_\alpha^{(k)-1} P_\alpha^{(k)}. \tag{6.1.28}$$

For any $A \in \mathfrak{B}(L^2(R_n))$, form the operator[2]

$$\varphi^{(k)}(A) = \sum_\alpha Q_\alpha^{(k)} A Q_\alpha^{(k)-1} P_\alpha^{(k)}.$$

Using (6.1.27) and (6.1.28), it is easily verified that $\varphi^{(k)}$ is a symmetric isomorphism of $\mathfrak{B}(L^2(R_n))$ onto $\mathfrak{A}^{(k)}$, such that, for every bounded Baire function f,

$$\varphi^{(k)}(f(U_0(x))) = f(U^{(k)}(x)), \qquad x \in R_n, \tag{6.1.29}$$

$$\varphi^{(k)}(f(V_0(x'))) = f(V^{(k)}(x')), \qquad x' \in R_n. \tag{6.1.30}$$

Taking $\psi = \varphi^{(2)}(\varphi^{(1)})^{-1}$, it follows at once that ψ fulfils the requirements of the theorem.

Now, suppose that ψ' is another isomorphism satisfying the conditions of the theorem. Consider the mapping

$$\varphi = (\varphi^{(2)})^{-1} \psi' \varphi^{(1)}. \tag{6.1.31}$$

Obviously, φ is a symmetric automorphism of $\mathfrak{B}(L^2(R_n))$, moreover, by virtue of (6.1.25), (6.1.26), (6.1.29), and (6.1.30), φ satisfies (6.1.24). Therefore, φ is the trivial automorphism of $\mathfrak{B}(L^2(R_n))$, whence, by (6.1.31), $\psi' = \psi$.]

3° Another Type of Representation

We shall now consider another type of representation which will be required later on. This type of representation differs from the Schrödinger representation in that it is based upon a Gaussian measure (which is only

[2] The sums appearing here and in (6.1.27), (6.1.28) are to be interpreted as strong limits.

quasi-invariant under translations) in R_n, rather than Lebesgue measure (which is invariant under translations).

Let $L^2(R_n, G)$ denote the totality of Lebesgue measurable complex-valued functions f on R_n such that

$$\int |f(u)|^2 \exp(-\| u \|^2)\, du < \infty.$$

$L^2(R_n, G)$ forms a Hilbert space with respect to the ordinary linear operations and the inner product

$$(f, g) = \frac{1}{\pi^{n/2}} \int f(u)\, \overline{g(u)} \exp(-\| u \|^2)\, du.$$

For $x, x' \in R_n$, define

$$(U(x)f)(u) = e^{i(u,x)} f(u),$$
$$(V(x')f)(u) = \exp\left[(u, x') - \frac{\| x' \|^2}{2}\right] f(u - x').$$

The system $\{U(x), V(x') \mid x, x' \in R_n\}$ clearly satisfies the Weyl commutation relations. Actually, this system is unitarily equivalent to the Schrödinger representation. In fact, such an equivalence is realized by the unitary operator

$$f(u) \to \frac{1}{\pi^{n/4}} \exp\left(-\frac{\| u \|^2}{2}\right) f(u)$$

from $L^2(R_n, G)$ onto $L^2(R_n)$. It follows that the representation $\{U(x), V(x') \mid x, x' \in R_n\}$ is also irreducible. In the present case, the corresponding self-adjoint operators p_ν and q_ν [see (6.1.7) and (6.1.8)] are defined as follows: the domain of p_ν is

$$\mathfrak{D}(p_\nu) = \{\varphi \mid \varphi \in L^2(R_n, G),\ u_\nu \varphi(u) \in L^2(R_n, G)\},$$

where $u = (u_1, \ldots, u_n)$, and, when $\varphi \in \mathfrak{D}(p_\nu)$,

$$(p_\nu \varphi)(u) = u_\nu \varphi(u). \tag{6.1.32}$$

The domain of q_ν is

$$\mathfrak{D}(q_\nu) = \Big\{\varphi \mid \varphi \in L^2(R_n, G),$$

$$\varphi(u) = \int^{u_\nu} \frac{d}{dt} \varphi(u_1, \ldots, u_{\nu-1}, t, u_{\nu+1}, \ldots)\, dt,$$

$$u_\nu \varphi(u) - \frac{\partial}{\partial u_\nu} \varphi(u) \in L^2(R_n, G)\Big\},$$

and, when $\varphi \in \mathfrak{D}(q_\nu)$,

$$(q_\nu \varphi)(u) = -iu_\nu \varphi(u) + i \frac{\partial}{\partial u_\nu} \varphi(u). \tag{6.1.33}$$

In fact, it is easily shown that (6.1.32) defines a self-adjoint operator, and, when $\varphi \in \mathfrak{D}(p_\nu)$,

$$(p_\nu \varphi, \psi) = \frac{1}{i} \frac{\partial}{\partial x_\nu} (U(0,\ldots, 0, x_\nu, 0,\ldots, 0)\varphi, \psi) \Big|_{x_\nu=0} .$$

Consequently, $U(x)$ and $\{p_1, \ldots, p_n\}$ are related as in (6.1.7). A similar argument applies to $\{q_1, \ldots, q_n\}$.

Let c_ν be the closure of the operator $(p_\nu + iq_\nu)/\sqrt{2}$. It is easily verified that the domain of c_ν is

$$\mathfrak{D}(c_\nu) = \Big\{\varphi \,\Big|\, \varphi \in L^2(R_n, G),$$

$$\varphi(u) = \int^{u_\nu} \frac{\partial}{\partial t} \varphi(u_1, \ldots, u_{\nu-1}, t, u_{\nu+1}, \ldots)\, dt,$$

$$2u_\nu \varphi(u) - \frac{\partial}{\partial u_\nu} \varphi(u) \in L^2(R_n, G)\Big\},$$

and, when $\varphi \in \mathfrak{D}(c_\nu)$,

$$(c_\nu \varphi)(u) = \sqrt{2}\, u_\nu \varphi(u) - \frac{1}{\sqrt{2}} \frac{\partial}{\partial u_\nu} \varphi(u). \tag{6.1.34}$$

The adjoint c_ν^* of c_ν is the closure of the operator $(p_\nu - iq_\nu)/\sqrt{2}$; its domain is

$$\mathfrak{D}(c_\nu^*) = \Big\{\varphi \,\Big|\, \varphi \in L^2(R_n, G), \frac{\partial}{\partial u_\nu} \varphi(u) \in L^2(R_n, G)\Big\},$$

and, when $\varphi \in \mathfrak{D}(c_\nu^*)$,

$$(c_\nu^* \varphi)(u) = \frac{1}{\sqrt{2}} \frac{\partial}{\partial u_\nu} \varphi(u). \tag{6.1.35}$$

A simple calculation yields the relation

$$c_k^* c_l - c_l c_k^* \subset \delta_{kl} I.$$

Consider the complete orthonormal system

$$\{h_k(u, 1); \quad k = (k_1, k_2, \ldots, k_n), \quad k_\nu = 0, 1, 2, \ldots\}$$

in $L^2(R_n, G)$ [see (5.4.15)]; hereafter, we shall simply write $h_k(u)$ in place of $h_k(u, 1)$. For each nonnegative integer m, let $H(m, n)$ be the finite-dimensional linear subspace of $L^2(R_n, G)$ spanned by the vectors $\{h_k(u) \mid |k| = m\}$. Then,

$$L^2(R_n, G) = \sum_{m=0}^{\infty} \oplus H(m, n).$$

Lemma 6.1.11. The operator c_ν maps $H(m, n)$ into $H(m+1, n)$, and

$$c_\nu h_k(u) = (k_\nu')^{1/2} h_{k'}(u),$$

where $k' = (k_1', \ldots, k_n')$ and $k = (k_1, \ldots, k_n)$ are related by

$$k_l' = \begin{cases} k_l, & \text{when } l \neq \nu, \\ k_\nu + 1, & \text{when } l = \nu. \end{cases} \tag{6.1.36}$$

The operator c_ν^* maps $H(m, n)$ into $H(m-1, n)$ [letting $H(-1, n) = (0)$], and

$$c_\nu^* h_k(u) = (k_\nu)^{1/2} h_{k'}(u),$$

where $k' = (k_1', \ldots, k_n')$ and $k = (k_1, \ldots, k_n)$ are related by

$$k_l' = \begin{cases} k_l, & \text{when } l \neq \nu, \\ \max(k_\nu - 1, 0), & \text{when } l = \nu. \end{cases} \tag{6.1.37}$$

PROOF. Note that the Hermite polynomials [see (5.4.11)] satisfy the relations

$$h_m'(x) = 2m h_{m-1}(x),$$

$$h_{m+1}(x) - 2x h_m(x) + 2m h_{m-1}(x) = 0.$$

Using these relations, (6.1.34) and (6.1.35), the assertions of the lemma follow immediately.]

We now form system $\{W(z), z \in C_n\}$, in accordance with (6.1.10).

Lemma 6.1.12. Let $k = (k_1, \ldots, k_n)$, $k' = (k_1', \ldots, k_n')$ be any two n-tuples of nonnegative integers, $n \geqslant 1$, and α any real number. Then

$$(W(e^{i\alpha} z) h_k(\cdot), h_{k'}(\cdot)) = (W(z) h_k(\cdot), h_{k'}(\cdot)) e^{i\alpha(|k| - |k'|)}. \tag{6.1.38}$$

PROOF. First, notice that, for any $\lambda, \lambda' \in C_n$,

$$(W(z) \exp(2(\lambda, t)), \exp(2(\bar{\lambda}', t))) = \frac{1}{\pi^{n/2}} \int \exp(-\varphi(t; z, \lambda, \lambda')) \, dt, \tag{6.1.39}$$

where

$$\varphi(t; z, \lambda, \lambda') = \| t \|^2 - (ix + y + 2\lambda + 2\lambda', t)$$
$$+ \frac{1}{2} \| y \|^2 + 2(\lambda, y) + \frac{i}{2} (x, y).$$

Calculating (6.1.39) by means of the Gaussian integral formula (5.1.5), we obtain

$$(W(z) \exp[-(\lambda, \bar{\lambda}) + 2(\lambda, t)], \exp[-(\bar{\lambda}', \lambda') + 2(\bar{\lambda}', t)])$$
$$= \exp\left[-\frac{(z, z)}{4} + 2(\lambda, \bar{\lambda}') + i(\lambda, \bar{z}) + i(\lambda', z)\right]. \qquad (6.1.40)$$

Now, the replacement of z by $e^{i\alpha}z$ in the right-hand member of (6.1.40) yields the same result as the replacement of λ by $\lambda e^{i\alpha}$ and λ' by $\lambda' e^{-i\alpha}$. Consequently,

$$(W(e^{i\alpha}z) \exp[-(\lambda, \bar{\lambda}) + 2(\lambda, t)], \exp[-(\bar{\lambda}', \lambda') + 2(\bar{\lambda}', t)])$$
$$= (W(z) \exp[-e^{2i\alpha}(\lambda, \bar{\lambda}) + 2e^{i\alpha}(\lambda, t)], \exp[- e^{2i\alpha}(\bar{\lambda}', \lambda') + 2e^{i\alpha}(\bar{\lambda}', t)]). \qquad (6.1.41)$$

Using (5.4.13), we expand both sides of (6.1.41) as power series in λ, λ'; comparing coefficients of like monomials $\lambda^k \lambda'^{k'}$, we obtain (6.1.38).]

4° Gradient Transformations

Let $\{W(z), z \in C_n\}$ be an irreducible family of unitary operators on a Hilbert space H, weakly continuous in the parameter z, and satisfying the commutation relations (6.1.11). Let U be a unitary operator on C_n. Form the family of unitary operators

$$W'(z) = W(Uz), \qquad z \in C_n$$

in H. Clearly, $\{W'(z), z \in C_n\}$ is also weakly continuous in z, and satisfies the commutation relations

$$W'(z)\, W'(z') = \exp\left[-\frac{i}{2} \Im(z, z')\right] W'(z + z'),$$

moreover, $\{W'(z), z \in C_n\}$ is also irreducible in H. Therefore, by virtue of Corollary 6.1.6, $\{W(z), z \in C_n\}$ and $\{W'(z), z \in C_n\}$ are unitarily equivalent, that is, there exists a unique[3] unitary operator $\Gamma(U)$ on H, such that

$$W(Uz) = \Gamma(U)\, W(z)\, \Gamma(U)^{-1}, \qquad z \in C_n\,.$$

In particular, let us consider the case in which U is of the form $e^{i\alpha}I : z \to e^{i\alpha}z$, $z \in C_n$, where α is a real number. In this case, $\Gamma(U)$ is called a *gradient transformation.*

Theorem 6.1.13. If $H = L^2(R_n, G)$ and $\{W(z)\}$ are the unitary operators introduced in 3⁰, then, for any real number α, we have

$$\Gamma(e^{i\alpha}I) = \sum_{n=0}^{\infty} e^{-i\alpha m} P_m\,, \tag{6.1.42}$$

where P_m denotes the projection of H onto $H(m, n)$.

PROOF. Since the operator defined by (6.1.42) has the property

$$\Gamma(e^{i\alpha}I)\, h_k(\cdot) = e^{-i\alpha|k|} h_k(\cdot),$$

it follows that (6.1.38) may be rewritten as

$$(\Gamma(e^{i\alpha}I)\, W(z)\, \Gamma(e^{i\alpha}I)^{-1}\, h_k\,,\, h_{k'}) = (W(e^{i\alpha}z)\, h_k\,,\, h_{k'}).$$

But $\{h_k\}$ is a complete orthonormal system in $L^2(R_n, G)$, hence

$$W(e^{i\alpha}z) = \Gamma(e^{i\alpha}I)\, W(z)\, \Gamma(e^{i\alpha}I)^{-1}.$$

This proves that (6.1.42) is, in fact, the required operator.]

§6.2. Quasi-Invariant Measures Applied to Representations of the Commutation Relations in Bose–Einstein Fields

1⁰ Representations of the Commutation Relations: Various Equivalent Formulations

We now turn to the consideration of the general case (including that of infinitely many degrees of freedom).

Definition 6.2.1. Let $\mathfrak{H}$ and $\mathfrak{H}'$ be real linear spaces, and let $B(x, x')(x \in \mathfrak{H}, x' \in \mathfrak{H}')$ be a nondegenerate real-valued bilinear functional on $(\mathfrak{H}, \mathfrak{H}')$, that is, the following conditions are satisfied.

(i) For any fixed $x \in \mathfrak{H}$, $B(x, x')$ is linear in x'; for any fixed $x' \in \mathfrak{H}'$, $B(x, x')$ is linear in x.

[3] *Translator's note:* That is, up to a numerical factor of unit modulus.

(ii) For any $x \in \mathfrak{H}$, $x \neq 0$, there exists an $x' \in \mathfrak{H}'$ such that $B(x, x') \neq 0$; conversely, for any $x' \in \mathfrak{H}'$, $x' \neq 0$, there exists an $x \in \mathfrak{H}$ such that $B(x, x') \neq 0$.

Then, $\Sigma = (\mathfrak{H}, \mathfrak{H}', B)$ is called a *single-particle* (*state vector*) *system.*

We shall have occasion to imbed $\mathfrak{H}'$ in $\mathfrak{H}^A$ as follows: for every $x' \in \mathfrak{H}'$, define a functional $f_{x'} \in \mathfrak{H}^A$ by

$$f_{x'}(x) = B(x, x'), \qquad x \in \mathfrak{H}.$$

Clearly, the correspondence $x' \to f_{x'}$ is an isomorphism of $\mathfrak{H}'$ onto a certain linear subspace of $\mathfrak{H}^A$. Thus, by identifying every x' with the corresponding $f_{x'}$, we may regard $\mathfrak{H}$ as a linear subspace of $\mathfrak{H}^A$.

Example 6.2.1. Let $\mathfrak{H}$ be a real linear space, and $\mathfrak{H}'$ a sufficient subspace of $\mathfrak{H}^A$. For any $x \in \mathfrak{H}$, $x' \in \mathfrak{H}'$, let $\mathfrak{B}(x, x') = x'(x)$. Then $(\mathfrak{H}, \mathfrak{H}', B)$ forms a single-particle system.

Example 6.2.2. Let $\mathfrak{H}$ be a real inner product space, let $\mathfrak{H}' = \mathfrak{H}$, and let $B(x, x')$ be the inner product on $\mathfrak{H}$. Then $(\mathfrak{H}, \mathfrak{H}, B)$ forms a single-particle system.

Definition 6.2.2. Let $\Sigma = (\mathfrak{H}, \mathfrak{H}', \mathfrak{B})$ be a single-particle system, and C the totality of complex numbers of unit modulus. Let $\Gamma(\Sigma) = \mathfrak{H} \times \mathfrak{H}' \times C$, and denote the elements of $\Gamma(\Sigma)$ by (x, x', α), $x \in \mathfrak{H}$, $x' \in \mathfrak{H}'$, $\alpha \in C$. Define a multiplication operation in $\Gamma(\Sigma)$ as follows:

$$(x, x', \alpha)(y, y', \beta) = (x + y, x' + y', \alpha\beta e^{-iB(y,x')}).$$

Then $\Gamma(\Sigma)$ forms a group, which we call the *group associated with the system* Σ. Let $S : \gamma \to S(\gamma)$ be a unitary representation of the group $\Gamma(\Sigma)$ in a complex Hilbert space H, such that

$$S(0, 0, \alpha) = \alpha I,$$

where I denotes the identity operator. Suppose that, for any pair of finite-dimensional linear subspaces $\mathfrak{M} \subset \mathfrak{H}$ and $\mathfrak{M}' \subset \mathfrak{H}'$, the mapping $\gamma \to S(\gamma)$ is weakly continuous on $\mathfrak{M} \times \mathfrak{M}' \times C$ with respect to the product topology induced by the ordinary Euclidean topologies on $\mathfrak{M}$, $\mathfrak{M}'$ and C. Then S is called a *canonical unitary representation* of $\Gamma(\Sigma)$ in H.

The above definition is equivalent to the following one.

Definition 6.2.3. Let $\Sigma = (\mathfrak{H}, \mathfrak{H}', B)$ be a single-particle system, and H a complex Hilbert space. Let $U : x \to U(x)$ and $V : x' \to V(x')$

be representations of $\mathfrak{H}$ and $\mathfrak{H}'$, respectively, in the group of unitary operators on H, such that the commutation relations

$$U(x)\, V(x') = e^{iB(x,x')} V(x')\, U(x), \qquad x \in \mathfrak{H}, \quad x' \in \mathfrak{H}' \tag{6.2.1}$$

are satisfied. Moreover, suppose that each of these representations is quasi-continuous, that is, if $x(x')$ is restricted to any finite-dimensional linear subspace of $\mathfrak{H}(\mathfrak{H}')$, then the mapping $x \to U(x)(x' \to V(x'))$ is weakly continuous (with respect to the Euclidean topology on the subspace). Then $\{U, V\}$ is called a (Weyl) *canonical system* (in H) over Σ.

In fact, given any canonical system $\{U, V\}$ over Σ, one may construct a canonical unitary representation of $\Gamma(\Sigma)$ in H, as follows: for any $\gamma = (x, x', \alpha)$, let

$$S(x, x', \alpha) = \alpha U(x)\, V(x').$$

We call S the *canonical unitary representation corresponding to* $\{U, V\}$. Conversely, given any canonical unitary representation S of $\Gamma(\Sigma)$ in H, let

$$U(x) = S(x, 0, 1), \qquad V(x') = S(0, x', 1).$$

Then $\{U(x), V(x') \mid x \in \mathfrak{H}, x' \in \mathfrak{H}'\}$ is a canonical system over Σ, to which S corresponds in the above sense.

Let $\{U, V\}$ be a canonical system. Choosing any $x \in \mathfrak{H}$, $x' \in \mathfrak{H}'$, we form the infinitesimal generators

$$p(x) = \frac{1}{i}\frac{d}{dt}\, U(tx)\Big|_{t=0}, \qquad q(x') = \frac{1}{i}\frac{d}{dt}\, V(tx')\Big|_{t=0} \tag{6.2.2}$$

of the one-parameter unitary groups $\{U(tx) \mid -\infty < t < \infty\}$ and $\{V(tx') \mid -\infty < t < \infty\}$, respectively (see §II.3, 2⁰). Consideration of the properties of these infinitesimal generators leads to the following definition.

Definition 6.2.3′. Let $\Sigma = (\mathfrak{H}, \mathfrak{H}', B)$ be a single-particle system and H a complex Hilbert space. Let $p : x \to p(x)$, $q : x' \to q(x')$ be linear mappings of $\mathfrak{H}$ and $\mathfrak{H}'$, respectively, into the set of self-adjoint operators in H. Suppose that, for all $x, y \in \mathfrak{H}$, $x', y' \in \mathfrak{H}'$, (i) $p(x)$ commutes[4] with

[4] Here, as in §6.1, the commutability of two (unbounded) self-adjoint operators $p(x)$ and $p(x')$ means that their spectral functions (i.e., resolutions of the identity) commute, and the "sum" of $p(x)$ and $p(x')$ means the closure of their linear sum (of course, we assume that this closure exists).

$p(y)$, (ii) $q(x')$ commutes with $q(y')$, and (iii) $p(x)$, $q(x')$ satisfy the commutation relation

$$q(x')\,p(x) - p(x)\,q(x') \subset iB(x, x')I. \tag{6.2.3}$$

Then, we say that $\{p(x), q(x') \mid x \in \mathfrak{H}, x' \in \mathfrak{H}'\}$ is an *infinitesimal canonical system* (in H) over Σ.

If $\{U, V\}$ is a Weyl canonical system, then, using Theorem 6.1.2 and the facts stated in §II.3, 2⁰, it is easily shown that the $\{p, q\}$ defined by (6.2.2) form an infinitesimal canonical system. Conversely, given any infinitesimal canonical system $\{p, q\}$, the unitary operators defined by

$$U(x) = e^{ip(x)}, \qquad V(x') = e^{iq(x')} \tag{6.2.4}$$

form a Weyl canonical system $\{U, V\}$.[5]

In what follows, we shall most often be concerned with the case mentioned in Example 6.2.2.

Let $\mathfrak{K}$ be a complex inner product space, and denote its inner product by (z, z'), $z, z' \in \mathfrak{K}$. Let $J : x \to x^*$ be a one-to-one mapping from $\mathfrak{K}$ onto $\mathfrak{K}$, satisfying the conditions

$$(x^*)^* = x, \qquad (\alpha x + \beta y)^* = \bar{\alpha} x^* + \bar{\beta} y^*, \quad \alpha, \beta \text{ complex numbers,}$$

and

$$(x^*, y^*) = (y, x).$$

Then J is called an *involution* in $\mathfrak{K}$. Let $\mathfrak{H} = \{x \mid x^* = x, x \in \mathfrak{K}\}$. Then $\mathfrak{H}$ forms a real inner product space with respect to the given inner product (x, y), $x, y \in \mathfrak{H}$. Moreover, every $z \in \mathfrak{K}$ can be uniquely expressed in the form

$$z = x + iy, \qquad x, y \in \mathfrak{H}.$$

$\mathfrak{K}$ is said to be the *complexification* of $\mathfrak{H}$. We note that involutions do exist in every complex inner product space $\mathfrak{K}$. In fact, one need only choose an arbitrary complete orthonormal system in $\mathfrak{K}$, and let $\mathfrak{H}$ denote the closure of the set of all real linear combinations of this system. Obviously, each $z \in \mathfrak{K}$ has a unique decomposition $z = x + iy$, $x, y \in \mathfrak{H}$, and we set

$$z^* = x - iy.$$

[5] *Translator's note:* The verification of equivalence between the notions of Weyl canonical system and infinitesimal canonical system (as defined by the author) involves certain technical difficulties. Since the definition is, in any case, an *ad hoc* one, perhaps it might be better to simply say that $\{p(x), q(x')\}$ is an infinitesimal canonical system provided that $\{e^{ip(x)}, e^{iq(x')}\}$ is a Weyl canonical system.

Then $z \to z^*$ is an involution. Furthermore, given any real inner product space $\mathfrak{H}$, there clearly exists a complex inner product space $\mathfrak{K}$ which is a complexification of $\mathfrak{H}$, and $\mathfrak{K}$ is unique up to an isomorphism which leaves $\mathfrak{H}$ invariant.

Now, if $\Sigma = (\mathfrak{H}, \mathfrak{H}, B)$, where $\mathfrak{H}$ is a real inner product space with the inner product B, we form the complexification $\mathfrak{K}$ of $\mathfrak{H}$, and call it the *state vector space* of Σ.

Definition 6.2.4. Let $\mathfrak{K}$ be a complex inner product space, with inner product (z, z'), $z, z' \in \mathfrak{K}$. Let $\{W(z) \mid z \in \mathfrak{K}\}$ be a family of unitary operators on a complex Hilbert space H, satisfying the von Neumann commutation relations

$$W(z)\, W(z') = \exp\left[-\frac{i}{2}\,\mathfrak{I}(z, z')\right] W(z + z').$$

Suppose that the mapping $z \to W(z)$, when restricted to any finite-dimensional linear subspace $\mathfrak{M}$, is weakly continuous (with respect to the Euclidean topology on $\mathfrak{M}$). Then we say that $\{W(z) \mid z \in \mathfrak{K}\}$ is a (von Neumann) *canonical system* over the state vector space $\mathfrak{K}$. If $\mathfrak{H}$ is a real inner product space, with inner product $B(x, x')$, and $\mathfrak{K}$ is the complexification of $\mathfrak{H}$, and if $\{U, V\}$ is a Weyl canonical system over $\Sigma = (\mathfrak{H}, \mathfrak{H}, B)$, form the system

$$W(x + iy) = \exp\left[-\frac{i}{2}\,B(x, y)\right] U(x)\, V(y), \qquad x, y \in \mathfrak{H}.$$

Then $\{W(z) \mid z \in \mathfrak{K}\}$ is a von Neumann canonical system. Conversely, if $\{W(z) \mid z \in \mathfrak{K}\}$ satisfies the conditions of Definition 6.2.3, choose any real linear subspace $\mathfrak{H}$ of $\mathfrak{K}$ such that (x, y) is real on $\mathfrak{H}$, and such that $\mathfrak{K}$ is the complexification of $\mathfrak{H}$. Let

$$U(x) = W(x), \qquad V(x') = W(ix'), \quad x, x' \in \mathfrak{H}.$$

Then, it is easily verified that $\{U, V\}$ constitutes a Weyl canonical system over $\Sigma = (\mathfrak{H}, \mathfrak{H}, B)$, $B(x, x') = (x, x')$.

2° The General Form of a Canonical System

Definition 6.2.5. Let $\{U_k, V_k\}$, $k = 1, 2$, be canonical systems, in the respective Hilbert spaces H_k ($k = 1, 2$), over the same single-particle system Σ. If there exists a unitary operator Q from H_1 onto H_2 such that

$$QU_1(x)Q^{-1} = U_2(x), \qquad QV_1(x')Q^{-1} = V_2(x') \tag{6.2.5}$$

for all $x \in \mathfrak{H}$, $x' \in \mathfrak{H}'$, then we say that the systems $\{U_k, V_k\}$, $k = 1, 2$, are *unitarily equivalent*.[6]

Insofar as their essential properties are concerned, any two unitarily equivalent canonical systems may be regarded as identical. Accordingly, we shall proceed to describe a certain type of concrete canonical system such that every canonical system is unitarily equivalent with one of this type, and thereby establish the general form of a canonical system.

Let $\Sigma = (\mathfrak{H}, \mathfrak{H}', B)$ be a single-particle system, and let $\{U, V\}$ be a canonical system over Σ in the complex Hilbert space H. As in §3.4, we let $\mathfrak{A}$ and $\mathfrak{C}$ be the weakly closed operator algebras generated by $\{U(x) \mid x \in \mathfrak{H}\}$ and $\{V(x') \mid x' \in \mathfrak{H}\}$, respectively. Clearly, both $\mathfrak{A}$ and $\mathfrak{C}$ are commutative.

Lemma 6.2.1. For every cardinal number n, there exists a projection operator $P_n \in \mathfrak{A}' \cap \mathfrak{C}'$ such that the restriction of $\mathfrak{A}$ to $P_n H$ has uniform multiplicity n. Moreover, $\sum_n P_n = I$.

PROOF. Applying Theorem 2.4.3 to the commutative weakly closed operator ring $\mathfrak{A}$, we see that there exists a unique system of projections $\{P_n\}$ satisfying all the requirements of the Lemma, except possibly the condition $P_n \in \mathfrak{C}'$. As in the proof of Theorem 3.4.1, we consider the mapping

$$T(x') : A \to V(x')AV(x')^{-1}, \qquad A \in \mathfrak{A},$$

where x' is an arbitrary element of $\mathfrak{H}'$. Using relation (6.2.1), it is easily proved that $T(x')\mathfrak{A} = \mathfrak{A}$. By the uniqueness of the projections $\{P_n\}$, it follows that $T(x')P_n = P_n$, whence $P_n \in \mathfrak{C}'$.]

In view of Lemma 6.2.1, we may, in the ensuing discussion, assume that $\mathfrak{A}$ has uniform multiplicity n in H. When this is the case, we shall say that $\{U(x) \mid x \in \mathfrak{H}\}$ has uniform multiplicity n. Also, we shall restrict our considerations to the case where H is separable, this being sufficient for applications to quantum field theory.

Theorem 6.2.2. Let $\Sigma = \{\mathfrak{H}, \mathfrak{H}', B\}$ be a single-particle system, and $\{U, V\}$ a canonical system over Σ in the separable Hilbert space H. Suppose that $\{U(x) \mid x \in \mathfrak{H}\}$ has uniform multiplicity n. Then, there exists a linear measure space $S = (\Omega, \mathfrak{B}, \mu)$, where Ω is a linear subspace of $\mathfrak{H}^A$, $\Omega \supset \mathfrak{H}'$, $\mathfrak{B}$ is the totality of weak Borel sets in Ω, $\mu(\Omega) < \infty$, and S is quasi-invariant under $\mathfrak{H}'$, such that $\{U, V\}$ is unitarily equivalent

[6] This simply amounts to the unitary equivalence, in the usual sense, of the corresponding canonical unitary representations of the group $\Gamma(\Sigma)$.

to the canonical system $\{\hat{U}, \hat{V}\}$ in $\mathfrak{L}_n^2(S)$ which is defined as follows. For any $\xi \in \mathfrak{L}_n^2(S)$,

$$\hat{U}(x)\,\xi(f) = e^{if(x)}\xi(f), \qquad x \in \mathfrak{H}, \tag{6.2.6}$$

$$\hat{V}(x')\,\xi(f) = z(f; x')\,\xi(f - x')\left(\frac{d\mu_{x'}(f)}{d\mu(f)}\right)^{1/2}, \qquad x' \in \mathfrak{H}'. \tag{6.2.7}$$

Here, for each $x' \in \mathfrak{H}'$, $z(f; x')$ is a unitary operator-valued measurable function[7] on S, such that, for any given pair of elements $x_1, x_2 \in \mathfrak{H}'$, the relations

$$z(f; x_1 + x_2) = z(f; x_1)\,z(f - x_1; x_2), \qquad z(f; 0) = 1 \tag{6.2.8}$$

hold for almost all f.

PROOF. Since H is separable, it is obvious that $n \leqslant \aleph_0$. Now, $\mathfrak{A}$ has uniform multiplicity n, that is, there exist closed linear subspaces H_α of H, $\alpha = 1, 2, \ldots, n$, such that $H = \sum_\alpha \oplus H_\alpha$, each H_α is invariant under $\mathfrak{A}$, each restriction $\mathfrak{A}_\alpha = \mathfrak{A} \mid H_\alpha$ is maximal commutative, and all the $\mathfrak{A}_\alpha$ are unitarily equivalent. Since H_α is separable, it follows from Corollary 2.4.9 that $\mathfrak{A}_\alpha$ has a cyclic element in H_α. Choose a system of cyclic elements $\{e_\alpha \in H_\alpha, \alpha = 1, 2, \ldots, n\}$ which correspond with one another under the unitary equivalences between the $\mathfrak{A}_\alpha$. Consider the function

$$\psi(x) = (U(x)\,e_\alpha, e_\alpha).$$

It is easily verified that $\psi(x)$ is a positive definite quasi-continuous function on $\mathfrak{H}$. By Theorem 4.3.5, there exists a linear measure space $S = (\Omega, \mathfrak{B}, \mu)$, where Ω is a linear subspace of $\mathfrak{H}^\Delta$, $\Omega \supset \mathfrak{H}'$ and $\mu(\Omega) < \infty$, such that

$$\psi(x) = \int_\Omega e^{if(x)}\,d\mu(f).$$

Proceeding as in the proof of Theorem 3.4.14, one can show that S is quasi-invariant under $\mathfrak{H}'$, and that there exists a unitary operator Q from H to $\mathfrak{L}_n^2(S)$ such that $\hat{U}(x) = QU(x)\,Q^{-1}$. The remaining assertions can then be established by following the proof of Theorem 3.4.5.]

Remark. Suppose we are given a linear measure space[8] $S = (\Omega, \mathfrak{B}, \mu)$, where $\mathfrak{H}' \subset \Omega \subset \mathfrak{H}^\Delta$, $\mathfrak{B}$ is the totality of weak Borel sets in Ω, $\mu(\Omega) < \infty$, and S is quasi-invariant under $\mathfrak{H}'$. Also, let there be given a system of

[7] That is, taking values in the group of unitary operators on n-dimensional complex Hilbert space (see Definition 2.4.4).

[8] *Translator's note:* Assuming only that Ω is a linear subspace of $\mathfrak{H}^\Delta$ and that $\mathfrak{B}$ is the totality of weak Borel sets in Ω, one can then deduce that the conditions of Definition 4.2.3 are necessarily satisfied.

unitary operator-valued measurable functions $z(f; x')$, $f \in \Omega$, $x' \in \mathfrak{H}'$, satisfying conditions (6.2.8). Then, it is easily verified that the operators $\{\hat{U}, \hat{V}\}$ defined by (6.2.6) and (6.2.7) constitute a canonical system[9] over Σ in $\mathfrak{L}_n{}^2(S)$. Thus, Theorem 6.2.2 gives the general form of a canonical system in the case of uniform multiplicity.

Corollary 6.2.3. Under the hypotheses of Theorem 6.2.2, (i) if $\mathfrak{H} = \mathfrak{H}'$ is a nuclear space, $B(x, x')$ is a continuous inner product on $\mathfrak{H}$, and $\mathfrak{M}$ is the completion of $\mathfrak{H}$ relative to $B(x, x')$, form the rigged Hilbert space

$$\mathfrak{H} \subset \mathfrak{M} \subset \mathfrak{H}^{\dagger}.$$

Suppose also that $x \to U(x)$ is weakly continuous relative to the topology of $\mathfrak{H}$. Then, we may choose $\Omega = \mathfrak{H}^{\dagger}$.

(ii) If $\mathfrak{H} = \mathfrak{H}'$ is an inner product space, with the inner product $B(x, x')$, and if $x \to U(x)$ is weakly continuous relative to the topology of $\mathfrak{H}$, then we may choose as Ω any Hilbert space containing $\mathfrak{H}$ such that the inclusion map of $\mathfrak{H}$ into Ω is a Hilbert–Schmidt type operator.

PROOF. In the proof of Theorem 6.2.2, instead of applying Theorem 4.3.5 to the positive definite continuous function ψ, we apply Corollaries 4.3.14 and 4.3.15, thereby deriving conclusions (i) and (ii),[10] respectively.]

The case of greatest interest to us is $n = 1$ and $z(f, x') \equiv 1$ (this is the case which suits the requirements of quantum field theory).

Theorem 6.2.4. Let $\Sigma = (\mathfrak{H}, \mathfrak{H}', B)$ be a single-particle system. Let Ω be a linear subspace of $\mathfrak{H}^{\Delta}$ containing $\mathfrak{H}'$, let $\mathfrak{B}$ be the totality of weak Borel sets in Ω, and let $(\Omega, \mathfrak{B}, \mu_k)$, $k = 1, 2$, be finite measure spaces which are quasi-invariant under $\mathfrak{H}'$. Suppose that $\{U_k, V_k\}$, $k = 1, 2$, are canonical systems over Σ in $L^2(\Omega, \mathfrak{B}, \mu_k)$, defined by

$$U_k(x)\, \xi(f) = e^{if(x)} \xi(f), \qquad x \in \mathfrak{H}, \tag{6.2.9}$$

$$V_k(x')\, \xi(f) = \xi(f - x') \left(\frac{d\mu_{kx'}(f)}{d\mu_k(f)} \right)^{1/2}, \qquad x' \in \mathfrak{H}'. \tag{6.2.10}$$

[9] *Translator's note:* It appears that some additional condition on S is necessary to ensure the weak continuity of $V(x')$ (on finite-dimensional subspaces of $\mathfrak{H}'$).

[10] *Translator's note:* Whether or not part (ii) of Corollary 6.2.3 is true, it is hard to see how it can be deduced from Corollary 4.3.15. If $\mathfrak{H}' = \mathfrak{H}$ is a real separable Hilbert space which is a dense linear subspace of another separable real Hilbert space Ω, and if the imbedding of $\mathfrak{H}$ into Ω is of Hilbert–Schmidt type, then by Corollary 5.3.3 it follows that there exists a finite measure space $(\Omega, \mathfrak{B}, \mu)$ which is quasi-invariant under $\mathfrak{H}$. Thus, one can obtain a canonical system (6.2.6), (6.2.7), but it is not clear why it should necessarily be equivalent to the *a priori* given canonical system $\{U, V\}$.

Then the measures μ_1 and μ_2 are equivalent if and only if the canonical systems $\{U_1, V_1\}$ and $\{U_2, V_2\}$ are unitarily equivalent.

PROOF. Assume that μ_1 and μ_2 are equivalent. Form a unitary operator Q, from $L^2(\Omega, \mathfrak{B}, \mu_1)$ to $L^2(\Omega, \mathfrak{B}, \mu_2)$, as follows:

$$(Q\xi)(f) = \left(\frac{d\mu_1(f)}{d\mu_2(f)}\right)^{1/2} \xi(f). \tag{6.2.11}$$

Then, using (6.2.9), (6.2.10), and (6.2.11), it is easy to check that (6.2.5) holds. Conversely, assume that Q is a unitary operator from $L^2(\Omega, \mathfrak{B}, \mu_1)$ to $L^2(\Omega, \mathfrak{B}, \mu_2)$ such that (6.2.5) holds. Write $\xi_0(f) = Q1$. Then, for any finite set of elements $x_1, x_2, \ldots, x_n \in \mathfrak{H}$ and any n complex numbers $\lambda_1, \ldots, \lambda_n$, we have

$$\int_\Omega \left| \sum_{k=1}^n \lambda_k \exp(if(x_k)) \right|^2 d\mu_1(f) = \left\| \sum \lambda_k U_1(x_k) 1 \right\|^2 = \left\| \sum \lambda_k U_2(x_k) Q1 \right\|^2$$
$$= \int_\Omega \left| \sum \lambda_k \exp(if(x_k)) \right|^2 \left| \xi_0(f) \right|^2 d\mu_2(f).$$

Introducing the finite measure

$$\mu_3(E) = \int_E \left| \xi_0(f) \right|^2 d\mu_2(f),$$

the foregoing relation may be written in the form

$$\int_\Omega \left| \sum_{k=1}^n \lambda_k \exp(if(x_k)) \right|^2 d\mu_1(f) = \int_\Omega \left| \sum_{k=1}^n \lambda_k \exp(if(x_k)) \right|^2 d\mu_3(f).$$

Now, since $\mathfrak{B}$ is the smallest σ-algebra with respect to which all the functions $\{f(x) \mid x \in \mathfrak{H}\}$ are measurable, the algebra

$$\mathscr{D} = \left\{ \sum_1^n \lambda_k \exp(if(x_k)) \;\middle|\; x_k \in \mathfrak{H}, \lambda_k \text{ complex numbers} \right\}$$

constitutes a determining set on $(\Omega, \mathfrak{B})$. Hence, by Corollary 1.1.6, $\mathfrak{D}$ is dense in $L^2(\Omega, \mathfrak{B}, \mu_1 + \mu_3)$. Consequently, for any $\xi \in L^2(\Omega, \mathfrak{B}, \mu_1 + \mu_3)$, we see that

$$\int_\Omega |\xi(f)|^2 d\mu_1(f) = \int_\Omega |\xi(f)|^2 d\mu_3(f).$$

In particular, letting ξ be the characteristic function of an arbitrary set $E \in \mathfrak{B}$, we obtain

$$\mu_1(E) = \int_E |\xi_0(f)|^2 \, d\mu_2(f).$$

Thus, $\mu_1 \ll \mu_2$. Similarly, $\mu_2 \ll \mu_1$. Therefore, μ_1 and μ_2 are equivalent.]

Definition 6.2.6. Let Σ be a single-particle system, and $\{U, V\}$ a canonical system over Σ in the complex Hilbert space H. If no nontrivial closed linear subspace of H is invariant under all the operators $U(x)$, $V(x')$,[11] then $\{U, V\}$ is said to be *irreducible*. [This is equivalent to the irreducibility of the corresponding canonical unitary representation of the group $\Gamma(\Sigma)$].

Let $\mathfrak{K}$ be a complex Hilbert space, and let $\mathfrak{U}$ denote the group of all unitary operators on $\mathfrak{K}$. In quantum field theory, one must consider not only a canonical system $\{W(z) \mid z \in \mathfrak{K}\}$ in some Hilbert space H, but also a unitary representation $\Gamma : U \to \Gamma(U)$ of $\mathfrak{U}$ in H. This representation has the property

$$\Gamma(U)\, W(z)\, \Gamma(U)^{-1} = W(Uz). \tag{6.2.12}$$

We note that, if $\{W(z) \mid z \in \mathfrak{K}\}$ is irreducible, then (6.2.12) uniquely determines $\Gamma(U)$ up to a scalar factor $\alpha(U)$, where $U \to \alpha(U)$ is a representation of $\mathfrak{U}$ in the group of complex numbers of unit modulus. In fact, suppose that $U \to \Gamma'(U)$ is another unitary representation of $\mathfrak{U}$ in H such that

$$\Gamma'(U)\, W(z)\, \Gamma'(U)^{-1} = W(Uz)$$

for all $z \in \mathfrak{K}$, and let $\Gamma_0(U) = \Gamma'(U)^{-1}\Gamma(U)$. Then

$$\Gamma_0(U)\, W(z) = W(z)\, \Gamma_0(U)$$

for all $z \in \mathfrak{K}$. Hence, $\Gamma_0(U) \in \{W(z) \mid z \in \mathfrak{K}\}'$. Since $\{W(z) \mid z \in \mathfrak{K}\}$ is irreducible, $\Gamma_0(U)$ must be of the form $\alpha(U)I$, where $\alpha(U)$ is a complex number of unit modulus.

Theorem 6.2.5. Let $\Sigma = (\mathfrak{H}, \mathfrak{H}', B)$ be a single-particle system. Let Ω be a linear subspace of $\mathfrak{H}^A$ such that $\mathfrak{H}' \subset \Omega$, let $\mathfrak{B}$ be the totality of weak Borel sets in Ω, and let $S = (\Omega, \mathfrak{B}, \mu)$ be a finite measure space

[11] Or, more concisely, $\{U(x), V(x') \mid x \in \mathfrak{H}, x' \in \mathfrak{H}'\}' = \{\lambda I \mid \lambda \text{ a complex number}\}$.

which is quasi-invariant under $\mathfrak{H}'$. Then the canonical system in $L^2(S)$ defined by

$$U(x)\,\xi(f) = e^{if(x)}\xi(f), \qquad x \in \mathfrak{H}, \tag{6.2.13}$$

$$V(x')\,\xi(f) = \xi(f - x')\left(\frac{d\mu_{x'}(f)}{d\mu(f)}\right)^{1/2}, \qquad x' \in \mathfrak{H}', \tag{6.2.14}$$

is irreducible if and only if S is ergodic with respect to the group of translations $\mathfrak{H}'$.

PROOF. Assume that $\{U, V\}$ is irreducible. If S were not ergodic, there would exist a quasi-invariant set $E \in \mathfrak{B}$ such that $\mu(E) > 0$, $\mu(\Omega - E) > 0$. Let $M = \{\xi \mid \xi(f) = 0$ for almost every $f \in E\}$. Then M is a closed linear subspace of $L^2(S)$ which is invariant under all the operators $\{U(x), V(x')\}$, and $L^2(S) \neq M \neq (0)$. This contradicts the irreducibility of $\{U, V\}$.

Conversely, assume that $\{U, V\}$ is not irreducible, so that H contains a nontrivial closed linear subspace M which is invariant under all the operators $U(x)$, $V(x')$. Let P denote the projection operator from H onto M; then $P \in \mathfrak{A}' \cap \mathfrak{C}'$. Now, since $\{e^{if(x)} \mid x \in \mathfrak{H}\}$ is a determining set on S, it follows by Lemma 2.4.4 that $\mathfrak{A}$ contains the multiplication algebra $\mathfrak{M}(S)$; but μ is finite, hence, by Lemma 2.4.10, $\mathfrak{M}(S)$ is maximal commutative, and therefore $\mathfrak{A} = \mathfrak{M}(S)$. Hence $\mathfrak{A}$ is maximal commutative, and so we have $P \in \mathfrak{A}' = \mathfrak{A} = \mathfrak{M}(S)$. Consequently, there exists a bounded measurable function $\eta(\cdot)$ on S such that P is the multiplication operator corresponding to $\eta(\cdot)$. Since P is a projection operator, it is easily seen that $\eta(\cdot)$ is equal almost everywhere to the characteristic function of some set $E \in \mathfrak{B}$. Moreover, using the fact that $P \in \mathfrak{C}'$, it is easy to deduce that E is quasi-invariant. Furthermore, since $I \neq P \neq 0$, $\mu(\Omega) \neq \mu(E) \neq 0$. Thus, S is not ergodic.]

Recall that, in the case of finitely many degrees of freedom (i.e., when $\mathfrak{H}$ is finite-dimensional), all irreducible canonical systems are (by Corollary 6.1.6) unitarily equivalent. However, in the case of infinitely many degrees of freedom, the situation becomes very complicated. For then, the number of mutually inequivalent ergodic measures is very great indeed (for example, there is at least a continuum of mutually inequivalent Gaussian measures on a countably dimensional space $\mathfrak{H}$!). Consequently, even if we confine ourselves to those having the relatively simple form (6.2.13)–(6.2.14), there are still a vast number of unitarily inequivalent irreducible canonical systems. In this context, the import of the following problem is clear.

Problem. Let Ω, $\mathfrak{H}$ be Hilbert spaces such that $\mathfrak{H}$ is a linear subspace of Ω and the operator imbedding $\mathfrak{H}$ into Ω is of Hilbert–Schmidt type. Determine the general form of all measures on $(\Omega, \mathfrak{B})$ which are quasi-invariant and ergodic with respect to $\mathfrak{H}$.[12]

We shall now express the von Neumann commutation relations in terms of quasi-invariant measures.

Let $\mathfrak{K}$ be a complex Hilbert space with inner product (z, z'), $z, z' \in \mathfrak{K}$, and write $|z| = (z, z)^{1/2}$. Now, $\mathfrak{K}$ may also be regarded as a real linear space, and is, in fact, a real Hilbert space with respect to the inner product

$$[z, z'] = \Re(z, z'), \qquad z, z' \in \mathfrak{K}.$$

We denote this real Hilbert space by $\mathfrak{H}$ to distinguish it from the underlying complex Hilbert space $\mathfrak{K}$.

Theorem 6.2.6. Let $S = (\Omega, \mathfrak{B}, \mu)$ be a standard quasi-invariant measure space associated with $\mathfrak{H}$ (see §4.2). For each $z \in \mathfrak{K}$, let $W(z)$ be the unitary operator on $L^2(S)$ defined by

$$(W(z)f)(\omega) = \exp\left[-\frac{i}{2}(iz)(\omega)\right]\left(\frac{d\mu_{-z}(\omega)}{d\mu(\omega)}\right)^{1/2} f(\omega + z), \qquad f \in L^2(S), \tag{6.2.15}$$

where $(iz)(\omega)$, $\omega \in \Omega$ denotes the quasi-linear functional[13] corresponding to the element iz of $\mathfrak{H}$. Then $\{W(z) \mid z \in \mathfrak{K}\}$ satisfies the von Neumann commutation relations.

PROOF. For any $z, z' \in \mathfrak{K}$, we have

$$(W(z)\, W(z')f)(\omega) = \exp\left(-\frac{i}{2}[(iz)(\omega) + (iz')(\omega + z)]\right) f(\omega + z + z') \left(\frac{d\mu_{-z-z'}(\omega)}{d\mu(\omega)}\right)^{1/2}. \tag{6.2.16}$$

Since $(iz')(\cdot)$ is the quasi-linear functional corresponding to iz', the relation

$$(iz')(\omega + z) = (iz')(\omega) + [iz', z] \tag{6.2.17}$$

[12] *Translator's note:* Although it is not explicitly stated, the author presumably intends that Ω, $\mathfrak{H}$ are real and separable, that $\mathfrak{H}$ is dense in Ω, that $\mathfrak{B}$ is the totality of weak Borel sets in Ω, and that only finite measures are to be considered.

[13] *Translator's note:* Unless some additional hypothesis is imposed (see Lemma 4.2.4 and the remark following it), these functionals need not be unique. However, a system (6.2.15) satisfying the von Neumann commutation relations can be constructed by choosing any subspace W of Ω_μ which maps isomorphically onto $\tilde{\mathfrak{H}}^\mu$ under the natural homomorphism (4.2.7), and, for each $z \in \mathfrak{H}$, letting $(iz)(\cdot)$ be unique antecedent of iz in W.

holds almost everywhere. But $[iz', z] = \mathfrak{R}(iz', z) = \mathfrak{I}(z, z')$, hence the required conclusion follows directly from (6.2.16) and (6.2.17).]

3° Algebras of Observables

Let $\Sigma = (\mathfrak{H}, \mathfrak{H}', B)$ be a single-particle system and $\{U, V\}$ a canonical system over Σ in the complex Hilbert space H. Let M be the weakly closed operator algebra generated by $\mathfrak{A} \cup \mathfrak{C}$. If $\{U, V\}$ is irreducible, then $M' = \{\lambda I \mid \lambda \text{ a complex number}\}$, hence $M = \mathfrak{B}(H)$. Suppose that $\{U_1, V_1\}$ is another irreducible canonical system over Σ, in another complex Hilbert space H_1; in general, there will not exist a symmetric isomorphism φ from $\mathfrak{B}(H)$ onto $\mathfrak{B}(H_1)$ such that $\varphi(U(x)) = U_1(x)$, $\varphi(V(x')) = V_1(x')$. For, by Lemma 6.1.8, such an isomorphism must be of the form $\varphi(A) = UAU^{-1}$, where U is a unitary operator, that is, $\{U, V\}$ and $\{U_1, V_1\}$ would have to be unitarily equivalent. Thus, it is apparent that the above definition of the algebra M is of little use. Accordingly, we proceed to reduce the size of M, in the following manner.

Let $\mathfrak{M}, \mathfrak{M}'$ be finite-dimensional linear subspaces of $\mathfrak{H}, \mathfrak{H}'$, respectively, and suppose that $B(x, x')$ is nondegenerate on $(\mathfrak{M}, \mathfrak{M}')$ (in particular, this implies that $\mathfrak{M}$ and $\mathfrak{M}'$ have the same dimension). We then say that the pair $(\mathfrak{M}, \mathfrak{M}')$ is *nondegenerate*.[14] For each nondegenerate pair of finite-dimensional subspaces $(\mathfrak{M}, \mathfrak{M}')$, let $W_{\mathfrak{M},\mathfrak{M}'}$ denote the weakly closed ring generated by the family of unitary operators $\{U(x), V(x') \mid x \in \mathfrak{M}, x' \in \mathfrak{M}'\}$ on H, and let $\mathfrak{W}$ be the smallest uniformly closed operator algebra which contains the union of all such rings $\mathfrak{W}_{\mathfrak{M},\mathfrak{M}'}$.

Definition 6.2.7. Let Σ be a single-particle system, $\{U, V\}$ a Weyl canonical system over Σ in the Hilbert space H, and $\{p, q\}$ the corresponding infinitesimal canonical system. Then, the operator ring $\mathfrak{W}$ described above is called the *concrete Weyl algebra* over Σ (on H) associated with $\{U, V\}$ (or with $\{p, q\}$). The operators in $\mathfrak{W}$ are called the *observables* of Σ.

Theorem 6.2.7. Let $\Sigma = (\mathfrak{H}, \mathfrak{H}', B)$ be a single-particle system, let $\{p_1, q_1\}$, $\{p_2, q_2\}$ be infinitesimal canonical systems over Σ in the Hilbert spaces H_1, H_2, respectively, and let $\mathfrak{A}_1$, $\mathfrak{A}_2$ be the associated concrete Weyl algebras. Then, there exists a unique symmetric isomorphism from $\mathfrak{A}_1$ onto $\mathfrak{A}_2$ such that, for any bounded Baire function f,

$$\varphi(f(p_1(x))) = f(p_2(x)), \qquad \varphi(f(q_1(x'))) = f(q_2(x')), \qquad x \in \mathfrak{H}, \quad x' \in \mathfrak{H}'.$$

[14] Note that, since $B(x, x')$ is nondegenerate on $(\mathfrak{H}, \mathfrak{H}')$, it follows that, for each finite-dimensional linear subspace $\mathfrak{M} \subset \mathfrak{H}$, there exists a finite-dimensional linear subspace $\mathfrak{M}' \subset \mathfrak{H}'$ such that the pair $(\mathfrak{M}, \mathfrak{M}')$ is nondegenerate.

PROOF. Let $(\mathfrak{M}, \mathfrak{M}')(\mathfrak{M} \subset \mathfrak{H}, \mathfrak{M}' \subset \mathfrak{H}')$ be any nondegenerate pair of finite-dimensional linear subspaces. Choose bases $\{e_1, ..., e_r\}$ and $\{e_1', ..., e_r'\}$ for $\mathfrak{M}$ and $\mathfrak{M}'$, respectively, such that, when $x = \sum_{\nu=1}^{r} x_\nu e_\nu \in \mathfrak{M}$ and $x' = \sum_{\nu=1}^{r} x_\nu' e_\nu' \in \mathfrak{M}'$,

$$B(x, x') = x_1 x_1' + \cdots + x_r x_r'.$$

For any $t = (t_1, ..., t_n)$, $s = (s_1, ..., s_n) \in R_n$, form the operators

$$U_k(t) = \exp(ip_k(e_1)\, t_1) \cdots \exp(ip_k(e_n)\, t_n),$$
$$V_k(s) = \exp(iq_k(e_1')\, s_1) \cdots \exp(iq_k(e_n')\, s_n),$$

$k = 1, 2$. Then $\{U_k(t) \mid t \in R_n\}$ and $\{V_k(s) \mid s \in R_n\}$ are weakly continuous unitary representations of R_n in H_k, and satisfy the commutation relations (6.1.9), $k = 1, 2$. Let $\mathfrak{W}^{(k)}_{\mathfrak{M},\mathfrak{M}'}$ denote the weakly closed operator ring on H_k generated by the family of operators $\{U_k(t), V_k(s) \mid t, s \in R_n\}$. By virtue of Theorem 6.1.10, there exists a unique symmetric isomorphism $\varphi_{\mathfrak{M},\mathfrak{M}'}$ from $\mathfrak{W}^{(1)}_{\mathfrak{M},\mathfrak{M}'}$ onto $\mathfrak{W}^{(2)}_{\mathfrak{M},\mathfrak{M}'}$ such that, for every bounded Baire function f,

$$\varphi_{\mathfrak{M},\mathfrak{M}'}(f(p_1(x))) = f(p_2(x)), \tag{6.2.18}$$

$$\varphi_{\mathfrak{M},\mathfrak{M}'}(f(q_1(x'))) = f(q_2(x')). \tag{6.2.19}$$

Let Λ denote the totality of nondegenerate pairs of finite-dimensional subspaces $(\mathfrak{M}, \mathfrak{M}')$. If $(\mathfrak{N}, \mathfrak{N}')$, $(\mathfrak{M}, \mathfrak{M}') \in \Lambda$ and $\mathfrak{M} \subset \mathfrak{N}$, $\mathfrak{M}' \subset \mathfrak{N}'$, then we write $(\mathfrak{M}, \mathfrak{M}') \prec (\mathfrak{N}, \mathfrak{N}')$; the set Λ is directed by this ordering relation. Now, $\mathfrak{W}^{(k)}$ is the uniform closure of the ring

$$\mathfrak{W}_0^{(k)} = \bigcup_{(\mathfrak{M},\mathfrak{M}') \in \Lambda} \mathfrak{W}^{(k)}_{\mathfrak{M},\mathfrak{M}'} .$$

Clearly, if $(\mathfrak{N}, \mathfrak{N}') \prec (\mathfrak{M}, \mathfrak{M}')$, then $\mathfrak{W}^{(k)}_{\mathfrak{N},\mathfrak{N}'} \subset \mathfrak{W}^{(k)}_{\mathfrak{M},\mathfrak{M}'}$. We assert that the restriction of $\varphi_{\mathfrak{M},\mathfrak{M}'}$ to $\mathfrak{W}^{(1)}_{\mathfrak{N},\mathfrak{N}'}$ is just $\varphi_{\mathfrak{N},\mathfrak{N}'}$. In fact, since (6.2.18) and (6.2.19) are, in particular, valid for all $x \in \mathfrak{N}$, $x' \in \mathfrak{N}'$, it can be seen from the proof of Theorem 6.1.10 that the restriction of $\varphi_{\mathfrak{M},\mathfrak{M}'}$ to $\mathfrak{W}^{(1)}_{\mathfrak{N},\mathfrak{N}'}$ is also a symmetric isomorphism from $\mathfrak{W}^{(1)}_{\mathfrak{N},\mathfrak{N}'}$ to $\mathfrak{W}^{(2)}_{\mathfrak{N},\mathfrak{N}'}$ such that (6.1.25) and (6.1.26) hold for all $x \in \mathfrak{N}$, $x' \in \mathfrak{N}'$. But Theorem 6.1.10 states that there exists just one such isomorphism, namely, $\varphi_{\mathfrak{N},\mathfrak{N}'}$.

We can now define a mapping φ_0 from $\mathfrak{W}_0^{(1)}$ to $\mathfrak{W}_0^{(2)}$, as follows. Given any $A \in \mathfrak{W}_0^{(1)}$, there exists $(\mathfrak{M}, \mathfrak{M}') \in \Lambda$ such that $A \in \mathfrak{W}_{\mathfrak{M},\mathfrak{M}'}$. Define

$$\varphi_0(A) = \varphi_{\mathfrak{M},\mathfrak{M}'}(A).$$

Then φ_0 is a symmetric isomorphism from $\mathfrak{W}_0^{(1)}$ onto $\mathfrak{W}_0^{(2)}$. Since $\mathfrak{W}_0^{(1)}$,

$\mathfrak{W}_0^{(2)}$ are dense in $\mathfrak{W}^{(1)}$, $\mathfrak{W}^{(2)}$, respectively, φ_0 can be extended to a symmetric isomorphism φ from $\mathfrak{W}^{(1)}$ onto $\mathfrak{W}^{(2)}$; by (6.2.18) and (6.2.19), φ satisfies the requirements of the theorem. Finally, the uniqueness of φ follows at once[15] from the uniqueness of the $\varphi_{\mathfrak{M},\mathfrak{M}'}$.]

4° The Characteristic Functional of a Canonical System

Definition 6.2.8. Let $\Sigma = (\mathfrak{H}, \mathfrak{H}, B)$ be a single-particle system, where $\mathfrak{H}$ is a real inner product space and $B(\cdot, \cdot)$ is the inner product on $\mathfrak{H}$. Let $\mathfrak{K}$ be the complexification of $\mathfrak{H}$, and let $\{W(z) \mid z \in \mathfrak{K}\}$ be a von Neumann canonical system over Σ in the Hilbert space H. Suppose there exists an $\eta_0 \in H$ such that $\{W(z)\,\eta_0 \mid z \in \mathfrak{K}\}$ generates H; then, the canonical system $\{W(z) \mid z \in \mathfrak{K}\}$ is said to be *cyclic*, η_0 is called a *cyclic element* for the system, and

$$\psi(z) = (W(z)\,\eta_0\,,\,\eta_0), \qquad z \in \mathfrak{K}$$

is called the *characteristic functional* of the canonical system (associated with η_0).

Following the proof of Lemma 2.4.6, we easily obtain the following result.

Lemma 6.2.8. Let Σ be a single-particle system, and let $\{W(z) \mid z \in \mathfrak{K}\}$ be a von Neumann canonical system over Σ in a separable Hilbert space H. Then, H can be decomposed into an orthogonal sum

$$H = \sum_{\xi \in \Xi} \oplus H_\xi\,,$$

where each H_ξ is invariant under the system $\{W(z) \mid z \in \mathfrak{K}\}$, and the restriction of $\{W(z) \mid z \in \mathfrak{K}\}$ to H_ξ is cyclic.

By virtue of the above lemma, we may henceforth restrict our considerations to cyclic canonical systems.

If two canonical systems are unitarily equivalent, and one of them is cyclic, then obviously the other is also cyclic, moreover, the characteristic functionals associated with corresponding cyclic elements are equal. Conversely, we have the following result.

Lemma 6.2.9. Let Σ be a single-particle system, and let $\{W_k(z) \mid z \in \mathfrak{K}\}$, $k = 1, 2$, be cyclic canonical systems in the respective Hilbert spaces H_k, $k = 1, 2$. Suppose that there exist cyclic elements

[15] *Translator's note:* If, in the statement of Theorem 6.2.7, one adds the condition that φ maps each $\mathfrak{W}^{(1)}_{\mathfrak{M},\mathfrak{M}'}$ onto $\mathfrak{W}^{(2)}_{\mathfrak{M},\mathfrak{M}'}$, then the uniqueness follows easily, as asserted. Otherwise, it seems difficult to prove.

η_k in H_k, $k = 1, 2$, such that the associated characteristic functionals are equal. Then the two canonical systems are unitarily equivalent.

PROOF. Let $\psi(z)$ denote the common characteristic functional associated with η_1 and η_2. Since

$$W_k(z)^* W_k(z') = \exp\left[\frac{i}{2}\Im(z, z')\right] W_k(z' - z) \tag{6.2.20}$$

for any $z, z' \in \mathfrak{K}$, it follows that

$$(W_1(z')\eta_1, W_1(z)\eta_1) = \exp\left[\frac{i}{2}\Im(z, z')\right]\psi(z' - z) = (W_2(z')\eta_2, W_2(z)\eta_2). \tag{6.2.21}$$

Let M_k be the linear hull of the set of vectors $\{W_k(z)\eta_k, z \in \mathfrak{K}\}$, $k = 1, 2$, and let Q be the linear mapping from M_1 onto M_2 defined by $QW_1(z)\eta_1 = W_2(z)\eta_2$. By (6.2.21), Q is an isometry, and since M_k is dense in H_k ($k = 1, 2$), Q can be uniquely extended to a unitary operator from H_1 onto H_2. Moreover,

$$\begin{aligned} &QW_1(z)Q^{-1}W_2(z')\eta_2 \\ &\quad = QW_1(z)W_1(z')\eta_1 \\ &\quad = \exp\left[-\frac{i}{2}\Im(z, z')\right]QW_1(z + z')\eta_1 = \exp\left[-\frac{i}{2}\Im(z, z')\right]W_2(z + z')\eta \\ &\quad = W_2(z)W_2(z')\eta_2. \end{aligned}$$

Consequently, for any $\xi \in M_2$, we have

$$QW_1(z)Q^{-1}\xi = W_2(z)\xi.$$

This proves that $QW_1(z)Q^{-1} = W_2(z)$, $z \in \mathfrak{K}$.]

Next, we establish a necessary and sufficient condition for a function to be the characteristic functional of some canonical system.

Lemma 6.2.10. Let $\mathfrak{K}$ be a complex inner product space and $\psi(z)$ a complex-valued function on $\mathfrak{K}$. Then, $\psi(z)$ is the characteristic functional of a von Neumann canonical system in some Hilbert space if and only if: (i) ψ is quasi-continuous, and (ii) the function

$$\psi(z_1 - z_2)\exp\left[-\frac{i}{2}\Im(z_1, z_2)\right], \qquad z_1, z_2 \in \mathfrak{K}$$

is a positive definite kernel, that is, for any finite set of complex numbers $\lambda_1, ..., \lambda_n$, and any n vectors $z_1, ..., z_n \in \mathfrak{K}$,

$$\sum_{l,m=1}^{n} \psi(z_l - z_m) \exp\left[-\frac{i}{2}\mathfrak{I}(z_l, z_m)\right] \lambda_l \bar{\lambda}_m \geqslant 0. \tag{6.2.22}$$

Proof. Suppose that ψ is the characteristic functional of a canonical system $\{W(z) \mid z \in \mathfrak{K}\}$. Then, using (6.2.20), we see that the left-hand side of (6.2.22) is equal to

$$\left\| \sum_{l=1}^{n} W(z_l)\lambda_l \right\|^2 \geqslant 0.$$

Thus, condition (ii) is satisfied. Furthermore, since $W(z)$ is weakly continuous on any finite-dimensional linear subspace of $\mathfrak{K}$, it follows immediately that ψ is quasi-continuous.

Conversely, assume that ψ satisfies conditions (i) and (ii). Let H_0 denote the totality of complex-valued functions on $\mathfrak{K}$ which vanish everywhere except possibly on some finite set of points $z_1, ..., z_n \in \mathfrak{K}$. Obviously, H_0 is a linear space relative to the ordinary linear operations. For any $f, g \in H_0$, define

$$(f, g) = \sum_{z, z' \in \mathfrak{K}} \psi(z - z') \exp\left[-\frac{i}{2}\mathfrak{I}(z, z')\right] f(z)\overline{g(z')}. \tag{6.2.23}$$

If we identify elements of H_0 whose difference f is such that $(f, f) = 0$, then (6.2.23) induces an inner product (f, g) on H_0. Let H be the completion of H_0 relative to this inner product. For each $z_0 \in \mathfrak{K}$, define an operator $W_0(z_0)$ on H_0 as follows:

$$(W_0(z_0)f)(z) = \exp\left[\frac{i}{2}\mathfrak{I}(z, z_0)\right] f(z - z_0). \tag{6.2.24}$$

It is easily verified that

$$(W_0(z_0)f, W_0(z_0)g) = (f, g).$$

In particular, if $(f, f) = 0$, then $(W_0(z_0)f, W_0(z_0)f) = 0$. Hence, $W_0(z_0)$ is an isometric operator from H_0 into H_0. But since $W_0(z_0)\, W_0(-z_0) = I$, it follows that $W_0(z_0)$ maps H_0 onto H_0, and can therefore be uniquely extended to a unitary operator $W(z_0)$ from H to H. Moreover, it is easy to calculate from (6.2.24) that the operators $W_0(z)$, and hence also $W(z)$, satisfy the von Neumann commutation relations.

Next, we must show that $W_0(z)$ is weakly continuous when z is restricted to any finite-dimensional linear subspace of $\mathfrak{K}$. Since H_0 is dense in H, it suffices to show that, if $f, g \in H_0$, then the function

$$(W_0(z)f, g) = \sum \psi(u - u') \exp\left[-\frac{i}{2}\mathfrak{I}(u, u' - z)\right] f(u - z)\overline{g(u')} \quad (6.2.25)$$

is quasi-continuous. Let

$$f(u) = \begin{cases} \lambda_l, & u = \sigma_l, \quad l = 1, 2, \ldots, n, \\ 0, & u \bar{\in} \{\sigma_1, \ldots, \sigma_n\}, \end{cases}$$

$$g(u) = \begin{cases} \mu_k, & u = \tau_k, \quad k = 1, 2, \ldots, m, \\ 0, & u \bar{\in} \{\tau_1, \ldots, \tau_m\}. \end{cases}$$

Then, (6.2.25) becomes

$$\sum_{l=1}^{n} \sum_{k=1}^{m} \psi(\sigma_l - \tau_k + z) \exp\left[-\frac{i}{2}\mathfrak{I}(\sigma_l + z, \tau_k - z)\right] \lambda_l \bar{\mu}_k . \quad (6.2.26)$$

Since ψ is quasi-continuous, it is clear that (6.2.26) is also quasi-continuous.

Finally, let $\eta \in H_0$ be defined by $\eta(0) = 1$, $\eta(z) = 0$ for $z \neq 0$. It is easily verified that η is a cyclic element for the system $\{W(z) \mid z \in \mathfrak{K}\}$, and that $(W(z)\eta, \eta) = \psi(z)$.]

§6.3. The Relation of Gaussian Measures and Rotationally Invariant Measures to Conventional Free-Field Systems

1° The Fock–Cook Free-Field System

We now proceed to formulate the Fock–Cook free-field system commonly used in quantum field theory. Let $\mathfrak{K}$ be an infinite dimensional (complex) Hilbert space. Let $\mathfrak{K}^{(0)}$ denote the one-dimensional Hilbert space formed by the totality of complex numbers with the inner product $(\lambda, \mu) = \lambda\bar{\mu}$, and let $\mathfrak{K}^{(n)}$ denote the tensor product $\prod_{\nu=1}^{n} \otimes \mathfrak{K}$ of n copies of $\mathfrak{K}$ (see Appendix II for the terminology of tensor product spaces and operators in such spaces); in particular, $\mathfrak{K}^{(1)} = \mathfrak{K}$. Let M_n be the space of symmetric tensors of order n over $\mathfrak{K}$ ($n \geqslant 1$), $M_0 = \mathfrak{K}^{(0)}$, and let

$$\mathfrak{L} = \sum_{n=0}^{\infty} \oplus \mathfrak{K}^{(n)}.$$

This is a Hilbert space. We consider the following closed linear subspace of $\mathfrak{L}$:

$$H = \sum_{n=0}^{\infty} \oplus M_n .$$

Let X denote the linear subspace of $\mathfrak{L}$ spanned by all the $\mathfrak{K}^{(n)}$, $n = 0, 1, \ldots$, and let $M = X \cap H$; M is called the *symmetric tensor algebra* over $\mathfrak{K}$.

For each $x \in \mathfrak{K}$, form the operator $C_1(x)$ on M, as follows: for $z \in M$, $z = z_0 + z_1 + \cdots + z_n$, $z_k \in \mathfrak{K}^{(k)}$, define

$$C_1(x)z = \sum_{k=0}^{n} (k+1)^{1/2} S_{k+1}(x \otimes z_k), \tag{6.3.1}$$

where $x \otimes z_k$ denotes the tensor product of the vector x and the tensor z_k of rank k, so that $x \otimes z_k \in \mathfrak{K}^{(k+1)}$. Clearly, $C_1(x)$ is a linear operator from M to M, and acts as a bounded linear operator from M_n to M_{n+1}.

We also form an operator $C_2(x)$ on M, defined as follows: for $z \in M$, $z = z_0 + z_1 + \cdots + z_n$, $z_k \in \mathfrak{K}^{(k)}$,

$$C_2(x)z = \sum_{k=1}^{\infty} \sqrt{k} z'_{k-1} , \tag{6.3.2}$$

where $z'_{k-1} \in M_{k-1}$ is such that, for all $y \in M_{k-1}$,

$$(z'_{k-1}, y) = (z_k, S_k(x \otimes y)). \tag{6.3.3}$$

It is easily seen that such a z'_{k-1} exists and is unique. For example, if $z = S_k(x_1 \otimes \cdots \otimes x_k)$, then

$$C_2(x)z = \frac{1}{\sqrt{k}} \sum_{\nu} (x_\nu , x) S_{k-1}(x_1 \otimes \cdots \otimes x_{\nu-1} \otimes x_{\nu+1} \cdots \otimes x_k).$$

$C_2(x)$ is also a linear operator from M to M, and acts as a bounded linear operator from M_n to M_{n-1} $(n \geqslant 1)$, with $C_2\mathfrak{K}^{(0)} = 0$.

The physical significance of these spaces and operators is as follows. M_n $(n \geqslant 1)$ represents the vector space of all n-particle states, and M_0 represents the vacuum vector space; $C_1(x)(C_2(x))$ represents the operator corresponding to the creation (annihilation) of a particle with wave-function x.

Lemma 6.3.1. $C_1(x)$ and $C_2(x)$ are related by

$$(C_1(x)\, z, y) = (z, C_2(x) y), \qquad z, y \in M. \tag{6.3.4}$$

PROOF. It suffices to prove that (6.3.4) holds for $z \in M_n$, $y \in M_{n'}$. Since $C_1(x) M_n \subset M_{n+1}$, $C_2(x) M_{n'} \subset M_{n'-1}$, it suffices, in fact, to prove (6.3.4) for $z \in M_k$, $y \in M_{k+1}$. But in this case, $C_1(x)z = (k+1)^{1/2} S_{k+1} x \otimes z$, hence, by (6.3.3),

$$(C_1(x)z, y) = ((k+1)^{1/2} S_{k+1} x \otimes z, y) = (z, C_2(x)y). \quad] \qquad (6.3.5)$$

By virtue of (6.3.5) and the mapping properties of $C_1(x)$ and $C_2(x)$, it is easily proved that the closure $C(x)$ of $C_1(x)$ exists, and that $C(x)^*$ [the adjoint of $C(x)$] is the closure of $C_2(x)$. We call $C(x)$, $C(x)^*$ the *creation* and *annihilation* operators, respectively, corresponding to x. Set

$$P(x) = \frac{1}{\sqrt{2}}(C(x) + C(x)^*), \qquad Q(x) = \frac{1}{i\sqrt{2}}(C(x) - C(x)^*).$$

Choose any real inner product subspace $\mathfrak{H}$ of $\mathfrak{K}$ such that $\mathfrak{K}$ is the complexification of $\mathfrak{H}$. We shall prove below that $\{P(x), Q(x') \mid x, x' \in \mathfrak{H}\}$ is an infinitesimal canonical system on $(\mathfrak{H}, \mathfrak{H}, B)$; we call it the *Fock–Cook system.*

We proceed to construct a canonical system of the form (6.2.13)–(6.2.14) which is unitarily equivalent with the Fock–Cook system. Let $S = (\Omega, \mathfrak{B}, N)$ be a standard Gaussian measure space, with parameter 1, associated with $\mathfrak{H}$, and let $\{x(\omega), x \in \mathfrak{H}\}$ be the corresponding standard Gaussian process.

We now define a linear mapping D of H into $L^2(S)$, as follows: arbitrarily choose a complete orthonormal system $\{g_m\}$ in $\mathfrak{H}$; then, by Lemma II.2.4 of Appendix II, the vectors of H may be expressed in the form

$$x = \sum_{k=(k_1,\cdots,k_n)} a_k S_{|k|}(g_1^{k_1} \otimes g_2^{k_2} \otimes \cdots \otimes g_n^{k_n})(|k|!/k!)^{1/2},$$

$$\sum |a_k|^2 = \|x\|^2 < \infty.$$

We define[16]

$$Dx(\omega) = \sum_{k=(k_1,\cdots,k_n)} a_k h_k(g_1(\omega),\ldots, g_n(\omega)).$$

By Lemma 5.4.6, we have

$$\| Dx \|^2 = \sum |a_k|^2 = \| x \|^2,$$

and, since $\{h_k\}$ is complete, $D(H) = L^2(S)$, thus, D is a unitary operator

[16] Here, as in §6.1, we denote $h_k(u_1, \ldots, u_n; 1)$ by $h_k(u_1, \ldots, u_n)$.

from H to $L^2(S)$. It can be shown that this unitary operator D is independent of the choice of the complete orthonormal system $\{g_m\}$.

We denote by $\mathfrak{P}$ the totality of polynomial functionals of the form $P(g_1(\omega),..., g_n(\omega))$ [where $P(t_1 ,..., t_n)$ is a polynomial in $t_1 ,..., t_n$]; then $\mathfrak{P}$ is a dense subspace of $L^2(S)$. Also, denote by $H(m)$ the closed linear subspace spanned by all functionals of the form $h_k(g_1(\omega),..., g_n(\omega))$, $|k| = m$; we call $H(m)$ the *m-particle space*. In fact,

$$H(m) = DM_m .$$

We now set up a representation of the commutation relations analogous to (6.2.13) and (6.2.14). Namely, if $f \in L^2(S)$, let

$$(U(x)f)(\omega) = e^{ix(\omega)}f(\omega), \tag{6.3.6}$$

$$(V(x')f)(\omega) = \exp\,[x'(\omega) - \tfrac{1}{2}(x', x')]\, f(\omega - x'). \tag{6.3.7}$$

Note that this $V(x')$ can also be written as

$$(V(x')f)(\omega) = \left(\frac{dN_{x'}(\omega)}{dN(\omega)}\right)^{1/2} f(\omega - x'). \tag{6.3.8}$$

From (6.3.6) and (6.3.8), it is easily seen that $\{U(x), V(x') \mid x, x' \in \mathfrak{H}\}$ satisfies the Weyl commutation relations (6.2.1), and that U, V are weakly continuous on any finite-dimensional linear subspace of $\mathfrak{H}$.

We note that, in analogy with (6.1.32), the operator

$$p(x) = \frac{1}{i}\frac{d}{dt}\, U(xt)\Big|_{t=0}$$

may be expressed as follows:

$$\begin{aligned} \mathfrak{D}(p(x)) &= \{\varphi \mid \varphi \in L^2(S),\, x(\cdot)\, \varphi(\cdot) \in L^2(S)\}, \\ p(x)\, \varphi(\omega) &= x(\omega)\, \varphi(\omega), \qquad \varphi \in \mathfrak{D}(p(x)). \end{aligned} \tag{6.3.9}$$

However, the expression corresponding to (6.1.33) is more complicated. Using again the complete orthonormal system $\{g_m\}$, and the notation of (6.2.2), we let $p_m = p(g_m)$, $q_m = q(g_m)$. Then, by (6.3.9), we have

$$\begin{cases} \mathfrak{D}(p_\nu) = \{\varphi \mid \varphi \in L^2(S),\, g_\nu(\cdot)\, \varphi(\cdot) \in L^2(S)\}, \\ p_\nu \varphi(\omega) = g_\nu(\omega)\, \varphi(\omega), \qquad \varphi \in \mathfrak{D}(p_\nu), \end{cases}$$

and, in analogy with (6.1.33), q_ν is the closure of the operator q_ν', where $\mathfrak{D}(q_\nu') = \mathfrak{P}$ and

$$q_\nu' \varphi(\omega) = -ig_\nu(\omega)\, \varphi(\omega) + i\frac{\partial}{\partial g_\nu}\varphi(\omega), \qquad \varphi \in \mathfrak{P}.$$

Set

$$c_\nu' = \frac{1}{\sqrt{2}}(p_\nu + iq_\nu), \qquad c_\nu^{*\prime} = \frac{1}{\sqrt{2}}(p_\nu - iq_\nu);$$

and let c_ν, $c_\nu{}^*$ be the closures of c_ν', $c_\nu{}^{*\prime}$, respectively. Then $c_\nu{}^*$ is the adjoint of c_ν, and, corresponding to Lemma 6.1.11, we have the following result.

Lemma 6.3.2. The operator c_ν maps $H(m)$ into $H(m+1)$, is bounded on $H(m)$, and satisfies

$$c_\nu h_k(g_1(\omega),\ldots, g_n(\omega)) = (k_\nu')^{1/2}\, h_{k'}(g_1(\omega),\ldots, g_n(\omega)),$$

where k and k' are related as in (6.1.36). Moreover, $c_\nu{}^*$ maps $H(m)$ into $H(m-1)$,[17] is bounded on $H(m)$, and satisfies

$$c_\nu{}^* h_k(g_1(\omega),\ldots, g_n(\omega)) = (k_\nu)^{1/2}\, h_{k'}(g_1(\omega),\ldots, g_n(\omega)),$$

where k and k' are related by (6.1.37).

Using (6.3.1), (6.3.2), Lemma 6.3.2, the boundedness of c_ν on $H(m)$ and the boundedness of $C(g_\nu)$ on M_m, one can calculate that

$$c_\nu = DC(g_\nu)\, D^{-1}.$$

Similarly,

$$c_\nu{}^* = DC(g_\nu)^*\, D^{-1}.$$

In general, writing $c(x) = (1/\sqrt{2})(p(x) + iq(x))$, we have

$$c(x) = DC(x)\, D^{-1},$$
$$c(x)^* = DC(x)^*\, D^{-1}.$$

Consequently,

$$p(x) = DP(x)\, D^{-1}, \qquad q(x) = DQ(x)\, D^{-1}.$$

Thus, since $\{p, q\}$ is an infinitesimal canonical system, it follows that $\{P, Q\}$ is an infinitesimal canonical system, moreover, the Weyl form of the Fock–Cook system is unitarily equivalent to (6.3.6)–(6.3.7), *via* the unitary transformation D. Since the standard Gaussian measure space is ergodic, it follows as in the proof of Theorem 6.2.5 that the Fock–Cook

[17] Here, $H(-1)$ is to be understood as meaning (0).

system is irreducible. Corresponding to Theorem 6.1.13, one can also prove the following result.

Theorem 6.3.3. Let P_m denote the projection of H onto M_m; then, for any real number α,

$$\Gamma(e^{i\alpha}I) = \sum_{m=0}^{\infty} e^{-im\alpha}P_m \tag{6.3.10}$$

is a unitary operator on H, moreover, if $\{W(z) \mid z \in \mathfrak{K}\}$ is the von Neumann form of the Fock–Cook system, then

$$\Gamma(e^{i\alpha}I)\, W(z)\, \Gamma(e^{i\alpha}I)^{-1} = W(e^{i\alpha}z).$$

In other words, (6.3.10) defines a gradient transformation on the Fock–Cook system.

It is easily seen that the Fock–Cook system is cyclic, with the vacuum state 1 as cyclic element. We now write out the characteristic functional of this system corresponding to the cyclic element 1. To do this, it suffices to consider the unitarily equivalent system defined by (6.3.6)–(6.3.7). Then

$$(W(z)1, 1) = \int_{\Omega} \exp\left[ix(\omega) + y(\omega) - \frac{1}{2}(y, y) - \frac{i}{2}(x, y)\right] dN(\omega), \tag{6.3.11}$$

and we may suppose that $x \neq 0$. Now, there exist mutually independent Gaussian variables $U(\omega)$, $V(\omega)$, each having mathematical expectation zero and variance $1/2$, and such that [18]

$$\begin{aligned} x(\omega) &= \| x \|\, U(\omega), \\ y(\omega) &= \left(y, \frac{x}{\| x \|}\right) U(\omega) + \left\| y - \left(y, \frac{x}{\| x \|}\right) \frac{x}{\| x \|} \right\| V(\omega). \end{aligned} \tag{6.3.12}$$

Substituting (6.3.12) into (6.3.11), we obtain

$$\begin{aligned} &(W(z)1, 1) \\ &\quad = \frac{1}{\pi} \iint_{-\infty}^{\infty} \exp\left[i \| x \| U + \left(y, \frac{x}{\| x \|}\right) U + \left\| y - \left(y, \frac{x}{\| x \|}\right) \frac{x}{\| x \|} \right\| V \right. \\ &\qquad \left. - \frac{1}{2}(y, y) - \frac{i}{2}(x, y) - (U^2 + V^2)\right] dU dV \\ &\quad = \exp[-\tfrac{1}{4}(\| x \|^2 + \| y \|^2)]. \end{aligned}$$

[18] $\| x \| = (x, x)^{1/2}$.

Therefore, the characteristic functional of the Fock–Cook system, corresponding to the vacuum state 1, is

$$(W(z)1, 1) = \exp(-\tfrac{1}{4}\| z \|^2). \tag{6.3.13}$$

We next proceed to formulate a class of canonical systems, similar to the Fock–Cook system, and also describable in terms of Gaussian measures.

Let $\mathfrak{K}$ be a complex Hilbert space, let $\mathfrak{H}$ be the real Hilbert space formed by $\mathfrak{K}$ with respect to the inner product $[z, z'] = \mathfrak{R}(z, z')$, and let $S_c = (\Omega, \mathfrak{B}, N_c)$ be a standard Gaussian measure space associated with $\mathfrak{H}$. Using (6.2.15), for each $z \in \mathfrak{K}$, we form the unitary operator

$$(W_c(z)f)(\omega) = \exp\Big[-\frac{i}{2}(iz)(\omega)\Big]\Big(\frac{dN_{c(-z)}(\omega)}{dN_c(\omega)}\Big)^{1/2} f(\omega + z), \qquad f \in L^2(S_c) \tag{6.3.14}$$

in the Hilbert space $L^2(S_c)$. Following the proof of Theorem 6.2.6, we see that $\{W_c(z) \mid z \in \mathfrak{K}\}$ satisfies the von Neumann commutation relations; moreover, one may verify that the correspondence $z \to W_c(z)$ is weakly continuous on any finite-dimensional subspace of $\mathfrak{K}$. Therefore, $\{W_c(z) \mid z \in \mathfrak{K}\}$ constitutes a canonical system. Furthermore, it is easily shown that it is irreducible[19] and cyclic, with 1 as cyclic element, and that the corresponding characteristic functional is

$$(W_c(z)1, 1) = \int_\Omega \exp\Big[-\frac{i}{2}(iz)(\omega) - \frac{1}{c}z(\omega) - \frac{1}{2c}(z, z)\Big]\, dN_c(\omega). \tag{6.3.15}$$

Since

$$[z, z] = [iz, iz] = \| z \|^2, \qquad [z, iz] = [iz, z] = \mathfrak{R}(z, iz) = 0,$$

the two Gaussian variables $z(\omega)$, $(iz)(\omega)$ are mutually independent, and (6.3.15) becomes

$$\begin{aligned}(W_c(z)1, 1) &= \frac{1}{\pi \| z \|^2}\iint_{-\infty}^{+\infty} \exp\Big[-\frac{i}{2}s - \frac{1}{c}t - \frac{1}{2c}\| z \|^2 - \frac{s^2 + t^2}{c\| z \|^2}\Big]\, ds\, dt \\ &= \exp\Big[-\frac{\| z \|^2}{8}\Big(\frac{c}{2} + \frac{2}{c}\Big)\Big].\end{aligned} \tag{6.3.16}$$

In particular, when $c = 2$, formula (6.3.16) reduces to (6.3.13). Thus,

[19] *Translator's note:* Details would be welcome, since the proof of Theorem 6.2.5 is not applicable in this case.

by virtue of Lemma 6.2.9, the Fock–Cook system is a special case of the type under consideration. We also note that, if one sets

$$k = \frac{1}{8}\left(\frac{c}{2} + \frac{2}{c}\right),$$

and writes $W_k(z)$ in place of $W_c(z)$ in (6.3.16), then (6.3.16) can be written in the form

$$(W_k(z)1, 1) = \exp(-\| z \|^2 k), \qquad \tfrac{1}{4} \leqslant k < \infty. \tag{6.3.17}$$

2° A More General Type of System

Definition 6.3.1. Let $\mathfrak{K}$ be a complex Hilbert space, let $\mathfrak{U}$ be the group of all unitary operators on $\mathfrak{K}$. Let $\{W(z) \mid z \in \mathfrak{K}\}$ be a canonical system over $\mathfrak{K}$ in the Hilbert space H, and let $\Gamma : U \to \Gamma(U)$ be a unitary representation of $\mathfrak{U}$ in H, satisfying the condition

$$\Gamma(U)\, W(z)\, \Gamma(U)^{-1} = W(Uz). \tag{6.3.18}$$

Suppose, moreover, that $\{W(z) \mid z \in \mathfrak{K}\}$ is cyclic and that there is a cyclic element η such that $\Gamma(U)\eta = \eta$ for all $U \in \mathfrak{U}$. Then we call $\{W(z) \mid z \in \mathfrak{K}\}$ a *conventional free-field system*, and η a *vacuum state vector*.

We observe that the system $\{W_c(z) \mid z \in \mathfrak{K}\}$ defined by (6.3.14) is a conventional free-field system. In fact, letting $\mathfrak{D}$ denote the linear space spanned by the totality of vectors of the form $W_c(z)1$, $z \in \mathfrak{H}$, it can be shown by methods similar to those used in §5.4 that $\mathfrak{D}$ is dense in $L^2(S_c)$. For each $U \in \mathfrak{U}$, we define a linear operator $\Gamma(U)$ on $\mathfrak{D}$ as follows:

$$\Gamma(U)\, W_c(z)1 = W_c(Uz)1. \tag{6.3.19}$$

By (6.2.20) and (6.3.16) we have, for any $z, z' \in \mathfrak{K}$,

$$\begin{aligned}
&(\Gamma(U)\, W_c(z')1, \Gamma(U)\, W_c(z)1) \\
&\quad = \exp\left[\frac{i}{2}\, \mathfrak{J}(Uz, Uz') - \frac{1}{8} \| U(z' - z)\|^2 \left(\frac{c}{2} + \frac{2}{c}\right)\right] \\
&\quad = \exp\left[\frac{i}{2}\, \mathfrak{J}(z, z') - \frac{1}{8} \| z' - z \|^2 \left(\frac{c}{2} + \frac{2}{c}\right)\right] \\
&\quad = (W_c(z')1, W_c(z)1).
\end{aligned}$$

Hence, $\Gamma(U)$ is an isometric mapping. Since $\Gamma(U)^{-1} = \Gamma(U^{-1})$, $\Gamma(U)$ can be uniquely extended to a unitary operator on $L^2(S_c)$; we also denote this extension by $\Gamma(U)$. From (6.3.19), it is easily seen that $U \to \Gamma(U)$ is a unitary representation of the group $\mathfrak{U}$, and that (6.3.18) holds. In this case, 1 is a vacuum state vector.

We shall now investigate the general form of the characteristic functional of a conventional free-field system corresponding to an arbitrary vacuum state vector. For this purpose, we first prove the following lemma, which provides a characterization of rotationally invariant measures.

Lemma 6.3.4. Let $\mathfrak{H}$ be an infinite dimensional real inner product space, and let $\psi(\xi)$, $\xi \in \mathfrak{H}$, be a positive definite quasi-continuous function on $\mathfrak{H}$ such that the value of $\psi(\xi)$ depends only upon $\| \xi \| = (\xi, \xi)^{1/2}$ [here $(\cdot, \cdot)$ denotes the inner product in $\mathfrak{H}$]. Then there exists a finite measure m on $[0, \infty)$ such that

$$\psi(\xi) = \int_0^\infty \exp(- \| \xi \|^2 t)\, dm(t), \qquad \xi \in H. \tag{6.3.20}$$

Proof. Since $\psi(\xi)$ depends only upon $\| \xi \|$, it may be written in the form $\varphi(\| \xi \|)$. Arbitrarily choose a unit vector e in $\mathfrak{H}$; then $\varphi(t) = \psi(te)$, $t \geqslant 0$, is a continuous function of t, hence $\psi(\xi) = \varphi(\| \xi \|)$ is a continuous functional on $\mathfrak{H}$. If we choose any orthonormal sequence of vectors $\{e_n, n = 1, 2, \ldots\}$ in $\mathfrak{H}$, then $\psi(\xi_1 e_1 + \cdots + \xi_n e_n)$ is a positive definite continuous function of the real variables $\xi_1, \ldots, \xi_n$. Let l be the totality of real number sequences, and let $\mathfrak{B}$ be the σ-algebra generated by the family of all Borel cylinders in l; by the Kolmogorov theorem (see Corollary 1.3.5′), there exists a finite measure P on $(l, \mathfrak{B})$ such that

$$\psi(\xi_1 e_1 + \cdots + \xi_n e_n) = \int_l \exp\Big(i \sum_{\nu=1}^{n} \xi_\nu x_\nu\Big)\, dP(x), \qquad x = (x_1, \ldots, x_n, \ldots). \tag{6.3.21}$$

Let Ω_n be the unit sphere in n-dimensional Euclidean space, and write

$$\rho = \Big(\sum_1^n \xi_\nu^2\Big)^{1/2}, \qquad \tau = \Big(\sum_{\nu=1}^n x_\nu^2\Big)^{1/2}.$$

Denote the coordinates of the points of Ω_n by $\omega = (\omega_1, \ldots, \omega_n)$; transforming to spherical coordinates, we have

$$\begin{aligned} \omega_1 &= \cos\varphi_1, & 0 &\leqslant \varphi_k \leqslant \pi, \quad 1 \leqslant k \leqslant n-2, \\ \omega_2 &= \sin\varphi_1 \cos\varphi_2, & 0 &\leqslant \varphi_{n-1} < 2\pi, \\ &\cdots \\ \omega_n &= \sin\varphi_1 \cdots \sin\varphi_{n-1}. \end{aligned}$$

The element of area on Ω_n is $dQ_n(\omega) = \sin^{n-2}\varphi_1 \cdots \sin\varphi_{n-2}\, d\varphi_1 \cdots d\varphi_{n-1}$, and the total area of Ω_n is $2\pi^{n/2}/\Gamma(n/2)$, which we denote by $|\Omega_n|$. If we set $\xi_\nu = \rho\omega_\nu$, $\nu = 1, 2, \ldots, n$, then the value of the integral

$$\frac{1}{|\Omega_n|}\int_{\Omega_n}\exp\Big(i\sum_1^n \xi_\nu x_\nu\Big)\, dQ_n(\omega) \tag{6.3.22}$$

depends only upon ρ and τ, hence, in the evaluation, we may choose $(x_1, \ldots, x_n) = (\tau, 0, \ldots, 0)$. Therefore, (6.3.22) becomes

$$\begin{aligned}&\frac{|\Omega_{n-1}|}{|\Omega_n|}\int_0^\pi \exp(i\rho\tau\cos\varphi_1)\sin^{n-2}\varphi_1\, d\varphi_1\\ &\quad = \frac{\Gamma(n/2)}{\Gamma\left(\frac{n-1}{2}\right)\sqrt{\pi}}\int_{-1}^{1}(1-t^2)^{(n-3)/2}\cos\rho\tau t\, dt, \qquad t = \cos\varphi_1 .\end{aligned}$$

Denote the above quantity by $\psi_n(\rho\tau)$. Applying the integral $\int_{\Omega_n}\cdots dQ_n(\omega)$ to both sides of (6.3.21), we obtain

$$\begin{aligned}\varphi(\rho) &= \frac{1}{|\Omega_n|}\int\varphi\Big(\rho\sum_\nu \omega_\nu e_\nu\Big)\, dQ_n(\omega) = \int_l \psi_n(\rho\tau)\, dP(x)\\ &= \int_0^\infty \psi_n(\rho\tau)\, d\sigma_n(\tau),\end{aligned} \tag{6.3.23}$$

where $\sigma_n(\tau) = P(\{x \mid \sum_{\nu=1}^n x_\nu^2 \leqslant \tau^2\})$. Make the substitution $\tau = \sqrt{n}u$ on the right-hand side of (6.3.23); since $\sigma_n(\sqrt{n}u)$, $n = 1, 2, \ldots$, is a uniformly bounded sequence of monotonically increasing nonnegative functions on $[0, \infty]$, it follows by Helly's theorem that there exists a subsequence $\{\sigma_{n_k}(\sqrt{n_k}u)\}$ which converges at all points of $[0, \infty]$ to a bounded monotonically increasing nonnegative function $\alpha(u)$. On the other hand, since[20]

$$\lim_{n\to\infty}\frac{\Gamma(n/2)}{\Gamma\left(\frac{n-1}{2}\right)(n\pi)^{1/2}} = \frac{1}{(2\pi)^{1/2}}$$

and, as $n \to \infty$,

$$\begin{aligned}&\sqrt{n}\int_{-1}^{1}(1-t^2)^{(n-3)/2}\cos\rho\tau\sqrt{n}\,t\, dt\\ &\quad = \int_{-\sqrt{n}}^{\sqrt{n}}\left(1-\frac{v^2}{n}\right)^{(n-3)/2}\cos\rho\tau v\, dv \to \int_{-\infty}^{\infty}\exp\left(-\frac{v^2}{2}\right)\cos\rho\tau v\, dv\\ &\quad = (2\pi)^{1/2}\exp\left(-\frac{\rho^2\tau^2}{2}\right),\end{aligned}$$

[20] Here, we use Stirling's formula $\Gamma(s+1) \sim (2\pi s)^{1/2} s^s e^{-s}$, $s \to \infty$.

therefore,

$$\lim_{n\to\infty} \psi_n(\sqrt{n}\,\rho\tau) = \exp\left(-\frac{\rho^2\tau^2}{2}\right).$$

Similarly, one can show that[21] there exist positive numbers c, c' such that $|\psi_n(\sqrt{n}\rho\tau)| \leqslant c\exp(-c'\rho^2\tau^2)$. Hence, taking $n = n_k$ in (6.3.23), and letting $k \to \infty$, we obtain

$$\varphi(\rho) = \int_0^\infty \exp\left(-\frac{\rho^2\tau^2}{2}\right) d\alpha(\tau) = \int_0^\infty \exp(-\rho^2 t)\, dm(t),$$

where $m(t) = \alpha((2t)^{1/2})$.]

If the characteristic function of a measure is of the type described by Lemma 6.3.4, then this measure is said to be *rotationally invariant.* It can be seen from (6.3.20) that any rotationally invariant measure is a superposition of Gaussian measures.[22]

Theorem 6.3.5. Let $\mathfrak{K}$ be a complex infinite-dimensional Hilbert space, $\{W(z) \mid z \in \mathfrak{K}\}$ a conventional free-field system, and η any vacuum state vector for this system. Then there exists a finite measure m on $[\frac{1}{4}, \infty)$ such that the characteristic functional of $\{W(z) \mid z \in \mathfrak{K}\}$ corresponding to η is

$$(W(z)\eta, \eta) = \int_{1/4}^\infty \exp(-\|z\|^2 t)\, dm(t). \tag{6.3.24}$$

Proof. Since $\Gamma(U)\eta = \eta$, the characteristic functional $\psi(z)$ has the following property: for $U \in \mathfrak{U}$, $z \in \mathfrak{K}$,

$$\begin{aligned}\psi(z) &= (W(z)\eta, \eta) = (W(z)\,\Gamma(U)^{-1}\eta, \Gamma(U)^{-1}\eta) \\ &= (W(Uz)\eta, \eta) = \psi(Uz).\end{aligned} \tag{6.3.25}$$

Now, if $\|z\| = \|z'\|$, there must be some $U \in \mathfrak{U}$ such that $z' = Uz$, hence, by (6.3.25), $\psi(z)$ depends only upon $\|z\|$. We choose any real linear subspace $\mathfrak{H}$ of $\mathfrak{K}$ such that $\mathfrak{H}$ is a real inner product space and $\mathfrak{K}$ is the complexification of $\mathfrak{H}$. If z is restricted to $\mathfrak{H}$, the function $\psi(z)$ still depends only upon $\|z\|$. Since $\mathfrak{H}$ is also infinite-dimensional, we know that there is a finite measure m on $[0, \infty)$ such that (6.3.20) holds for all $\xi \in \mathfrak{H}$. But since $\psi(\xi)$, on the entire space $\mathfrak{K}$, depends only upon $\|\xi\|$,

[21] *Translator's note:* Actually, it suffices to show that the convergence of $\psi_n(\sqrt{n}\,\rho\tau)$ is uniform with respect to τ.

[22] See Umemura [1] for further details.

it follows that (6.3.20) holds for all $\xi \in \mathfrak{K}$. Using the condition in Lemma 6.2.10 and carrying out a rather intricate calculation, it can be proved that[23]

$$m([0, \tfrac{1}{4})) = 0,$$

whence we obtain (6.3.24).]

APPENDIX

I

BACKGROUND MATERIAL ON TOPOLOGICAL GROUPS AND LINEAR TOPOLOGICAL SPACES

§I.1. Pseudometrics, Convex Functions, and Pseudonorms

Definition I.1.1. Let R be a set, and let $\rho(x, y)$, $x, y \in R$, be a real-valued function on $R \times R$, satisfying the following conditions: (i) $\infty > \rho(x, y) \geqslant 0$, $\rho(x, x) = 0$; (ii) $\rho(x, y) \leqslant \rho(x, z) + \rho(y, z)$. Then ρ is called a *pseudometric* on R. If R is also a group, and $\rho(y, x) = \rho(x^{-1}y, e)$ for all $x, y \in R$, then ρ is said to be a *left-invariant pseudometric* on R. *Right-invariance* is similarly defined.

Let $\{\rho_\alpha, \alpha \in \mathfrak{A}\}$ be a family[1] of pseudometrics on R. The topology generated by the totality of sets of the form

$$\{x \mid \rho_{\alpha_k}(x, y) < \epsilon, k = 1, 2, \ldots, n\}, \qquad \alpha_k \in \mathfrak{A}, \quad \epsilon > 0$$

[1] Naturally, this subsumes the case in which $\mathfrak{A}$ contains only a single index. This case will not be singled out for special attention in the ensuing discussion.

is called the topology *induced* by $\{\rho_\alpha, \alpha \in \mathfrak{A}\}$. The set R, provided with this topology, is said to be a *pseudometric space.*

If R is a group, and ρ is a (left and right) invariant pseudometric on R, then R, given the topology induced by ρ, becomes a topological group.

Definition I.1.2. Let G be a group, and let $N(g)$, $g \in G$, be a real-valued function on G, satisfying the following conditions: (i) $\infty > N(g) \geqslant 0$, $N(e) = 0$; (ii) $N(yx^{-1}) \leqslant N(x) + N(y)$. Then N is called a *convex function* (or, in some literature, a *pseudonorm*) on G.

It is easily verified that any convex function also satisfies the condition: (iii) $N(x) = N(x^{-1})$. Given a convex function N on a group G, we introduce the function

$$\rho(g, h) = N(h^{-1}g) \tag{I.1.1}$$

on $G \times G$. It is easily verified that ρ is a left-invariant pseudometric on G; we call it the *pseudometric corresponding to the convex function* N. Conversely, given a left-invariant pseudometric ρ on G, the formula

$$N(g) = \rho(g, e)$$

defines a convex function on G, and ρ is just the pseudometric corresponding to N. A similar discussion applies to right-invariant pseudometrics.

Suppose that $\{N_\alpha, \alpha \in \mathfrak{A}\}$ is a family of convex functions on a group G, and that $\{\rho_\alpha, \alpha \in \mathfrak{A}\}$ is the corresponding family of (left-invariant) pseudometrics. In such a case, the topology induced by $\{\rho_\alpha, \alpha \in \mathfrak{A}\}$ may also be called the *topology induced by* $\{N_\alpha, \alpha \in \mathfrak{A}\}$.

Definition I.1.3. Let R be a linear space (which is, of course, a group with respect to addition) and let N be a real-valued function on R. (i) If N is a convex function and, for any $x \in R$ and number t,

$$N(tx) = |t| N(x),$$

then N is said to be a *pseudonorm* on R. (ii) If N is a convex function and satisfies the condition

$$N(tx) \leqslant N(x), \qquad |t| \leqslant 1,$$

then N is called a *homogeneous convex function*, and the corresponding pseudometric is said to be a *homogeneous pseudometric*. (iii) If N satisfies the conditions: $0 \leqslant N(h) \leqslant \infty$, $N(g + h) \leqslant N(g) + N(h)$, $N(th) = |t| N(h)$, then N is called a *convex functional.*

Proceeding directly from the definition of a linear topological space, one can easily prove the following lemma.

Lemma I.1.1. Let R be a linear space, ρ an invariant pseudometric on R. Suppose that ρ also satisfies the following conditions: (i) if $\{\lambda_n\}$ is a sequence of numbers converging to zero, and $x \in R$, then

$$\lim_{n\to\infty} \rho(\lambda_n x, 0) = 0;$$

(ii) if $\{\lambda_n\}$ is a bounded sequence of numbers, $\{h_n\} \subset R$, and

$$\lim_{n\to\infty} \rho(h_n, 0) = 0,$$

then $\lim_{n\to\infty} \rho(\lambda_n h_n, 0) = 0$. Then, R is a linear topological space relative to the topology induced by ρ.

We call this kind of linear topological space a *linear pseudometric space.* In particular, if the pseudometric ρ is a metric, we call R a *linear metric space.*

Notice that, if ρ is homogeneous, then condition (ii) of Lemma I.1.1 is automatically satisfied.

The next lemma provides a method by which, starting from certain invariant pseudometrics, one may construct pseudometrics satisfying the conditions of Lemma I.1.1.

Lemma I.1.2. Let $\mathfrak{G}$ be a real linear space, and let $M(h)$, $h \in \mathfrak{G}$ be a nonnegative bounded functional on $\mathfrak{G}$, satisfying the following conditions:(i) for any $h_1, h_2 \in \mathfrak{G}$, $M(h_1 + h_2) \leqslant M(h_1) + M(h_2)$; (ii) $M(0) = 0$, and, for any $h \in \mathfrak{G}$, $M(th)$ is a continuous function of t, $-\infty < t < \infty$; (iii) for any $h \in \mathfrak{G}$, $M(-h) = M(h)$. Let

$$R(h) = \left[\int_0^\infty e^{-t} M(th)^2 \, dt\right]^{1/2}, \qquad h \in \mathfrak{G}.$$

Then, $\mathfrak{G}$ is a linear pseudometric space with respect to the pseudometric $R(h_1 - h_2)$, $h_1, h_2 \in \mathfrak{G}$.

PROOF. By virtue of condition (i), we clearly have

$$R(h_1 + h_2) \leqslant R(h_1) + R(h_2), h_1, h_2 \in \mathfrak{G}.$$

By condition (iii), we have $R(h) = R(-h)$. Now, take any $h \in \mathfrak{G}$, and let $\{t_n\}$ be any sequence of real numbers converging to zero. If M is an upper bound for $M(h)$, we have

$$\begin{aligned} R(t_n h)^2 = R(|t_n| h)^2 &\leqslant \int_0^\lambda e^{-t} M(t\,|t_n|\,h)^2 \, dt + M^2 e^{-\lambda} \\ &\leqslant \max_{0 \leqslant t \leqslant \lambda |t_n|} M(th)^2 + M^2 e^{-\lambda}. \end{aligned}$$

Using condition (ii), we deduce that $\lim_{n\to\infty} R(t_n h) = 0$.

Next, suppose that $\{h_n\} \subset \mathfrak{G}$, $R(h_n) \to 0$, and let $\{t_n\}$ be any bounded sequence of real numbers. Then,

$$R(t_n h_n) \leqslant R((|\, t_n \,| + 1) h_n) + R(h_n). \tag{I.1.2}$$

Let c be a positive constant such that $(|\, t_n \,| + 1) \leqslant c$ for all n, and let λ be an arbitrary positive number. Then,

$$\begin{aligned} R((|\, t_n \,| + 1)\, h_n)^2 &= \int_0^\infty \frac{1}{|\, t_n \,| + 1} \exp\left(-\frac{t}{|\, t_n \,| + 1}\right) M(th_n)^2\, dt \\ &\leqslant \int_0^\infty \exp\left(-\frac{1}{c}\, t\right) M(th_n)^2\, dt \\ &\leqslant e^\lambda \int_0^\lambda e^{-t} M(th_n)^2\, dt + cM^2 \exp\left(-\frac{1}{c}\lambda\right) \\ &\leqslant e^\lambda R(h_n)^2 + cM^2 \exp\left(-\frac{1}{c}\lambda\right). \end{aligned} \tag{I.1.3}$$

Since $R(h_n) \to 0$, it follows from (I.1.2) and (I.1.3) that

$$\overline{\lim_{n\to\infty}}\, R(t_n h_n)^2 \leqslant cM^2 \exp\left(-\frac{1}{c}\lambda\right).$$

Letting $\lambda \to \infty$, we conclude that $\lim_{n\to\infty} R(t_n h_n) = 0$. Thus, by Lemma I.1.1, $\mathfrak{G}$ is a linear pseudometric space relative to the pseudometric $R(h_1 - h_2)$, $h_1, h_2 \in \mathfrak{G}$.]

§I.2. Some Properties of Semicontinuous Functions

1° The Concept of a Semicontinuous Function

First, we introduce the following notations. If $f(x)$ is a real-valued function on a set M, and B a subset of M, we write

$$S(f, B) = \sup_{x\in B} f(x), \qquad I(f, B) = \inf_{x\in B} f(x).$$

Now, suppose that M is also a topological space, and $x_0 \in M$. Let $\mathscr{V}_{x_0}$ denote the neighborhood system at the point x_0. The numbers

$$S(x_0, f) = \inf_{V\in\mathscr{V}_{x_0}} S(f, V), \qquad I(x_0, f) = \sup_{V\in\mathscr{V}_{x_0}} I(f, V)$$

are called, respectively, the *upper* and *lower bounds of f at x_0*. If $S(x_0, f) = f(x_0)$, we say that f is *upper semicontinuous at x_0*; if $I(x_0, f) = f(x_0)$, we say that f is *lower semicontinuous at* x_0. If f is upper (lower) semicontinuous at every point $x_0 \in M$, we say that f is *upper (lower) semicontinuous on* M.

For any point $x \in M$, we write $\omega(x, f) = S(x, f) - I(x, f)$.

Lemma I.2.1. For each positive number c, the set $\{x \mid \omega(x, f) < c\}$ is open.

PROOF. Suppose $\omega(x_0, f) < c$. Then, $S(x_0, f) < +\infty$, $I(x_0, f) > -\infty$. Hence, given any positive number $\epsilon < \frac{1}{2}(c - \omega(x_0, f))$, there exists a neighborhood V of x_0 such that

$$S(f, V) < S(x_0, f) + \epsilon, \qquad I(f, V) > I(x_0, f) - \epsilon.$$

Therefore, if $x \in V$, then

$$\omega(x, f) \leqslant S(f, V) - I(f, V) < \omega(x_0, f) + 2\epsilon < c.$$

Thus, $V \subset \{x \mid \omega(x, f) < c\}$.]

Lemma I.2.2. Let M be a topological space of the second category and f an upper semicontinuous function on M such that, for every point $x_0 \in M$, $I(x_0, f) > -\infty$. Then M contains at least one continuity point of f.

PROOF. Let c be any positive number. We assert that the open set $O_c = \{x \mid \omega(x, f) < c\}$ is dense in M. In fact, given any $x_0 \in M$, we have $I(x_0, f) > -\infty$, hence there must exist a $V \in \mathscr{V}_{x_0}$ such that $I(f, V) > -\infty$. Therefore, for any neighborhood U of x_0 such that $U \subset V$, we have $I(f, U) > -\infty$. Obviously, we may assume that $I(f, U) < \infty$, and hence that

$$f(x) < I(f, U) + c \tag{I.2.1}$$

for some point $x \in U$. On the other hand,

$$I(f, U) \leqslant I(x, f) \leqslant f(x) = S(x, f). \tag{I.2.2}$$

From (I.2.1) and (I.2.2), we get

$$\omega(x, f) = f(x) - I(x, f) \leqslant f(x) - I(f, U) < c,$$

that is, U intersects O_c. Thus, O_c is dense in M. The set $\{x \mid \omega(x, f) \geqslant c\}$, being the complement of the dense open set O_c, is nowhere dense. Since M is of the second category, it follows that $M \neq \bigcup_{n=1}^{\infty} \{x \mid \omega(x, f) \geqslant 1/n\}$. Hence, the set $\bigcap_{n=1}^{\infty} O_{1/n} = \{x \mid \omega(x, f) = 0\}$ is nonempty. But the points of this set are just the continuity points of f.]

Lemma I.2.3. Let $\mathfrak{G}$ be a linear topological space of the second category, and let $M(h)$ be a nonnegative, lower semicontinuous functional on $\mathfrak{G}$, satisfying the following conditions.

(i) For any $h_1, h_2 \in \mathfrak{G}$, $M(h_1 + h_2) \leqslant M(h_1) + M(h_2)$.
(ii) For any $h \in \mathfrak{G}$, $M(h) = M(-h)$.
(iii) For any $h \in \mathfrak{G}$, $\lim_{n\to\infty} M(h/n) = 0$.

Then, $M(h)$ is continuous on $\mathfrak{G}$.

PROOF. Given any positive number ϵ, form the set $E_\epsilon = \{h \mid M(h) \leqslant \epsilon\}$. Since $M(h)$ is lower semicontinuous, E_ϵ is closed. Now, by condition (iii), for any $h \in \mathfrak{G}$, there exists an integer n such that $M(h/n) \leqslant \epsilon$, that is, $h \in nE_\epsilon$. Hence,

$$\mathfrak{G} = \bigcup_{n=1}^{\infty} nE_\epsilon .$$

Therefore, since $\mathfrak{G}$ is of the second category, there exists an n such that nE_ϵ is not a nowhere dense set. Moreover, $\mathfrak{G}$ being a linear topological space, the mapping $h \to h/n$ is a homeomorphism of $\mathfrak{G}$ onto itself, consequently, E_ϵ itself cannot be nowhere dense. As E_ϵ is closed, it follows that E_ϵ contains a nonempty open set V. Let $U = \{x - y \mid x, y \in V\}$; U is a neighborhood of 0 in G. By condition (i), when

$$h = x - y \in U (x, y \in V),$$

we have

$$M(h) \leqslant M(x) + M(y) \leqslant 2\epsilon. \tag{I.2.3}$$

Now, for any $x_0 \in \mathfrak{G}$, $x_0 + U$ is a neighborhood of x_0, and if $x \in x_0 + U$, then, by condition (i) and (I.2.3),

$$| M(x) - M(x_0)| \leqslant M(x - x_0) \leqslant 2\epsilon.$$

This proves that $M(h)$ is continuous.]

2° Sequences of Lower Semicontinuous Convex Functionals on Banach Spaces

We limit the statement and proof of the following result to the case which finds application in the present book. The reader is referred to Gel'fand and Vilenkin [1] for a more general formulation.

Theorem I.2.4. Let Φ be a Banach space, let $\{p_n(\varphi), n = 1, 2,\ldots\}$ be a sequence of lower semicontinuous (but not necessarily everywhere finite) convex functionals on Φ, and suppose that, for every $\varphi \in \Phi$, there is an n such that $p_n(\varphi) < \infty$. Then, there exists an n_0 such that

$p_{n_0}(\varphi)$ is finite and continuous on Φ. In other words, there exists an n_0 and a positive number $c < \infty$, such that

$$p_{n_0}(\varphi) \leqslant c \| \varphi \|, \qquad \varphi \in \Phi,$$

where $\| \cdot \|$ denotes the norm in Φ.

PROOF. Consider the closed sets $A_n = \{\varphi \mid p_n(\varphi) \leqslant 1\}$ in Φ. For each $\varphi \in \Phi$, there is an n such that $p_n(\varphi) < \infty$, and hence, for some positive integer k, $p_n(\varphi) \leqslant k$, that is, $\varphi \in kA_n$. Thus,

$$\Phi = \bigcup_{k,n=1}^{\infty} (kA_n).$$

Now Φ, being complete, is of the second category, hence, for some k and n, kA_n is not a nowhere dense set. Therefore, since kA_n is closed, it contains an open ball $\{\varphi \mid \| \varphi - \varphi_0 \| < \epsilon\}$. Hence, if $\psi \in \Phi$, $\| \psi \| < \epsilon/k$, then $\varphi_0 + k\psi \in \{\varphi \mid \| \varphi - \varphi_0 \| < \epsilon\} \subset kA_n$, whence it follows that $\psi + (1/k)\,\varphi_0 \in A_n$, and so

$$p_n(\psi) \leqslant \frac{1}{2}\left(p_n\left(\psi + \frac{1}{k}\varphi_0\right) + p_n\left(\psi - \frac{1}{k}\varphi_0\right)\right) \leqslant 1.$$

Therefore, for any $\psi \in \Phi$, we have

$$p_n(\psi) \leqslant \frac{k}{\epsilon} \| \psi \|. \quad]$$

§I.3 Countably Hilbert Spaces and Rigged Hilbert Spaces

1° Countably Normed Spaces

Definition I.3.1. Let Φ be a linear space and $\{\| \varphi \|_n, n = 1, 2, \ldots\}$ a sequence of pseudonorms on Φ. Introducing the topology determined by the pseudometric

$$\rho(\varphi, \psi) = \sum_{n=1}^{\infty} \frac{1}{2^n} \frac{\| \varphi - \psi \|_n}{1 + \| \varphi - \psi \|_n}, \qquad \varphi, \psi \in \Phi, \tag{I.3.1}$$

Φ becomes a linear pseudometric space. We say that Φ is a *countably pseudonormed space*, with the *topology determined by the sequence of pseudonorms* $\{\| \varphi \|_n, n = 1, 2, \ldots\}$.

We may, without loss of generality, assume that

$$\| \varphi \|_1 \leqslant \| \varphi \|_2 \leqslant \cdots \leqslant \| \varphi \|_n \leqslant \cdots. \tag{I.3.2}$$

For, we may replace $\{\| \varphi \|_n, n = 1, 2,...\}$ by

$$| \varphi |_n = \max_{1 \leqslant \nu \leqslant n} \| \varphi \|_\nu ;$$

the pseudonorms $\{| \varphi |_n\}$ satisfy condition (I.3.2), moreover, it is easy to verify that the pseudometrics derived from the respective sequences $\{| \varphi |_n\}$ and $\{\| \varphi \|_n\}$, in accordance with (I.3.1), define the same topology.

Also, notice that, in the topology of Definition I.3.1, a sequence $\{\varphi_n\} \subset \Phi$ converges to $\varphi \in \Phi$ if, and only if, for every natural number k,

$$\lim_{n \to \infty} \| \varphi_n - \varphi \|_k = 0.$$

Similarly $\{\varphi_n\}$ is a fundamental sequence in Φ if, and only if, for every k,

$$\lim_{m,n \to \infty} \| \varphi_m - \varphi_n \|_k = 0.$$

Definition I.3.2. Let Φ be a linear space, and let $\| \cdot \|_1$ and $\| \cdot \|_2$ be two norms on Φ. Suppose that, for any sequence $\{\varphi_n\} \subset \Phi$ which is fundamental with respect to both $\| \cdot \|_1$ and $\| \cdot \|_2$, we have $\| \varphi_n \|_1 \to 0$ if and only if $\| \varphi_n \|_2 \to 0$. We then say that the two norms $\| \cdot \|_1$ and $\| \cdot \|_2$ are *compatible.*

Let Φ be a linear space, $\{\| \cdot \|_n, n = 1, 2,...\}$ a sequence of norms on Φ, satisfying condition (I.3.2), and such that any two norms $\| \cdot \|_m$ and $\| \cdot \|_n$ are compatible. Then Φ, with the metric topology defined by $\{\| \cdot \|_n\}$ in accordance with (I.3.1), is called a *countably normed space.*

Obviously, countably normed spaces form a special class of countably pseudonormed spaces.

Now, construct the completion Φ_1 of Φ with respect to the norm $\| \cdot \|_1$. Consider those elements $\varphi \in \Phi_1$ for which there exists a sequence $\{\varphi_n\} \subset \Phi$ such that

$$\lim_{m,n \to \infty} \| \varphi_n - \varphi_m \|_2 = 0, \qquad \lim_{n \to \infty} \| \varphi - \varphi_n \|_1 = 0. \tag{I.3.3}$$

Denote the totality of such elements φ by Φ_2; clearly, Φ_2 is a linear subspace of Φ_1. For any $\varphi \in \Phi_2$, choose a sequence $\{\varphi_n\} \subset \Phi$ satisfying (I.3.3). Then, (I.3.3) implies that $\lim_{m,n\to\infty} | \| \varphi_n \|_2 - \| \varphi_m \|_2 | = 0$. We define

$$\| \varphi \|_2 = \lim_{n \to \infty} \| \varphi_n \|_2. \tag{I.3.4}$$

The value of $\| \varphi \|_2$ is independent of the choice of the sequence $\{\varphi_n\}$. In fact, suppose that $\{\psi_n\} \subset \Phi$ is another sequence such that

$$\lim_{m,n \to \infty} \| \psi_m - \psi_n \|_2 = 0, \qquad \lim_{n \to \infty} \| \psi_n - \varphi \|_1 = 0. \tag{I.3.5}$$

Then, by (I.3.3) and (I.3.5), the sequence $\{\psi_n - \varphi_n\}$ is fundamental with respect to $\|\cdot\|_2$, and hence, by condition (I.3.2), also fundamental with respect to $\|\cdot\|_1$, moreover, $\lim_{n\to\infty} \|\psi_n - \varphi_n\|_1 = 0$. Since $\|\cdot\|_2$ and $\|\cdot\|_1$ are compatible, it follows that $\lim_{n\to\infty} \|\psi_n - \varphi_n\|_2 = 0$. Therefore, $\lim_{n\to\infty} \|\psi_n\|_2 = \lim_{n\to\infty} \|\varphi_n\|_2$. In particular, when $\varphi \in \Phi$, we may choose $\varphi_n = \varphi$, $n = 1, 2,\dots$, so that the norm defined by (I.3.4) coincides with the original norm $\|\varphi\|_2$. Furthermore, if $\varphi \in \Phi_2$ and $\{\varphi_n\} \subset \Phi$ satisfies (I.3.3), then

$$\|\varphi - \varphi_n\|_2 = \lim_{m\to\infty} \|\varphi_m - \varphi_n\|_2,$$

hence,

$$\lim_{n\to\infty} \|\varphi - \varphi_n\|_2 = 0. \tag{I.3.6}$$

This shows that Φ is dense in Φ_2. Moreover, we assert that Φ_2 is complete with respect to the norm $\|\cdot\|_2$. In fact, suppose that $\{\psi_n\} \subset \Phi_2$ is fundamental with respect to $\|\cdot\|_2$. Since Φ is dense in Φ_2, we may choose elements $\varphi_n \in \Phi$ such that

$$\|\psi_n - \varphi_n\|_2 < \frac{1}{n}, \qquad n = 1, 2,\dots . \tag{I.3.7}$$

Then, $\{\varphi_n\}$ is also fundamental with respect to $\|\cdot\|_2$, and hence fundamental with respect to $\|\cdot\|_1$. Consequently, there exists a $\varphi \in \Phi_1$ such that (I.3.3) holds. But this means that $\varphi \in \Phi_2$, and it follows as before that (I.3.6) holds. Hence, in view of (I.3.7), we get

$$\lim_{n\to\infty} \|\varphi - \psi_n\|_2 = 0,$$

which proves that Φ_2 is complete. Thus, Φ_2 is a completion of Φ with respect to the norm $\|\cdot\|_2$. By repetition of the preceding arguments, we arrive at the following result.

Lemma I.3.1. Let Φ be a countably normed space, with norm sequence $\{\|\cdot\|_n, n = 1, 2,\dots\}$. Then, a completion Φ_n of Φ with respect to $\|\cdot\|_n$ can be chosen for every n in such a way that

$$\Phi_1 \supset \Phi_2 \supset \cdots \supset \Phi_n \supset \cdots \supset \Phi.$$

Lemma I.3.2. Φ is complete if and only if

$$\Phi = \bigcap_{n=1}^{\infty} \Phi_n . \tag{I.3.8}$$

Proof. Assume that Φ is complete. Choose any $\varphi \in \bigcap_{n=1}^{\infty} \Phi_n$. Since Φ

is dense in Φ_n with respect to the norm $\|\cdot\|_n$, there exists, for every n, an element $\varphi_n \in \Phi$ such that

$$\|\varphi - \varphi_n\|_n < \frac{1}{n}.$$

Then, for $l, m \geqslant n$, we have

$$\begin{aligned}\|\varphi_l - \varphi_m\|_n &\leqslant \|\varphi - \varphi_l\|_n + \|\varphi - \varphi_m\|_n \\ &\leqslant \|\varphi - \varphi_l\|_l + \|\varphi - \varphi_m\|_m < \frac{1}{l} + \frac{1}{m}.\end{aligned}$$

Hence, the sequence $\{\varphi_l\}$ is fundamental with respect to $\|\cdot\|_n$ for every n, that is, $\{\varphi_l\}$ is a fundamental sequence in Φ. Therefore, there exists a $\psi \in \Phi$ such that

$$\lim_{l\to\infty} \|\psi - \varphi_l\|_n = 0,$$

from which it follows that $\psi = \varphi$, and hence $\varphi \in \Phi$. Thus, (I.3.8) holds.

Conversely, assume that (I.3.8) holds. If $\{\varphi_m\}$ is a fundamental sequence in Φ, then $\{\varphi_m\}$ is also a fundamental sequence in Φ_n (with respect to $\|\cdot\|_n$) for every n. By the completeness of Φ_n, there exists $\varphi^{(n)} \in \Phi_n$ such that

$$\lim_{m\to\infty} \|\varphi_m - \varphi^{(n)}\|_n = 0. \tag{I.3.9}$$

But $\Phi_{n+1} \subset \Phi_n$, hence $\varphi^{(n+1)} \in \Phi_n$, moreover,

$$\begin{aligned}\|\varphi^{(n+1)} - \varphi^{(n)}\|_n &\leqslant \|\varphi_m - \varphi^{(n+1)}\|_n + \|\varphi_m - \varphi^{(n)}\|_n \\ &\leqslant \|\varphi_m - \varphi^{(n+1)}\|_{n+1} + \|\varphi_m - \varphi^{(n)}\|_n \to 0.\end{aligned}$$

Therefore, $\varphi^{(n)} = \varphi^{(n+1)}$, $n = 1, 2,\ldots$, so that all the elements $\varphi^{(n)}$ are identical and may be denoted by φ. Since $\varphi \in \bigcap_{n=1}^{\infty} \Phi_n$, and (I.3.8) holds, we have $\varphi \in \Phi$. Moreover, by (I.3.9), the sequence $\{\varphi_m\}$ converges to φ in Φ. Thus, Φ is complete.]

If $m \geqslant n$, then Φ_m is a subspace of Φ_n. Thus, regarding Φ_n and Φ_m as Banach spaces with norms $\|\cdot\|_n$ and $\|\cdot\|_m$, respectively, we may define a linear operator $T_n{}^m$, from Φ_m to Φ_n, by

$$T_n{}^m : \varphi \to \varphi, \qquad \varphi \in \Phi_m.$$

It follows from (I.3.2) that $T_n{}^m$ $(m \geqslant n)$ is bounded; we call it the *imbedding operator* from Φ_m to Φ_n.

A complete countably normed space will be called a *countably Banach space.* If all the norms $\|\cdot\|_n$ which define a countably normed space Φ are induced by inner products $(\varphi, \psi)_n$, that is,

$$\|\varphi\|_n = [(\varphi, \varphi)_n]^{1/2},$$

then we call Φ a *countably inner product space.* A complete countably inner product space is called a *countably Hilbert space.*

Example I.3.1. Let Δ denote the finite closed interval $[a, b]$ on the real line. Let $K(\Delta)$ denote the totality of infinitely differentiable complex-valued functions $\varphi(x)$ on the real line which vanish outside of Δ. Obviously, $K(\Delta)$ forms a linear space with respect to the ordinary linear operations. For any $\varphi, \psi \in K(\Delta)$, we define

$$(\varphi, \psi)_n = \sum_{\nu=0}^{n} \int_{\Delta} \varphi^{(\nu)}(t)\, \overline{\psi^{(\nu)}(t)}\, dt, \qquad n = 1, 2, \ldots, \tag{I.3.10}$$

where $\varphi^{(0)}$ means just φ itself. We write $\|\varphi\|_n = [(\varphi, \varphi)_n]^{1/2}$.

Clearly, $K(\Delta)$, with the sequence of norms $\{\|\cdot\|_n, n = 1, 2, \ldots\}$, forms a countably inner product space.

Consider those functions φ, defined on the real line, which vanish outside of Δ, have an absolutely continuous derivative of order $n - 1$, and such that $\varphi^{(n)} \in L^2(\Delta)$. Denote the totality of such functions φ by $K_n(\Delta)$. Define an inner product on $K_n(\Delta)$ by formula (I.3.10). Clearly, $K_n(\Delta)$, with the ordinary linear operations and the inner product (I.3.10), forms a Hilbert space. In fact, $K_n(\Delta)$ is just the completion of $K(\Delta)$ with respect to the norm $\|\cdot\|_n$. Moreover, $K_1(\Delta) \supset K_2(\Delta) \supset \cdots \supset K_n(\Delta) \supset \cdots$ and

$$K(\Delta) = \bigcap_{n=1}^{\infty} K_n(\Delta),$$

consequently, $K(\Delta)$ is a countably Hilbert space.

2° Spaces of Continuous Linear Functionals

Let Φ be a countably normed space, with norm sequence $\{\|\cdot\|_n, n = 1, 2, \ldots\}$, and, for every n, let Φ_n be the completion of Φ with respect to the norm $\|\cdot\|_n$. Then Φ_n is, of course, a Banach space; we denote its conjugate space by $\Phi_n^\dagger$, and the norm in $\Phi_n^\dagger$ by

$$\|F\|_{-n} = \sup_{\varphi} \frac{|F(\varphi)|}{\|\varphi\|_n}, \qquad F \in \Phi_n^\dagger.$$

If $m \geqslant n$, any element of $\Phi_n^\dagger$, when restricted to Φ_m, is also continuous with respect to the norm $\|\cdot\|_m$, and similarly, when restricted to Φ, is continuous with respect to the topology of Φ. Thus, denoting by $\Phi^\dagger$ the linear space consisting of all continuous linear functionals on Φ, we have

$$\Phi_1^\dagger \subset \Phi_2^\dagger \subset \cdots \subset \Phi_n^\dagger \subset \cdots \subset \Phi^\dagger, \tag{I.3.11}$$

moreover, we assert that

$$\Phi^\dagger = \bigcup_{n=1}^{\infty} \Phi_n^\dagger. \tag{I.3.12}$$

In fact, if $F \in \Phi^\dagger$, then there is a positive number ϵ such that $\varphi \in \Phi$, $\rho(\varphi, 0) < \epsilon$ implies $|F(\varphi)| < 1$. Choose n_0 sufficiently large, so that $\sum_{n=n_0+1}^{\infty} (1/2^n) < \epsilon/2$, and choose a positive number $\delta < 1$, such that $\delta/(1+\delta) < \epsilon/2$. Then, when $\|\varphi\|_{n_0} < \delta$, we have

$$\rho(\varphi, 0) \leqslant \frac{\delta}{1+\delta} \sum_{n=1}^{n_0} \frac{1}{2^n} + \sum_{n=n_0+1}^{\infty} \frac{1}{2^n} < \epsilon,$$

whence $|F(\varphi)| < 1$. Therefore, $|F(\varphi)| \leqslant (1/\delta)\|\varphi\|_{n_0}$ for all $\varphi \in \Phi$, and hence F can be extended to a bounded linear functional $F \in \Phi_{n_0}^\dagger$. This proves (I.3.12).

Now, let Φ be a countably inner product space, with the topology defined by the sequence of inner products $\{(\varphi, \psi)_n,\ n = 1, 2, \ldots\}$. Let (φ, ψ) be another inner product on Φ, which is continuous with respect to the topology of Φ, in other words, there exists a positive number ϵ, such that, if $\varphi, \psi \in \Phi$, $\rho(\varphi, 0) < \epsilon$, $\rho(\psi, 0) < \epsilon$, then $|(\varphi, \psi)| < 1$. It then follows from the preceding discussion that there is a positive number δ, and a natural number n_0, such that $|(\varphi, \psi)| < 1$ whenever $\|\varphi\|_{n_0} < \delta$, $\|\psi\|_{n_0} < \delta$. Consequently,

$$|(\varphi, \psi)| \leqslant M \|\varphi\|_{n_0} \|\psi\|_{n_0}, \qquad \varphi, \psi \in \Phi, \tag{I.3.13}$$

where $M = 1/\delta^2$.

Forming the completion of Φ with respect to the new inner product (φ, ψ), we obtain a Hilbert space $H \supset \Phi$. We may, in the usual manner, identify H with its own conjugate space $H^\dagger$. The restriction to Φ of any element ξ of $H^\dagger$ $(= H)$ is a linear functional on Φ, defined explicitly by $\varphi \to (\varphi, \xi)$. Moreover, by (I.3.13),

$$|(\varphi, \xi)| \leqslant (\varphi, \varphi)^{1/2} (\xi, \xi)^{1/2} \leqslant [M(\xi, \xi)]^{1/2} \|\varphi\|_{n_0},$$

hence $\xi : \varphi \to (\varphi, \xi)$, $\varphi \in \Phi$, is a continuous linear functional on Φ, in fact, $\xi \in \Phi^{\dagger}_{n_0}$. Thus, we obtain an imbedding of H in $\Phi^{\dagger}$. The triple of spaces

$$\Phi \subset H \subset \Phi^{\dagger} \tag{I.3.14}$$

is called a *rigged Hilbert space*.

Example I.3.2. Define an inner product on $K(\Delta)$ by

$$(\varphi, \psi) = \int_{\Delta} \varphi(t)\, \overline{\psi(t)}\, dt.$$

The completion of $K(\Delta)$ with respect to (φ, ψ) is just $L^2(\Delta)$, and $K(\Delta)^{\dagger}$ is the space of generalized functions on Δ. Thus, we obtain the rigged Hilbert space $K(\Delta) \subset L^2(\Delta) \subset K^{\dagger}(\Delta)$.

Example I.3.3. Let $K = \bigcup_{n=1}^{\infty} K(\Delta_n)$, $\Delta_n = [-n, n]$. We specify that a sequence $\{\varphi_n\}$ converges to φ in K, provided that, for some k, $\{\varphi_n\} \subset K(\Delta_k)$, $\varphi \in K(\Delta_k)$, and that $\{\varphi_n\}$ converges to φ in the topology of $K(\Delta_k)$ (as defined in Example I.3.1). Accordingly, we say that a functional f on K is continuous provided that $f(\varphi_n) \to f(\varphi)$ whenever $\{\varphi_n\}$ converges to φ in the above sense. It can be proved that K is the exact union of the sequence of spaces $\{K(\Delta_n), n = 1, 2, \ldots\}$ (see Definition 4.3.5).

APPENDIX

II

BACKGROUND MATERIAL ON FUNCTIONAL ANALYSIS IN HILBERT SPACES

§II.1. Operators of Hilbert–Schmidt Type, Nuclear Operators, and Equivalence Operators

1° Basic Properties of Hilbert–Schmidt Operators and Nuclear Operators

In this subsection, we restrict our considerations to Hilbert spaces. Let H and G be two Hilbert spaces, T a completely continuous linear operator from H to G. Let T^* denote the adjoint of T. Then T^*T is a completely continuous self-adjoint linear operator from H to H, moreover, for any $\varphi \in H$,

$$(T^*T\varphi, \varphi) = (T\varphi, T\varphi) \geqslant 0,$$

that is, T^*T is a positive operator. According to the spectral resolution theorem for completely continuous self-adjoint operators, there is an orthonormal system of eigenvectors $\{e_n\}$ of T^*T, with corresponding eigenvalues $\lambda_n{}^2 > 0$, such that, for any $\varphi \in H$,

$$T^*T\varphi = \sum_{n=1}^{\infty} \lambda_n{}^2(\varphi, e_n)\, e_n\,. \tag{II.1.1}$$

Let $g_n = (1/\lambda_n)\, Te_n$. Then,

$$(g_m, g_n) = \frac{1}{\lambda_n\lambda_m}(Te_n, Te_m) = \frac{1}{\lambda_n\lambda_m}(T^*Te_n, e_m) = \delta_{n,m}\,,$$

where $\delta_{n,m}$ is the Kronecker delta, that is, $\{g_n\}$ is an orthonormal system in G. Now, for any $\varphi \in H$, there is a vector $u \perp \{e_n\}$ such that

$$\varphi = \sum (\varphi, e_n)\, e_n + u.$$

Since $u \perp \{e_n\}$, $u \perp T^*Tu$, that is, $(Tu, Tu) = 0$. Hence,

$$T\varphi = \sum (\varphi, e_n)\, Te_n\,,$$

or,

$$T\varphi = \sum_{n=1}^{\infty} \lambda_n(\varphi, e_n)\, g_n\,, \tag{II.1.2}$$

where $\lambda_n > 0$ and $\lim_{n\to\infty} \lambda_n = 0$.

Definition II.1.1. If, in (II.1.1), we have $\sum_{n=1}^{\infty} \lambda_n{}^2 < \infty$, then the operator T is said to be of *Hilbert–Schmidt type* (written briefly as H-S type); if $\sum_{n=1}^{\infty} \lambda_n < \infty$, then T is said to be a *nuclear operator*.

Obviously, every nuclear operator is of H-S type, but an operator of H-S type is not necessarily nuclear. Any continuous linear operator of finite rank (i.e., having a finite-dimensional range) is, of course, nuclear.

Lemma II.1.1. Let T be a bounded linear operator from a Hilbert space H to a Hilbert space G. Then T is of H-S type if and only if T^*T is nuclear.

PROOF. The necessity of the condition is obvious. Conversely, if T^*T is nuclear, then there exist numbers $\lambda_n > 0$, $\sum \lambda_n{}^2 < \infty$, and an orthonormal system $\{e_n\}$ in H, such that (II.1.1) holds. Hence, (II.1.2) also holds. It then follows that T is completely continuous and is of H-S type.]

Lemma II.1.2. Let T be a completely continuous linear operator from a Hilbert space H to a Hilbert space G. Then, T is of H-S type if and only if there exists a positive number M such that, for any orthonormal system $\{\varphi_\nu\}$ in H,

$$\sum_{\nu=1}^{\infty} \| T\varphi_\nu \|^2 \leqslant M < \infty. \tag{II.1.3}$$

PROOF. Assume that T is of H-S type. Then, there exist orthonormal systems $\{e_\nu\}$ and $\{g_\nu\}$ in H and G, respectively, such that (II.1.2) holds. By (II.1.2) we have, for any $\varphi \in H$,

$$\| T\varphi \|^2 = \sum_{\nu=1}^{\infty} \lambda_\nu^2 \, |(\varphi, e_\nu)|^2. \tag{II.1.4}$$

Let $\{\varphi_n\}$ be an arbitrary orthonormal system in H. Then, by the Bessel inequality,

$$\sum_n |(\varphi_n, e_\nu)|^2 \leqslant \| e_\nu \|^2 = 1. \tag{II.1.5}$$

Using (II.1.4) and (II.1.5), we get

$$\sum_n \| T\varphi_n \|^2 = \sum_\nu \sum_n \lambda_\nu^2 \, |(\varphi_n, e_\nu)|^2 \leqslant \sum_\nu \lambda_\nu^2.$$

Thus, choosing $M = \sum \lambda_\nu^2 < \infty$, we obtain (II.1.3).

Conversely, suppose there exists an $M < \infty$ such that (II.1.3) holds for every orthonormal system $\{\varphi_\nu\}$ in H. Form the decomposition (II.1.2), and choose $\varphi_\nu = e_\nu$. Then, since $Te_\nu = \lambda_\nu g_\nu$, it follows from (II.1.3) that $\sum \lambda_\nu^2 < \infty$.]

Remark. Actually, it suffices to assume that T is a bounded linear operator, and that (II.1.3) holds for every orthonormal system, in order to deduce that T is completely continuous, and, consequently, of H-S type.

Lemma II.1.3. If H, G are Hilbert spaces and T is a H-S type operator from H to G, then T^* is a H-S type operator from G to H.

PROOF. First, since T is completely continuous, T^* is also completely continuous. Now, let $\{\psi_n\}$ be any orthonormal system in G. Choose a complete orthonormal system $\{\varphi_\nu\}$ in H. Then, by the Parseval equality and the Bessel inequality, we get

$$\sum_n \| T^*\psi_n \|^2 = \sum_n \sum_\nu |(T^*\psi_n, \varphi_\nu)|^2$$

$$= \sum_n \sum_\nu |(\psi_n, T\varphi_\nu)|^2 \leqslant \sum_\nu \| T\varphi_\nu \|^2.$$

By Lemma II.1.2, there is a positive number M such that (II.1.3) holds, hence,

$$\sum \| T^*\psi_n \|^2 \leqslant M.$$

Again applying Lemma II.1.2, it follows at once that T^* is of H-S type.]

Lemma II.1.4. The totality of H-S type operators, from a fixed Hilbert space H to a fixed Hilbert space G, forms a linear space with respect to addition and scalar multiplication. If T is a H-S type operator from H to G, B a bounded linear operator from a Hilbert space H_1 to H, and C a bounded linear operator from G to a Hilbert space G_1, then the operators TB and CT are of H-S type.

PROOF. Using Lemma II.1.2, it is easily seen that the H-S type operators from H to G form a linear space. Next, suppose that the operator T, from H to G, is of H-S type, and that the linear operator C, from G to G_1, is bounded. Given any orthonormal system $\{\varphi_\nu\}$ in H, we have, by (II.1.3),

$$\sum \| CT\varphi_\nu \|^2 \leqslant \| C \|^2 \sum \| T\varphi_\nu \|^2 \leqslant \| C \|^2 M.$$

Again, by Lemma II.1.2, it follows that CT is of H-S type.

Finally, if B is a bounded linear operator from H_1 to H, then, since T^* is of H-S type, it follows, by the result just proved, that B^*T^* is of H-S type. Using Lemma II.1.3, we conclude that $TB = (B^*T^*)^*$ is of H-S type.]

In the present book, we make use of the following sufficient condition for a bounded self-adjoint operator to be of H-S type.

Theorem II.1.5. Let B be a bounded self-adjoint operator in a Hilbert space H. Suppose there exists a positive number $K < \infty$, such that, for any finite set of vectors $\eta_1, \ldots, \eta_n$ in H satisfying the conditions

$$(\eta_k, \eta_l) = \delta_{kl}, \qquad k, l = 1, 2, \ldots, n, \tag{II.1.6}$$

$$(B\eta_k, \eta_l) = 0, \qquad k \neq l, \quad k, l = 1, 2, \ldots, n, \tag{II.1.7}$$

the inequality

$$\sum_{k=1}^{n} (B\eta_k, \eta_k)^2 \leqslant K \tag{II.1.8}$$

always holds. Then, B is of H-S type.

PROOF. (i) We shall first prove that, if λ_0 belongs to the spectrum of B, and $\lambda_0 \neq 0$, then λ_0 is an isolated point of the spectrum. We may assume that $\lambda_0 > 0$. Let $\{E_\lambda, a \leqslant \lambda \leqslant b\}$ be the resolution of the identity corresponding to B. If λ_0 were not an isolated point, there would exist a sequence of points $\{\lambda_n\}$, in the spectrum of B, converging to λ_0. For definiteness, we assume that the sequence $\{\lambda_n\}$ is monotonic increasing, and that $\lambda_1 > \lambda_0/2$. Choose any sequence of numbers $\{\mu_n\}$ such that

$$\frac{\lambda_0}{2} < \lambda_1 < \mu_1 < \lambda_2 < \mu_2 < \cdots < \lambda_n < \mu_n < \cdots < \lambda_0 .$$

Since every λ_n belongs to the spectrum of B, we have $E_{\mu_{n+1}} - E_{\mu_n} \neq 0$, $n = 1, 2, \ldots$. Choose vectors $\eta_n \in (E_{\mu_{n+1}} - E_{\mu_n})H$, $\| \eta_n \| = 1$. Then, for any n, the vectors $\eta_1, \ldots, \eta_n$ satisfy conditions (II.1.6) and (II.1.7) [because the subspaces $(E_{\mu_{n+1}} - E_{\mu_n})H$ are mutually orthogonal and invariant under B]. On the other hand,

$$\begin{aligned}(B\eta_k, \eta_k) &= \int_{\mu_k}^{\mu_{k+1}} \lambda \, d(E_\lambda \eta_k, \eta_k) \geqslant \frac{\lambda_0}{2} \int_{\mu_k}^{\mu_{k+1}} d(E_\lambda \eta_k, \eta_k) \\ &= \frac{\lambda_0}{2} \| \eta_k \|^2. \end{aligned} \qquad \text{(II.1.9)}$$

From (II.1.8) and (II.1.9), we get

$$n \left(\frac{\lambda_0}{2} \right)^2 \leqslant \sum_{k=1}^{n} (B\eta_k, \eta_k)^2 \leqslant K.$$

But, since n is arbitrary, this is impossible. Hence, we conclude that every nonzero point of the spectrum of B is isolated, and is therefore an eigenvalue of B.

(ii) The eigenspace H_{λ_0}, corresponding to any nonzero eigenvalue λ_0 of B, is finite dimensional. For otherwise, there would exist infinitely many mutually orthogonal unit vectors $\{\eta_k\}$ in H_{λ_0}; for every n, the vectors $\eta_1, \ldots, \eta_n$ would satisfy (II.1.6) and (II.1.7) (since $B_{\eta_k} = \lambda_0 \eta_k$). But then (II.1.8) becomes

$$n\lambda_0^2 \leqslant K, \qquad n = 1, 2, \ldots,$$

which is impossible.

It follows from (i) and (ii) that B is completely continuous.

(iii) Let $\mu_1, \mu_2, \ldots, \mu_n, \ldots$ denote the totality of nonzero eigenvalues of B (repeated according to their multiplicity), and $\eta_1, \eta_2, \ldots, \eta_n, \ldots$ a corresponding orthonormal system of eigenvectors. Thus,

$$B\varphi = \sum_{k=1}^{\infty} \mu_k (\varphi, \eta_k) \, \eta_k , \qquad \text{(II.1.10)}$$

moreover, for any n, the vectors $\eta_1, \ldots, \eta_n$ satisfy conditions (II.1.6) and (II.1.7). Hence, by virtue of (II.1.8),

$$\sum_{k=1}^{n} \mu_k^2 \leqslant K \qquad \text{(II.1.11)}$$

for any n. Comparing (II.1.10) with (II.1.2), we conclude from (II.1.11) that B is of H-S type.]

Repeating the line of reasoning used in the proof of Theorem II.1.5, we can also obtain the following result.

Theorem II.1.6. Let B be a bounded self-adjoint operator in a Hilbert space H. Suppose there exists a positive number $K < \infty$, such that, for any finite set of vectors $\eta_1, \ldots, \eta_n$ satisfying conditions (II.1.6) and (II.1.7), the inequality

$$\sum_{k=1}^{n} |(B\eta_k, \eta_k)| \leqslant K \qquad \text{(II.1.12)}$$

holds. Then B is a nuclear operator.

Remark. In Theorem II.1.6 (Theorem II.1.5), one need only assume that inequality (II.1.12) [inequality (II.1.8)] holds for every finite set of vectors $\eta_1, \ldots, \eta_n$ belonging to some dense linear subspace of H and satisfying conditions (II.1.6), (II.1.7); the conclusion of the Theorem still remains valid.

Let B be a nuclear operator from H to G, and write

$$\| B \|_1 = \sup \sum_{k=1}^{\infty} |(B\xi_k, \eta_k)|,$$

where the supremum is taken over all the orthonormal systems $\{\xi_k\}$ and $\{\eta_k\}$ in H and G, respectively. It is easily verified that $\| B \|_1 < \infty$. The number $\| B \|_1$ is called the *trace norm* of B.

The following theorem gives another important relation between H-S type operators and nuclear operators.

Theorem II.1.7. The product of two H-S type operators is a nuclear operator.

Proof. Let H_1, H_2, H_3 be Hilbert spaces and $B : H_2 \to H_3$, $C : H_1 \to H_2$ operators of H-S type. Then $T = BC$ is a completely continuous operator from H_1 to H_3. By (II.1.2), there exist orthonormal systems of vectors $\{e_\nu\}$ in H_1 and $\{g_n\}$ in H_3, and numbers $\lambda_n > 0$, such that

$$T\varphi = \sum \lambda_n(\varphi, e_n) g_n .$$

We must prove that

$$\sum_{n=1}^{\infty} \lambda_n < \infty.$$

In fact,

$$\begin{aligned}\lambda_n &= (Te_n, g_n) = (BCe_n, g_n) = (Ce_n, B^*g_n) \\ &\leqslant \tfrac{1}{2}(\| Ce_n \|^2 + \| B^*g_n \|^2).\end{aligned}$$

By Lemma II.1.3, B^* is of H-S type, hence, by Lemma II.1.2, we have

$$\sum \lambda_n \leqslant \frac{1}{2}\Big(\sum_n \| Ce_n \|^2 + \sum \| B^*g_n \|^2\Big) < \infty,$$

as required.]

Remark. It can also be proved that any nuclear operator can be resolved into the product of two H-S type operators.

Next, we give another useful sufficient condition for an operator to be nuclear.

Theorem II.1.8. Let T be a linear operator from a Hilbert space H to a Hilbert space G. If there exist families of vectors $\{h_n\}$ in H and $\{\psi_n\}$ in G, such that T has the representation

$$T\varphi = \sum (\varphi, h_n) \psi_n , \qquad \varphi \in H,$$

and

$$\sum \| h_n \| \| \psi_n \| < \infty, \tag{II.1.13}$$

then T is a nuclear operator.

PROOF. In view of (II.1.13), it may be assumed that $\{h_n\}$ and $\{\psi_n\}$ are countable. Form the operator

$$T_n\varphi = \sum_{\nu=1}^{n} (\varphi, h_\nu) \psi_\nu , \qquad \varphi \in H.$$

Then T_n is a continuous linear operator of finite rank, and is therefore completely continuous. Moreover, since

$$\|(T - T_n)\varphi \| = \Big\| \sum_{\nu=n+1}^{\infty} (\varphi, h_\nu) \psi_\nu \Big\| \leqslant \sum_{\nu=n+1}^{\infty} \| h_\nu \| \| \psi_\nu \| \| \varphi \|,$$

it follows by (II.1.13) that

$$\lim_{n\to\infty} \| T - T_n \| = 0.$$

Hence, T is completely continuous. Representing T in the form (II.1.2), and using the Bessel inequality, we get

$$\sum_n \lambda_n = \sum_n |(Te_n, g_n)| \leqslant \sum_\nu \sum_n |(e_n, h_\nu)(\psi_\nu, g_n)|$$

$$\leqslant \sum_\nu \Big(\sum_n |(e_n, h_\nu)|^2\Big)^{1/2} \Big(\sum_n |(\psi_\nu, g_n)|^2\Big)^{1/2}$$

$$\leqslant \sum_\nu \| h_\nu \| \cdot \| \psi_\nu \| < \infty.$$

Therefore, T is a nuclear operator.]

2° Equivalence Operators

The concept of an equivalence operator, as defined below, was first introduced by Feldman [1] in connection with the equivalence of Gaussian measures.

Definition II.1.2. Let H, K be Hilbert spaces and A a linear operator from H to K, having the following properties.

(i) A is a homeomorphism from H onto K.

(ii) $A^*A - I$ is a H-S type operator on H (here, I denotes the identity operator in H).

Then A is said to be an *equivalence operator* (from H onto K).

Note that condition (ii) in the above definition may be replaced by

(ii)′ $(A^*A)^{1/2} - I$ is a H-S type operator on H.

In fact, if A is an equivalence operator, then $A^*A - I = T$ is a self-adjoint H-S type operator. Let T be represented in the form

$$T\varphi = \sum \lambda_n(\varphi, e_n)\, e_n, \qquad \sum \lambda_n^2 < \infty, \quad \lambda_n > -1,$$

where $\{e_n\}$ is an orthonormal system in H. Then, one easily calculates that

$$((A^*A)^{1/2} - I)\varphi = \sum((1 + \lambda_n)^{1/2} - 1)(\varphi, e_n)\, e_n .$$

But $(1 + \lambda_n)^{1/2} - 1$ is a real number, moreover $\sum ((1 + \lambda_n)^{1/2} - 1)^2 \leqslant$

$\sum \lambda_n^2 < \infty$. Hence, (ii)′ holds. Conversely, assume that (ii)′ holds, and write $S = (A^*A)^{1/2} - I$. Then

$$A^*A - I = (I + S)^2 - I = 2S + S^2,$$

which shows that (ii) holds.

Theorem II.1.9. The inverse and adjoint of an equivalence operator are equivalence operators; the product of two equivalence operators is an equivalence operator.

PROOF. Let A be an equivalence operator from H onto K, and B an equivalence operator from G onto H. Obviously, AB satisfies condition (i) in the above definition. Now, set

$$A^*A = I + T, \qquad Q = (B^*B)^{1/2} = I + S.$$

By conditions (ii) and (ii)′, both T and S are of H-S type.

Suppose that $x \in G$ and $Qx = 0$. Then $\| Bx \|^2 = (B^*Bx, x) = (Q^2x, x) = 0$, but since B^{-1} exists, it follows that $x = 0$. This shows that -1 is not an eigenvalue of S. Therefore, since S is completely continuous, -1 is a regular point of S, so that $(I + S)^{-1}$ is defined and bounded on the entire space G. Hence the operator $V = BQ^{-1}$ is one-to-one from G onto H, moreover, for any $x \in G$,

$$(Vx, Vx) = (BQ^{-1}x, BQ^{-1}x) = (B^*BQ^{-1}x, Q^{-1}x) = (x, x).$$

Therefore, V is a unitary operator from G onto H, and so we have

$$\begin{aligned}(AB)^*(AB) &= QV^*A^*AVQ = Q(I + V^*TV)Q \\ &= I + (2S + V^*TV + SV^*TV + V^*TVS + S^2 + SV^*TVS).\end{aligned}$$

By virtue of Lemma II.1.4, the above expression shows that $(AB)^*(AB) - I$ is of H-S type, that is, AB is an equivalence operator. The other assertions of the theorem can be proved in a similar fashion.]

3° Nuclear Spaces

Definition II.1.3. Let Φ be a countably Hilbert space, with the sequence of inner products $\{(\cdot, \cdot)_n\}$. Let Φ_n be the completion of Φ with respect to $(\cdot, \cdot)_n$, and, for $m \geqslant n$, let T_n^m be the imbedding operator from Φ_m to Φ_n. Suppose that, for every n, there is an $m \geqslant n$ such that T_n^m is a nuclear operator. Then Φ is called a *nuclear space*.

Since the product of two H-S type operators is a nuclear operator (see Theorem II.1.7), it is easily seen that a countably Hilbert space Φ

is a nuclear space if and only if, for every n, there is an $m \geqslant n$ such that $T_n{}^m : \Phi_m \to \Phi_n$ is of H-S type.

Furthermore, since the product of a bounded linear operator and a nuclear (H-S type) operator is again a nuclear (H-S type) operator, it is easily seen that a necessary and sufficient condition for a countably Hilbert space to be nuclear is the existence of a sequence of natural numbers

$$n_1 < n_2 < \cdots < n_k < \cdots$$

such that every $T_{n_k}^{n_{k+1}} : \Phi_{n_{k+1}} \to \Phi_{n_k}$ is a nuclear (or H-S type) operator.

Example II.1.1. $K(\Delta)$ (see Example I.3.1) is a nuclear space.

In fact, we shall prove that the imbedding operator

$$T_n^{n+1} : K_{n+1}(\Delta) \to K_n(\Delta)$$

is of H-S type for every n. To simplify notation, we shall suppose that $\Delta = [0, 2\pi]$. Let $K_n'(\Delta)$ denote the totality of complex-valued functions on Δ which have an absolutely continuous derivative of order $n - 1$ and a quadratically integrable derivative of order n (defined almost everywhere). Then $K_n'(\Delta)$ forms a Hilbert space with respect to the ordinary linear operations and the inner product $(\cdot, \cdot)_n$, moreover, $K_n(\Delta)$ is a closed linear subspace of $K_n'(\Delta)$. Let P_n denote the projection operator from $K_n'(\Delta)$ onto $K_n(\Delta)$, A_n the imbedding operator from $K_{n+1}'(\Delta)$ to $K_n'(\Delta)$, and C_{n+1} the imbedding operator from $K_{n+1}(\Delta)$ to $K_{n+1}'(\Delta)$. Clearly,

$$T_n^{n+1} = P_n A_n C_{n+1} .$$

Since C_{n+1} and P_n are bounded linear operators, we need only prove that A_n is of H-S type, whence it will follow, in view of Lemma II.1.4, that T_n^{n+1} is of H-S type.

Consider the orthonormal system

$$e_m(t) = \frac{e^{imt}}{(2\pi \sum_{\nu=0}^{n+1} m^{2\nu})^{1/2}}, \qquad m = 0, \pm 1, \pm 2, \ldots$$

in $K_{n+1}'(\Delta)$, and the orthonormal system

$$g_m(t) = \frac{e^{imt}}{(2\pi \sum_{\nu=0}^{n} m^{2\nu})^{1/2}}, \qquad m = 0, \pm 1, \pm 2, \ldots$$

in $K_n'(\Delta)$. Let $\varphi \in K_{n+1}'(\Delta)$; expansion of φ in a Fourier series shows that

$$A_n \varphi = \sum_m \lambda_m (\varphi, e_m)_{n+1} g_m \tag{II.1.14}$$

where

$$\lambda_m = \Big(\sum_{\nu=0}^{n} m^{2\nu} \Big/ \sum_{\nu=0}^{n+1} m^{2\nu}\Big)^{1/2}. \tag{II.1.15}$$

In fact, (II.1.14) and (II.1.15) follow from the expansion

$$\varphi(s) = \frac{1}{2\pi} \sum_m \int_0^{2\pi} \varphi(t)\, e^{-imt}\, dt\, e^{ims}$$

and the formula

$$(\varphi, e_m)_{n+1} = \Big(\sum_{\nu=0}^{n+1} m^{2\nu} \Big/ 2\pi\Big)^{1/2} \int_0^{2\pi} \varphi(t)\, e^{-imt}\, dt,$$

which is derived using integration by parts. From (II.1.15), it is easily verified that $\sum \lambda_m^2 < \infty$. Therefore, A_n is of H-S type.]

§II.2. Tensor Products of Hilbert Spaces

The definition of a tensor product of finite-dimensional spaces can likewise be extended to the infinite-dimensional case. In the present book, we shall be concerned only with tensor products of inner product spaces (see von Neumann [2]).

Definition II.2.1. Let $\mathfrak{H}_1, \ldots, \mathfrak{H}_n$ be vector spaces (e.g., over the complex numbers). Consider those functionals $\Phi(f_1, \ldots, f_n)$, $f_\nu \in \mathfrak{H}_\nu$, $\nu = 1, 2, \ldots, n$, which satisfy the conditions

(i) $\Phi(\ldots, \alpha f_\nu, \ldots) = \bar{\alpha}\Phi(\ldots, f_\nu, \ldots)$, α any complex number, and

(ii) $\Phi(\ldots, f_\nu + g_\nu, \ldots) = \Phi(\ldots, f_\nu, \ldots) + \Phi(\ldots, g_\nu, \ldots)$, for every index $\nu = 1, 2, \ldots, n$. The totality of such functionals forms a linear space with respect to the usual operations of addition and multiplication by scalars; denote this linear space by $\prod_{\nu=1}^{n} \otimes \mathfrak{H}_\nu$.

Next, we further assume that each $\mathfrak{H}_\nu$ is an inner product space. Given any set of vectors $f_\nu^0 \in \mathfrak{H}_\nu$, $\nu = 1, 2, \ldots, n$, we denote the functional $\Phi(f_1, \ldots, f_n) = \prod_{\nu=1}^{n} (f_\nu^0, f_\nu)$ by the symbol $\prod_{\nu=1}^{n} \otimes f_\nu^0$, or, alternatively, $f_1^0 \otimes \cdots \otimes f_n^0$. The totality of functionals of the form

$$\Phi = \sum_{p=1}^{l} \prod_{\nu=1}^{n} \otimes f_{\nu,p}^0, \qquad f_{\nu,p}^0 \in \mathfrak{H}_\nu \tag{II.2.1}$$

constitute a linear space, which we denote by $\prod_{\nu=1}^{\prime n} \otimes \mathfrak{H}_\nu$. Given any other element of the form (II.2.1), say

$$\Psi = \sum_{q=1}^{m} \prod_{\nu=1}^{n} \otimes g_{\nu,q}^0, \qquad g_{\nu,q}^0 \in \mathfrak{H}_\nu, \tag{II.2.2}$$

we define

$$(\Phi, \Psi) = \sum_{p=1}^{l} \sum_{q=1}^{m} \prod_{\nu=1}^{n} (f^0_{\nu,p}, g^0_{\nu,q}). \tag{II.2.3}$$

Lemma II.2.1. The number (Φ, Ψ) defined by (II.2.3) depends only upon the functionals Φ and Ψ, and is independent of the particular representations (II.2.1) and (II.2.2). Moreover, (Φ, Ψ) is an inner product on $\prod'^{n}_{\nu=1} \otimes \mathfrak{H}_\nu$.

PROOF. Choose any two representations of Φ. Subtracting one from the other, we obtain another representation of the form (II.2.1), but in this case the left-hand side $\Phi = 0$. Hence, to prove that (II.2.3) is independent of the representation of Φ, it suffices to show that the right-hand side of (II.2.3) is zero whenever the functional $\Phi = 0$. But (II.2.3) may be written as

$$(\Phi, \Psi) = \sum_{q=1}^{m} \Phi(g^0_{1,q}, ..., g^0_{n,q}), \tag{II.2.4}$$

which vanishes whenever $\Phi = 0$. Thus, (II.2.3) is indeed independent of the representation of Φ. Noting the fact that $(\Phi, \Psi) = \overline{(\Psi, \Phi)}$, it follows at once that (II.2.3) is also independent of the representation (II.2.2) of Ψ.

It is obvious that (Φ, Ψ) depends linearly upon Φ. Next, we prove that $(\Phi, \Phi) \geqslant 0$ for all Φ. Observe that, for any finite set of complex numbers $\xi_1, ..., \xi_l$,

$$\sum_{p,q=1}^{l} (f^0_{\nu,p}, f^0_{\nu,q})\, \xi_p \bar{\xi}_q = \Big\| \sum_{p=1}^{l} f^0_{\nu,p} \xi_p \Big\|^2 \geqslant 0,$$

that is, the matrix $((f^0_{\nu,p}, f^0_{\nu,q}))_{p,q=1,2,...,l}$ is nonnegative definite. Therefore, this matrix can be expressed as a sum of l matrices[1] $(\alpha^{(s)}_{\nu,p} \bar{\alpha}^{(s)}_{\nu,q})_{p,q=1,2,...,l}$, $s = 1, 2, ..., l$, whence we have

$$(\Phi, \Phi) = \sum_{1 \leqslant s_\nu \leqslant l} \sum_{p,q=1}^{l} \prod_{\nu=1}^{n} \alpha^{(s_\nu)}_{\nu,p} \bar{\alpha}^{(s_\nu)}_{\nu,q}$$

$$= \sum_{1 \leqslant s_\nu \leqslant l} \Big| \sum_{p=1}^{l} \prod_{\nu=1}^{n} \alpha^{(s_\nu)}_{\nu,p} \Big|^2 \geqslant 0. \tag{II.2.5}$$

[1] Let $X = (x_{pq})_{p,q=1,2,...,l}$ be any nonnegative definite $l \times l$ matrix. Then there exists an $l \times l$ unitary matrix $U = (u_{p,q})$ such that $X = U\Lambda U^*$, where Λ is a diagonal matrix whose diagonal entries are the eigenvalues $\lambda_1, ..., \lambda_l (\lambda_\nu \geqslant 0)$ of X. Consequently, X is the sum of l matrices of rank one, that is,

$$((u_{p,\nu}(\lambda_\nu)^{1/2})(\bar{u}_{q,\nu}(\lambda_\nu)^{1/2}))_{p,q=1,2,...,l}, \qquad \nu = 1, 2, ..., l.$$

From (II.2.5), we deduce at once that the Schwarz inequality

$$|(\Phi, \Psi)| \leqslant ((\Phi, \Phi)(\Psi, \Psi))^{1/2}$$

is valid. Consequently, if $\Phi \in \prod_{\nu=1}^{\prime n} \otimes \mathfrak{H}_\nu$, $(\Phi, \Phi) = 0$, then $(\Phi, \Psi) = 0$ for every $\Psi \in \prod_{\nu=1}^{\prime n} \otimes \mathfrak{H}_\nu$. In particular, choosing $\Psi = \prod_{\nu=1}^{n} \otimes f_\nu$, $f_\nu \in \mathfrak{H}_\nu$, we get

$$\Phi(f_1, \ldots, f_n) = \left(\Phi, \prod_{\nu=1}^{n} \otimes f_\nu\right) = 0,$$

hence $\Phi = 0$. Thus, we have proved that (Φ, Ψ) is an inner product.]

Throughout the following, $\prod_{\nu=1}^{\prime n} \otimes \mathfrak{H}_\nu$ will be regarded as an inner product space with respect to the inner product (II.2.3), and we shall write

$$\| \Phi \| = (\Phi, \Phi)^{1/2}.$$

It is easily verified that

$$\left\| \prod_{\nu=1}^{n} \otimes f_\nu \right\| = \prod_{\nu=1}^{n} \| f_\nu \|.$$

Moreover, since

$$\Phi(f_1, \ldots, f_n) = \left(\Phi, \prod_{\nu=1}^{n} \otimes f_\nu\right)$$

for any $\Phi \in \prod_{\nu=1}^{\prime n} \otimes \mathfrak{H}_\nu$, it follows that

$$| \Phi(f_1, \ldots, f_n)| \leqslant \| \Phi \| \prod_{\nu=1}^{n} \| f_\nu \|. \tag{II.2.6}$$

Lemma II.2.2. Let $\{\Phi_s\}$, $s = 1, 2, \ldots$ be a fundamental sequence in $\prod_{\nu=1}^{\prime n} \otimes \mathfrak{H}_\nu$. Then, there exists a functional $\Phi \in \prod_{\nu=1}^{n} \circledast \mathfrak{H}_\nu$, such that

$$\Phi(f_1, \ldots, f_n) = \lim_{s\to\infty} \Phi_s(f_1, \ldots, f_n)$$

for every n-tuple of vectors $f_\nu \in \mathfrak{H}_\nu$, $\nu = 1, 2, \ldots, n$.

PROOF. By (II.2.6), we have

$$| \Phi_s(f_1, \ldots, f_n) - \Phi_{s'}(f_1, \ldots, f_n)| \leqslant \| \Phi_s - \Phi_{s'} \| \prod_{\nu=1}^{n} \| f_\nu \|.$$

Consequently, for any fixed n-tuple $f_1, \ldots, f_n$, $\lim_{s\to\infty} \Phi_s(f_1, \ldots, f_n)$ exists; denote this limit by $\Phi(f_1, \ldots, f_n)$. It is easily verified that the functional $\Phi(f_1, \ldots, f_n) \in \prod_{\nu=1}^{n} \circledast \mathfrak{H}_\nu$.]

Let $\prod_{\nu=1}^{n} \otimes \mathfrak{H}_\nu$ be the totality of functionals Φ in $\prod_{\nu=1}^{n} \circledast \mathfrak{H}_\nu$ which satisfy the following condition: there exists a fundamental sequence $\{\Phi_s\}$ in $\prod_{\nu=1}^{\prime n} \otimes \mathfrak{H}_\nu$ such that

$$\Phi(f_1, \dots, f_n) = \lim_{s\to\infty} \Phi_s(f_1, \dots, f_n) \tag{II.2.7}$$

for every n-tuple $f_\nu \in \mathfrak{H}_\nu$, $\nu = 1, 2, \dots, n$. When this condition is satisfied, we say that $\{\Phi_s\}$ is a *defining sequence* for Φ. If $\Psi \in \prod_{\nu=1}^{n} \otimes \mathfrak{H}_\nu$, and $\{\Psi_s\}$ is a defining sequence for Ψ, we define the inner product of Φ and Ψ by

$$(\Phi, \Psi) = \lim_{n\to\infty} (\Phi_s, \Psi_s). \tag{II.2.8}$$

It is easily verified that the above limit exists, is independent of the choice of defining sequences for Φ and Ψ, and does, in fact, define an inner product on $\prod_{\nu=1}^{n} \otimes \mathfrak{H}_\nu$. Moreover, from Lemma II.2.2, the reader can easily deduce the following result.

Theorem II.2.3. $\prod_{\nu=1}^{n} \otimes \mathfrak{H}_\nu$ forms a Hilbert space with respect to the inner product (II.2.8).

In fact, $\prod_{\nu=1}^{n} \otimes \mathfrak{H}_\nu$ is a completion of the inner product space $\prod_{\nu=1}^{\prime n} \otimes \mathfrak{H}_\nu$.

Definition II.2.2. The Hilbert space $\prod_{\nu=1}^{n} \otimes \mathfrak{H}_\nu$ is called the *tensor product* of the inner product spaces $\mathfrak{H}_1, \dots, \mathfrak{H}_n$.

Example II.2.1. Let $\Omega_\nu = (G_\nu, \mathfrak{B}_\nu, P_\nu)$, $\nu = 1, 2, \dots, n$, be a family of measure spaces. For every ν, let $L^2(\Omega_\nu)$ be the Hilbert space formed by the totality of quadratically integrable functions on Ω_ν with the usual linear operations and the inner product

$$(f, g) = \int_{G_\nu} f(\omega_\nu)\, \overline{g(\omega_\nu)}\, dP_\nu(\omega_\nu).$$

Let $\Omega = (G, \mathfrak{B}, P)$ be the product of the measure spaces $\Omega_1, \dots, \Omega_n$. Then, any $\Phi \in L^2(\Omega)$ may be regarded as a functional $\Phi(f_1, \dots, f_n)$, $f_\nu \in L^2(\Omega_\nu)$, $\nu = 1, 2, \dots, n$, namely,

$$\Phi(f_1, \dots, f_n) = \int_\Omega \Phi(\omega)\, \overline{f_1(\omega_1)} \cdots \overline{f_n(\omega_n)}\, dP(\omega),$$

$$\omega = (\omega_1, \dots, \omega_n).$$

In accordance with this identification, $\prod_{\nu=1}^{n} \otimes f_\nu$ is just the function $f_1(\omega_1) \cdots f_n(\omega_n)$ in $L^2(\Omega)$. It is easily seen that the tensor product $\prod_{\nu=1}^{n} \otimes L^2(\Omega_\nu)$ is, in this case, simply $L^2(\Omega)$.

We now consider tensor products of identical factors $\mathfrak{H}_1 = \cdots = \mathfrak{H}_\nu = \mathfrak{H}$. In this case, we write $\mathfrak{H}^{(n)}$ in place of $\prod_{\nu=1}^{n} \otimes \mathfrak{H}_\nu$; the elements of $\mathfrak{H}^{(n)}$ are called *tensors of order n over* $\mathfrak{H}$. Let Σ_n denote the symmetric group of degree n, that is, the group of all permutations of the symbols $1, 2,\ldots, n$. If $\pi \in \Sigma_n$, we denote the result of applying the permutation π to the symbols $1,\ldots, n$ by $\pi(1),\ldots, \pi(n)$, respectively. Now, form a linear operator $V^{(n)}(\pi)$ in $\prod_{\nu=1}^{\prime n} \otimes \mathfrak{H}_\nu$, as follows: if

$$\Phi = \sum_{p=1}^{l} \prod_{\nu=1}^{n} \otimes f_\nu^{(p)}, f_\nu^{(p)} \in \mathfrak{H}_\nu ,$$

then

$$V^{(n)}(\pi)\Phi = \sum_{p=1}^{l} \prod_{\nu=1}^{n} \otimes f_{\pi(\nu)}^{(p)} .$$

It is easily seen that $V^{(n)}(\pi)$ is a unitary operator from $\prod_{\nu=1}^{\prime n} \otimes \mathfrak{H}_\nu$ onto itself. Since $\prod_{\nu=1}^{\prime n} \otimes \mathfrak{H}_\nu$ is dense in $\mathfrak{H}^{(n)}$, $V^{(n)}(\pi)$ can be uniquely extended to a unitary operator from $\mathfrak{H}^{(n)}$ onto itself. Furthermore, the correspondence $\pi \to V^{(n)}(\pi^{-1})$ is a unitary representation of the symmetric group in the Hilbert space $\mathfrak{H}^{(n)}$ (for the definition of a unitary representation, see §II.3).

Consider the bounded linear operator

$$S_n = \frac{1}{n!} \sum_{\pi \in \Sigma_n} V^{(n)}(\pi)$$

in $\mathfrak{H}^{(n)}$. Actually, S_n is a projection operator. In fact, since $V^{(n)}(\pi)^* = V^{(n)}(\pi^{-1})$, we have

$$S_n^* = \frac{1}{n!} \sum_{\pi} V^{(n)}(\pi^{-1}) = S_n .$$

On the other hand, for any $\pi \in \Sigma_n$, $\{\pi\pi' \mid \pi' \in \Sigma_n\} = \{\pi'\pi \mid \pi' \in \Sigma_n\} = \Sigma_n$. Consequently,

$$V^{(n)}(\pi)\, S_n = \frac{1}{n!} \sum_{\pi'} V^{(n)}(\pi\pi') = S_n = S_n V^{(n)}(\pi). \tag{II.2.9}$$

Therefore, since Σ_n is of order $n!$, we obtain

$$S_n^2 = \frac{1}{n!} \sum_{\pi} V^{(n)}(\pi)\, S_n = S_n .$$

Formula (II.2.9) also shows that every vector of $M_n = S_n\mathfrak{H}^{(n)}$, the range of S_n, is invariant with respect to all the operators $V^{(n)}(\pi)$,

$\pi \in \Sigma_n$. The elements of M_n are called *symmetric tensors* of order n, and the projection S_n is called the *symmetrization operator*. M_n, being a closed subspace of a Hilbert space $\mathfrak{H}^{(n)}$, is itself a Hilbert space with respect to the inner product (Φ, Ψ).

Example II.2.2. If, in Example II.2.1, we take $\Omega_1 = \Omega_2 = \cdots = \Omega_n$, then the symmetric tensors of order n are just the symmetric functions $f(\omega)$ in $L^2(\Omega)$, that is, those which satisfy

$$f(\omega_1, \ldots, \omega_n) = f(\omega_{\pi(1)}, \ldots, \omega_{\pi(n)})$$

for all permutations $\pi \in \Sigma_n$.

Lemma II.2.4. Let $\{g_n\}$ be a complete orthonormal system in the Hilbert space $\mathfrak{H}$. Then,[2]

$$\left(\frac{n!}{k_1! k_2! \cdots k_l!}\right)^{1/2} S_n(g_1^{k_1} \otimes \cdots \otimes g_l^{k_l}), \qquad 0 \leqslant k_\nu, \quad \sum_{\nu=1}^{l} k_\nu = n \tag{II.2.10}$$

forms a complete orthonormal system in the corresponding space M_n of symmetric tensors of order n.

PROOF. Clearly, (II.2.10) is an orthonormal system; we need only prove that it is complete. Suppose that $g \in M_n$ is orthogonal to all of the vectors (II.2.10). Then

$$(g, S_n(g_1^{k_1} \otimes \cdots \otimes g_l^{k_l})) = (S_n g, g_1^{k_1} \otimes \cdots \otimes g_l^{k_l}) = 0.$$

But every vector in $\mathfrak{H}^{(n)}$ is a limit of linear combinations of vectors of the form $g_1^{k_1} \otimes \cdots \otimes g_l^{k_l}$, moreover, $S_n g = g$. Consequently, g is orthogonal to the whole of $\mathfrak{H}^{(n)}$, that is, $g = 0$. This shows that (II.2.10) is complete.]

§II.3. Unitary Representations of Groups

1° General Concepts

Definition II.3.1. Let $\mathfrak{G}$ be a group, H a Hilbert space. Suppose that, to every element $g \in \mathfrak{G}$, there corresponds a unitary operator $U(g)$

[2] Here, $g_1^{k_1}$ represents $\underbrace{g_1 \otimes \cdots \otimes g_1}_{k_1 \text{ factors}}$.

on H, and that, for any $g_1, g_2 \in \mathfrak{G}$, we have $U(g_1)\, U(g_2) = U(g_1 g_2)$. Then the correspondence

$$U : g \to U(g), \qquad g \in \mathfrak{G}$$

is called a *unitary representation* of the group $\mathfrak{G}$ in the Hilbert space H. If we let $\mathfrak{U}$ denote the totality of unitary operators in H, then a unitary representation U of $\mathfrak{G}$ in H is simply a homomorphism of $\mathfrak{G}$ into $\mathfrak{U}$. If there exists no nontrivial closed linear subspace of H which is invariant under all the operators $U(g)$, $g \in \mathfrak{G}$, then the representation U is said to be *irreducible*.

Clearly, U is irreducible if and only if $\{U(g) \mid g \in G\}'$ (see §2.3, 2°) consists of just the operators λI, where λ is a number and I is the identity operator in H.

Now, suppose that $\mathfrak{G}$ is a topological group, and U is a unitary representation of $\mathfrak{G}$ in a Hilbert space H. If, for every fixed pair of vectors φ, $\psi \in H$, the function

$$(U(h)\varphi, \psi)$$

is continuous on $\mathfrak{G}$, then the representation U is said to be *weakly continuous*. In other words, U is a weakly continuous representation provided that, when $\mathfrak{U}$ is given the weak topology (see §2.3, 1°), the correspondence $U : \mathfrak{G} \to \mathfrak{U}$ is a continuous homomorphism.

If, for every fixed $\varphi \in H$, the correspondence

$$h \to U(h)\varphi, \qquad h \in \mathfrak{G}$$

is a continuous mapping from $\mathfrak{G}$ to H (relative to the norm of H), then the representation U is said to be *strongly continuous*. In other words, the representation U is strongly continuous provided that U is a continuous mapping from $\mathfrak{G}$ to $\mathfrak{U}$ when the latter is given the strong topology (see §2.3, 1°).

Obviously, any strongly continuous unitary representation is also weakly continuous. Actually, the converse is also true.

Lemma II.3.1. A weakly continuous unitary representation is also strongly continuous.

PROOF. Let U be a weakly continuous unitary representation of a topological group $\mathfrak{G}$ in a Hilbert space H. Using the definition of a unitary representation, one readily calculates that, for any $\varphi \in H$, $g_0, g \in \mathfrak{G}$,

$$\|(U(g) - U(g_0))\varphi\|^2 = 2\mathfrak{R}\{(\varphi, \varphi) - (U(g_0^{-1}g)\varphi, \varphi)\}. \qquad \text{(II.3.1)}$$

Let ϵ be an arbitrary positive number. Since U is weakly continuous, there exists a neighborhood V of the identity e in $\mathfrak{G}$, such that

$$|(U(h)\varphi, \varphi) - (U(e)\varphi, \varphi)| < \frac{\epsilon}{2} \tag{II.3.2}$$

for every $h \in V$. But $(U(e)\varphi, \varphi) = (\varphi, \varphi)$, hence, if $g \in g_0 V$ (which is a neighborhood of g_0), it follows from (II.3.1) and (II.3.2) that

$$\|(U(g) - U(g_0))\varphi\| < \epsilon.$$

Therefore, U is strongly continuous.]

Let $U_k : g \to U_k(g)$, $g \in G$, $k = 1, 2$, be unitary representations of a group G in the respective Hilbert spaces H_1 and H_2. If there exists a unitary operator Q from H_1 onto H_2 such that $U_2(g) = QU_1(g)\,Q^{-1}$, $g \in G$, then U_1 and U_2 are said to be *unitarily equivalent*.

2° One-parameter Groups of Unitary Operators

Let H be a Hilbert space, and $\{U(t), -\infty < t < \infty\}$ a family of unitary operators in H such that

$$U(t_1 + t_2) = U(t_1)\,U(t_2), \qquad -\infty < t_1, t_2 < \infty, \quad U(0) = I. \tag{II.3.3}$$

Then, $\{U(t), -\infty < t < \infty\}$ is called a *one-parameter group*[3] *of unitary operators* in H. In other words, saying that $\{U(t), -\infty < t < \infty\}$ is a one-parameter group of unitary operators in H means that the correspondence $t \to U(t)$ is a unitary representation of R (the additive group of real numbers) in H. If this representation is strongly (or weakly) continuous, then we say that the one-parameter group of unitary operators $\{U(t), -\infty < t < \infty\}$ is *strongly* (or *weakly*) *continuous*.

If, for every fixed pair of vectors $\varphi, \psi \in H$,

$$(U(t)\varphi, \psi), \qquad -\infty < t < \infty$$

is a Lebesgue measurable function of t, then $\{U(t), -\infty < t < \infty\}$ is said to be *weakly measurable*. Clearly, any weakly continuous one-parameter group of unitary operators is weakly measurable. Conversely, we have the following result.

Lemma II.3.2. In a separable Hilbert space, any weakly measurable one-parameter group of unitary operators is weakly continuous.

[3] The term "parameter" in this designation refers, of course, to the real variable t, $-\infty < t < \infty$.

This result is due to von Neumann. Because the proof is rather lengthy and involves techniques not closely related to the subject of the present book, it will not be given here. A proof may be found in, for example, Guan Zhao-zhi [2].

For the proof of the following lemma, see, for example, Riesz and Sz.-Nagy [1].

Lemma II.3.3. Let $\{U(t), -\infty < t < \infty\}$ be a strongly continuous one-parameter group of unitary operators in a Hilbert space. Then, there exists a resolution of the identity $\{P_\lambda, -\infty < \lambda < \infty\}$ such that

$$U(t) = \int_{-\infty}^{\infty} e^{it\lambda}\, dP_\lambda \,.$$

The result of Lemma II.3.3 can be further elaborated, as follows. We define a linear operator A in H: the domain of A is to be

$$\mathfrak{D}(A) = \left\{x \,\middle|\, \int_{-\infty}^{\infty} \lambda^2\, d\,\| P_\lambda x \|^2 < \infty\right\},$$

and, when $x \in \mathfrak{D}(A)$,

$$Ax = \int_{-\infty}^{\infty} \lambda\, dP_\lambda x.$$

We then write

$$U(t) = e^{itA}, \tag{II.3.4}$$

and call A the *infinitesimal generator* of the one-parameter group of unitary operators $\{U(t), -\infty < t < \infty\}$. Actually, the right-hand side of (II.3.4) can be interpreted as

$$e^{itA} = \lim_{n\to\infty} \exp(itA(P_n - P_{-n})) = \lim_{n\to\infty} \sum_{k=0}^{\infty} \frac{(itA(P_n - P_{-n}))^k}{k!}\,.$$

Moreover, it can also be proved that, when $x \in \mathfrak{D}(A)$,

$$Ax = \lim_{t\to 0} \frac{U(t) - I}{it}\, x,$$

or, written more concisely,

$$A = \frac{1}{i}\frac{d}{dt}\, U(t)\Big|_{t=0}\,.$$

NOTES AND REFERENCES TO THE LITERATURE

Chapter I

§1.1. Some of this material is common knowledge. Regarding 3° and 5°, see Segal [3]. The notion of a pseudo-σ-finite measure (in 6°) was introduced, for convenience, by the author. Regarding Lemmas 1.1.13, 1.1.14 and Corollary 1.1.15, see Naĭmark [1]. Lemma 1.1.16 (7°) is due to Prohorov (see Gel'fand and Vilenkin [1]). Regarding 8°, see Doob [1].

§1.2. See Segal [1].

§1.3. The generalization of the classical theorem of Kolmogorov [1], in terms of projective limits, is due to the author.

§1.4. These results are due to Kakutani [1], except for Lemma 1.4.2, which is due to the author.

Chapter II

§2.1. For the basic ideas and properties of topological algebras, see Naĭmark [1] or Xia Dao-xing [2]. Regarding the representation of symmetric multiplicative linear functionals as points of an infinite product space (Lemma 2.1.1), see Xia Dao-xing [1].

§2.2. These results were proved for normed algebras by Gel'fand and Naĭmark, but the discussion given here applies to a still more general context. Lemma 2.2.1, Corollary 2.2.2, Lemma 2.2.3, and Theorem 2.2.4 are due to the author (see Xia Dao-xing [2]); the methods used are entirely different from those of Gel'fand and Naĭmark. In particular, the use of the Kolmogorov Theorem to study positive functionals, and to relate the theory of positive functionals more intimately with measure theory, is due to the author (see Xia Dao-xing [1]).

§2.3. Some of the concepts introduced here are due to von Neumann; see von Neumann [2], [3], [4], [5], [6], [7], [8], or Naĭmark ([1], Chapter 8).

§2.4. See Segal [1], [2].

Chapter III

§3.1. Lemma 3.1.3, for the special case where G is a finite-dimensional space, can be found in Gel'fand and Vilenkin [1]. Theorem 3.1.13 was first proved by Yan Shao-zong [1], using a method due to Girsanov and Mityagin [1]. Lemma 3.1.31 was also proved by Yan Shao-zong [1], using a different method. Theorem 3.1.32, concerning ergodic measures, was first proved by Segal [5] (although he stated the result in a somewhat different form). All the other results in §3.1 are due to the author; some of these may be found in Xia Dao-xing [5], [9], while some are published here for the first time.

§3.2. The basic notions concerning characters, and certain results for locally compact groups, are well known (see, e.g., Naĭmark [1]). Aside from these, the theory in §3.2 is due to the author; some of the results are published here for the first time, others may be found in Xia Dao-xing [5]–[9].

§3.3. Lemma 3.3.3 and Theorem 3.3.4 (concerning positive definite functions on discrete groups), as well as Example 3.3.2 and Lemma 3.3.9 (concerning locally compact groups) are all well known (see, e.g., Naĭmark [1]). The remaining results are due to the author, and were previously published for the case where there exists a finite quasi-invariant measure (see Xia Dao-xing [6]).

§3.4. The theory of Fourier transforms on locally compact groups is well known (see, e.g., Naĭmark [1]). Theorem 3.4.8 (for kth order cyclic measures) has been discussed by Ou Rui-sen [2]. The remaining results in this section are due to the author. Regarding other research along these lines, see Ou Rui-sen [2] and Shen Hai-yu [2].

Chapter IV

§4.1. Except for Example 4.1.2, all the results in this section are due to the author, and are published here for the first time.

§4.2. As regards linear functionals, the basic concepts and propositions given here are well known. Regarding quasi-linear functionals, the results in 2° and 3° are due to the author (some may be found in Xia Dao-xing [8]). Lemmas 4.2.16 and 4.2.17 are due to Erohin (see Minlos [1]). Theorem 4.2.18 is also due to the author, but Corollary 4.2.19 was first proved by Shen Hai-yu, while Corollary 4.2.20 was first proved by Mityagin [1], using a geometrical argument. Example 4.2.1 was first given by Chen Jing-jian in his university graduation thesis. Yan Shao-zong [3] used the L^2-Fourier transform to solve a certain type of functional variational equation, similar to that which occurs in quantum field theory.

§4.3. The notion of a cylinder measure was introduced by Gel'fand (see Gel'fand and Vilenkin [1]). Lemma 4.3.7 is due to Mityagin [1], Corollary 4.3.9 to Erohin (see Minlos [1]), and Theorem 4.3.12 to the author (see Xia Dao-xing [3]). For another approach to the representation of continuous positive-definite functions, see Xia Dao-xing [3], Ou Rui-sen [1], and Shen Hai-yu [1]. Concerning the material in 5°, see Gel'fand and Vilenkin [1]. The notion of a generalized stochastic process was first formulated by Gel'fand (see Gel'fand and Vilenkin [1]); concerning the results in 6°, see also Xia Dao-xing [3] and Yan Shao-zong [2].

Chapter V

§5.1. Concerning the general properties of Gaussian measures, see Doob [1]. Theorem 5.1.4 is due to the author; Feldman [2] attempted to give an example of an ergodic measure not equivalent to any Gaussian measure, but the proof of Lemma 6 in Feldman's paper is open to question (because the assertion that $C^k \to 0$ lacks adequate justification), and so the general validity of his example remains uncertain. More extensive discussions of Wiener processes may be found in Doob [1] and Paley and Wiener [1]; Koval'čik [1] gives a fairly comprehensive bibliography on the subject.

§5.2. The fundamental criterion for the equivalence of Gaussian measures (Theorem 5.2.5) was first obtained by Feldman [1] and

Hájek [1]. The proof given here by the author is a modification of that given by Rozanov [1]; it avoids the use of relatively complicated devices such as entropy, and so on.[1] Concerning Example 5.2.1, see Xia Dao-xing [4]. A number of survey type articles on the equivalence of Gaussian measures may be found in the proceedings of the Symposium on Time Series Analysis, Brown University, 1962 (Edited by M. Rosenblatt), Wiley, New York, 1963.

§5.3. The notion of a standard Gaussian measure was introduced, for convenience, by the author. Theorem 5.3.5 summarizes results obtained by Minlos (see Gel'fand and Vilenkin [1]) and Mityagin [1], but some parts of the present proof were devised by the author. Example 5.3.1 is due to Ou Rui-sen [1] and Chen Jing-jian. Theorem 5.3.7 is due to the author. As for the elementary properties of Wiener measure, these are, of course, well-known.

§5.4. The general results presented in this subsection are written here for the first time in their present form, but are not essentially new; they originate in certain results of Cameron and Martin [1] and Segal [4] concerning Wiener and Gaussian measures.

Chapter VI

§6.1. Theorem 6.1.1 is due to Wielandt [1]; Theorems 6.1.3, 6.1.4 and Corollaries 6.1.5, 6.1.6 are due to von Neumann [1]. The remaining results of this section originated in the work of Segal [3], [6].

§6.2. Regarding 1° and 3°, see Segal [6], [7] and [8]. Theorem 6.2.6, as stated here, is due to the author.

§6.3. Lemma 6.3.4 was obtained by Umemura [1], using Bernstein's theorem; the proof given here is due to the author. Theorem 6.3.5 was first obtained by Segal [8]; again, the proof given here is due to the author.

The remaining results in §6.2 and §6.3 are similar to those obtained by Segal ([4], [5], [6], [7], [8]), using the notion of a weak distribution, but here, they are formulated in terms of quasi-invariant measures. Gel'fand used quasi-invariant measures on conjugate spaces of nuclear spaces to construct a special class of unitary representations of the commutation relations (see Gel'fand and Vilenkin [1]).

[1] *Translator's note:* For a very elementary and easily comprehensible proof, see Brody [1].

BIBLIOGRAPHY

Brody, E. J.[1]

[1] An Elementary Proof of the Gaussian Dichotomy Theorem, *Z. Wahrscheinlichkeitstheorie und Verw. Gebiete* **20** (1971), 217-226.

Brush, S. G.

[1] Functional Integrals and Statistical Physics, *Rev. Modern Phys.* **33** (1961), 79–92.

Cameron, R. H., and Martin, W. T.

[1] The Orthogonal Development of Nonlinear Functionals in Series of Fourier–Hermite Functionals, *Ann. of. Math.* **48** (1947), 385–392.

[2] Fourier–Wiener Transforms of Functionals belonging to L_2 over the Space C, *Duke Math. J.* **14** (1947), 99–107.

Chen, Jian-gong

[1] Real Function Theory (in Chinese), Kexue Chubanshe, Peking, 1958.

Doob, J. L.

[1] "Stochastic Processes." Wiley, New York, 1953.

Feldman, J.

[1] Equivalence and Perpendicularity of Gaussian Processes, *Pacific J. Math.* **8** (1958), 699–707.

Gel'fand, I. M.

[1] Some Problems in Analysis and Differential Equations (in Russian), *Uspehi Mat.*

[1] *Translator's note:* Reference added by translator.

Nauk. **14** (1959), No. 3 (87), 3–19. [English transl.: *Amer. Math. Soc. Transl.*, Ser. 2, **26** (1963), 201–219.]

Gel'fand, I. M., and Vilenkin, N. Ya.

[1] "Generalized Functions," Vol. 4 (in Russian). Moscow, 1961. English translation: Academic Press, New York, 1964.

Girsanov, I. V., and Mityagin, B. S.

[1] Quasi-invariant Measures in Linear Topological Spaces (in Russian), *Naučn. Dokl. Vysš. Školy Fiz.-Mat. Nauki* **2** (1959), 5–9.

Guan Zhao-zhi

[1] "Essentials of the Theory of Topological Spaces" (in Chinese), Kexue Chubanshe, Peking, 1958.

[2] "Lectures on Functional Analysis" (in Chinese), Gaodeng Jiaoyu Chubanshe, Peking, 1958.

Hájek, J.

[1] On a Property of Normal Distributions of an Arbitrary Stochastic Process (in Russian), *Czechoslovak Math. J.* **8** (1958), 610–618. [English transl.: "Selected translations in Mathematical Statistics and Probability," pp. 245–252. American Mathematical Society, Providence, Rhode Island, 1961.]

Halmos, P. R.

[1] "Measure Theory." Van Nostrand, Princeton, New Jersey, 1950.

Kakutani, S.

[1] On Equivalence of Infinite Product Measures, *Ann. of Math.* **49** (1948), 214–224.

Kolmogorov, A. N.

[1] "Foundations of Probability Theory" (in Russian), Moscow–Leningrad, 1936. [English transl.: Chelsea, New York, 1950.]

Koval'čik, I. M.

[1] The Wiener Integral (in Russian), *Uspehi Mat. Nauk.* **18** (1963), No. 1 (109), 97–134.

Mityagin, B. S.

[1] A Remark on Quasi-invariant Measures (in Russian), *Uspehi Mat. Nauk.* **16** (1961), No. 5 (101), 191–193.

Naĭmark, M. A.

[1] "Normed Rings" (in Russian), Moscow, 1958. [English transl.: Noordhoff, Groningen, 1959.]

Ou Rui-sen

[1] Measures on the Spaces l^p (in Chinese), *Fudan Daxue Xuebao (Journal of Fudan University)* **8** (1963), 39–46.

[2] Measures on Banach Spaces (in Chinese), Doctoral dissertation, Fudan University, 1964.

Paley, R. E. A. C., and Wiener, N.

[1] Fourier Transforms in the Complex Domain, *Amer. Math. Soc. Colloq. Publ.* **19** (1934), Chap. 9.

Riesz, F., and Sz.-Nagy, B.

[1] "Leçons d'analyse fonctionnelle," Budapset, 1952. [English transl.: "Functional Analysis," Ungar, New York, 1955.]

Rozanov, Yu. A.

[1] On the Density of one Gaussian Measure with Respect to Another (in Russian), *Teor. Verojatnost. i Primenen.* **7** (1962), 84–89.

Sazonov, V. V.

[1] On Characteristic Functionals (in Russian), *in* "Proceedings of the Sixth All-Union Conference on Probability Theory and Mathematical Statistics, pp. 455–

462, Vilnius, 1960. Gosudarstv. Izdat. Političesk i Naučn. Lit. Litovsk. S.S.R., Vilnius, 1962.

Segal, I. E.

[1] Equivalences of Measure Spaces, *Amer. J. Math.* **73** (1951), 275–313.

[2] Decompositions of Operator Algebras II, *Mem. Amer. Math. Soc.* **9** (1951), 1–66.

[3] Abstract Probability Spaces and a Theorem of Kolmogoroff, *Amer. J. Math.* **76** (1954), 721–732.

[4] Tensor Algebras over Hilbert Spaces, I, *Trans. Amer. Math. Soc.* **81** (1956), 106–134.

[5] Distributions in Hilbert Spaces and Canonical Systems of Operators, *Trans. Amer. Math. Soc.* **88** (1958), 12–41.

[6] Foundations of the Theory of Dynamical Systems of Infinitely Many Degrees of Freedom, I, *Kgl. Danske. Videnskab. Selskab. Mat.–Fys. Medd.* **31** (1959), No. 12, 1–33.

[7] Foundations of the Theory of Dynamical Systems of Infinitely Many Degrees of Freedom, II, *Canadian J. Math.* **13** (1961), 1–18.

[8] Foundations of the Dynamics of Infinite Systems, III: Mathematical Characterization of the Physical Vacuum for a Linear Bose–Einstein Field, *Illinois J. Math.* **6** (1962), 500–523.

Shen Hai-yu

[1] Measures on the Spaces L^p (in Chinese), *Fudan Daxue Xuebao (Journal of Fudan University)* **8** (1963), 205–220.

[2] Measures on Banach Spaces and Groups (in Chinese), Doctoral dissertation, Fudan University, 1964.

Umemura, Y.

[1] Rotationally Invariant Measures in the Dual Space of a Nuclear Space, *Proc. Japan Acad.* **38** (1962), 15–17.

von Neumann, J.

[1] Die Eindeutigkeit der Schrödingerschen Operatoren, *Math. Ann.* **104** (1928), 570–578.

[2] On a Certain Topology for Rings of Operators, *Ann. Math.* **37** (1936), 111–115.

[3] (with F. J. Murray) On Rings of Operators, *Ann. Math.* **37** (1936), 116–229.

[4] On Rings of Operators II, *Trans. Amer. Math. Soc.* **41** (1937), 208–248.

[5] On Rings of Operators III, *Ann. of Math.* **41** (1940), 94–161.

[6] On Some Algebraical Properties of Operator Rings, *Ann. of Math.* **44** (1943), 709–715.

[7] On Rings of Operators IV, *Ann. of Math.* **44** (1943), 716–808.

[8] On Rings of Operators. Reduction Theory, *Ann. Math.* **50** (1949), 401–485.

Wielandt, H.

[1] Über die Unbeschränktheit der Operatoren der Quantenmechanik, *Math. Ann.* **121** (1949), 21.

Xia Dao-xing (Hsia Tao-hsing, Sya Do-šin)

[1] Positive Functionals on Algebras (in Russian), *Dokl. Akad. Nauk SSSR.* **121** (1958), 233–235.

[2] On Semi-normed Rings with Involution (in Russian), *Dokl. Akad. Nauk SSSR.* **124** (1959), 1223–1225; *Izv. Akad. Nauk SSSR. Ser. Mat.* **23** (1959), 509–528.

[3] Measures on Functional Spaces (in Chinese), *Fudan Daxue Xuebao (Journal of Fudan University)* **7** (1962), 121–131, **8** (1963), 261–262. English transl.: *Sci. Sinica* **12** (1963), 479–494.

[4] On the Equivalence of a Class of Gaussian Measures (in Chinese), *Fudan Daxue Xuebao (Journal of Fudan University)* **8** (1963), 151–161.

[5] On Linear Topological Spaces with Quasi-invariant Measures (in Chinese), *Fudan Daxue Xuebao (Journal of Fudan University)* **8** (1963), 273–277.

[6] Positive Definite Functions on Commutative Topological Groups with Quasi-invariant Measures (in Chinese), *Shuxue Xuebao* **14** (1964), 340–352. English transl.: *Chinese Math.-Acta* **5** (1964), 369–381.

[7] On Dual Quasi-invariant Measures (in Chinese), *Shuxue Xuebao* **14** (1964), 680–688. English transl.: *Chinese Math.-Acta* **6** (1965), 70–78.

[8] On Some Properties of Quasi-invariant Measures (in Chinese), *Shuxue Xuebao* **15** (1965), 81–90. English transl.: *Chinese Math.-Acta* **6** (1965), 382–392.

[9] On Quasi-invariant Measures in Linear Spaces (unpublished).

Xia Dao-xing, Wu Zhuo-ren, and Yan Shao-zong

[1] "Theory of Functions of a Real Variable and Essentials of Functional Analysis" (in Chinese). Shanghai Kexue Jishu Chubanshe, Shanghai, 1963.

Yan Shao-zong

[1] A Note Concerning Quasi-invariant Measures (unpublished).

[2] On Generalized Random Processes (in Chinese), *Fudan Daxue Xuebao (Journal of Fudan University)* **9** (1964).

[3] On a Class of Functional Variational Equations (unpublished).

INDEX

Pure and Applied Mathematics

A Series of Monographs and Textbooks

Editors **Paul A. Smith and Samuel Eilenberg**

Columbia University, New York

1: Arnold Sommerfeld. Partial Differential Equations in Physics. 1949 (Lectures on Theoretical Physics, Volume VI)

2: Reinhold Baer. Linear Algebra and Projective Geometry. 1952

3: Herbert Busemann and Paul Kelly. Projective Geometry and Projective Metrics. 1953

4: Stefan Bergman and M. Schiffer. Kernel Functions and Elliptic Differential Equations in Mathematical Physics. 1953

5: Ralph Philip Boas, Jr. Entire Functions. 1954

6: Herbert Busemann. The Geometry of Geodesics. 1955

7: Claude Chevalley. Fundamental Concepts of Algebra. 1956

8: Sze-Tsen Hu. Homotopy Theory. 1959

9: A. M. Ostrowski. Solution of Equations and Systems of Equations. Second Edition. 1966

10: J. Dieudonné. Treatise on Analysis. Volume I, Foundations of Modern Analysis, enlarged and corrected printing, 1969, Volume II, 1970. Volume III, in preparation

11: S. I. Goldberg. Curvature and Homology. 1962.

12: Sigurdur Helgason. Differential Geometry and Symmetric Spaces. 1962

13: T. H. Hildebrandt. Introduction to the Theory of Integration. 1963.

14: Shreeram Abhyankar. Local Analytic Geometry. 1964

15: Richard L. Bishop and Richard J. Crittenden. Geometry of Manifolds. 1964

16: Steven A. Gaal. Point Set Topology. 1964

17: Barry Mitchell. Theory of Categories. 1965

18: Anthony P. Morse. A Theory of Sets. 1965

19: Gustave Choquet. Topology. 1966

20: Z. I. Borevich and I. R. Shafarevich. Number Theory. 1966

21: José Luis Massera and Juan Jorge Schaffer. Linear Differential Equations and Function Spaces. 1966

22: Richard D. Schafer. An Introduction to Nonassociative Alegbras. 1966

23: Martin Eichler. Introduction to the Theory of Algebraic Numbers and Functions. 1966

24: Shreeram Abhyankar. Resolution of Singularities of Embedded Algebraic Surfaces. 1966

25: François Treves. Topological Vector Spaces, Distributions, and Kernels. 1967
26: Peter D. Lax and Ralph S. Phillips. Scattering Theory. 1967.
27: Oystein Ore. The Four Color Problem. 1967
28: Maurice Heins. Complex Function Theory. 1968
29: R. M. Blumenthal and R. K. Getoor. Markov Processes and Potential Theory. 1968
30: L. J. Mordell. Diophantine Equations. 1969
31: J. Barkley Rosser. Simplified Independence Proofs: Boolean Valued Models of Set Theory. 1969
32: William F. Donoghue, Jr. Distributions and Fourier Transforms. 1969
33: Marston Morse and Stewart S. Cairns. Critical Point Theory in Global Analysis and Differential Topology. 1969
34: Edwin Weiss. Cohomology of Groups. 1969
35: Hans Freudenthal and H. De Vries. Linear Lie Groups. 1969
36: Laszlo Fuchs. Infinite Abelian Groups: Volume I, 1970. Volume II, in preparation
37: Keio Nagami. Dimension Theory. 1970
38: Peter L. Duren. Theory of H^p Spaces. 1970.
39: Bodo Pareigis. Categories and Functors. 1970
40: Paul L. Butzer and Rolf J. Nessel. Fourier Analysis and Approximation: Volume 1, One-Dimensional Theory. 1971
41: Eduard Prugovečki. Quantum Mechanics in Hilbert Space. 1971
42: D. V. Widder: An Introduction to Transform Theory. 1971
43: Max D. Larsen and Paul J. McCarthy. Multiplicative Theory of Ideals. 1971
44: Ernst-August Behrens. Ring Theory. 1972
45: Morris Newman. Integral Matrices. 1972
46: Glen E. Bredon. Introduction to Compact Transformation Groups. 1972
47: Werner Greub, Stephen Halperin, and Ray Vanstone. Connections, Curvature, and Cohomology: Volume I, De Rham Cohomology of Manifolds and Vector Bundles. 1972
48: Xia Dao-xing. Measure and Integration Theory of Infinite-Dimensional Spaces: Abstract Harmonic Analysis. 1972
49: Ronald G. Douglas. Banach Algebra Techniques in Operator Theory. 1972

In preparation

E. R. Kolchin. Differential Algebra and Algebraic Groups

T. Benny Rushing. Topological Embeddings

Willard Miller, Jr. Symmetry Groups and Their Applications